Annual Review of

INFORMATION SCIENCE AND TECHNOLOGY

Annual Review of
INFORMATION SCIENCE AND TECHNOLOGY

Volume 39 • 2005
Blaise Cronin, Editor

Published on behalf of the
American Society for Information Science and Technology
by Information Today, Inc.

Medford, New Jersey

ISBN: 1-57387-209-1
ISSN: 0066-4200
CODEN: ARISBC
LC No. 66-25096

Published and distributed by
Information Today, Inc.
143 Old Marlton Pike
Medford, NJ 08055-8750

On behalf of

The American Society for Information Science and Technology
1320 Fenwick Lane, Suite 510
Silver Spring, MD 20910-3602, U.S.A.

Information Today, Inc. Staff
Publisher: Thomas H. Hogan, Sr.
Editor-in-Chief: John B. Bryans
Managing Editor: Amy Holmes
Proofreader: Pat Hadley-Miller
Graphics Department Director:
  M. Heide Dengler
Cover Designer: Victoria Stover
Book Designers: Kara Mia Jalkowski,
  Lisa M. Boccadutre

ARIST Staff
Editor: Blaise Cronin
Associate Editor: Debora Shaw
Copy Editors: Dorothy Pike,
  Joy Hanson, Tiana Tew
Indexer: Amy Novick

Contents

SECTION I
Information Retrieval

SECTION II
Technology and Systems

v

Introduction

Blaise Cronin

ARIST is going the way of so many things American: supersizing. Truth be told, the beefiness of this year's volume is more a matter of chance than design: good news for subscribers, not such good news for our publisher. Last year I described some of the challenges associated with editing an annual review and the near impossibility of knowing with certitude what the end product will look like. A great deal can happen in the almost two years that constitute the gestation period of the typical *ARIST* chapter. Life has a knack of cocking a snoot at contracts, deadlines, and good intentions, and even when the Fates are smiling there is the occasional disappointment: *parturiunt montes nascitur ridiculus mus*. More often than not, though, the mountains labor to magisterial effect and the mice are few and far between. But, ultimately, it is for you, dear reader, to dispense and withhold the encomia. I will, however, say this: Volume 39 has something for just about everyone. The topic range is wide, almost indecently so—a seductive perquisite of editorship. Each year one adds new ingredients and experiments cautiously with the overall form of the volume, safe in the knowledge that readers and reviewers alike will not be slow to provide constructive feedback. With volume 39 we inaugurate a cumulative index to *ARIST*, beginning with volume 1 in 1966, available online at http://www.asis.org/Publications/ARIST.search.html.

This year we have 14 chapters grouped into five sections: Information Retrieval; Technology and Systems; Social Informatics; National Intelligence; Theory. We begin with reviews of recent research on information retrieval (IR), a staple of *ARIST* over the years, in both controlled document collections and Web environments. These chapters explore new approaches—statistical language modeling and fusion techniques, respectively—to problems old and new. They are written by a doyen of the field (Bruce Croft) and two relative neophytes (Xiaoyong Liu and Kiduk Yang) continuing an *ARIST* tradition of showcasing both established and lesser known names. The third chapter in the opening section is by Mike Thelwall and colleagues. Although not strictly speaking an IR chapter, it is as comfortable in this section as it is likely to be in any other. Aficionados of classical information science will know that it is 35 years since Alan Pritchard (1969) coined the term "bibliometrics"; today, with rapid advances in hyperlinking, new measures of association and intellectual interaction are emerging, hence the proliferation of neologisms such as "webometrics." But longstanding issues of validity and reliability remain, whether the links being counted are citational or hyper in character. An important role for *ARIST* is thus to ensure that we do not develop an ahistorical mindset or overlook the contributions of information science's pioneers as we scamper to embrace the new. But I shall have more to say on this subject next year in the introduction to what will be the fortieth volume of *ARIST*.

Section II, "Technology and Systems," provides a snapshot of developments in three rather different areas of research and practice. The chapter by Bin Zhu and Hsinchun Chen on information visualization methods, techniques, and applications follows closely on the heels of the review by Börner, Chen, and Boyack (2002). But this is a fast moving and important domain—witness the recent launch of the journal *Information Visualization* (http://www.palgrave-journals.com/ivs)—and research findings have potentially wide practical relevance across many scientific fields and disciplines, not least bioinformatics, the topic of Gerald Benoît's chapter. The term "informatics" (coined in 1968 by A. I. Mikhailov, as best I can tell) has emerged from relative East European seclusion to become one of the most widely used programmatic descriptors in existence; a quick search on Google generated more than 2.5 million hits. Colleges, schools, and departments of informatics, having erupted like acne across the face of higher education in the U.S. and farther afield, can incorporate everything from hydro, to medical, to quantum informatics. As the routine computational demands of scientific (and other) disciplines intensify—think of genomics or astrophysics—a new hybrid species is required, individuals who combine domain knowledge with advanced information technology capability. One is reminded of the over-used adage, a picture is worth a thousand words. Faced with petabytes of data, scientists are struggling to extract meaning and significance, and this has turbocharged research in information visualization, data mining, and related areas, as the chapters by Zhu and Chen and Benoît make abundantly clear.

"As ye sow, so shall ye reap." The tools for data generation, capture, and analysis may be in place, or in the process of being developed, but that is not the end of the story. What are we to do with the mass of information in circulation? How are data and evidence to be recorded and preserved? What, in fact, is an electronic record? This is the question posed here by Anne Gilliland-Swetland. These issues extend well beyond the world of big science and infrastructural computing to include, for example, humanistic scholarship, archival management, and routine business practice. And it is not just a matter of building bigger and better mousetraps; the creation and stewardship of electronic records raises many at times complex public policy and legal issues, as Gilliland-Swetland tells us.

"Social informatics" is a truly protean label, one about which I remain ambivalent. I could equally well have used the heading "Socio-Technical Systems Design and Analysis," which would effectively signify the common orientation of the three chapters contributing to this section, but I have gone instead with "informatics" in memory of my late colleague Rob Kling, whose championing of the term, dogged efforts to institutionalize the embryonic field, and decades-long, intellectual contributions to the research literature on the social dimensions of computing meant so much to so many. Rob established the Center for Social Informatics at Indiana University well before the "informatics" label had achieved fashionable status. Incidentally, his last *ARIST* chapter was published posthumously in volume 38.

Systems are designed for use by people. But people belong to different cultures, broadly and variously understood. What, asks Ewa Callahan, should interaction designers know about culture so that they can build culturally sensitive and effective interfaces? In some respects this is an almost intractable problem, albeit less in terms of artifactual design or engineering considerations than the fundamental issue of defining what culture actually means. Her chapter nonetheless raises a number of substantive topics and areas in need of systematic research, but rightly eschews definitive answers.

We use digital technologies and online fora for a variety of reasons, ranging from straightforward information seeking to looking for social and emotional support in chat rooms and other online settings. Caroline Haythornthwaite and Christine Hagar provide some insights into the social dimensions of Web use and the ways in which we move between the virtual and physical worlds. As Sproull and Patterson (2004, p. 34) note: "Participation in electronic communities often begins with a search for information. Rarely does anyone wake up in the morning and say, 'I think I'll join an electronic community today.' But the results of a search for information often lead individuals to them. Today millions of people worldwide participate in hundreds of thousands of voluntary electronic discussion groups and communities." The world of the Web is now part of our everyday world. Further insights into patterns of use of the Web, this time specifically by children and teens, are provided by Andrew Large, although his primary focus is instrumental usage in educational contexts. For more detailed information and benchmark data on trends in Web and Internet use in general, I recommend the Pew Internet and American Life Project (http://www.pewinternet.org) and also the UCLA

Internet Report (http://ccp.ucla.edu/pages/internet-report.asp), both of which are cited in the chapter.

I am writing this Introduction a matter of days after the March 11th terrorist bombings in Madrid, and although national intelligence has become a topic of great personal interest to all of us for obvious reasons, that is not the motivation for including it in the present volume. As both Lee Strickland and I (here exercising editorial *droit de seigneur*, as it were) point out, there are longstanding ties between the worlds of intelligence and information science, both intellectually and professionally. That is why I first introduced the topic to *ARIST* in the shape of Philip Davis's (2002) chapter on intelligence and information warfare, and why I expect to include further contributions on the subject in the years ahead. By way of an aside, two of this year's contributors, Lee Strickland and Herbert Snyder, have intelligence backgrounds, as does at least one of our advisory board members, Glynn Harmon. Again, this will hardly come as a surprise to historians of Anglo-American information science; our professional and academic ranks include some who, for example, served with the OSS (Office of Strategic Services) in the U.S. or with the code breakers at Bletchley Park in the U.K.

A recent issue of the journal *Social Epistemology* was devoted to the topic of social capital. Arnaldo Bagnasco's (2003, p. 359) article in that compilation opened as follows: "I could say the same thing about social capital that St. Augustine of Hippo said about time: I know exactly what it is, but if I start thinking about it I don't know any more." One knows the feeling only too well. The lead chapter in the Theory section is by Elisabeth Davenport and Herbert Snyder, no strangers to *ARIST*. They don't cite St. Augustine on time nor Supreme Court Justice Potter Stewart on pornography (the "I know it when I see it" approach); instead they dissect many of the claims and assumptions embedded in the sprawling literature on the subject of social capital, a construct popularized by Robert Putnam (2000) in his best-seller, *Bowling Alone*, and assess its potential utility in an age of networked organizations.

Julian Warner attempts something that I don't think has been tried previously in the pages of *ARIST*: the application of Marx's claims about physical labor to information systems and intellectual labor. He analyzes concepts of syntactic and semantic labor that have been both implicit and to some extent explicit in information theory. It is not unreasonable to surmise that Warner's interest in Marxist thinking (which can be found elsewhere in his writings) may have been sparked or reinforced by Terry Eagleton, the noted critical theorist and author of the hugely successful *Literary Theory* (1983), with whom he studied at the University of Oxford. I mention this biographical bagatelle because it provides a nice segue to the final chapter of the volume, written by Ron Day on the subject of post-structuralism, one of the major paradigms of twentieth-century literary theory. I am more than ready to concede that Dead Germans and sesquipedalian Frenchmen may not be everyone's cup of tea, least of all regular *ARIST* readers, but Ron Day's chapter raises fundamentally important questions about the limits of traditional theorizing in the domains of library and information science. Moreover, and this will surely appeal to those of an historical bent, he invokes Jesse Shera (1970), who recognized early on the need for the field to explore general problems of knowledge and information.

We know that *ARIST* is rarely read cover to cover; it is a resource to be dipped into selectively and at leisure. It may well be that those who devour Yang's account of Web-based IR will ignore the light of Day, and those who tarry with Warner's exegesis on information labor will bypass Benoît's bioinformatics chapter. But the seriously open-minded reader will discover many obvious and not-so-obvious intertextualities across these 14 chapters. Devotees of *ARIST* will also find that not a few of the issues raised in the present volume have been foreshadowed in earlier volumes, even if the resonances are at times understated or unacknowledged. I am certainly not suggesting that there is nothing new under the sun, it's just that we—whether as readers or authors—being so consumed with the immediate and the emerging, sometimes fail to see the threads that wind back over the years and through the cumulating pages of *ARIST*.

References

Bagnasco, A. (2003). Social capital in changing capitalism. *Social Epistemology*, 17(4), 359–380.

Börner, K., Chen, C., & Boyack, K. (2002). Visualizing knowledge domains. *Annual Review of Information Science and Technology*, 37, 179–255.

Davis, P. H. J. (2002). Intelligence, information technology, and information warfare. *Annual Review of Information Science and Technology*, 36, 313–352.

Eagleton, T. (1983). *Literary theory: An introduction*. Minneapolis: University of Minnesota Press.

Pritchard, A. (1969). Statistical bibliography or bibliometrics? *Journal of Documentation*, 25(4), 348–349.

Putnam, R. (2000). *Bowling alone: The collapse and revival of American community*. New York: Simon & Schuster.

Shera, J. (1970). *Sociological foundations of librarianship*. New York: Asia Publishing House.

Sproull, L., & Patterson, J. F. (2004). Making information cities livable. *Communications of the ACM*, 47(2), 33–37.

Acknowledgments

Many individuals are involved in the production of *ARIST*. I should like to acknowledge the contributions of both our Advisory Board members and the outside reviewers. Their names are listed in the pages that follow. The bulk of the copyediting and bibliographic checking has been carried out by Joy Hanson, Dorothy Pike, and Tiana Tew, for which I am most grateful. As always, Debora Shaw did what a good associate editor is supposed to do.

ARIST Advisory Board

Judit Bar-Ilan
Hebrew University of Jerusalem, Israel

Micheline Beaulieu
University of Sheffield, UK

Nicholas J. Belkin
Rutgers University, New Brunswick, USA

David C. Blair
University of Michigan, Ann Arbor, USA

Christine L. Borgman
University of California at Los Angeles, USA

Terrence A. Brooks
University of Washington, Seattle, USA

Ian Cornelius
University College Dublin, Ireland

Elisabeth Davenport
Napier University, Edinburgh, UK

Susan Dumais
Microsoft Research, Redmond, USA

Abby Goodrum
Syracuse University, USA

E. Glynn Harmon
University of Texas at Austin, USA

Leah A. Lievrouw
University of California at Los Angeles, USA

Katherine W. McCain
Drexel University, Philadelphia, USA

Charles Oppenheim
Loughborough University, UK

Chapter Reviewers

Philip Agre
Phil Bantin
Judit Bar-Ilan
Peter Bath
David Blair
Christine Borgman
Katy Börner
Ian Cornelius
Elisabeth Davenport
Ronald Day
Luciana Duranti
Steve Fuller
Abby Goodrum
Melissa Gross
Dennis Groth
Glynn Harmon
Djoerd Hiemstra

Peter Ingwersen
Ray Larson
Loet Leydesdorff
Leah Lievrouw
Peter Lyman
Craig Norman
Charles Oppenheim
Greg Paris
Jennifer Preece
Edie Rasmussen
Alice Robbin
Yvonne Rogers
Victor Rosenberg
Ben Shneiderman
Nancy Van House
Gary Wiggins

Contributors

David A. Baldwin is currently Supervisor of Stacks and Shelving at the Enoch Pratt Free Library, Baltimore, and has previously served in other management positions at the Baltimore County Historical Society, the Maryland Historical Society, the Library of Virginia, and both the Texas State Archives and Library. Before his career in archives and libraries, Mr. Baldwin worked in the broadcasting and recording industries for more than ten years, including KLBJ AM/FM in Austin, Texas. He received his M.L.S. from the University of Maryland in 2003 and also holds M.A., B.A., and B.S. degrees from the University of Texas.

Gerald Benoît is an Assistant Professor at the Simmons College Graduate School of Library and Information Science. His research interests are in Web-based systems development, human–computer interaction, visualization, and the role philosophy plays in creating, using, and testing systems. In addition, Dr. Benoît lectures in computer science and bioinformatics. He is also a consultant in digital library and archives creation in the Boston area.

Lennart Björneborn is an Assistant Professor at the Royal School of Library and Information Science, Copenhagen, Denmark. He obtained his Ph.D. in Webometrics from the Royal School in 2004 with a thesis on *Small-World Link Structures across an Academic Web Space – a Library and Information Science Approach*. His research interests are concerned with Webometrics, link structure analysis, small-world connectivity, distributed knowledge organization, serendipitous information finding, and networked creativity stimulation. He has published papers on Webometrics with Peter Ingwersen in *Scientometrics* and the *Journal of the American Society for Information Science and Technology*.

Ewa Callahan is a Ph.D. candidate in information science at Indiana University. She earned her M.A. in history with specialization in archives at Jagiellonian University, Cracow, Poland in 1993, and then moved into the field of information science, receiving her M.I.S. at Indiana University in 1998. She is currently working on her dissertation, examining similarities and differences in Web design among world cultures. Her research and teaching interests are in the areas of international issues in information science, human–computer interaction, information visualization, and the development and use of information technologies in post-communist countries.

Hsinchun Chen is McClelland Professor of M.I.S. at the University of Arizona. He received the Ph.D. degree in Information Systems from New York University in 1989. He is author of four books and more than 150 articles covering digital libraries, intelligence analysis, medical informatics, semantic retrieval, search algorithms, knowledge management, and Web computing. He serves on the editorial board of *Journal of the American Society for Information Science and Technology, ACM Transactions on Information Systems*, and *Decision Support Systems*. Dr. Chen is founding director of the Artificial Intelligence Laboratory and Hoffman E-Commerce Laboratory. The Artificial Intelligence Laboratory, which houses more than 40 researchers, has received more than $15 million in research funding from the NSF, NIH, NLM, NIJ, CIA, and other agencies over the past 10 years. Dr. Chen has received numerous industry awards in education and research including: AT&T Foundation Award, SAP Award, and the Andersen Consulting Professor of the Year Award.

W. Bruce Croft is a Distinguished Professor in the Department of Computer Science at the University of Massachusetts, Amherst, which he joined in 1979. In 1992, he founded the Center for Intelligent Information Retrieval (CIIR). Professor Croft has been Editor-in-Chief of the *ACM Transactions on Information Systems* and on the editorial boards of six other journals. He is a Fellow of ACM, and received the ASIST Research Award in 2000. In 2003, he received the Gerald Salton Award from the ACM SIGIR.

Elisabeth Davenport was appointed Professor of Information Management at Napier University in 1998. She is currently head of the Social Informatics Research Group in the School of Computing and a Board Member of the International Democracy Centre. She is also a Visiting Scholar in the School of Library and Information Science at Indiana University. Dr. Davenport has recently been principal investigator for two projects funded by the European Commission that explored social capital in networked enterprises. She has published widely in the areas of information strategy, knowledge management, social informatics, and ethnographic methods in the workplace. With her co-author, Herbert Snyder, she is currently exploring entity-based models for management accounting in information service work.

Ronald E. Day is the author of *The Modern Invention of Information: Discourse, History, and Power* (Southern Illinois University, 2001). He is the author of numerous articles which apply humanities research and techniques to problems in information science and, more generally, to information culture. Many of his articles have been published in the *Journal of the American Society for Information Science and Technology*. He is an Assistant Professor at Wayne State University.

Anne Gilliland-Swetland is an Associate Professor and Director of the Center for Information as Evidence in the Department of Information Studies, Graduate School of Education and Information Studies at the University of California, Los Angeles. Her research interests relate to the design and evaluation of record keeping and cultural information systems, metadata creation and management, and the use of primary sources in K-12 education. Her current research activities include InterPARES 2, the Museums and the Online Archive of California Evaluation (MOACII), MUSTICA, and the Clever Recordkeeping Metadata Project. She is a Fellow of the Society of American Archivists.

Christine Hagar is a doctoral student at the Graduate School of Library and Information Science, University of Illinois at Urbana-Champaign. She has worked in U.K. academic and public libraries and taught at both British and Polish library and information science schools. Her main area of research is the impact of information and communication technologies on communities during crisis situations. She also has an interest in knowledge sharing in international nongovernment development organizations, particularly indigenous knowledge. She has worked on projects funded by the U.K. Department for International Development and The British Council.

Caroline Haythornthwaite is an Associate Professor at the Graduate School of Library and Information Science, University of Illinois at Urbana-Champaign. Her research examines what interactions are important for members of online work and learning communities, and how computer media support such interactions. Her studies have examined: social networks and media use by academic researchers, distance learners, and inter-disciplinary scientists; online communication and community; and processes of collaboration, knowledge co-construction, and technology definition in research teams. Her work appears in *The Information Society*, *Journal of Computer-Mediated Communication*, *New Media & Society* and as chapters in various books. She recently co-edited with Barry Wellman *The Internet in Everyday Life* (Blackwell, 2002), and with Michelle M. Kazmer *Learning, Culture and Community in Online Education* (Peter Lang, 2004).

Marlene Justsen was born and educated in Europe where she majored in history and international relations at both the undergraduate and graduate levels. Her publications have examined ethnic tensions in Eastern Europe, considered theories of change within international relations, and analyzed the election and presidency of John F. Kennedy. She is a candidate for her M.L.S. at the University of Maryland with a focus on archival administration and records management.

Andrew Large is CN-Pratt-Grinstad Professor of Information Studies at McGill University (Canada) and the former Director of McGill's Graduate School of Library and Information Studies. His major research interests in recent years have been the information-seeking behavior in electronic environments of elementary school students and design issues relating to information systems intended for this user community. He co-edits the quarterly journal, *Education for Information*, and is on the editorial boards of several other journals. He has authored or edited twelve books, published extensively in the journal literature and acted as consultant for a wide range of international organizations.

Xiaoyong Liu is a Ph.D. student in the Center for Intelligent Information Retrieval of the Department of Computer Science at the University of Massachusetts, Amherst. Her research interests include language modeling, text retrieval, text representation, and question answering.

Herbert Snyder is Associate Professor of accounting and information systems at North Dakota State University. He received his Ph.D. from Syracuse University and taught at Indiana University prior to joining the faculty at NDSU. He has published in the areas of scholarly communication, accounting for information assets, and white-collar crime. During 2003 he was a Fulbright scholar in Ukraine. Prior to entering the academic world, Dr. Snyder worked as a fraud investigator and an intelligence office for the U.S. Army.

Lee S. Strickland is a career attorney and intelligence officer with the United States government and has been a member of the Senior Intelligence Service since 1986. He has also held an appointment as a Visiting Professor at the College of Information Studies, University of Maryland since 1999. His teaching efforts there have led to the development of a number of courses including Legal Issues in Managing Information, Information and the War on Terrorism, Information Policy, and Competitive Intelligence. He publishes frequently on various information policy issues (e.g., intellectual property, national security, and privacy) for ARMA, ASIST, GIQ, and others. He is a graduate of the University of Central Florida (B.S. in Mathematics, magna cum laude), the University of Virginia (Master of Computer Science), and the University of Florida (Juris Doctor with honors) and is also admitted to practice law in the District of Columbia, the state of Virginia, and various federal courts.

Mike Thelwall is a Reader in Computer Science in the School of Computing, and a member of the Research Institute for Advanced Technologies, both at the University of Wolverhampton, U.K. He obtained his Ph.D. in Pure Mathematics from the University of Lancaster, U.K. He develops and applies text and hyperlink analysis methods in a variety of contexts, from computer science to sociology. Thelwall is part of the European Union project WISER—Web Indicators for Science, Technology & Innovation Research. He is on the editorial board of two journals and two book series, and has published sixty-six refereed journal articles.

Liwen Vaughan is an Associate Professor with the Faculty of Information and Media Studies, University of Western Ontario, Canada. Dr. Vaughan's current research interests focus on Webometrics and Web search engine evaluation. She has conducted Web data mining projects and published papers on topics such as Web citation and Web hyperlink analysis. She is the principal investigator of a major research project entitled "Mining the Web for Business Intelligence through Link Structure Analysis." Dr. Vaughan's main areas of teaching include research methods and statistics, informetrics, information retrieval, and database management systems.

Julian Warner is a faculty member in information science at the School of Management and Economics, in the Queen's University of Belfast. He has a B.A. and M.A. in English Language and Literature from the University of Newcastle upon Tyne, a D.Phil. in English Literature from the University of Oxford, and an M.A. in Librarianship from the University of Sheffield. He has been a visiting scholar at the University of California at Berkeley and the University of Illinois. His books include *From Writing to Computers* (Routledge, 1994), *Information, Knowledge, Text* (Scarecrow, 2001), and *Humanizing Information Technology* (Scarecrow, 2004). The themes of his chapter on labor in information systems are to be developed further.

Kiduk Yang is an Assistant Professor of Information Science at Indiana University. He began a career as an application programmer and systems developer after receiving his B.S. in computer science. While working for more than fourteen years as an IT professional, he earned a master's degree and Ph.D. in Information Science at the University of North Carolina, Chapel Hill. His principal research area is information retrieval with an emphasis on leveraging human knowledge for information discovery on the Web. He is currently investigating the integration of knowledge organization with text and link analysis to enhance Web searching.

Bin Zhu received her Ph.D. in Management Information Systems from the University of Arizona. She is an Assistant Professor in the information systems department at Boston University. Her current research interests include human–computer interaction, information visualization, computer-mediated communication, and knowledge management systems. She has been a lead author for papers that have appeared in the *Journal of the American Society for Information Science*, *IEEE Transaction on Image Processing*, and *D-Lib Magazine*. Her research also received an IBM faculty award in 2003.

About the Editor

Blaise Cronin is the Rudy Professor of Information Science at Indiana University, Bloomington, where he has been Dean of the School of Library and Information Science for thirteen years. From 1985 to 1991 he held the Chair of Information Science and was Head of the Department of Information Science at the University of Strathclyde Business School in Glasgow. He has also held visiting professorships at the Manchester Metropolitan University and Napier University, Edinburgh. Professor Cronin is the author of numerous research articles, monographs, technical reports, conference papers, and other publications. Much of his research focuses on collaboration in science, scholarly communication, citation analysis, the academic reward system, and cybermetrics—the intersection of information science and social studies of science. He has also published extensively on topics such as information warfare, information and knowledge management, competitive analysis, and strategic intelligence. Professor Cronin sits on a number of editorial boards, including *Journal of the American Society for Information Science and Technology*, *Scientometrics*, *Cybermetrics*, and *International Journal of Information Management*.

Professor Cronin has extensive international experience, having taught, conducted research, or consulted in more than 30 countries: clients have included the World Bank, NATO, Asian Development Bank, UNESCO, U.S. Department of Justice, Brazilian Ministry of Science and Technology, European Commission, British Council, Her Majesty's Treasury, Hewlett-Packard Ltd., British Library, Commonwealth Agricultural Bureaux, Chemical Abstracts Service, and Association for Information Management. He has been a keynote or invited speaker at scores of conferences, nationally and internationally. Professor Cronin was a founding director of Crossaig, an electronic publishing start-up in Scotland, which was acquired in 1992 by ISI (Institute for Scientific Information) in Philadelphia. He was educated at Trinity College Dublin (M.A.) and the Queen's University of Belfast (Ph.D., D.S.Sc.). In 1997, he was awarded the degree Doctor of Letters (D.Litt., *honoris causa*) by Queen Margaret University College, Edinburgh, for his scholarly contributions to information science.

About the Associate Editor

Debora Shaw is a Professor at Indiana University, Bloomington and also Executive Associate Dean of the School of Library and Information Science. Her research focuses on information organization and information seeking and use. Her work has been published in the *Journal of the American Society for Information Science and Technology*, the *Journal of Documentation*, *Online Review*, and *First Monday*, among others. She serves on the editorial board of the *Journal of Educational Resources in Computing*.

Dr. Shaw served as President of the American Society for Information Science & Technology (1997), and has also served on the Society's Board of Directors. She has been affiliated with *ARIST* as both a chapter author and as indexer over the past 18 years. Dr. Shaw received bachelor's and master's degrees from the University of Michigan and the Ph.D. from Indiana University. She was on the faculty at the University of Illinois before joining Indiana University.

Information Retrieval

CHAPTER 1

Statistical Language Modeling for Information Retrieval

Xiaoyong Liu and W. Bruce Croft
University of Massachusetts

Introduction

This chapter reviews research and applications in statistical language modeling for information retrieval (IR), which has emerged within the past several years as a new probabilistic framework for describing information retrieval processes. Generally speaking, statistical language modeling, or more simply language modeling (LM), involves estimating a probability distribution that captures statistical regularities of natural language use. Applied to information retrieval, language modeling refers to the problem of estimating the likelihood that a query and a document could have been generated by the same language model, given the language model of the document either with or without a language model of the query.

The roots of statistical language modeling date to the beginning of the twentieth century when Markov tried to model letter sequences in works of Russian literature (Manning & Schütze, 1999). Zipf (1929, 1932, 1949, 1965) studied the statistical properties of text and discovered that the frequency of words decays as a power function of each work's rank. However, it was Shannon's (1951) work that inspired later research in this area. In 1951, eager to explore the applications of his newly founded information theory to human language, Shannon used a prediction game involving n-grams to investigate the information content of English text. He evaluated n-gram models' performance by comparing their cross-entropy on texts with the true entropy estimated using predictions made by human subjects. For many years, statistical language models have been used primarily for automatic speech recognition. Since 1980, when the first significant language model was proposed (Rosenfeld, 2000), statistical language modeling has become a fundamental component of speech recognition, machine translation, and spelling correction. It has also proven useful for natural language processing tasks such as natural language generation and summarization. In 1998 it was introduced to information retrieval and has opened up new ways of thinking about the retrieval process.

The first uses of the language modeling approach for IR focused on its empirical effectiveness using simple models. In the basic approach, a query is considered to have been generated from an "ideal" document that satisfies the information need. The system's job is then to estimate the likelihood of each document in the collection being the ideal document and rank the documents accordingly. This query-likelihood retrieval model, first proposed by Ponte and Croft (1998), and later described in terms of a "noisy channel" model by Berger and Lafferty (1999), has produced results that are at least comparable to the best retrieval techniques previously available. The basic model has been extended in a variety of ways. For example, documents have been modeled as mixtures of topics (Hofmann, 1999b) and phrases as well as single words have been used as the basis for analysis (Song & Croft, 1999). Progress has also been made in understanding the formal underpinnings of the statistical language modeling approach and in comparing it to traditional probabilistic approaches. This research found various connections and also identified important differences. Recent work has developed sophisticated models that are more closely related to traditional approaches. For example, a language model that explicitly models relevance has been proposed by Lavrenko and Croft (2001), and Lafferty and Zhai (2001a) have developed a risk-minimization framework based on Bayesian decision theory. Successful applications of the LM approach to a number of retrieval tasks have also been reported, including cross-lingual retrieval (Lavrenko, Choquette, & Croft, 2002; Xu, Weischedel, & Nguyen, 2001) and distributed retrieval (Si, Jin, Callan, & Ogilvie, 2002; Xu & Croft, 1999). Research carried out by a number of groups has confirmed that the language modeling approach is a theoretically attractive and potentially very effective probabilistic framework for studying information retrieval problems (Croft & Lafferty, 2003). This empirical success and the overall potential of the approach have given rise to the LEMUR[1] project.

As a new family of probabilistic retrieval models, language models for IR share the theoretical foundations underlying general probabilistic IR. Numerous authors have contributed to the theoretical discussions of probabilistic retrieval including Maron and Kuhns (1960), Cooper and Maron (1978), Cooper (1995), Robertson and Sparck Jones (1976), Robertson (1977), van Rijsbergen (1979, 1992), and Sparck Jones, Walker, and Robertson (2000a, 2000b). Well known probabilistic IR models include the Robertson and Sparck Jones model (1976), the Croft and Harper model (1979), the Fuhr model (1989), and the inference network model (Turtle & Croft, 1991). Detailed treatment of these earlier probabilistic IR theories and approaches is beyond the scope of this chapter. Excellent accounts of this work can be found in van Rijsbergen (1979) and Baeza-Yates and Ribeiro-Neto (1999); readers are encouraged to consult them for more information.

This review brings together contemporary research on statistical language modeling and smoothing for retrieval of written text. The review does not cover language-modeling techniques that were developed for speech recognition and other language technologies but have not been applied to text retrieval. It also limits the discussion of earlier probabilistic IR work for which there is a wealth of literature. This is the first *ARIST* review of statistical language modeling for IR, but Rosenfeld (2000) reviewed language modeling techniques for speech recognition and other domains and Chen and Goodman (1998) provided a survey of smoothing techniques developed for those domains.

The rest of the chapter is organized as follows: first we introduce statistical language modeling in more detail and overview major LM techniques for IR. The next section discusses various smoothing strategies used in language models for IR. We then draw comparisons between LM and traditional probabilistic IR approaches. Applications of LM to various retrieval tasks are discussed in the following section. The review concludes with some observations regarding future research directions.

Language Models for IR

The formulation of language models has been based on probability theory. There are a number of theories of what probability means, and the differences can have an effect on how probabilistic models are interpreted. A good discussion of the various theories can be found in Good (1950). A statistical language model is a probability distribution over all possible sentences or other linguistic units in a language (Rosenfeld, 2000). It can also be viewed as a statistical model for generating text. The task of language modeling, in general, is to answer the question: How likely is it that the ith word in a sequence would occur given the identities of the preceding $i - 1$ words? In most applications of language modeling, such as speech recognition and information retrieval, the probability of a sentence is decomposed into a product of n-gram probabilities.

Let's assume that S denotes a specified sequence of k words,

$$S = w_1, w_2, \ldots, w_k$$

An n-gram language model considers the word sequence S to be a Markov process with probability

$$P_n(S) = \prod_{i=1}^{k} P(w_i \mid w_{i-1}, w_{i-2}, w_{i-3}, \ldots, w_{i-n+1})$$

where n refers to the order of the Markov process. When $n = 2$ we call it a bigram language model, which is estimated using information about the co-occurrence of pairs of words. In the case of $n = 1$, we call it a unigram language model, which uses only estimates of the probabilities of individual words. For applications such as speech recognition or machine translation, word order is important and higher-order (usually trigram) models are used. In information retrieval, the role of word order is less clear and unigram models have been used extensively.

To establish the word n-gram language model, probability estimates are typically derived from frequencies of n-gram patterns in training data. It is common that many possible word n-gram patterns do not appear in the actual data used for estimation, even if the size of the data is huge and the value of n is small. As a consequence, for rare or unseen events basing the likelihood estimates directly on counts becomes problematic. This is often referred to as the data sparseness problem. Smoothing is used to address this problem and is an important part in any language model. We discuss it in detail in the next section. Here, we focus our discussion on the conceptual similarities and differences of various language models.

Evaluation of language models in other domains has typically been done using a measure called "perplexity" (Manning & Schütze, 1999). This measure is directly related to entropy. Entropy measures the average uncertainty present for a random variable (Cover & Thomas, 1991). The more knowledge or structure a model captures, the lower the uncertainty, or entropy, will be. Models with lower entropy can therefore be considered better. In ad hoc IR, performance of retrieval models has most often been evaluated based on precision and recall. The average precision measure combines precision and recall into a single number summary. Baeza-Yates and Ribeiro-Neto (1999) give a good discussion of these measures and their appropriateness. In order for the performance of language models to be directly comparable to that of other retrieval models, researchers have taken the average precision measure as the method of choice for evaluation. Throughout this chapter our discussions of retrieval performance will refer to average precision.

Query-Likelihood Model

The basic approach for using language models for IR assumes that the user has a reasonable idea of the terms likely to appear in the "ideal" document that can satisfy his or her information need and that the query terms the user chooses can distinguish the "ideal" document from the rest of the collection (Ponte & Croft, 1998). The query is thus generated as the piece of text representative of the "ideal" document. The task of the system is then to estimate, for each of the documents in the collection, which is most likely to be the ideal document. That is, we calculate:

$$\arg\max_{D} P(D \mid Q) = \arg\max_{D} P(Q \mid D) P(D) \tag{1}$$

where Q is a query and D is a document. The prior probability $P(D)$ is usually assumed to be uniform, and a language model $P(Q|D)$ is estimated for every document. In other words, we estimate a probability distribution over words for each document and calculate the probability that the query is a sample from that distribution. Documents are ranked according to this probability. This is generally referred to as the *query-likelihood* retrieval model and was first proposed by Ponte and Croft (1998). In their paper, Ponte and Croft take a *multivariate Bernoulli* approach to approximate $P(Q|D)$. They represent a query as a vector of binary attributes, one for each unique term in the vocabulary, indicating the presence or absence of terms in the query. The number of times that each term occurs in the query is not captured. A couple of assumptions are behind this approach. The first is the *binary* assumption: All attributes are binary. If a term occurs in the query, the attribute representing the term takes the value of 1. Otherwise, it takes the value of 0. The second is the *independence* assumption: Terms occur independently of one another in a document. These assumptions are the same as those underlying the *binary independence* model proposed in earlier probabilistic IR work (Robertson & Sparck Jones, 1976; van Rijsbergen, 1977). Based on these assumptions, the query likelihood $P(Q|D)$ is thus formulated as the product of two probabilities—the probability of producing the query terms and the probability of not producing other terms.

$$P(Q|D) = \prod_{w \in Q} P(w|D) \prod_{w \notin Q} (1.0 - P(w|D)) \qquad (2)$$

where $P(w|D)$ is calculated by a nonparametric method that makes use of the average probability of w in documents containing it and a risk factor. For nonoccurring terms, the global probability of w in the collection is used instead. It is worth mentioning that collection statistics such as term frequency and document frequency are integral parts of the language model and not used heuristically as they are in traditional probabilistic and other approaches. In addition, document length normalization does not have to be done in an ad hoc manner as it is implicit in the calculation of the probabilities. This approach to retrieval, although very simple, has demonstrated superior performance to traditional probabilistic retrieval using the Okapi-style tf*idf weighting (Robertson, Walker, Sparck Jones, Hancock-Beaulieu, & Gatford, 1995) on Text REtrieval Conference (TREC) test collections.[2] An 8.74 percent improvement in performance (measured in average precision) is reported in the paper. This finding is important because with few heuristics the simple language model can do at least as well as one of the most successful probabilistic retrieval models previously available with heuristic tf*idf weighting.

In contrast to Ponte and Croft's approach, Hiemstra (1998); Miller, Leek, and Schwartz (1999); and Song and Croft (1999) employ a *multinomial* view of the query generation process. They treat the query Q as a sequence of independent terms (i.e., $Q = q_1, ..., q_m$), taking into account possible multiple occurrences of the same term. The "ordered sequence of terms assumption" behind this approach states that both queries and documents are defined by an ordered sequence of terms (Hiemstra, 1998). A query of length k is modeled by an ordered sequence of k random variables, one for each term occurrence in the query. Although this assumption is not usually made in traditional probabilistic IR work, it has been essential for many statistical natural language processing tasks (e.g., speech recognition). Based on this assumption, the query probability can be obtained by multiplying the individual term probabilities.

$$P(Q \mid D) = \prod_{i=1}^{m} P(q_i \mid D) \qquad (3)$$

where q_i is the ith term in the query. Although they employ different theoretical derivations, these models all arrive at a similar way of computing $P(w \mid D)$ (with w denoting any term)—combining a component estimated from the document and one from the collection by linear interpolation (we refer to this formulation henceforth as the probability-weighted model in the rest of this review).

$$P(w \mid D) = \lambda P_{document}(w \mid D) + (1 - \lambda) P_{collection}(w) \qquad (4)$$

where λ is a weighting parameter between 0 and 1. This can also be viewed as a combination of information from a local source, i.e., the document, and a global source, i.e., the collection. The differences between those models reside in how $P_{document}(w \mid D)$ and $P_{collection}(w)$ are estimated. Hiemstra (1998) relates his model to the well-known tf*idf formulation by approximating $P_{document}(w \mid D)$ using the maximum likelihood of term w appearing in the document D, which can be thought of as based on term frequency, and estimating $P_{collection}(w)$ using document frequency information. That is,

$$P_{document}(w \mid D) \cong P_{ml}(w \mid D) = \frac{c(w, D)}{\sum_{w \in D} c(w', D)} \qquad (5)$$

where $c(w, D)$ is the number of times w occurs in D and $\sum_{w \in D} c(w', D)$ is the total number of tokens in D. And

$$P_{collection}(w) \cong \frac{df(w)}{\sum_{w \in V} df(w')} \qquad (6)$$

where V is the vocabulary and $df(w)$ is the document frequency of term w, i.e., the number of documents in which term w appears. A pilot experiment on the Cranfield test collection shows that this model outperforms the traditional vector space model with tf*idf and cosine normalization.

Miller et al. (1999) use a two-state hidden Markov model (HMM) with one state representing the choice of a word directly from the document and the other state representing the choice of a word from general English. $P_{document}(w \mid D)$ is the output distribution for the "document" state, which is estimated in the same way as in equation (5). To approximate the probability distribution of the "general English" state, the sample distribution of the entire document collection is used and $P_{collection}(w)$ is estimated by the maximum likelihood of term w occurring in the collection

$$P_{collection}(w) \cong P_{ml}(w \mid C) = \frac{c(w, C)}{\sum_{w' \in V} c(w', C)} \qquad (7)$$

where $c(w, C)$ is the number of times w occurs in the entire collection, V is the vocabulary, and $\sum_{w' \in V} c(w', C)$ is the total number of tokens in the collection. If we substitute equations (5) and (7) into equation (4), we get the following estimate: $\qquad (8)$

$$P(w \mid D) = \lambda \frac{c(w, D)}{\sum_{w' \in D} c(w', D)} + (1 - \lambda) \frac{c(w, C)}{\sum_{w' \in V} c(w', C)}$$

This is the basic formulation of the HMM model proposed by Miller et al., often referred to as the simple language model, which has been used as the baseline language model in several studies (Jin, Hauptmann, & Zhai, 2002; Lavrenko & Croft, 2001; Liu & Croft, 2002). Retrieval experiments on TREC test collections show that the simple two-state system can do dramatically better than the tf*idf measure. This work also demonstrates that commonly used IR techniques such as relevance feedback as well as prior knowledge of the usefulness of documents can be incorporated into language models. For example, the authors modify their basic model (in equation [8]) to permit: 1) incorporating automatic relevance feedback by re-estimation of $P(w \mid D)$ with additional information from top-ranked documents; 2) adding a third, document-dependent bigram state to the HMM; 3) weighting the importance of different query sections[3] and using the weights as part of the model; and 4) varying the document prior $P(D)$ (as in equation [1]) according to features that are predictive of the usefulness of the document such as source, length, and average word length. The refined system, whether using just one technique or all four, has had further performance gains; using all techniques together is found to be superior to using any single one.

Rather than using maximum likelihood estimates for computing probabilities in equation (4), Song and Croft (1999) propose using Good-Turing estimates. Their basic model is the unigram model presented by equation (4) with Good-Turing estimates. In their second model, the unigram model is combined with a bigram model through linear interpolation. They compare the performance of the combined model with their basic model, the Ponte and Croft (1998) model, and the language models with INQUERY—a probabilistic retrieval system based on the inference net model (Turtle, 1990). Experiments on TREC data sets show that the results of language models are comparable to those of INQUERY. However, their basic and combined models produce similar results and the improvement over the Ponte and Croft model is marginal.

Statistical Translation Model

Taking a different angle, Berger and Lafferty (1999) view a query as a distillation or translation from a document. The query generation process is described in terms of a "noisy channel" model. To determine the relevance of a document to a query, their model estimates the probability that the query would have been generated as a translation of that document. Documents are then ranked according to these probabilities. More specifically, the mapping from a document term w to a query term q is achieved by estimating translation models $t(q \mid w)$. Using translation models, the retrieval model becomes

$$P(Q \mid D) = \prod_{i=1}^{m} \sum_{w} t(q_i \mid w) P(w \mid D)$$

A notable feature of this model is an inherent query expansion component and its capability of handling the issues of synonymy (multiple terms having similar meanings) and polysemy (the same term having multiple meanings). However, as the translation models are context independent, their ability to handle the ambiguity of word senses is limited.

Significant improvements over the baseline language model through the use of translation models have been reported, but this approach is not without its weaknesses: the need of a large collection of training data for estimating translation probabilities and inefficiency in ranking documents.

Building upon the ideas of Berger and Lafferty (1999), Jin et al. (2002) propose constructing language models of document titles and determining the relevance of a document to a query by estimating the likelihood that the query would have been the title for the document. The title of a document is viewed as a translation from that document and the title language model is regarded as an approximate language model of the query. Jin et al. (2002) first estimate a translation model by using all the document-title pairs in a collection. The translation model is then used for mapping a regular document language model to a title language model. In the final step, the title language model estimated for each document is used to compute the query likelihood, and documents are ranked accordingly. It has been shown empirically that the title language model outperforms the simple language model (given in equation [8]) as well as the traditional Okapi method.

Risk Minimization Framework

Lafferty and Zhai (2001a, 2001b) and Zhai (2002) have developed a risk minimization framework based on Bayesian decision theory. In this framework, queries and documents are modeled using statistical language models, user preferences are modeled through loss functions, and retrieval is cast as a risk minimization problem. The similarity between a document and a query is measured by the Kullback-Leibler (KL) divergence between the document model and the query model.

$$KL(Q \parallel D) = \sum_{w \in V} P(w \mid Q) \log \frac{P(w \mid Q)}{P(w \mid D)} \qquad (9)$$

One important advantage of this framework over previous approaches is its capability of modeling not only documents but also queries directly though statistical language models. This makes it possible to set retrieval parameters automatically and improve retrieval performance through utilization of statistical estimation methods, which is not typically done with traditional retrieval methods. This framework bears resemblance to classical probabilistic retrieval models and can accommodate the existing language models proposed by Ponte and Croft (1998) and others. Lafferty and Zhai (2001a) also introduce the idea of estimating expanded query language models for which they use a Markov chain method to help overcome the limitations of the translation models used by Berger and Lafferty (1999). In subsequent work Zhai and Lafferty (2002) suggest using two-stage language models to capture explicitly different influences of the query and document collection on the optimal setting of retrieval parameters. In the first stage, a document language model is estimated independent of the query. In the second stage, query likelihood is computed according to a query language model, which is based on the estimated document language model from the first stage and a query background language model. This approach is similar to the original query likelihood approach by Ponte and Croft (1998) in that it involves both estimation of a document language model and computation of the query likelihood. The difference lies in whether the query likelihood is computed directly using the estimated document model (as is done in the original approach) or using a query model that is based on the estimated language model (as is done in the two-stage approach). A two-stage smoothing method is developed in this approach to set retrieval parameters completely automatically. Empirical evaluations indicate that the two-stage smoothing method consistently gives performance that is comparable with or better than the best obtainable by a single-stage smoothing method, which is usually achieved by an exhaustive search through the whole parameter space.

Relevance Model

Instead of attempting to model the query generation process, Lavrenko and Croft (2001) explicitly model relevance and put forward a novel technique that estimates a relevance model from the query alone, with no training data. Conceptually, the relevance model is a description of an information need or, alternatively, a description of the topic area associated with the information need. It is assumed that, given a collection of documents and a user query Q, there exists an unknown relevance model R that assigns the probabilities $P(w \mid R)$ to the word occurrence in the relevant documents. The relevant documents are random samples from the distribution $P(w \mid R)$. Both the query and the documents are samples from R. The essence of their model is to estimate $P(w \mid R)$. Let $P(w \mid R)$ denote the probability that a word sampled at random from a relevant document would be the word w. If we know what documents are relevant, estimation of these probabilities would be straightforward; but in a typical retrieval environment we are not given any examples of relevant documents. Lavrenko and Croft (2001) and Lavrenko et al. (2002) suggest a reasonable way to approximate $P(w \mid R)$ by using the joint probability of observing the word w together with query words $q_1, ..., q_m$ ($Q = q_1, ..., q_m$):

$$P(w \mid R) \approx P(w \mid Q) = \frac{P(w, q_1 \dots q_m)}{P(q_1 \dots q_m)} = \frac{P(w, q_1 \dots q_m)}{\sum_{v \in vocabulary} P(v, q_1 \dots q_m)} \quad (10)$$

Two methods of estimating the joint probability $P(w, q_1, ..., q_m)$ are described by Lavrenko and Croft (2001). Both methods assume the existence of a set U of underlying source distributions from which $w, q_1, ..., q_m$ could have been sampled. They differ in their independence assumptions. *Method 1* assumes that all query words and the words in relevant documents are sampled from the same distribution, thus w and $q_1, ..., q_m$ are mutually independent once we pick a source distribution M from U. If we assume U to be the universe of our unigram language models, one for each document in the collection, we get:

$$P(w, q_1...q_m) = \sum_{M \in U} P(M)P(w, q_1...q_m | M) = \sum_{M \in U} P(M) \left[P(w | M) \prod_{i=1}^{m} P(q_i | M) \right] \quad (11)$$

where $P(M)$ denotes some prior distribution over the set U, which is usually taken to be uniform, and $P(w|M)$ specifies the probability of observing w if we samples a random word from M. $P(w|M)$ is computed using equation (8). *Method 2* assumes that the query words $q_1, ..., q_m$ are independent of each other but are dependent on w. That is

$$P(w, q_1...q_m) = P(w) \prod_{i=1}^{m} P(q_i | w)$$

The conditional probability $P(q_i|w)$ can be estimated by calculating the expectation over the universe U of our unigram models:

$$P(q_i | w) = \sum_{M_i \in U} P(q_i | M_i) P(M_i | w)$$

Here again an assumption is made that q_i is independent of w once a source distribution M_i is picked. However, the difference from the assumption made in *Method 1* is that each q_i is now allowed to have a separate M_i. Although *Method 2* is shown to be less sensitive to the choice of the universe of distributions U and slightly more effective for retrieval, the relative simplicity and decomposability of *Method 1* has often made it the method of choice for estimation when the relevance model is used (Cronen-Townsend, Zhou, & Croft, 2002; Lavrenko et al., 2002; Liu & Croft, 2002). Lavrenko et al. (2002) employ the KL divergence between the relevance model and the document model to rank documents. Documents with smaller divergence from the relevance model are considered more relevant. The relevance model presents a natural incorporation of query expansion into the language model framework.

Significantly improved retrieval performance has been reported by Lavrenko and Croft (2001) over a simple baseline language model similar to that of equations (3) and (8).

Smoothing Strategies

Virtually all language models developed for IR to date use some form of an n-gram. To derive n-gram probabilities from corpora, one natural solution is to use maximum likelihood (ML) estimation, that is,

$$P_{ml}(w_n \mid w_1, w_2, \ldots, w_{n-1}) = \frac{c(w_1, w_2, \ldots, w_n)}{c(w_1, w_2, \ldots, w_{n-1})}$$

where $c(w_1, w_2, \ldots w_n)$ is the raw counts of the occurrences of the word sequence $w_1, w_2, \ldots w_n$ in the corpus. For unigram, the maximum likelihood of a term w is simply the number of occurrences of w in the corpus divided by the total number of tokens in the corpus. The name maximum likelihood estimate comes from the fact that it does not waste any probability mass on unseen events but rather maximizes the probability of observed events based on the training data while subject to the normal stochastic constraints (Manning & Schütze, 1999). However, it is common that many possible word n-gram patterns would not appear in the actual data, even if there is a very large amount of data and the value of n is small. This causes the ML probability estimates of the unseen n-grams to be zero, which is rather extreme—the fact that we have not seen an n-gram does not make its occurrence impossible. Even worse is the fact that these zero probabilities propagate. For example, the basic LM approach to IR attempts to find the document D that maximizes the probability $P(Q \mid D)$ for a given query Q; and that probability is generally computed by multiplying the probabilities of individual query terms. If a document is missing one or more of the query terms, it will receive a probability of zero even though it may be highly relevant. This is generally known as the sparse data problem, sometimes referred to as the zero-frequency problem (Witten & Bell, 1991). Smoothing is used to address this problem and has become an indispensable part of any language model. Smoothing is so called because the techniques developed for this purpose tend to make the distributions more uniform, by pushing low probabilities (including zero probabilities) upward while balancing out the adjustments by pushing high probabilities downward, so that the total probability mass still sums to one. Although the sparse data problem is the most prominent reason for smoothing in LM for IR, it is not the only reason. Smoothing plays other, subtler roles including combining multiple knowledge sources (Lavrenko, 2000, 2002; Ogilvie, 2000) and accommodating generation of common words in a query (Zhai & Lafferty, 2001a). Several studies have shown that smoothing has a very strong impact on the performance of LM-based retrieval systems (Ponte, 2001; Zhai & Lafferty, 2001b).

Many smoothing techniques have been proposed, mostly for speech recognition tasks (Chen & Goodman, 1998). These include: 1) correcting the ML estimates by pretending each n-gram occurs δ times more than it does with $0 < \delta \leq 1$ (Jeffreys, 1948; Johnson, 1932; Laplace, 1995; Lidstone, 1920); 2) discounting or scaling all non-zero ML estimates by a small constant and distributing the probability mass so gained uniformly over unseen events (Ney & Essen, 1993; Ney, Essen & Kneser, 1994); 3) interpolating probability estimates from different models[4] (e.g., n-grams of different orders) provided that the contribution of each is weighted so that the result is another probability distribution (Chen & Goodman, 1996; Jelinek & Mercer, 1980); and 4) recursively backing off to lower order n-grams (Katz, 1987; Kneser & Ney, 1995; Ney et al., 1994). Note that back-off models of 4) can be thought of as a special case of the general linear interpolation model of 3) if the weights of the interpolated components are chosen so that their value is 0 except for the weight of the model that would have been chosen using the back-off method, which has the value 1.

As an alternative to ML, the Good-Turing estimate (Good, 1953) is often used (usually not by itself but in combination with other techniques) and has been central to many smoothing techniques. There are a number of other methods: the Witten-Bell smoothing (Bell, Cleary, & Witten, 1990; Witten & Bell, 1991), which can be viewed as an instance of interpolation-based methods; the Kneser-Ney smoothing (Kneser & Ney, 1995), which is an extension of the discounting methods; and Bayesian smoothing with Dirichlet priors (MacKay & Peto, 1995) or beta prior (Nadas, 1984), to name a few. Detailed technical treatments of these methods can be found in Jelinek (1997) and Chen and Goodman (1998). Among the smoothing strategies used by IR researchers, many are borrowed from the speech community, but there has also been considerable development where smoothing has been considered specifically for the IR domain. Different situations call for different strategies.

Parameter Smoothing

In the LM approach to IR, one has to infer a language model $P(w \mid D)$ for each document D, which is a probability distribution of all possible words w (e.g., in the vocabulary) in that document. Clearly, a document by itself is too small a sample for which to derive good ML estimates. To produce a reasonable estimated document language model, Ponte and Croft (1998) and Ponte (1998) employ a variety of smoothing techniques. For observed terms, their ML estimates are adjusted by a factor based on the average probability of a term in documents containing it and the associated risk (which is a geometric distribution). For unseen terms, the estimates simply back off to the collection probabilities. As an improvement to the basic method, a histogram estimator is used for low frequency terms.

The two-state HMM developed by Miller et al. (1999) can be viewed as a two-part mixture model. Although the motivations are different, the effect is to adjust the estimates of the query terms by using additional information from the collection. Song and Croft (1999) use the Good-Turing estimate in place of the ML estimate for the document model and then interpolate each document model with a background (or collection) model to produce the final document model (equation [4]). In addition, they consider term pairs and interpolate the unigram document model with a bigram model, using the intuition that phrases would aid retrieval.

Differing from these approaches in which the background model is estimated from relative term frequencies in collection, Hiemstra (1998) uses relative document frequencies of terms for smoothing. He argues that the background model so constructed is, in effect, a probabilistic interpretation of the traditional idf heuristics and that the final smoothed document model achieves the effect of traditional tf*idf weighting with document length normalization.

In other work Hiemstra (2002) proposes to model the importance of the query terms explicitly. Previous approaches have treated each query term as equally important; thus a single smoothing parameter λ is used for all query terms (see equations [3] and [4]). Hiemstra (2002) suggests using a different smoothing parameter λ_i for each query term q_i; thus the model becomes

$$P(Q \mid D) = \prod_{i=1}^{m} (\lambda_i P_{document}(q_i \mid D) + (1 - \lambda_i) P_{collection}(q_i))$$

Relevance feedback can be done with this model by boosting the λ values of the query terms that appear in the relevant documents while deemphasizing those terms that do not.

Jin et al. (2002) construct title language models based on the idea of statistical machine translation. The use of translation models in their approach is theoretically motivated, but it also plays a role in smoothing. Jin et al. (2002) first treat the titles as translations of documents and train a translation model based on the document-title pairs in the whole collection. Instead of being estimated based solely on the observations of the document titles, the title language model of a document is now estimated by applying the learned translation model to the document. The translation model helps alleviate the data sparseness problem to some degree, but this is not enough. The training data are still sparse given the large number of parameters involved. To cope with this, Jin et al. (2002) extend the standard learning algorithms of the translation models by adding special parameters to model the self-translation probabilities of words.

Zhai and Lafferty (2001b) study a number of interpolation-based approaches to smoothing, including Jelinek-Mercer smoothing, Bayesian smoothing with Dirichlet priors, and absolute discounting, as well as their back-off versions. Several large and small TREC collections are used for evaluating these methods. They find that different situations call for different approaches to smoothing, that retrieval performance is generally sensitive to smoothing parameters, and that the effect of smoothing is strongly dependent on the type of queries. They explain their empirical results by suggesting that smoothing plays two roles: one is to overcome the sparse data problem, and the other is to help generate common words in a query so as to cause query terms to be weighted in a similar fashion as the idf heuristics (Zhai & Lafferty, 2001a). Motivated by decoupling the two different roles of smoothing, Zhai and Lafferty (2002) develop a two-stage smoothing method. In the first stage, the document language model is smoothed with a Dirichlet prior with the collection model, and in the second stage, the smoothed document model is further interpolated with a query background model. An important advantage of this smoothing method is that it allows fully automatic estimation of the parameters. It is shown to be quite effective compared to a single-stage smoothing with an exhaustive parameter search on the test data.

Semantic Smoothing

To move beyond parameter smoothing, which plays a role similar to traditional term weighting, researchers have begun to look at semantic smoothing—another role that smoothing plays. As we pointed out earlier in this section, smoothing (more specifically linear interpolation or mixture models) can be used to combine knowledge sources. Most language models in IR can be generalized to the form of a mixture of a sparse topic model and a background model. The smoothing strategies we have discussed so far approximate the topic model by counting words in a sample of text (e.g., document). What if we have more information about the topic than just the document? Semantic smoothing aims to adjust the probability estimates of terms by exploiting the context of words, for example, relevant documents, so that terms that are closely related to the original topic will attract higher probabilities. The translation model proposed by Berger and Lafferty (1999) captures semantic relations between words based on term co-occurrences. However, the semantic relations between words are captured not by the language model but by the translation model, as they are learned from synthetic query/document pairs rather than the co-occurrences within the document collection.

Hofmann (1999a, 1999b) discusses how latent classes or topic models can be incorporated into the language modeling framework. These topic models provide a form of smoothing based on reducing the dimensionality of the corpus. Peters (2001) proposes a similar approach based on clustering. Lavrenko (2000) hypothesizes that he could achieve semantic smoothing by using a zone of closely related text samples. He estimates a contextual model based on the texts in the zone and interpolates this contextual model with the document model and the background model. He chooses to use Witten-Bell smoothing for estimating the weights of different models. The main problem with this approach is that retrieval performance is extremely sensitive to the number of text samples in the zone. Ogilvie (2000) encounters similar problems when he smoothes document models with models of their nearest neighbors.

Lafferty and Zhai (2001a) and Lavrenko and Croft (2001) propose using a weighted mixture of top-ranked documents from the query to approximate a topic model. Lafferty and Zhai (2001a) make use of a Markov chain method to assign weights to documents whereas Lavrenko and Croft (2001) employ an estimate of a joint probability from observing the query words together with any word in the vocabulary. Both methods have enabled large improvements in retrieval performance over models that do not use semantic smoothing, but again they are very sensitive to the number of top-ranked documents that are used for probability estimation. Lavrenko (2002) explores the possibility of automatically finding the optimal subset of documents to construct an optimal mixture model. Two types of mixture models—set-based and weighted— are considered. He proves that it is not feasible to compute set-based optimal mixture models and proposes a gradient descent procedure for estimating weighted mixture models. Retrieval experiments indicate that the weighted mixture models are relatively insensitive to the number of top-ranked documents used in the estimation.

Comparisons with Traditional Probabilistic IR Approaches

The language modeling approach has introduced a new family of probabilistic models to IR. Several researchers have attempted to relate this new approach to the traditional probabilistic IR approaches and compare performance. The first workshop on language modeling and information retrieval, held at Carnegie Mellon University from May 31 to June 1, 2001, has facilitated this discussion (Croft, Callan, & Lafferty, 2001).

Sparck Jones and Robertson (2001) examine the notion of relevance in the traditional probabilistic approach (probabilistic modeling, or PM) and the new language modeling approach, pointing out that two distinctions should be made between the two approaches. The first, and what they call *surface* distinction, is that although in both approaches a good match on index keys between a document and a query implies relevance, relevance figures explicitly in PM but is never mentioned in LM. The second, and the more important difference, is that the underlying principle of LM is to identify *the* ideal document that generates the query rather than a list of relevant documents. Thus, once this ideal document is recovered, retrieval stops. Because of this, they argue, it is difficult to describe important processes such as relevance feedback in existing LM approaches. Lafferty and Zhai (2001a, 2001b) and Lavrenko and Croft (2001) address these issues directly and suggest new forms of LM to retrieval that are more closely related to the traditional probabilistic approach. Lafferty and Zhai (2001b) argue that, in the traditional probabilistic approach proposed by Robertson and Sparck Jones (1976), a document could be thought of as generated from a query using a binary latent variable that indicates whether the document is relevant to the query. They show through mathematical derivations that, if a similar binary latent variable is introduced to LM, these two methods are on equal footing in terms of the relevance ranking principle and interpretation of the ranking process. However, this does not mean that PM and LM are just versions of each other. The differences go beyond a simple application of Bayes's law. They point out that document length normalization is a critical issue in PM but not in LM. Another difference is that in LM we have more data for estimating a statistical model than in PM, which is the advantage of "turning the problem around." Both the risk-minimization framework suggested by Lafferty and Zhai (2001a, 2001b) and the relevance model suggested by Lavrenko and Croft (2001) move away from estimating the probability of generating query text (the query-likelihood model) to estimating the probability of generating document text (document-likelihood) or comparing query and document language models directly. Greiff (2001) suggests that the main contributions of LM to IR lie in the recognition of parameter estimation's importance in modeling and in the treatment of term frequency as the manifestation of an underlying probability distribution rather than as the probability of word occurrence itself. Zhai and Lafferty (2002) point out that traditional IR models rely heavily on ad hoc parameter tuning to achieve satisfactory performance whereas in LM, statistical estimation methods can be used to set parameters automatically.

Hiemstra and de Vries (2000) relate LM to traditional approaches by comparing Hiemstra's model (1998) with the tf*idf term weighting and the combination with relevance weighting as done in the BM25 algorithm. They conclude that LM and PM have important similarities in that LM provides a probabilistic interpretation and justification of the tf*idf weighting and suggests why, in combination with relevance weighting in BM25, it is effective. Fuhr (2001) shows how the LM approach can be related to other probabilistic retrieval models (Wong & Yao, 1995) in the framework of uncertain inference.

Applications

The language modeling approach was initially developed for ad hoc retrieval and found to be very effective. Soon after, successful applications of this approach to other retrieval tasks were reported, including relevance feedback, distributed IR, cross-lingual IR, quantification of query ambiguity, and passage retrieval.

Query Ambiguity

Predicting query performance and thus dealing gracefully with ambiguous queries have long been an interest and challenge in information retrieval. An often-used example of ambiguous queries is "bank," where the context in which the word appears could be financial institutions, river bank, or a flight maneuver, among others. Given such a query, IR systems need more user information than just the query words to resolve the ambiguity. One obvious solution is relevance feedback. That is, the retrieval system treats all queries (regardless of their degree of ambiguity) in the same way initially and then refines the document list based on user clarification of the initial presentation. However, relying on relevance feedback to solve the problem may not always be realistic. A more effective alternative is to determine if each query is ambiguous and ask users specific questions about ambiguous queries. From the beginning vague queries will be handled differently from clear ones. Croft, Cronen-Townsend, and Lavrenko (2001); Cronen-Townsend and Croft (2002); and Cronen-Townsend et al. (2002) have developed a "clarity" measure within the language model framework to quantify query ambiguity with respect to a collection of documents without relevance information. In this approach, a query receives a non-negative "clarity" value based on how different its associated language model is from the language model of the collection, with zero meaning that the query is maximally ambiguous—its associated language model is indistinguishable from the collection language model. Formally, the "clarity" measure is defined as the Kullback-Liebler divergence between the query distribution and the collection distribution (Croft, Cronen-Townsend, & Lavrenko, 2001; Cronen-Townsend et al., 2002).

$$clarity \equiv KL(Q \parallel Coll) = \sum_{w \in V} P(w \mid Q) \log \frac{P(w \mid Q)}{P_{coll}(w)}$$

where w is any term, Q is the query, $Coll$ is the collection, V is the vocabulary of the collection, $P_{coll}(w)$ is the relative frequency of the term w in the collection, and $P(w \mid Q)$ is the query language model. Two types of language models, namely, the probability-weighted models and relevance models, have been used in their work for creating the query language model. In the probability-weighted approach, the query language model is taken to be

$$P(w \mid Q) \cong \sum_{D \in Ret} P(w \mid D) P(D \mid Q)$$

where D is a document, Q is a query, and Ret is the set of retrieved documents. $P(D \mid Q)$ is the Bayesian inversion of $P(Q \mid D)$ with uniform document priors. Individual document models $P(w \mid D)$ are estimated using equation (8) and $P(Q \mid D)$ is estimated using equations (3) and (8). In the relevance model approach, the query language model is given by $P(w \mid R)$ as defined in equations (10) and (11). Croft, Cronen-Townsend, and Lavrenko (2001) observe that the two types of query language models produce similar ranking when ordering queries based on their clarity scores. Cronen-Townsend and Croft (2002) and Cronen-Townsend et al. (2002) demonstrate that the clarity measure is highly correlated with the retrieval performance of the query and show how thresholds can be automatically determined to identify queries with poor language models.

Relevance Feedback

The basic approach to relevance feedback has been to modify the query using words from top-ranked or identified relevant documents. One way to use LM for this task is to build a language model for the top-ranked or relevant documents and augment the query with words that have a relatively high log ratio of the probability of occurring in the model for relevant documents against the probability of occurring in the background (collection) model. This is the approach taken by Ponte (2000). In his work, the language model of the relevant documents is simply the sum of the individual document models. In the HMM system used by Miller et al. (1999), query expansion is achieved by adding to the initial query the words appearing in two or more of the top N retrieved documents and adjusting the HMM transition probabilities through training over queries. Croft, Cronen-Townsend, and Lavrenko (2001) view the query as a sample of text from a model of the information need. They hypothesize that users could also be represented as a mixture of topic language models generated from previous interactions and other sources. The task then boils down to estimating a language model associated with the query. Once we have the query language model, retrieval is straightforward—we can rank documents either according to the likelihood of their being generated by the query model (Lavrenko & Croft, 2001) or directly based on their similarity to the query model (Lafferty & Zhai, 2001a). Lavrenko and Croft (2001) develop relevance-based language models to approximate query models; Lafferty and Zhai (2001a) make use of a Markov chain method.

Distributed IR

The major difference between distributed information retrieval and centralized retrieval is that distributed IR typically uses multiple collections at different sites so each site maintains its own index, whereas centralized retrieval uses a centralized index for both indexing and retrieval. Therefore, one has to solve the problems of 1) resource selection, 2) ad hoc retrieval, and 3) results merging, when doing IR in a distributed environment. Language modeling can be used in any of these steps or applied as a general integrated framework. Xu and Croft (1999) apply language modeling to resource selection. The basic idea is to group documents into clusters, each cluster representing a topic. A language model is built on the word usage in a topic (cluster). To determine which topics are best for the query, the KL divergence is used to measure how well a topic model predicts the query. Based on this general idea, three different methods of organizing a distributed retrieval system are tested, all showing improvements over the traditional method of distributed retrieval with heterogeneous collections. Si et al. (2002) present an LM-based framework that integrates all three subtasks of distributed IR. Query-based sampling is first applied to acquire language model descriptions of individual databases. In the resource selection step, databases are ranked according to the likelihood of a given query being generated from each of them. In the retrieval step, a language model-based retrieval algorithm is used. In the result-merging step, the document scores are recomputed to remove the possible bias caused by different collection statistics. The authors demonstrate through experiments that a simple language model can be used effectively in an intranet distributed IR environment.

Cross-Lingual IR

The goal of cross-lingual IR is to find documents in one language for queries in another. A straightforward adaptation of an LM approach to this task is to view the query in one language as generated from a document in another. Berger and Lafferty (1999) treat query generation as a translation process. Although their model has been used only for monolingual retrieval so far, it can easily accommodate a cross-lingual environment. Hiemstra and de Jong (1999) extend the model proposed by Hiemstra (1998; also given in equations [4] through [6]) to incorporate statistical translation for use in cross-lingual retrieval (in their work, Dutch queries on English document collections). By assuming the independence of the translation of a term and the document from which it is drawn, and also assuming each query term q_i has k_i possible English translations t_{ij} ($1 \leq j \leq k_i$), they end up with

$$P(Q \mid D) = P(q_1 \ldots q_m \mid D) = \prod_{i=1}^{m} \sum_{j=1}^{k_i} P(q_i \mid t_{ij}) P(t_{ij} \mid D)$$

where $P(t_{ij} \mid D)$ is computed using equations (4), (5), and (6). Note that the probability estimates of query terms in the source language are, in effect, smoothed by using the background model built on a document collection in the target language. Xu and Weischedel (2000) and Xu et al. (2001) suggest that if the background model can be built directly using a document collection in the source language, then the noise introduced by translation can be avoided. Motivated by this suggestion, they present an extension of the monolingual HMM model proposed by Miller et al. (1999) to use for querying Chinese (or Spanish) document collections with English queries. The computation involves the following expression

$$P(Q_e \mid D_c) = \prod_{i=1}^{m}(\lambda P(q_i \mid GE) + (1 - \lambda) \sum_{c \in Chinese} P(c \mid D)P(q_i \mid c))$$

where Q_e stands for an English query, D_c stands for a Chinese document, GE stands for general English, and c is any word in Chinese. Their cross-lingual system achieves roughly 90 percent of monolingual performance in retrieving Chinese documents and 85 percent in retrieving Spanish documents. Lavrenko et al. (2002) apply the relevance model to a similar task with new estimation methods that are adapted to cross-lingual retrieval. The model starts with a query in the source language and directly estimates the model of relevant documents in the target language, which is different from other models in that it does not attempt to translate either the documents or the query. Retrieval performance comparable to other models is observed.

Passage Retrieval

Given that early work on language modeling for IR has been entirely document based, Liu and Croft (2002) address the question whether language models would be feasible for passage retrieval. In their work, a probability-weighted, query-likelihood model and a relevance model are used. After experimenting with different types of passages (e.g., half-overlapped windows and arbitrary passages at different passage sizes) and various ways of constructing the language models for doing passage retrieval, they find that all passage retrieval runs produce results comparable with, and sometimes significant improvements over, full-length document retrieval when using language models. A second finding is that passage retrieval can provide more reliable performance than full-length document retrieval in the language model framework, especially when using relevance models.

Incorporating Prior Knowledge

It is common in IR that documents are treated as being equally likely to be relevant; that is, the prior probability *P(D)* is assumed to be uniform (see also equation [1]). However, in practice, some documents with certain properties may be more likely to be relevant for a given task. An example is document length. If these properties are known and can be exploited by the retrieval system, they should help improve retrieval performance. Traditional IR approaches have attempted to incorporate prior knowledge, but the process was rather heuristic and not based on formal models. The LM approach to IR, in contrast, has provided a natural and principled way of incorporating knowledge. Several papers describe efforts in using prior knowledge with the LM framework. Hiemstra and Kraaij (1999) show that document length is helpful for ad hoc retrieval, especially for short queries. Miller et al. (1999) combine several features in their document priors, including document source, length, and average word length. Small improvement over uniform priors is observed. Zhu (2001) discusses how to derive features beyond bags of words in language models via discriminative techniques. Kraaij, Westerveld, and Hiemstra (2002) use several Web page features such as the number of inlinks and the form of the URL as prior knowledge for an LM-based entry page retrieval system. They show that language models can accommodate such knowledge in an almost trivial manner, by estimating the posterior probability of a page, given its features and using this posterior probability as the document prior in the language model. Evaluations demonstrate that significant performance improvements over the baseline LM system can be obtained through proper use of prior probabilities. Li and Croft (2003) show that time serves as a useful prior for queries that favor very recent documents and those that have more relevant documents within a specific period in the past. Depending on the type of query, either an exponential distribution or a normal distribution is used to replace the uniform prior probability in a query-likelihood model and a relevance model.

Research Directions

Statistical language modeling has brought many opportunities to IR. First of all, it provides us with a formal method of reasoning about the design and empirical performance of an IR system. Many heuristic techniques introduced in the past can now be explained and accommodated by this new framework. A case in point is the tf*idf weighting used in traditional IR systems. Second, because language models have been in use for nearly thirty years in other language processing tasks such as automatic speech recognition, considerable experience has been gained and lessons learned that could be leveraged by the IR community. For example, smoothing has been studied extensively in the speech community and many smoothing methods have been developed. Researchers in the IR community have started looking at what is available there and either adapting existing models to retrieval tasks or developing new techniques based on them. Third, the statistical language modeling approach applies naturally to a variety of information system technologies, such as ad hoc and distributed retrieval, cross-lingual IR, summarization, filtering, topic tracking and detection, and, possibly, question answering. Preliminary successes have been reported in many of these areas. Because of the advantages LM offers, it is considered the most promising framework for advancing information retrieval to meet future challenges (Allan, Aslam, Belkin, Buckley, Callan, Croft, et al., 2003).

One long-term challenge of IR is to provide global information access to users so that information needs expressed in any language can be satisfied by information from many sources, encoded in any language. This calls for the development of massively distributed, cross-lingual retrieval systems. Existing techniques in distributed IR, cross-lingual IR, and possibly data fusion can easily lend themselves to the design of such a system, but simply combining these techniques would not result in a satisfactory solution. Therefore, to cope with this challenge, current LM methods must be extended to provide a unified retrieval framework that leverages techniques from multiple fields.

Another long-term challenge of IR is to capture information about the user and context of the information retrieval process and to integrate models of users into retrieval models. In current IR systems, little has been done to achieve this; models of the user are weak if not missing from retrieval models. As a result, current IR systems resort to a user-generic approach in which different users with the same query will be provided with the same results. Although this approach has proven good enough for the average user, it leaves much room for improvement in retrieval effectiveness for individual users. One way for LM to help IR move beyond this limitation is to represent the user by a probability distribution of interests (e.g., words, phrases, or topics), actions (e.g., browsing behavior), and annotations/judgments (Allan et al., 2003). Knowledge about the user-related context (e.g., user type, background, and personalization) and task-related context (e.g., genre, level, authority, and subject domain) can be encoded in the priors of the language models.

As stepping stones toward achieving these long-term goals, there are some shorter-term challenges with which LM must cope. Current LM techniques must be extended to incorporate diverse data sources and multiple forms of evidence. In Web searching, for instance, it is a challenge to exploit all sorts of evidence, including the Web structure, metadata, and user context information, in order to find high-quality information for users. This requires LM to allow more precise representation of information needs, integration of structural evidence, and incorporation of linguistic/semantic knowledge. For question answering (QA), the standard approach has been locating documents or passages with candidate answers, then integrating multiple passages and multiple data sources, including structured and unstructured text, to provide the final answer. Current statistical language models have proven useful for document/passage retrieval, but are not adequate for finding exact answers. A promising direction seems to be to extend language models to include structured patterns which are used rather heuristically in existing systems.

New LM techniques also need to be developed to support more advanced retrieval tasks as well as provide more integrated models for current tasks. For example, retrieval would benefit from adding structure (e.g., proximity operators) to the query. The challenge for LM is to study how the structure can be represented in probabilistic terms. In cross-lingual IR, most current LM approaches make use of a probabilistic model for translation and a language model for retrieval. However, the two components are only loosely coupled and independence assumptions are made for each of them. One possible way to improve this is to explore models with loosened independence assumptions, thus enabling the direct use of contextual information on words and phrases for translation. LM techniques that allow for a tighter integration of the translation model and the retrieval model are likely to be explored. In addition, better mechanisms for relevance feedback may be necessary. Techniques developed for ad hoc retrieval have been successfully employed in distributed IR, but developing a theoretically grounded model specific to distributed IR could provide a unified model for the metasearch problem (data fusion and collection fusion).

As LM for IR grows, existing models will continue to be refined, parameter estimation procedures will be improved, and the relationships among the various modeling approaches will be examined more carefully. Interest in applying more smoothing techniques and combining language models for multiple topics and collections is likely to continue. For example, probabilistic latent semantic indexing (Hofmann, 1999b) seems to be a promising technique to be used for topic models. Models based on clustering techniques may also be explored. We can expect more sophisticated language modeling techniques to be developed that allow for increasingly integrated representations across documents, collections, languages, topics, queries, and users.

Acknowledgments

This review was supported in part by the Center for Intelligent Information Retrieval, in part by SPAWARSYSCEN-SD grant number N66001-02-1-8903, and in part by Advanced Research and Development Activity under contract number MDA904-01-C-0984. Any opinions, findings and conclusions, or recommendations expressed in this material are the authors' and do not necessarily reflect those of the sponsors.

Endnotes

1. This is a collaborative project of the University of Massachusetts and Carnegie Mellon University, with the sponsorship of the Advanced Research and Development Activity in Information Technology (ARDA), to develop an open-source toolkit for language modeling in information retrieval. The toolkit is available at http://www.cs.cmu.edu/~lemur.

2. TREC, co-sponsored by the National Institute of Standards and Technology (NIST) and the Defense Advanced Research Projects Agency (DARPA), supports research in the information retrieval community by providing infrastructure (such as realistic test collections and appropriate evaluation procedures) for large-scale evaluation of various (text) retrieval methods. It also serves as a forum for the exchange of research ideas and

for the discussion of research methodology. More information can be found at: http://trec.
nist.gov.

3. Miller et al. (1999) take queries from TREC topics. A TREC topic typically consists of
three sections: Title, Description, and Narrative, with increasing details about a given
subject.

4. Another name for interpolation-based models is mixture models.

References

Allan, J., Aslam, J., Belkin, N., Buckley, C., Callan, J., Croft, W. B., et al. (2003). Challenges
in information retrieval and language modeling [Special issue]. *SIGIR Forum, 37*(1).

Baeza-Yates, R., & Ribeiro-Neto, B. (1999). *Modern information retrieval.* New York: ACM
Press Series/Addison Wesley.

Bell, T. C., Cleary, J. G., & Witten, I. H. (1990). *Text compression.* Englewood Cliffs, NJ:
Prentice Hall.

Berger, A., & Lafferty, J. (1999). Information retrieval as statistical translation. *Proceedings
of the 22nd ACM SIGIR Annual International Conference on Research and Development
in Information Retrieval,* 222–229.

Chen, S. F., & Goodman, J. (1996). An empirical study of smoothing techniques for language
modeling. *Proceedings of the 34th Annual Meeting of the Association for Computational
Linguistics,* 310–318.

Chen, S. F., & Goodman, J. (1998). *An empirical study of smoothing techniques for language
modeling* (Technical Report TR-10-98). Boston, MA: Center for Research in Computing
Technology, Harvard University.

Cooper, W. S. (1995). Some inconsistencies and misidentified modeling assumptions in probabilistic information retrieval. *ACM Transactions on Information Systems, 13*(1),
100–111.

Cooper, W. S., & Maron, M. E. (1978). Foundations of probabilistic and utility-theoretic
indexing. *Journal of the ACM, 25*(1), 67–80.

Cover, T. M., & Thomas, J. A. (1991). *Elements of information theory.* New York: Wiley.

Croft, W. B., Callan, J., & Lafferty, J. (2001). Workshop on language modeling and information retrieval. *SIGIR Forum, 35*(1), 4–6.

Croft, W. B., Cronen-Townsend, S., & Lavrenko, V. (2001). Relevance feedback and personalization: A language modeling perspective. *Proceedings of the DELOS-NSF Workshop
on Personalization and Recommender Systems in Digital Libraries,* 49–54.

Croft, W. B., & Harper, D. J. (1979). Using probabilistic models of document retrieval without relevance information. *Journal of Documentation, 35,* 285–295.

Croft W. B., & Lafferty, J. (Eds.). (2003). *Language modeling for information retrieval.*
Dordrecht, The Netherlands: Kluwer.

Cronen-Townsend, S., & Croft, W. B. (2002). Quantifying query ambiguity. *Proceedings of
the Human Language Technology 2002 Conference,* 94–98.

Cronen-Townsend, S., Zhou, Y., & Croft, W. B. (2002). Predicting query performance.
*Proceedings of the 25th ACM SIGIR Annual International Conference on Research and
Development in Information Retrieval,* 299–306.

Fuhr, N. (1989). Models for retrieval with probabilistic indexing. *Information Processing &
Management, 25*(1), 55–72.

Fuhr, N. (2001). Language models and uncertain inference in information retrieval. In J.
Callan, W. B. Croft, & J. Lafferty (Eds.), *Proceedings of the Workshop on Language
Modeling and Information Retrieval* (pp. 6–11). Pittsburgh, PA: Carnegie Mellon
University.

Good, I. J. (1950). *Probability and the weighting of evidence.* London: Charles Griffin.

Good, I. J. (1953). The population frequencies of species and the estimation of population parameters. *Biometrika, 40*, 237–264.

Greiff, W. (2001). Is it the language model in language modeling? In J. Callan, W. B. Croft, & J. Lafferty (Eds.), *Proceedings of the Workshop on Language Modeling and Information Retrieval* (pp. 26–30). Pittsburgh, PA: Carnegie Mellon University.

Hiemstra, D. (1998). A linguistically motivated probabilistic model of information retrieval. *Proceedings of the Second European Conference on Research and Advance Technology for Digital Libraries (ECDL)*, 569–584.

Hiemstra, D. (2002). Term-specific smoothing for the language modeling approach to information retrieval: The importance of a query term. *Proceedings of the 25th ACM SIGIR Annual International Conference on Research and Development in Information Retrieval*, 35–41.

Hiemstra, D., & de Jong, F. (1999). Disambiguation strategies for cross-language information retrieval. *Proceedings for the Third European Conference on Research and Advanced Technology for Digital Libraries*, 274–293.

Hiemstra, D., & de Vries, A. (2000). *Relating the new language models of information retrieval to the traditional retrieval models* (CTIT Technical Report TR-CTIT-00-09). Enschede, The Netherlands: Centre for Telematics and Information Technology, University of Twente.

Hiemstra, D., & Kraaij, W. (1999). Twenty-one at TREC-7: Ad-hoc and cross-language track. In E. M. Voorhees & D. K. Harman (Eds.), *The Seventh Text Retrieval Conference (TREC7)* (pp. 174–185). Gaithersberg, MD: National Institute of Standards and Technology.

Hofmann, T. (1999a). From latent semantic indexing to language models and back. In J. Callan, W. B. Croft, & J. Lafferty (Eds.), *Proceedings of the Workshop on Language Modeling and Information Retrieval* (pp. 42–46). Pittsburgh, PA: Carnegie Mellon University.

Hofmann, T. (1999b). Probabilistic latent semantic indexing. *Proceedings of the 22nd ACM SIGIR Annual International Conference on Research and Development in Information Retrieval*, 50–57.

Jeffreys, H. (1948). *Theory of probability*. Oxford, UK: Clarendon Press.

Jelinek, F. (1997). *Statistical methods for speech recognition*. Cambridge, MA: MIT Press.

Jelinek, F., & Mercer, R. L. (1980, May). Interpolated estimation of Markov source parameters from sparse data. In S. Gelsema & L. N. Kanal (Eds.), *Proceedings of the Workshop on Pattern Recognition in Practice* (pp. 381–397). Amsterdam: North-Holland.

Jin, R., Hauptmann, A. G., & Zhai, C. (2002). Title language model for information retrieval. *Proceedings of the 25th ACM SIGIR Annual International Conference on Research and Development in Information Retrieval*, 42–47.

Johnson, W. E. (1932). Probability: Deductive and inductive problems. *Mind, 41*, 421–423.

Katz, S. M. (1987). Estimation of probabilities from sparse data for the language model component of a speech recognizer. *IEEE Transactions on Acoustics, Speech and Signal Processing, 35*(3), 400–401.

Kneser, R., & Ney, H. (1995). Improved backing-off for m-gram language modeling. *Proceedings of the IEEE International Conference on Acoustics, Speech and Signal Processing: Vol. I* (pp. 181–184). Detroit, MI: IEEE.

Kraaij, W., Westerveld, T., & Hiemstra, D. (2002). The importance of prior probabilities for entry page search. *Proceedings of the 25th ACM SIGIR Annual International Conference on Research and Development in Information Retrieval*, 27–34.

Lafferty, J. & Zhai, C. (2001a). Document language models, query models, and risk minimization for information retrieval. *Proceedings of the 24th ACM SIGIR Annual International Conference on Research and Development in Information Retrieval*, 111–119.

Lafferty, J., & Zhai, C. (2001b). Probabilistic IR models based on document and query generation. In J. Callan, W. B. Croft, & J. Lafferty (Eds.), *Proceedings of the Workshop on Language Modeling and Information Retrieval* (pp. 1–5). Pittsburgh, PA: Carnegie Mellon University.

Laplace, P. S. (1995). *Philosophical essay on probabilities*. New York: Springer-Verlag.

Lavrenko, V. (2000). *Localized smoothing of multinomial language models* (CIIR Technical Report IR-222). Amherst: University of Massachusetts.

Lavrenko, V. (2002). Optimal mixture models in IR. *Proceedings of the 24th European Colloquium on IR Research*, 193–212.

Lavrenko, V., Choquette, M., & Croft, W. B. (2002). Cross-lingual relevance models. *Proceedings of the 25th ACM SIGIR Annual International Conference on Research and Development in Information Retrieval*, 175–182.

Lavrenko, V., & Croft, W. B. (2001). Relevance-based language models. *Proceedings of the 24th ACM SIGIR Annual International Conference on Research and Development in Information Retrieval*, 120–127.

Lidstone, G. J. (1920). Note on the general case of the Bayes-Laplace formula for inductive or a posteriori probabilities. *Transactions of the Faculty of Actuaries, 8*, 182–192.

Li, X., & Croft, W. B. (2003). Time-based language models. *Proceedings of the Twelfth International Conference on Information and Knowledge Management (CIKM'03)*, 469–475.

Liu, X., & Croft, W. B. (2002). Passage retrieval based on language models. *Proceedings of the Eleventh International Conference on Information and Knowledge Management (CIKM'02)*, 375–382.

MacKay, D. J. C., & Peto, L. C. B. (1995). A hierarchical Dirichlet language model. *Natural Language Engineering, 1*(3), 1–19.

Manning, C. D., & Schütze, H. (1999). *Foundations of statistical language processing*. Cambridge, MA: MIT Press.

Maron, M. E., & Kuhns, J. L. (1960). On relevance, probabilistic indexing and information retrieval. *Journal of the ACM, 7*, 216–244.

Miller, D., Leek, T., & Schwartz, R. (1999). A hidden Markov model information retrieval system. *Proceedings of the 22nd ACM SIGIR Annual International Conference on Research and Development in Information Retrieval*, 214–221.

Nadas, A. (1984). Estimation of probabilities in the language model of the IBM speech recognition system. *IEEE Transactions on Acoustics, Speech and Signal Processing, 32*(4), 859–861.

Ney, H., & Essen, U. (1993). Estimating "small" probabilities by leaving-one-out. *Eurospeech'93: Vol. 3*, 2239–2242.

Ney, H., Essen, U., & Kneser, R. (1994). On structuring probabilistic dependences in stochastic language modeling. *Computer Speech and Language, 8*, 1–38.

Ogilvie, P. (2000). *Nearest neighbor smoothing of language models in IR*. Unpublished manuscript.

Peters, J. (2001). Semantic text clusters and word classes: The dualism of mutual information and maximum likelihood. In J. Callan, W. B. Croft, & J. Lafferty (Eds.). *Proceedings of the Workshop on Language Modeling and Information Retrieval* (pp. 55–59). Pittsburgh, PA: Carnegie Mellon University.

Ponte, J. (1998). *A language modeling approach to information retrieval*. Unpublished doctoral dissertation, University of Massachusetts, Amherst.

Ponte, J. (2000). Language models for relevance feedback. In W. B. Croft (Ed.), *Advances in information retrieval: Recent research from the CIIR* (pp. 73–95). Boston: Kluwer Academic.

Ponte, J. (2001). Is information retrieval anything more than smoothing? In J. Callan, W. B. Croft, & J. Lafferty (Eds.), *Proceedings of the Workshop on Language Modeling and Information Retrieval* (pp. 37–41). Pittsurgh, PA: Carnegie Mellon University.

Ponte, J., & Croft, W. B. (1998). A language modeling approach to information retrieval. *Proceedings of the 21st ACM SIGIR Annual International Conference on Research and Development in Information Retrieval,* 275–281.

Robertson, S. E. (1977). The probability ranking principle in IR. *Journal of Documentation, 33,* 294–304.

Robertson, S. E., & Sparck Jones, K. (1976). Relevance weighting of search terms. *Journal of the American Society for Information Science, 27*(3), 129–146.

Robertson, S. E., Walker, S., Sparck Jones, K., Hancock-Beaulieu, M. M., & Gatford, M. (1995). Okapi at TREC-3. *Proceedings of the Third Text REtrieval Conference (TREC-3),* 109–126.

Rosenfeld, R. (2000). Two decades of statistical language modeling: Where do we go from here? *Proceedings of the IEEE, 88*(8), 1270–1278.

Shannon, C. E. (1951). Prediction and entropy of printed English. *Bell System Technical Journal, 30,* 50–64.

Si, L., Jin, R., Callan, J., & Ogilvie, P. (2002). Language modeling framework for resource selection and results merging. *Proceedings of the Eleventh International Conference on Information and Knowledge Management (CIKM'02),* 391–397.

Song, F., & Croft, W. B. (1999). A general language model for information retrieval. *Proceedings of the 22nd ACM SIGIR Annual International Conference on Research and Development in Information Retrieval,* 279–280.

Sparck Jones, K. & Robertson, S. (2001). LM vs. PM: Where is the relevance? In J. Callan, W. B. Croft, & J. Lafferty (Eds.), *Proceedings of the Workshop on Language Modeling and Information Retrieval* (pp. 12–15). Pittsburgh, PA: Carnegie Mellon University.

Sparck Jones, K., Walker, S., & Robertson, S. E. (2000a). A probabilistic model of information retrieval: Development and comparative experiments: Part 1. *Information Processing & Management, 36*(6), 779–808.

Sparck Jones, K., Walker, S., & Robertson, S. E. (2000b): A probabilistic model of information retrieval: Development and comparative experiments: Part 2. *Information Processing & Management, 36*(6), 809–840.

Turtle, H. R. (1990). *Inference networks for document retrieval.* Unpublished doctoral dissertation, University of Massachusetts, Amherst.

Turtle, H., & Croft, W. B. (1991). Efficient probabilistic inference for text retrieval. In A. Lichnerowicz (Ed.), *Proceedings of RIAO-91, 3rd International Conference Recherche d'Informations Assistee par Ordinateur* (pp. 644–661). Amsterdeam: Elsevier.

van Rijsbergen, C. J. (1977). A theoretical basis for the use of co-occurrence data in information retrieval. *Journal of Documentation, 33*(2), 106–119.

van Rijsbergen, C. J. (1979). *Information retrieval.* London: Butterworths. Retrieved January 2, 2004, from http://www.dcs.gla.ac.uk/Keith/Preface.html

van Rijsbergen, C. J. (1992). Probabilistic retrieval revisited. *The Computer Journal, 35*(3), 291–298.

Witten, I. H., & Bell, T. C. (1991). The zero-frequency problem: Estimating the probabilities of novel events in adaptive text compression. *IEEE Transactions on Information Theory, 37*(4), 1085–1094.

Wong, S., & Yao, Y. (1995). On modeling information retrieval with probabilistic inference. *ACM Transactions on Information Systems, 13*(1), 38–68.

Xu, J., & Croft, W. B. (1999). Cluster-based language models for distributed retrieval. *Proceedings of the 22nd ACM SIGIR Annual International Conference on Research and Development in Information Retrieval,* 254–261.

Xu, J., & Weischedel, R. (2000). TREC-9 cross-lingual retrieval at BBN. *Proceedings of the Ninth Text REtrieval Conference (TREC-9)*, 106–116.

Xu, J., Weischedel, R., & Nguyen, C. (2001). Evaluating a probabilistic model for cross-lingual retrieval. *Proceedings of the 24th ACM SIGIR Annual International Conference on Research and Development in Information Retrieval*, 105–110.

Zhai, C. (2002). *Risk minimization and language modeling in text retrieval*. Unpublished doctoral dissertation, Carnegie Mellon University, Pittsburgh, PA.

Zhai, C., & Lafferty, J. (2001a). Dual role of smoothing in the language modeling approach. In J. Callan, W. B. Croft, & J. Lafferty (Eds.), *Proceedings of the Workshop on Language Models for Information Retrieval* (pp. 31–36). Pittsburgh, PA: Carnegie Mellon University.

Zhai, C., & Lafferty, J. (2001b). A study of smoothing methods for language models applied to ad hoc information retrieval. *Proceedings of the 24th ACM SIGIR Annual International Conference on Research and Development in Information Retrieval*, 334–342.

Zhai, C., & Lafferty, J. (2002). Two-stage language models for information retrieval. *Proceedings of the 25th ACM SIGIR Annual International Conference on Research and Development in Information Retrieval*, 49–56.

Zhu, J. X. (2001). Language model feature induction via discriminative techniques. In J. Callan, W. B. Croft, & J. Lafferty (Eds.), *Proceedings of the Workshop on Language Modeling and Information Retrieval* (pp. 47–54). Pittsburgh, PA: Carnegie Mellon University.

Zipf, G. K. (1929). Relative frequency as a determinant of phonetic change. *Harvard Studies in Classical Philology, 40*, 1–95.

Zipf, G. K. (1932). *Selected Studies of the principle of relative frequency in language*. Cambridge, MA: Harvard University Press.

Zipf, G. K. (1949). *Human behavior and the principle of least effort*. Cambridge, MA: Addison Wesley.

Zipf, G. K. (1965). *The psycho-biology of language: An introduction to dynamic philology*. Cambridge, MA: MIT Press.

CHAPTER 2

Information Retrieval on the Web

Kiduk Yang
Indiana University

Introduction

How do we find information on the Web? Although information on the Web is distributed and decentralized, the Web can be viewed as a single, virtual document collection. In that regard, the fundamental questions and approaches of traditional information retrieval (IR) research (e.g., term weighting, query expansion) are likely to be relevant in Web document retrieval.[1] Findings from traditional IR research, however, may not always be applicable in a Web setting. The Web document collection—massive in size and diverse in content, format, purpose, and quality—challenges the validity of previous research findings that are based on relatively small and homogeneous test collections. Moreover, some traditional IR approaches, although applicable in theory, may be impossible or impractical to implement in a Web setting. For instance, the size, distribution, and dynamic nature of Web information make it extremely difficult to construct a complete and up-to-date data representation of the kind required for a model IR system.

To further complicate matters, information seeking on the Web is diverse in character and unpredictable in nature. Web searchers come from all walks of life and are motivated by many kinds of information needs. The wide range of experience, knowledge, motivation, and purpose means that searchers can express diverse types of information needs in a wide variety of ways with differing criteria for satisfying those needs. Conventional evaluation measures, such as precision and recall, may no longer be appropriate for Web IR, where a representative test collection is all but impossible to construct.

Finding information on the Web creates many new challenges for, and exacerbates some old problems in, IR research. At the same time, the Web is rich in new types of information not present in most IR test collections. Hyperlinks, usage statistics, document markup tags, and collections of topic hierarchies such as Yahoo! (http://www.yahoo.com) present an opportunity to leverage Web-specific document characteristics in novel ways that go beyond the term-based retrieval framework of traditional IR. Consequently, researchers in Web IR have reexamined the findings from traditional IR research to discover which conventional text retrieval approaches may be applicable in Web settings, at the same time exploring new approaches that can accommodate Web-specific characteristics.

Web data still consist mostly of text, which means that various text retrieval tools (e.g., term weighting, term similarity computation, query expansion) may be useful in Web IR. As the name implies, Web documents are heavily interconnected. Link analysis approaches, such as PageRank, HITS, and CLEVER, treat hyperlinks as implicit recommendations about the documents to which they point (Chakrabarti, Dom, Raghavan, Rajagopalan, Gibson, Kleinberg, 1998; Kleinberg, 1998; Page, Brin, Motwani, & Winograd, 1998). Some researchers (Pirolli, Pitkow, & Rao, 1996; Schapira, 1999), as well as commercial search services (e.g., DirectHit,[2] Alexa: http://www.alexa.com) have also looked at ways to exploit Web user statistics (e.g., counting link clicks or page browsing time). Another approach popularized by Web search services, although less explored in research, is the organization of quality-filtered Web data into a topic hierarchy. Due to the volume and diversity of data, traditional automatic classification approaches do not seem to fare well in a Web setting, so most such topic hierarchies are constructed manually (e.g., Yahoo!).

Two of the most promising Web IR tools, namely, Google and CLEVER, seem to be using combinations of these retrieval techniques. CLEVER combines topic-dependent link analysis techniques called Hyperlink Induced Topic Search (HITS) with term similarity techniques from text retrieval, although Google seems to employ a smorgasbord of techniques to obtain high-performance text retrieval influenced by a universal link analysis score called PageRank. Both Google and CLEVER not only leverage heavily off the implicit human judgments embedded in hyperlinks, but also improve on link analysis by combining it with text retrieval techniques.

By way of contrast, there seems to be a shortage of techniques that utilize the considerable body of explicit human judgment (e.g., Web directories[3]) in combination with hyperlinks and text contents of Web documents. Most IR research dealing with knowledge organization focuses on automatic clustering and classification of documents. There is little research that investigates how hierarchical knowledge bases like Yahoo! can be integrated with text retrieval and link analysis techniques in Web IR.

Scope

Information discovery on the Web is challenging. The complexity and richness of the Web search environment call for approaches that extend conventional IR methods to leverage rich sources of information on the Web. We first review findings from research that investigates characteristics of the Web search environment; then we examine key ideas and approaches in Web IR research. Although the emphasis in this chapter is on Web retrieval strategies, a review of research that explores the classification of Web documents is included to highlight the importance of the knowledge organization approach to information discovery. We also examine research that combines traditional and Web-specific IR methods.

Sources of Information

Web IR is an active research area with participants from a variety of disciplines, such as computer science, library and information science, and human–computer interaction. Research results are published in a wide range of journals, including *Information Processing & Management* and the *Journal of the American Society for Information Science and Technology*, as well as many conference proceedings, such as the International World Wide Web Conference (IW3C), the Association for Computing Machinery (ACM) conferences, and the Text REtrieval Conference (TREC). A wide spectrum of research in Web IR can be found in a variety of ACM conferences such as the Annual International Special Interest Group on Information Retrieval (SIGIR) Conference on Research and Development in Information Retrieval and the Hypertext and Hypermedia (HT) Conference; the TREC conference provides a common ground where cutting-edge Web IR approaches are investigated in a standardized environment.

In addition to individual research findings, a number of overviews of Web IR issues have been published. Two of the most recent are the *ARIST* chapters by Rasmussen (2003) on indexing and retrieval from the Web, which reviews research on indexing and ranking functions of search engines on the Web, and Bar-Ilan's (2004) on the use of Web search engines in information science research, which examines research into the design and use of search engines from both social and application-centered perspectives.

Research in Web IR

The main focus of early IR research was on the development of retrieval strategies for relatively small, static, and homogeneous text corpora. The Web, however, contains massive amounts of dynamic, heterogeneous, and hyperlinked information. Furthermore, information seeking on the Web is much more diverse and unpredictable than in traditional IR contexts. Consequently, determining the applicability of traditional IR methods to Web IR should begin with a consideration of the contextual differences between the two worlds. The findings of traditional IR are based primarily on experiments conducted with small, stable document collections and sets of relatively descriptive and specific queries. Web IR, on the other hand, must deal with mostly short and unfocused queries posed against a massive collection of heterogeneous, hyperlinked documents that change dynamically. One focus of Web IR research has thus been the characteristics of the Web search environment (notably, document characteristics and searcher behavior).

Characteristics of the Web Search Environment

Studies of the Web search environment are of two main types: those that attempt to characterize the Web by content and structural analysis of sample sets of documents obtained by a Web crawler (Bray, 1996; Broder et al., 2000; Lawrence & Giles, 1998, 1999a, 1999b; Woodruff, Aoki, Brewer, Gauthier, & Rowe, 1996), and those that attempt to characterize searcher behavior via user surveys (Kehoe, Pitkow, Sutton, Aggarwal, & Rogers, 1999) or analysis of search engine logs (Jansen, Spink, Bateman, & Saracevic, 1998; Jansen, Spink, & Saracevic, 1998, 2000; Silverstein, Henzinger, Marais, & Moricz, 1998; Spink, Wolfram, Jansen, & Saracevic, 2001).

Early studies of Web characteristics were based on relatively small samples and were therefore limited in scope. Moreover, the Web has been growing and changing at a significant rate since its inception, so findings from early studies are outdated and unrepresentative of the current search environment. Tracking Web characterization research over time, however, might give us insight into the evolution of the Web and thus lead to a better understanding of its underlying nature.

Characteristics of Web Documents

One of the first large-scale Web studies was conducted by a Berkeley research project group called Inktomi[4] (Woodruff et al., 1996), who examined various characteristics of Web documents obtained by the Inktomi crawler in 1995. The study reported the usage pattern of hypertext mark-up language (HTML) tags as well as various statistics on Web documents, such as the average document size after HTML tag extraction (4.4 Kbytes), the average number of tags per document (71), and the average number of unique tags per document (11). The most prevalent HTML tag was the title tag (used by 92 percent of documents), and the most frequent attribute was hypertext reference (HREF) (used by 88 percent of documents; on average 14 times per document). The study also observed rapid changes in the Web by comparing the Web crawl from July to October 1995 (1.3 million unique HTML documents) with that of November 1995 (2.6 million unique HTML documents). The number of documents doubled in one month and many of the most popular documents (i.e., those with the highest indegree[5]) in the first crawl disappeared in the second crawl.

In a related study, Bray (1996) examined the content and structure of 1.5 million Web documents in 1995. In addition to the document statistics (size, tag count, etc.), which were comparable to those of the Inktomi group, Bray characterized Web documents by their connectivity using measures called visibility (indegree) and luminosity (outdegree). The most visible documents in Bray's sample consisted of home pages of well-known universities, organizations, and companies; the most luminous sites were Web indexes such as Yahoo!. Bray also observed that the majority (80 percent) of documents in his sample did not contain any outlinks to external sites and had only "a few" (1 to 10) inlinks from external sites.[6] This implies that the Web is held together by a relatively few hubs (i.e., documents with high outdegree).

Despite Bray's observation about the irregular connectivity pattern of the Web, the impression of the Web as a network of densely interconnected communities persisted, not least with proponents of link-based search strategies (Kleinberg, 1998; Page et al., 1998). However, findings from a more recent study of Web structure, based on three sets of experiments between May 1999 and October 1999 with two AltaVista crawls of more than 200 million pages and 1.5 billion links, seem to validate Bray's picture of the Web (Broder et al., 2000). Broder's study found that only about 28 percent of the Web pages[7] are "strongly connected," and it takes an average of 16 link traversals to move from one strongly connected site to another, which suggests that the Web may not be as tightly connected as previously thought (Barabási, 2003). If such findings hold for the whole Web, it would pose yet another challenge for Web search engines, which collect most of their data by following hyperlinks. Incidentally, Broder's study also verified an earlier observation regarding the power law phenomenon of indegree (Kumar, Raghavan, Rajagopalan, & Tomkins, 1999), which estimated the probability that a page has in-degree k to be roughly $1/k^2$.

Findings from Broder's study, which are based on the graph analysis of the Web link structure, have been supported by Lawrence and Giles (1998, 1999a, 1999b), who conducted a series of studies analyzing Web content. According to their most recent study (Lawrence & Giles, 1999a, 1999b), the estimated size of the publicly indexable Web[8] as of February 1999 was 15 terabytes of data comprising 800 million pages scattered over 3 million Web servers; this more than doubled their earlier estimate of 320 million pages in December 1997 (Lawrence & Giles, 1998).

According to the Internet Archive Project (http://www.archive.org), which has been building a digital library of the Internet by compiling snapshots of publicly accessible Internet sites since 1996 (Kahle, 1997), the Web had reached 1 billion pages as of March 2000 with page content changing at the rate of 15 percent per month. Given the accelerated growth of the Web and its estimated 4 billion indexable pages (Glover, Tsioutsiouliklis, Lawrence, Pennock, & Flake, 2002), it is not surprising that an up-to-date study of Web document characteristics is lacking.

Information Seeking on the Web

Thus far we have considered the characteristics of Web documents, which is only one aspect of the Web search environment. Information seeking is the other, no less crucial than document collection in terms of its influence on the retrieval process. The considerable body of literature on information seeking is beyond the scope of this chapter. Instead, we focus on studies of search patterns and user characteristics on the Web.

The most recent user survey conducted by the Graphics, Visualization and Usability (GVU) Center at Georgia Tech (Kehoe et al., 1999) found that most Web users access the Internet on a daily basis (79 percent) and use search engines to find information (85 percent). The survey is now several years old, but the trend appears to be continuing, as evidenced by the 670 million searches logged daily by seven major Web search engines (250 million by Google).[9]

User studies, such as the GVU survey, provide demographic data about searchers. In order to study patterns of search behavior, however, researchers have examined system transaction logs that record actual search sessions. Findings from one of the first large-scale studies analyzing an Excite search engine log[10] suggested that Web search behavior patterns are rather different from those in traditional IR contexts (Jansen, Spink, Bateman, & Saracevic, 1998). By examining the content and use of queries (e.g., search terms, use of search features), the study found that Web searchers tend to use very short queries and exert minimum effort in evaluating or refining their searches. About 30 percent of queries in the study were single term queries, with an average query length of 2.35 terms. The study also reported that most searchers browsed only the first page of search results and did not engage in search refinement processes such as relevance feedback or query reformulation. In a related study using the same data, Jansen, Spink, and Saracevic (1998) conducted a failure analysis of queries (i.e., incorrect query construction) and found that Web searchers not only used advanced query features sparingly, but also tended to use them incorrectly.

Silverstein et al. (1998) conducted a much larger study of Web query patterns, analyzing 280 gigabytes of AltaVista query logs collected over a period of 43 days. The data consisted of 1 billion entries in AltaVista transaction logs from August 2, 1998, to September 13, 1998, presumably less affected by short-term query trends than might be the case with smaller snapshots of query data. The findings were consistent with earlier studies in that Web searchers used short queries, tended to look at the first few results only, and seldom modified the query. The study reported that 77 percent of all search sessions contained only one query; and 85 percent of the time only the top 10 results were examined. The searchers' relatively low investment in the process was further evidenced in the short average query length of 2.35 words, most of which were intended to be phrases. Except for a few very common queries (mostly sex-related), over two thirds of all queries were unique, implying that information needs on the Web are highly particular (or at least expressed in diverse ways).

Wolfram, Spink, Jansen, and Saracevic (2001) investigated trends in Web search behavior by comparing findings from two follow-up studies of Excite search logs in 1997 (Jansen et al., 2000) and 1999 (Spink et al.,

2001), each of which contained more than 1 million queries from more than 200,000 Excite users. They found that Web searchers tended to invest little effort in both query formulation and result evaluation, which confirmed patterns reported in earlier studies.

Large-scale studies of search engine logs describe characteristics of an average searcher on the Web, but do not provide detailed information on individual searchers. Although small in scale, and thus more anecdotal than conclusive, studies that examine the characteristics of searchers in context are nevertheless illuminating. For example, Pollock and Hockley (1997) found that Internet-naïve users had a poor grasp of the concept of iterative searching and relevance ranking. They had difficulty formulating good queries, either because they did not understand what were likely to be good-quality differentiators or because they did not realize that contextual terms should be included. At the same time, they expected results to be clear and organized, and considered anything less than a perfect match to be an outright failure.

Hölscher and Strube (2000), who investigated the search behavior of expert and novice Web users, found that expert Web users exhibited more complex behaviors than average Web searchers observed in previous studies. For example, they engaged in various search enhancement strategies, such as query reformulation, exploration of advanced search options, and a combination of browsing and querying. In general, they seemed to be much more willing to go the extra mile to satisfy their information needs, as further evidenced by the longer query (average 3.64 words) and a tendency to review more search results. Furthermore, Hölscher and Strube found that searchers with a high level of domain knowledge used a variety of terminology in their queries and took less time evaluating documents than their counterparts, who often got stuck in unsuccessful query reformulation cycles resulting from ineffective query modifications.

The profile of the average Web searcher that emerges from the Excite and AltaVista studies (Jansen, Spink, & Saracevic, 1998; Jansen et al., 2000; Silverstein et al., 1998; Spink et al., 2001; Wolfram et al., 2001) seems to be consistent with assumptions about information seeking in electronic environments described by Marchionini (1992). Web searchers in general do not want to engage in an involved retrieval process. Instead, they expect immediate answers while expending minimum effort. Web or domain experts, however, tend to engage in more elaborate efforts to satisfy their information needs. This difference in the degree of involvement with the retrieval process is in all likelihood influenced by the level of search skill and domain knowledge. These help reduce the cognitive load required for engaging in the various steps of the retrieval process. The important point here is not how expert and novice searchers differ, but why. If we believe that humans seek the path of least resistance, then reducing the cognitive load for users should be one of the primary goals of IR system design. Marchionini (1992) describes an appropriate information system as one that combines and integrates the information-seeking functions to help users clarify their problems and find solutions.

Information seeking on the Web not only deviates from traditional IR expectations of searcher behavior, but also encompasses a wider range of information needs than previously investigated. In his exploration of Web search taxonomy, Broder (2002) discovered that information sought on the Web was often not informational. In fact, Broder's analysis of AltaVista user surveys and search logs,[11] which revealed three broad types of Web search, found informational queries to account for slightly less than 50 percent of all searches, followed by transactional queries (about 30 percent) (e.g., shopping, downloading), and navigational queries (about 20 percent), which are used to reach a particular site (e.g., a home page).

Detailed examination of the Web highlights some of the potential problems with applying traditional IR approaches. It is practically impossible for any one search engine to construct a comprehensive and current index, due to the Web's massive size and dynamic nature. Moreover, there is relatively small overlap in search engines' coverage. Thus, combining retrieval results from multiple search engines offers considerable potential advantage. The variable nature of both documents and information needs on the Web hints at the potential for specialized search systems that cater to specific situations, or flexible systems that can respond appropriately to a variety of situations. The evidence of minimalist approaches to information seeking on the Web suggests the need for support features that can shift the cognitive burden from the user to the system.

We must go beyond conventional approaches to IR and devise new techniques that fit the Web environment, adopting established techniques as appropriate. One approach is the use of Web-specific features, such as HTML and link structure (Google) or usage statistics (Direct Hit) to identify relevant documents. Another example is the use of manually constructed hierarchical categories like Yahoo!, where the application of age-old cataloging principles allows users to find high-quality information by browsing a predefined, topical taxonomy without having to formulate their information need explicitly in the language of the search system. Both of these approaches—the use of topical hierarchies and leveraging of Web-specific information—are important aspects of Web IR.

Web IR Approaches

Much of IR research focuses on developing strategies for identifying documents "relevant" to a given query. In traditional IR, evidence of relevance is typically mined from information in the text, ranking documents according to their estimated degree of relevance using such measures as term similarity or term occurrence probability. On the Web, however, information is not confined to the textual content of documents. For example, it is relatively easy to collect Web document metadata, such as usage statistics and file characteristics (e.g., size, date), which can be used to supplement term-based estimation of document relevance. Hyperlinks, being by far the most prominent source of evidence in Web documents, have been the subject of numerous studies exploring retrieval strategies based on link exploitation.

The advent of link-based approaches predates the Web and can be traced back to citation analysis and hypertext research (Borgman & Furner, 2002). Two measures of document similarity based on citations were proposed in bibliometrics (White & McCain, 1989): bibliographic coupling (Kessler, 1963), which is the number of documents cited by both document p and q, and cocitation (Small, 1973), the number of documents that cite both p and q. A successful deployment of these measures was demonstrated by Shaw (1991a, 1991b), who used a combination of text similarity, bibliographic coupling, and cocitation as part of a graph-based clustering algorithm to improve retrieval performance.

In hypertext research, both citations and hyperlinks have been used for clustering and searching. Rivlin, Botafogo, and Shneiderman (Botafogo, Rivlin, & Shneiderman, 1992; Rivlin, Botafogo, & Shneiderman, 1994) used connectivity and compactness measures based on node distance to identify clusters as well as link analysis to enhance relevance ranking. Weiss et al. (1996) defined similarity measures based on link structure, generalized from cocitation and bibliographic coupling to permit long chains of reference. Using citations or links to cluster documents is closely related to conventional clustering approaches that group together documents with similar content. Instead of identifying related documents (Willett, 1988), passages (Salton & Buckley, 1991), or terms (Sparck Jones, 1971) based on their textual similarity, however, link-based clustering approaches rest on the assumption that documents cited/linked together many times (i.e., cocitation) or documents with many common citations/links (i.e., bibliographic coupling) are likely to be related.

Leveraging Hyperlinks

Use of hyperlinks to enhance searching is also based on the notion that hyperlinks connect related documents and thus can provide additional information. Link-based retrieval strategies in hypertext research explore various methods of enriching local document content with the content of hyperlinked documents. In the case of hypertext, where the collection usually consists of homogeneous documents on a single topic that are linked together for the purpose of citation, external content introduced by hyperlinks tends to be of high quality and useful. On the Web, however, where hyperlinks connect documents of varying quality and content for various purposes (Kim, 2000), document enrichment via hyperlink can sometimes introduce noise and degrade retrieval performance.

One way to address such problems is to categorize hyperlinks by purpose and usefulness, so that they may be utilized appropriately.

Discussions of hyperlink classification, however, have not moved beyond explicit link typing (Baron, Tague-Sutcliffe, & Kinnucan, 1996; Kopak, 1999; Shum, 1996; Trigg & Weiser, 1983) or rudimentary attempts at link type identification based on visualization of text similarity relationships between documents (Allan, 1996; Salton, Buckley, & Allan, 1994). Allan's (1996) automatic link typing strategy, for example, involves determination of the density and pattern of similarity between document subparts (i.e., paragraphs), and may not be best suited for Web IR. It seems clear that there are various patterns in hypertext linking (Bernstein, 1998), but whether those patterns can be automatically identified for effective retrieval remains to be seen.

Although hypertext approaches look beyond immediate document content, they still mine for evidence in the text of neighboring documents. Web IR, on the other hand, goes beyond the textual content of the document corpus and leverages various sources of information such as link structure, usage patterns, and manually constructed topic hierarchies. Herein lies one of the fundamental differences between Web IR and traditional IR. Whereas traditional IR depends solely on the text of documents, Web IR goes beyond the textual content and utilizes implicit human judgments about documents, whether embedded in hyperlinks, user statistics, or Web directories.

One of the earliest attempts to adapt the traditional IR model was conducted by Croft (1993), who demonstrated the successful incorporation of hypertext links into the Inference Network model. The Inference Network is a probabilistic retrieval model based on a Bayesian inference network, where nodes represent documents and queries and directed edges represent dependency relations between nodes. In a text-based implementation of the Inference Network model, the middle layer of nodes consists of terms whose dependencies to outer layer nodes (i.e., documents, query) are computed based on term occurrence probabilities. In a link-based implementation, where the existence of a hyperlink is taken as evidence of a dependency between linked documents, the hyperlink evidence is incorporated into the term-based network structure by adding to each document the dependency relations of new terms introduced by the linked documents and strengthening the relations of existing terms shared by the linked documents. Croft's use of hyperlinks results in increased importance of terms contained in linked documents, which can be problematic if the quality of hyperlinks is poor.

A more discriminatory method of leveraging hyperlinks was proposed by Frei and Stieger (1995). Instead of blindly following all hyperlinks to obtain additional information about a document, Frei and Stieger annotated each link with a content-specific link description comprising terms common to the source and destination of a link. They followed only those links whose query-link description similarities were above some threshold. Each time a link is traversed, Frei and Stieger's algorithm updates the Retrieval Status Value (RSV), the sum of query-document similarity scores weighted by link distance with which retrieval results are ranked.

Computation of RSV, as described in the equation below, is not only similar in essence to the link-based document enrichment strategy proposed by Marchiori (1997), but also resembles in form subsequent link-based algorithms that propagate information through hyperlinks:

$$RSV_{d+1} = RSV_d + w_d * RSV_{current}, \qquad (1)$$

where $RSV_{current}$ is the sum of similarity between query and documents at distance d+1, and w_d is a propagation factor dependent on the navigation distance.

Marchiori describes a mathematical model, in which the information propagated through hyperlinks is scored with a recursive formula based on exponentially fading textual content of the linked documents. Marchiori coins the term "hyper information" to denote the information provided by hyperlinks and suggests that hyper information could work on top of "local" textual information to produce a more "global" score for a Web document. Marchiori's *HyperSearch* algorithm scores a document's information content by adding to the textual information of a document p its hyper information, which is computed by summing the textual information of documents reachable from p by recursively following outlinks, diminished by a damping factor that decays exponentially with link distance from p. If we denote hyper information of p with $HI(p)$, textual information of p by $TI(p)$, and the damping factor by F ($0 < F < 1$), $HI(p)$ can be expressed as follows:

$$HI(p) = F^*TI(p_1) + F^2{}^*TI(p_2) + \ldots + F^k{}^*TI(p_k) \qquad (2)$$

At the heart of this approach is the notion that a document can be enriched with the textual contents of linked documents. Because linked documents are often linked in turn, the computation becomes recursive, with the provision for fading information propagation based on link distance. The informative content of a Web document ideally should involve all the linked documents, but Marchiori fixes an arbitrary link distance limit k in his model for reasons of practicality; he uses the HyperSearch algorithm in practice to compute the relevance of a document to a given query by propagating the relevance scores of documents within link distance k.

The idea of information propagation via links is extended by Page et al. (1998), who developed a method for assigning a universal rank to Web pages based on a weight-propagation algorithm called PageRank. Instead of propagating textual information backwards through outlinks of a fixed size neighborhood as in HyperSearch, PageRank propagates the PageRank scores forward through inlinks of the entire Web. This recursive definition of PageRank differs sharply from other link-based methods in that it arrives at a global measure of a Web page without taking into account any textual information. In other words, the PageRank score of a Web page is influenced by neither the page itself nor any potential query, but is based solely on the aggregate measure of human-judged importance implied in each hyperlink.

Page et al. start with the notion of counting backlinks (i.e., indegree) to assess the importance of a Web page, but point out that simple indegree does not always correspond to importance; thus they arrive at propagation of importance through links, where a page is important if the sum of the importance of its backlinks is high. This idea is captured in the PageRank formula as follows:

$$R(p) = d \cdot \frac{1}{T} + (1-d) \cdot \sum_{i=1}^{k} \frac{R(p_i)}{C(p_i)} \qquad (3)$$

where T is the total number of pages on the Web, d is a damping factor, $C(p)$ is the outdegree of p, and p_i denotes the inlinks of p. $R(p)$ can be calculated iteratively, starting with all $R(p_i)$ values equal to 1 and repeating computations until the values converge. This calculation corresponds to computing the principal eigenvector of the link matrix of the Web.[12] When PageRank is scaled so that $\sum R(p) = 1$, it can be thought of as a probability distribution over Web pages. In that light, $R(p)$ can be interpreted as a weighting function that estimates the probability that a Web surfer will arrive at page p from some starting point by a series of forward link traversals as well as occasional jumps to random pages. By incorporating into the formula the damping factor d, which represents the probability that p will be arrived at randomly instead of via link traversal, PageRank models the behavior of the "random" Web surfer who walks the Web by following the hyperlinks for a finite amount of time before going on to something unrelated.

The underlying assumption of PageRank is the notion that a link from page p_i to page p signifies the recommendation of p by the author of p_i. By aggregating all such recommendations recursively over the entirety of the Web, where each recommendation is weighted by its importance and normalized by its outdegree, PageRank arrives at an objective measure of importance from subjective determinations of importance scattered over the Web. By the same token, PageRank can be said to measure a collective notion of importance, otherwise known as "authority" in Web IR. One may challenge the "conferred authority" assumption about links and argue that a linked page is popular rather than important or authoritative. But, then, even journal citations sometimes reflect an author's deference or preferential treatment toward the cited document rather than careful judgment (Cronin, Snyder, & Atkins, 1997). The debate surrounding popularity and authority is important; it has been argued that the recursive and exhaustive nature of PageRank tends to favor authority over popularity because it is more likely that important pages will link to other important pages than popular pages will link to other popular pages.

There are several ways PageRank can be used in Web IR. It can be used as a part of a search engine's ranking mechanism to improve term-based retrieval results by giving preference to more important and central pages (Brin & Page, 1998). It can also be used to guide a Web crawler (Cho, Garcia-Molina, & Page, 1998) or help to find representative pages for page clusters. When employed by a search engine, PageRank works well with broad queries that return large numbers of documents by identifying a few high-quality (i.e., important, popular) documents. Because PageRank is a global measure based on collective opinions, its effectiveness depends largely on the coverage on which the computations are made. Google, for example, computes PageRank based on link analysis of almost 600 million pages (1 billion URLs), a large chunk of the publicly indexable Web.

Kleinberg's (1998) HITS algorithm considers both inlinks and outlinks to identify mutually reinforcing communities of "authority" and "hub" pages. HITS defines "authority" as a page that is pointed to by many good hubs and defines a "hub" as a page that points to many good authorities. Mathematically, these circular definitions can be expressed as follows:

$$a(p) = \sum_{q \to p} h(q) \,,00 \qquad (4)$$

$$00 \qquad h(p) = \sum_{p \to q} a(q) \,. \qquad (5)$$

These equations define the authority weight $a(p)$ and the hub weight $h(p)$ for each page p, where $p \to q$ denotes "page p has a hyperlink to page q."

Although HITS embraces the link analysis assumption that equates a hyperlink with a human judgment, conferring authority on the pages pointed to, it differs from other link-based approaches in several ways. Instead of simply counting the number of links, HITS calculates the value of page p based on the aggregate values of pages that point to p or are pointed to by p, in a fashion similar to PageRank. HITS, however, differs from PageRank in three major regards. First, it takes into account the contributions from both inlinks and outlinks to compute two separate measures of a page's value, namely authority and hub scores, instead of a single measure of importance. Second, HITS measures pages' values dynamically for each query, rather than assigning global scores regardless of the query. Third, HITS scores are computed from a relatively small subset rather than the totality of the Web.

Unique to HITS is the premise that the Web contains mutually reinforcing communities (hubs and authorities) on sufficiently broad topics. To identify these communities, HITS starts with a root set S of text-based search engine results in response to a query about some topic, expands S to a base set T with the inlinks and outlinks of S, eliminates links between pages with the same domain name in T to define the graph G, runs the iterative algorithm (equations 4 and 5) on G until convergence, and returns a set of documents with high $h(p)$ weights (hubs) and another set with high $a(p)$ weights (authorities). The iterative algorithm works as follows: Starting with all weights initialized to 1, each step of the iterative algorithm computes $h(p)$ and $a(p)$ for every page p in T, normalizes each of them so that the sum of the squares equals 1, and repeats until the weights stabilize. It can be shown that the authority weights at convergence correspond to the principal eigenvalues of $A^T A$ and hub weights correspond to those of AA^T, where A is the link matrix of the base set T.[13] Typically, convergence occurs in 10 to 50 iterations for T consisting of about 5,000 Web pages, expanded from the root set S of 200 pages while constrained by the expansion limit of 50 inlinks per page.

The base set T often contains multiple distinct communities (sets of hubs and authorities), which turn out to be document clusters of sorts with different meanings (e.g., "jaguar"), different contexts (e.g., recall), standpoints (e.g., abortion), or simply varying degrees of "relevance" to the query topic. The most densely linked community, which is returned by the iterative algorithm of HITS, is called the principal community; others are called nonprincipal communities. Nonprincipal communities, which can be identified by finding the nonprincipal eigenvectors of $A^T A$ and AA^T, can not only contain relevant documents when the principal community misses the mark, but also reveal interesting information about the fine structure of a Web community (Kumar et al., 1999).

In addition to finding a set of high-quality pages on a topic and hubs pointing to many such pages, HITS can be used to find similar pages to a given page p by starting with a root set S of pages that point to p (as opposed to obtaining S by querying a search engine) and returning authority pages in the community to which p belongs. Of course, the target page p should have enough inlinks with which to construct S of sufficient size. The advantage of the HITS method of finding similar pages over text-based methods is that it can locate pages classified together by Web authors (hub creators) whether there is any text overlap or not.

Although HITS is query-dependent in that it begins with a root set of documents returned by a search engine in response to a query, the textual contents of pages are considered only in the initial step of obtaining that root set, after which the algorithm simply propagates weight over links without regard to the relevance of pages to the original topic. In other words, once HITS locates a topic neighborhood, it is guided by the link structure alone. Consequently, HITS can drift away from relevant documents if the neighborhood contains a community, T, of documents with a higher link density. This phenomenon is called "diffusion." It has been observed most frequently in response to a specific query with a large, generalized topic presence, when the algorithm converges to the community on the generalized topic instead of focusing on the original topic. The diffusion effect, also known as "topic drift" (Bharat & Henzinger, 1998) or the "Tightly-Knit Community (TKC)" effect (Lempel & Moran, 2001), is perhaps the most serious weakness of HITS. As we shall see in the next section, several researchers have investigated ways to compensate for the diffusion effect by tempering link analysis with content analysis.

Enhancing Link Analysis with Content Analysis

Chakrabarti, Dom, Raghavan, et al. (1998) extended HITS in the ARC (Automatic Resource Compiler) of the CLEVER project by incorporating the text around links into the computation of hub and authority weights. The idea that the text surrounding the links pointing to p is descriptive of the content of p had been recognized for some time (McBryan, 1994), but fusing textual content with the link-based framework of HITS was an innovative step. The ARC algorithm is largely identical to HITS except that the root expansion is by 2 links (i.e., expand S by all pages that are 2-link distance away from S), and the use of anchor text similarity weights $w(p,q)$ in hub and authority weight computations.[14] $w(p,q)$ is defined as $1 + n(t)$, where $n(t)$ is the number of query terms in the window of 50 bytes around a link (i.e., 50 bytes before <a href> and after </a>) from p to q. Anchor window size was determined by examining the occurrence distribution of the term "Yahoo!" around http://www.yahoo.com anchor in 5,000 pages, where most occurrences fell within the 50-byte window.

As the name indicates, the main goal of ARC was to devise a mechanism that could automatically compile and maintain resource lists similar to those provided by Yahoo!. To test the effectiveness of ARC in that regard, Chakrabarti et al. compared ARC with both Yahoo! and Infoseek in an experiment where volunteers evaluated the retrieval results of 27 broad topic queries. The queries, constructed by choosing words or short phrases representative of pages in both Yahoo! and Infoseek directories, were submitted to AltaVista to obtain the root sets for ARC and submitted to Yahoo! and Infoseek to find matching category pages. Volunteers were first asked to use the resulting list from each search engine for 15 to 30 minutes as a starting point to learn about the topic in any way they choose and then asked to assign three scores to each search engine.

Volunteers used a 10-point scale to rate the accuracy (how topically focused the list was), comprehensiveness (how broadly the list covered the topic), and overall value (how helpful the list was in locating valuable pages). Although the results of the experiment found no statistically significant difference, the average score ratios[15] showed ARC to be comparable with Infoseek and only marginally worse than Yahoo!.

Chakrabarti, Dom, Gibson, et al. (1998) continued the development of the HITS-based method in the CLEVER project by making improvements to ARC. Although still focusing on the main objective of topic distillation, where the objective is to derive a small number of high-quality Web pages most representative of the topic specified by a query, CLEVER incorporates more detailed information about links between pages. Based on the assumption that pages on the same logical Web site were authored by the same organization or individual, CLEVER varies the weight of links according to the location of their endpoints, so that intrasite links confer less authority than intersite links. Although Kleinberg (1998) had suggested eliminating intrahost links in his original paper on HITS, it is not clear whether this was implemented in ARC. A host or Web site, determined by the root portion of a URL as suggested by Kleinberg, may contain within it many distinct authors (e.g., America Online), so weighting links based on location may be a better strategy than eliminating intrahost links altogether.

Another significant feature of CLEVER is based on the observation that some good hub pages contain within them sections of subtopical link clusters. Such hub pages could cause problems for queries focusing on the subtopics by blurring the topic boundaries of authority pages. For example, a good hub page on Web IR, containing links to pages on text-based techniques and link-based techniques, may return Kleinberg's paper on HITS along with Salton's paper on term weighting in response to a query on link analysis. Unfortunately, the exact detail of how CLEVER addresses this problem is not provided in the paper, which mentions only that CLEVER uses "interesting (physically contiguous) sections of Web pages to determine good hubs or authorities" (Chakrabarti, Dom, Gibson, et al., 1998, p.15). It might be that CLEVER partitions hub pages into physically contiguous sections based on some heuristic using HTML tags and/or anchor texts. CLEVER also reduces the scores of near-duplicate hubs in order to keep them from unduly dominating the computation.

In an experiment conducted to test the performance of the new and improved CLEVER, Chakrabarti, Dom, Gibson, et al. (1998) compared the system's precision with that of Yahoo! and AltaVista on 27 benchmark topics used in the ARC experiment. The precision measure was based on relevance judgments made by 37 users, who evaluated 30 documents per topic (10 each from Yahoo! and AltaVista, five hubs and five authorities from CLEVER, merged and sorted alphabetically) as bad, fair, good, or fantastic, based on how useful they were in learning about the topic. Unlike in the ARC experiment, the source of documents was hidden from the user and no time limit was imposed. Comparison of precision results, where "good" and "fantastic" documents were considered relevant, showed that Yahoo! and CLEVER tied in 31 percent of cases and CLEVER did better in 50 percent of topics.

Bharat and Henzinger (1998) explored methods to augment HITS with full content analysis of documents going beyond the use of anchor text. Their research was motivated by failure analyses of HITS, where the results of query topics with poor performance were analyzed to better understand HITS's weaknesses. In addition to the obvious finding that the neighborhood graph induced by a base set T must be densely connected for HITS to be effective, examinations of the problem neighborhood graphs revealed three additional instances where HITS could fail.

The first failure situation, called "mutually reinforcing relationships between hosts" (Bharat & Henzinger, 1998, p.104), was caused by certain linkage patterns that unduly influenced HITS computation. A set of documents A on one host ($host1$) all pointing to a single document b on another host ($host2$) will drive up the hub scores of A and the authority score of b. Conversely, a single document a on $host1$ pointing to a set of documents B on $host2$ will drive up the hub score of a and authority scores of B. When documents on the same host are authored by the same person, which is a common occurrence, such linkage patterns gives undue weight to the opinion of one person, thus violating the notion of collaborative judgment inherent in HITS. The second problem was caused by automatically generated links in documents created with authoring tools, which often insert commercial or proprietary links that have nothing to do with perceived authority. The third problem observed in the neighborhood graph of failed results is probably the most common and serious failing of HITS. Bharat and Henzinger often found that hubs and authorities returned by HITS were not relevant to the query topic because the computation drifted away from the query topic toward the topic areas that were more densely linked. Kleinberg, who discussed HITS's tendency to focus on the generalized topic of a specific query, described this as "diffusion."

To overcome the HITS problems they observed, Bharat and Henzinger combined the connectivity analysis of HITS with content analysis and also strengthened the connectivity analysis by weighting links based on linkage pattern. The purpose of link weights, called "edge weights" by Bharat and Henzinger, is to combat mutually reinforcing relationships by giving fractional weights to edges (links) connecting multiple nodes (documents) on one host to a single node on another host. CLEVER's use of link weights arises from the same underlying assumption that documents on the same host implies single authorship, but the weighting is based on the location of link endpoints rather than the pattern of linkage.[16] The edge weights essentially normalize the contribution of authorship by dividing the contribution of each page by the number of pages created by the same author. To achieve this effect, the HITS formulae are modified as follows:

$$a(p) = \sum_{q \to p} h(q) \times auth\_wt(q,p) \, 00, \qquad (6)$$

$$h(p) = \sum_{p \to q} a(q) \times hub\_wt(p,q) \, 00. \qquad (7)$$

In these equations, $auth\_wt(q,p)$ is $1/m$ for page q, whose host has m documents pointing to p, and $hub\_wt(p,q)$ is $1/n$ for page q, which is pointed by n documents from the host of p. According to Bharat and Henzinger, application of this modified algorithm successfully eliminated all the instances of mutually reinforcing relationships they had observed.

Because the topic drift problem is caused by HITS's attraction to densely linked documents that are not relevant to the query, Bharat and Henzinger reasoned that the solution lay in eliminating nonrelevant documents from consideration or regulating the influence of documents based on their relevance to the query. One of the key tasks of such an approach, of course, is measuring the relevance of documents with respect to a given query, for which most link-based methods are ill equipped. Consequently, Bharat and Henzinger adapted a fusion strategy (BHITS[17]), where text-based methods of computing the relevance scores of documents were combined with the link-based methods of HITS in order to temper the contribution of links with documents' relevance to a query. It turned out that Bharat and Henzinger's method of combining connectivity and content analysis was an effective remedy for the problem caused by auto-generated links as well as the topic drift problem, because both instances involved influences of nonrelevant documents that were modulated by the method.

The first step of BHITS is determining a document's relevance to the query topic, where the relevance is estimated by text similarity between the document and the query. In keeping with Bharat and Henzinger's contention that the query topic is often broader than the query itself in the context of topic distillation, where the aim is to find quality documents related to the topic of the query rather than to find documents that match the query precisely, BHITS defines a broad query by concatenating the first 1,000 words from each document in the root set S (i.e., top 200 search results from a search engine). In BHITS, $tf*idf$ weights[18] are used for terms in both documents and the query and the similarity score is computed by the cosine vector similarity formula to compensate for length variations in documents (Salton & Buckley, 1988).

Once the relevance weights of all documents in the base set T have been computed, they can be used either to eliminate nonrelevant documents entirely or to regulate the link influence based on the relevance of the document in which the link occurs. There is no precise point that separates relevant documents from nonrelevant documents, so eliminating nonrelevant documents—or pruning nonrelevant nodes, in BHITS terminology—involves setting a relevance threshold. BHITS implements three different methods of determining the relevance threshold: in one the threshold is the median relevance weight of the base set T, in the second the threshold is the median relevance weight of the root set S, and in the third the threshold is one tenth of the maximum relevance weight.

Regulating the node influence, as BHITS calls it, is an attempt to reduce the influence of less relevant nodes on the scores of their neighbors by introducing the relevance weight, $rel\_wt(q)$, into the computation of hub and authority scores in the following manner:

$$a(p) = \sum_{q \to p} h(q) \times rel\_wt(q)\, 00, \qquad (8)$$

$$h(p) = \sum_{p \to q} a(q) \times rel\_wt(q)\, 00. \qquad (9)$$

These formulae, modulating the node influence based on its relevance weight, can be combined easily with the edge weights (equations 6 and 7).

Bharat and Henzinger also explored partial content analysis (PCA) approaches in order to reduce the content analysis cost of downloading thousands of pages. Based on the assumption that not all nodes are equally influential in deciding the outcome of HITS, PCA methods attempt to prune the most highly connected, nonrelevant nodes that dominate the computation. As a first step, relevance weights are computed for the 30 most highly connected documents in the root set against a reduced query vector consisting of the first 1,000 words of those 30 documents and the pruning threshold is set at the 25th percentile. Bharat and Henzinger suggest two pruning strategies: degree-based pruning selects influential nodes based on their indegree and outdegree (4*indegree + outdegree) and iterative pruning selects nodes to prune based on HITS computations. In degree-based pruning, the relevance weights of the top 100 influential nodes are computed, nodes with relevance weights below the threshold are pruned, and 10 iterations of edge-weighted HITS are run on the reduced graph. In iterative pruning, pruning occurs in each of 10 iterations of edge-weighted HITS by eliminating documents below the threshold from the top-5 ranking documents. The rationale for using the top-5 documents to prune is as follows: pruning top-ranking documents is sufficient to combat topic-drift because other documents supported in the ranking by high-ranking, pruned documents will be affected as well by their mutually reinforcing relationships.

To evaluate the effectiveness of the methods outlined so far, Bharat and Henzinger devised an experiment comparing 10 combinations of their methods with the baseline HITS using 28 queries from the ARC experiment (Chakrabarti, Dom, Raghavan, et al., 1998).[19] The evaluation metrics used were precision and relative recall at 5 and 10 documents, where relative recall was computed by dividing the number of relevant documents retrieved by the total number of relevant documents in the pool of top-10 documents retrieved by all systems. Three volunteers, who independently evaluated a pool of the top 14 documents from all systems, made the relevance judgments.

The results of the experiment showed that the edge weights alone improved precision over HITS (26 percent for authorities and 23 percent for hubs); adding regulation and pruning each improved precision further (10 percent each for authorities, 10 percent by pruning and under 1 percent by regulation for hubs), but combining regulation and pruning did not give any additional improvement. Recall exhibited similar patterns. The performances of partial-content analysis systems were comparable with those of full-content analysis systems. In fact, the precision of the iterative pruning system was highest overall. Bharat and Henzinger suggest that the good performance of partial content analysis systems is due to their ability to avoid pruning noninfluential but useful (i.e., connected to good hubs and authorities) documents with low relevance weights. It is also possible that good hubs with little content other than links may be pruned when matched against a fully expanded query but manage to remain in the computation when matched against a shorter query.

Lempel and Moran (2001) take another approach to combat what they call the TKC effect (i.e., diffusion effect, topic drift) of HITS. Their Stochastic Approach for Link-Structure Analysis (SALSA) algorithm is similar in principle to PageRank in that it considers "random" walks on graphs derived from link structure; but in its implementation it follows the HITS approach of identifying topic-driven neighborhood graphs but replaces the iterative algorithm of the mutual reinforcement approach with a noniterative stochastic approach to identify hubs and authorities. The authority and hub scores computed by the SALSA algorithm turn out to be equivalent to the normalized sum of weights of inlinks and outlinks, or simple count of inlinks and outlinks when link contributions are not differentiated (i.e., the link is unweighted).

Because SALSA is a noniterative algorithm, the effect of link density on hub and authority computation is less profound than with HITS. SALSA in essence uses the indegree as the measure of authority, in direct contrast to Kleinberg's contention that the contribution of links should be differentiated by iterative link propagation. SALSA does differentiate the link contributions in the case of weighted links, but not with link propagation. Instead, Lempel and Moran (2001) suggest that links may be weighted by such factors as query-anchor text similarity or link location in a page.

Perhaps the most important reason for the claimed effectiveness of SALSA is its careful filtering of links in the link graph formulation stage. Lempel and Moran, proposing that filtering out "noninformative" links is crucial in link analysis, eliminate 38 percent of links to arrive at a high-quality link graph by ignoring related-domain links (e.g., http://www.yahoo.com and http://shopping.yahoo.com), cgi scripts, and advertisement links[20] in addition to intradomain links—the only type of link excluded from link analysis in HITS. Lempel and Moran suggest that link differentiation by link propagation is not as important when the link graph is of high quality. In other words, the "noise" in the link graph, which is suppressed by iterative mutual reinforcement computation of HITS, is eliminated in the link graph identification phase, thus making the end results comparable. The overpowering effect of the high-quality link graph is also reported by Amento, Terveen, and Hill (2000), who compared the effectiveness of indegree, PageRank, and authority scores on Yahoo! documents and found no significant performance differences.

Li, Shang, and Zhang (2002) extend HITS even further to address what they call the "small-in-large-out" problem, which is the tendency of a hub page with a large in-to-outdegree ratio to dominate the HITS computation. In their investigation of HITS-based algorithms, they observed that a hub page with a small indegree and a large outdegree in a tightly knit link community would achieve a substantial hub score and thus affect the high authority scores of its outlinks, regardless of their relevance to the query. The authors' solution to this problem is to assign high inlink weights when a small-in-large-out page exists in the root set so that such pages will not unduly dominate the HITS computation. Their experiment, comparing HITS, BHITS, and BHITS with small-in-large-out weighting (WHITS), showed significant performance improvement by WHITS over both HITS and BHITS.

Leveraging Implicit Links

Link analysis approaches such as HITS and PageRank aim to measure the relative value of a Web page by mining the human thought involved in creating hyperlinks. These algorithms capture collective judgments reflected in hyperlink structures and use these to rank Web pages. Hyperlinks, however, are not the only type of links that can be leveraged in Web IR. Implicit links, such as those found in a bibliographic citation index, a hierarchy of URLs, or the structure of documents, can be used in various ways to help find information on the Web. Lawrence, Giles, and Bollacker (1999), for example, specifically mined citation links in Web pages to build an effective search engine called CiteSeer that specializes in finding scientific research papers on the Web.

CiteSeer uses a method called "autonomous citation indexing" to compile a citation index automatically by extracting citations and the context of those citations into a database. The first step for any search engine is obtaining documents of interest. Even general purpose search engines conduct selective crawls to keep their indices current, because the entire Web is simply too large to crawl in a reasonable amount of time. For a special purpose search engine that targets documents of a specific type, efficient and effective identification of such documents is crucial for its successful deployment.

Instead of using a selective crawler guided by machine learning heuristics (Chakrabarti, van der Berg, & Dom, 1999; Cho et al., 1998), CiteSeer starts by querying multiple search engines with terms such as postscript, portable document format (PDF), technical report, conference proceedings, publication, and paper. Then CiteSeer converts postscript and PDF documents to text using PreScript from the New Zealand Digital Library project[21] or pstotext from the DEC Virtual Paper research project,[22] and filters them by eliminating documents without reference/bibliography sections. Having thus found the seed set of scientific research papers, CiteSeer can build its database by selectively crawling the Web to locate the cited documents. CiteSeer also monitors mailing lists and newsgroups, keeps in contact with publishers, and accepts notifications by authors to supplement its collection of scientific literature; CiteSeer's full-text and citation indexes are continually updated using the incremental indexing method (Brown, Callan, & Croft, 1994). Once the documents have been collected, CiteSeer applies a series of heuristics to extract the citations as well as the context in which they occur.

In addition to providing full Boolean search with proximity support, CiteSeer finds related documents by several methods. It uses similarity scoring based on tf^*idf weights to find documents with similar textual content by string distance comparison of article headers[23] to find similar headers, and the CCIDF measure to find articles with similar citations. CCIDF stands for Common Citation x Inverse Document Frequency, which is computed by summing up common citations weighted by their inverse frequency of citation. The IDF component of CCIDF works in similar fashion as idf in the tf^*idf measure by downplaying the importance of citations that occur frequently in the collection (i.e., citations to highly cited documents).

An experimental system called ParaSite by Spertus (1997) employs "implicit" link analysis methods that examine URLs and document structure. URLs are leveraged by ParaSite in several ways. ParaSite uses the file hierarchy of URLs to determine relationships (e.g., parent, child, sibling) between pages, examines domain names of URLs to group together related pages (e.g., http://www.ai.mit.edu and http://www.mit.edu), and considers stereotypical names in URLs to infer page types (e.g., hometown.aol.com). ParaSite also examines the structure within a document in order to mine information about links. For example, it uses link proximity with respect to document hierarchy to gauge the strength of link relationships. In other words, ParaSite considers links within a same list item (i.e., <LI> tag) to be most closely related, and those within a same list (i.e., between <OL> and </OL>, or <UL> and </UL> tags) to be more closely related than those located outside list boundaries. Consideration of document structure also comes into play in locating information about links. For example, general information about a group of links in a list can be found in the headers and text preceding a list, and specific information about an individual link can be found in the text surrounding and anchoring a link.

Spertus describes three potential applications—finding moved pages, person finder, and finding related pages—that use the information extracted from implicit link analysis. Pages whose URLs have changed may be found by link-based heuristics. One method is to check inlinks of moved pages, which may reveal updated URLs. Another heuristic is to remove portions of URLs until a valid URL is found and then search downward in the URL hierarchy. Finding a person's home page by querying a search engine with his or her name is not always successful. The page of interest may not yet be indexed by search engines, it may not contain the actual name, or the correct name may not be known. For example, finding someone named Albert who teaches Physics at Princeton University by searching within the pages of the Princeton physics department would be a lot more efficient than brute force searching of the entire Web. When the full name is known, searching the anchor text may prove quite effective (e.g., "anchor: Bill Clinton"). Related pages can be identified by finding pages pointed to by links in close proximity to one another.

Finding Related Pages

Finding related pages by link proximity as proposed by Spertus (1997) can be thought of as an extension of cocitation analysis and collaborative filtering (Shardanand & Maes, 1995), which matches users with other "like-minded" users, where "like-mindedness" is indicated by correlations among user ratings of items.

A more sophisticated approach to finding related pages was proposed by Dean and Henzinger (1999), who described two link-based algorithms that identify related Web pages, one of which is based on HITS and the other on cocitation analysis. Their objective was to find quickly a set of high-quality Web pages that addresses the same topic as the page specified by a query URL. Their Companion algorithm extends HITS in several ways. First, it uses the same edge weight devised by Bharat and Henzinger (equations 6 and 7) to reduce the influence of pages that all reside on one host. Second, it compresses duplicate and near-duplicate pages as well as eliminating a stoplist of URLs with very high indegree (e.g., Microsoft, Yahoo!) to keep them from dominating the hub and authority computation. Third, the base set T includes not only the parents (inlinks) and children (outlinks) of the query page, but also its siblings (i.e., pages that share a child with the query page). Fourth, it considers link location in a page to determine which pages to include.

To build the "vicinity graph" induced by the base set T, the Companion algorithm starts by including m randomly chosen parents and the first n children of the query page u, and then includes the siblings of u by getting k children of each parent of u immediately before and after its link to u as well as l parents of each child of u (excluding u) with highest indegree. Dean and Henzinger found that large m (2,000) and small k (8) values work better than moderate values, because the large m prevents undue influence being exercised by a single parent page and small k is enough to capture links on a similar topic, which tend to be clustered together on a page. Near-duplicate pages are defined as pages with more than 10 links, 95 percent of which are common. When duplicate pages are detected, they are replaced with a page consisting of the union of links in duplicate pages.

The Cocitation algorithm, another of Dean and Henzinger's methods of finding related pages, finds siblings of the query page with the highest degree of cocitation (i.e., pages most frequently cocited with the query page). Siblings, as seen above, are pages with a common parent; the degree of cocitation is the number of common parents for a given pair of pages. The Cocitation algorithm builds a vicinity graph that includes only the siblings of the query page u by following m randomly chosen parents of u and including k children of each of them immediately before and after its link to u. The algorithm then determines the degree of cocitation with u for every sibling and returns 10 pages with highest scores.

The effectiveness of the Companion algorithm (CM) and Cocitation algorithm (CC) were compared in a user study with Netscape's (NS) "What's Related,"[24] which combines link, usage, and content information to determine relationships between Web pages. The evaluation metrics for comparison were precision at a fixed rank, average precision (the sum of precisions at ranks with relevant documents divided by the number of relevant documents retrieved), and overall average precision (the average precision for all queries divided by the number of queries). Eighteen volunteers made binary[25] relevance judgments on randomly ordered pools of 30 documents retrieved by the three systems (10 by each system) for self-supplied query URLs, with an instruction that a page had to be both relevant and high quality to be scored as 1. There was a total of 59 URLs, 37 of which had results returned by all three systems.

All three algorithms did well with highly linked query URLs, although answers by NS were about broader topics than those produced by the other systems. CM had the highest precision at rank 10 (73 percent better than NS for 59 queries, 40 percent better than NS for 37 queries), followed by CC (51 percent better than NS for 59 queries, 22 percent better than NS for 37 queries). In general, CM and CC substantially outperformed NS at all ranks, with CM performing better than CC on average. Sign Tests and Wilcoxson Sums of Ranks Tests for each pair of algorithms showed that differences between CM and NS, and between CC and NS, were statistically significant. There was a large overlap in the answers of CM and CC due to the similarity in vicinity graph construction methods, but relatively little overlap between NS and CM or CC, which is consistent with previous research findings regarding the overlap of results among different search systems (Harman, 1994; Katzer, McGill, Tessier, Frakes, & Das-Gupta, 1982; Lawrence & Giles, 1998).

Other Link-Based Approaches

Link-based approaches use link structure to assess documents in ranking the retrieval results. There are other ways to leverage hyperlinks, such as visualizing the link structure (Carriere & Kazman, 1997; Mukherjea & Foley, 1995) or querying the Web as a relational database (Arocena, Mendelzon, & Mihaila, 1997; Mendelzon, Mihaila, & Milo, 1996). Although these are interesting Web IR approaches in their own right, only a cursory review will be included here as they are peripheral to the main thrust of this chapter.

Carriere and Kazman (1997) examined a link-based method for visualizing as well as ranking search engine query results. They constructed the WebQuery system, which queries a search engine and expands the returned results by inlinks and outlinks in a manner similar to HITS. WebQuery's resemblance to HITS, however, ends there. It ignores both link directionality and link importance, simply ranking pages in the expanded result set by the sum of their indegree and outdegree. Although WebQuery's search method is designed to augment text-based retrieval by finding relevant documents that do not contain query terms and to filter out uninteresting documents by ordering the search results by the degree of connectivity, it falls short of more sophisticated link analysis approaches such as HITS or PageRank. WebQuery's visualization approach, on the other hand, is a significant attempt to capitalize on humans' innate pattern recognition abilities by allowing the searchers to interact with visual representations of retrieved documents and their interconnectivity, so that they may find documents of interest more efficiently.

WebSQL by Arocena et al. (1997) demonstrates another approach to exploiting link structure between pages by using structured queries and a relational database representation of the Web. WebSQL queries multiple search engines with a uniform query interface integrating text-based retrieval with link-topology based queries. In the relational model of the Web proposed by WebSQL, Web pages are represented in a relational database containing document attributes (e.g., URL, title, text) as well as link attributes (e.g., base, href, label) so that queries specifying both content and link related features can be satisfied by combinations of attribute matching and selective link traversal. Florescu, Levy, and

Mendelzon (1998) provide a more detailed examination of the database approach to Web IR.

Mining Usage Data

In addition to hyperlinks, two other major sources of information provide human judgments on the value of Web documents. One source is usage data collected by Web servers and search engines that monitor surfer and searcher actions; the other is Web directories that contain a considerable body of human knowledge and judgment in the form of topic hierarchies. Despite the fact that commercial search engines (e.g., Google, DirectHit) have readily embraced usage data mining,[26] there has been scant research on exploiting Web user data until recently.

Schapira (1999) describes an experimental system called Pluribus, which reranks search engine results by "user popularity" scores based on accumulated user selection frequencies of search results. Pluribus was designed to test the hypothesis that retrieval performance could be improved by the implicit "relevant feedback" in users' collective past actions on search results. The basic idea of Pluribus is encapsulated in its ranking formula, which increments the scores of documents frequently selected by searchers and decreases the scores of documents infrequently selected by searchers. The user popularity score $U(p)$, as seen in the equation below, consists of the selection count $S(p)$, which is the number of times searchers selected a page p, and the expected selection count $E(p)$, which is the rank-based number of times p should have been selected (0.6 for the top 25 percent, 0.3 for the middle 50 percent, 0.1 for the bottom 25 percent of the search engine results), and the base relevance score $R(p)$ returned by a search engine.

$$U(p) = R(p) + 100*S(p) - 100*E(p) \qquad (10)$$

With this equation, low-ranking documents will rise rapidly as they are selected by users, but their rate of ascent will slow as they near the top; and conversely, high-ranking documents not selected by users will fall rapidly. One potential problem is the formula's tendency to overpower the original text-base ranking with exaggerated contributions from selection counts as the number of selections increases.

The effectiveness of Pluribus hinges on the validity of its assumptions: First, user selection/nonselection of a document indicates relevance/irrelevance. Second, users consistently select relevant documents and ignore irrelevant documents. Third, the same query represents the same information need. Similar assumptions underpin collaborative filtering (Shardanand & Maes, 1995), a successful example of which is a recommender system (Resnik & Varian, 1997) that suggests music, films, books, and other products and services to users based on examples of their likes and dislikes. In the context of collaborative filtering, however, users' evaluations of items are explicitly assigned in the form of scores, ranks, or opinions, rather than implied in the form of link clicks.

Estimating relevance by counting link clicks can be problematic. Users can select a document only to decide it is not relevant, or skip known relevant documents they have seen before. Consideration of page browsing time and link traversal count (i.e., the number of links traversed for a page) may be needed for a more accurate measure of user-based relevance. Another weakness of Pluribus is its dependence on query overlap (i.e., queries must be repeated), which is necessary for learning to occur. Pluribus also does nothing to improve the recall of the original search engine results; all it does is improve precision by reranking documents based on user popularity.

Cui, Wen, Nie, and Ma (2002) also explored the idea of link click counts as implicit relevance judgments, but they leveraged the usage data for probabilistic query expansion instead of direct manipulation of document scores. After computing correlations between query terms and document terms based on 4.8 million query sessions and 42,000 documents from the Encarta Web site (http://encarta.msn.com), they expanded queries with highly correlated terms to improve retrieval performance. They compared their approach with Local Context Analysis query expansion (Xu & Croft, 1996), which leverages both corpus-wide term co-occurrence statistics and pseudo-relevance feedback. Cui et al. found that their method improved retrieval performance substantially. Their query expansion by usage data mining improves on the Pluribus approach because the massive quantity of usage data compensates for potential false positives (i.e., user clicks that are not an indication of relevance), and the implicit relevance judgments embedded in usage data are further distilled into term correlation probabilities, which enable high-quality query expansion that can boost the overall retrieval performance, rather than simple reranking of the initial retrieval results.

In a related study, Zhou, Chen, Shi, Zhang, and Wu (2001) analyzed Web logs to compute the "Associate Degree" (AD), which is essentially a conditional probability of users visiting page q from page p. Although their application of AD was limited to increasing the link connectivity of Web sites by adding implicit links described by high AD, it is not difficult to envision AD as a basis to derive a user-based popularity measure of a Web page. One simple approach, given a large quantity of usage data, would be to apply a link analysis formula, such as PageRank, to the weighted directed graph of user visiting patterns. Such a measure could be either query-dependent (if AD is computed from search engine logs) or query-independent (if AD is computed from Web server logs) and computed solely from usage data or in combination with hyperlink data. Miller et al. (Miller, Rae, Schaefer, Ward, LoFaro, & Farahat, 2001) describe such an approach, a modification of the HITS algorithm based on the idea that more frequently followed links should play a larger role in determining authority scores by employing a usage-weighted link matrix.[27]

Although Davison's (2002) strategy for prefetching Web pages is designed to speed up page access by improved Web caching, it deserves mention here because he estimates the degree of association between pages based on user visiting patterns with a new twist. Davison's approach differs from that of Zhou et al. in that he uses the content similarity of page q to a set of pages leading to q as an association measure between pages p and q. Such a set consists of page p, which explicitly links q, and pages the user visited prior to p during the same user session.

Last but certainly not least, Xue et al.'s (Xue, Zeng, Chen, Ma, Zhang, & Lu, 2003) study of implicit link analysis for small Web searches raises some interesting notions. Observing the frequent failure of link analysis methods in TREC experiments (Hawking, 2001; Hawking & Craswell, 2002), Xue et al. contend that link analysis applied to small Web collections is ineffective due to the truncated link structure; they suggest that link analysis should be applied to an implicit link structure constructed from user access patterns instead of the incomplete hyperlink structure of a small Web collection. Their approach to small Web search extends previous work on Web usage data mining by incorporating both the probabilistic approach to implicit link construction (Zhou et al., 2001) and usage-based modification of link analysis (Miller et al., 2001). Xue et al.'s implicit link construction method is based on conditional probabilities associated with pairs of pages visited in user sessions and produces implicit links with associated weights, which are then used to modify the link matrix from the PageRank formula.

To test their approach on a small Web search, Xue et al. mined 336,000 implicit links collected over a four month period from the UC Berkeley Web server log and compared their modified PageRank approach to full-text, PageRank, modified HITS, and DirectHit searches. Because the effectiveness of the modified PageRank method they proposed is directly affected by the quality of implicit link construction, Xue et al. had seven human subjects evaluate random subsets of 375 implicit and 290 explicit links. They found that 67 percent of implicit and 39 percent of explicit links were recommendation links. This finding suggests that implicit link construction can not only help complete the link structure but also raise the quality of links by reducing the link noise (e.g., navigational links). The search results, evaluated using precision at rank 30 and degree of authority (proportion of authority pages) at rank 10, showed the proposed method outperforming all other methods by significant margins.

The basic idea of mining usage data for user-based relevance seems to hold promise for Web IR, especially when combined with other sources of evidence. The 670 million searches logged by major search engines per day suggest that usage data is a potential source of evidence rich with possibilities.

Web IR Experiments in TREC

TREC has been a fertile ground for IR experimentation using large-scale test collections (Harman, 1994; Voorhees, 2003). In addition to demonstrating the continuing viability of statistical IR approaches, TREC has also shown that fine tuning of IR systems to the data and the task at hand can improve retrieval performance. Over the years, TREC has expanded its scope beyond the realm of text retrieval to areas such as cross-language retrieval, interactive retrieval, video retrieval, and Web retrieval; TREC is actively investigating various nontraditional IR issues such as user–system interaction and link analysis, as well as testing the applicability of traditional methods to nontraditional settings.

The Web IR experiment of TREC, called the Web track, initially investigated the same ad hoc retrieval task undertaken with plain text documents. Although many TREC participants explored methods of leveraging nontextual sources of information, such as hyperlinks and document structure, the general consensus among the early Web track participants was that link analysis and other nontextual methods did not perform as well as content-based retrieval methods (Gurrin & Smeaton, 2001; Hawking, Craswell, Thistlewaite, & Harman, 1999; Hawking, Voorhees, Craswell, & Bailey, 2000; Savoy & Rasolofo, 2001).

There has been much speculation as to why link analysis, which showed much promise in previous research and has been so readily embraced by commercial search engines, did not prove more useful in Web track experiments. Speculation pointed to potential problems with the Web track's test collection, inadequate link structure of truncated Web data (Savoy & Picard, 1998; Singhal & Kaszkiel, 2001), relevance judgments that penalize link analysis by not counting the hub pages as relevant (Voorhees & Harman, 2000) and boost content analysis by counting multiple relevant pages from the same site as relevant (Singhal & Kaszkiel, 2001), and to queries too detailed and specific to be representative of real-world Web searches (Singhal & Kaszkiel, 2001).

In a case study of TREC algorithms, Singhal and Kaszkiel (2001) suggest that the Web track task is not representative of real-world Web searches, which range from locating a specific site to finding high-quality sites on a given topic. In an experiment that compared retrieval performances of TREC's content-based, ad hoc algorithms to the search results of several Web search engines in a more "realistic" environment—designed specifically to address the problems with the Web track experiments[28]—Singhal and Kaszkiel found that TREC content-based systems performed much worse than Web search engines utilizing link analysis and suggested that content-based methods may retrieve documents about a given topic, but not necessarily the documents users are looking for.

In a related study, Craswell, Hawking, and Robertson (2001) used link anchor text to find the main entry point of a specific Web site (i.e., home page). In a carefully controlled experiment, where all the retrieval system parameters were identical except for the document content that consisted of either textual content or inlink anchor texts of a Web page, they found that using anchor text was twice as effective as using document content in what they called the site-finding task.

In an effort to address the criticisms and problems associated with the early Web track experiments, TREC abandoned the ad hoc Web retrieval task in 2002 in favor of topic distillation and named-page finding tasks, and replaced its earlier Web test collection of randomly selected Web pages with a larger and potentially higher quality, domain-specific collection.[29] The topic distillation task in TREC-2002 entailed finding a short, comprehensive list of pages that were good information resources; the named-page finding tasks involved locating a specific page, the name of which was described by the query (Craswell & Hawking, 2003; Hawking & Craswell, 2002).

Adjustment of the Web track environment generated renewed interest in retrieval approaches that leverage Web-specific sources of evidence, such as link structure and document structure. For the home page finding task, where the objective is to find the entry page of a specific site described by the query, the Web page's URL characteristics, such as type and length, as well as the anchor text of the Web page's inlinks, proved to be useful sources of information (Hawking & Craswell, 2002). In the named-page finding task, which is similar to the home page finding task except that the target page described by the query is not necessarily the entry point of a Web site but may be any specific page on the Web, the use of anchor text still proved to be an effective strategy, although the use of URL characteristics did not work as well as it did in the home page finding task (Craswell & Hawking, 2003).

In the topic distillation task, the anchor text still seemed to be a useful resource, especially as a way to boost the performance of content-based methods via fusion (i.e., result merging), although its utility level fell much below that achieved in named-page finding tasks (Craswell & Hawking, 2003; Hawking & Craswell, 2002). Site compression strategies, which attempt to select the "best" pages of a given site, were another common theme in the topic distillation task, once again demonstrating the importance of fine-tuning the retrieval system to the task (Amitay, Carmel, Darlow, Lempel, & Soffer, 2003b; Zhang, Song, et al., 2003). Interestingly, link analysis (e.g., PageRank, HITS variations) has not yet proven effective and the content-based method still seems to be dominant in the Web track. In fact, the two best results in the TREC-2002 topic distillation task were achieved by the baseline systems that used only content-based methods (MacFarlane, 2003; Zhang, Song, et al., 2003).

The two tasks in the TREC-2003 Web track were slight modifications of tasks in TREC-2002. The 2003 topic distillation task is described as finding relevant "home pages" given a broad query, which introduces bias in favor of home page finding approaches. For the second Web track task, the named-page finding task was combined with a home page finding task in a blind mix of 150 home-page queries and 150 named-page queries (Craswell, Hawking, Wilkinson, & Wu, 2003). There were only a few relevant documents for topic distillation queries in 2003 (average 10.32 relevant documents per topic), so R-precision, which is the precision at rank n, where n is the number of relevant documents for a given topic, was used to evaluate the topic distillation results instead of precision at rank 10, as had been done in the previous year. The evaluation metrics for home/named-page finding results were mean reciprocal rank, (the sum of reciprocal rank of the first correct answer for all queries divided by the number of queries), and the success rate, (the proportion of queries with correct answers at top-10 ranks).

The home page bias of topic distillation task in TREC-2003 prompted many participants to combine their topic distillation approaches with strategies found to be effective in the home-page finding task (Kraaij, Westerveld, & Hiemstra, 2002). Tomlinson (2003), for example, used URL information[30] to identify potential home pages and boost their retrieval scores. Jijkoun et al. (2003) combined the results of body text, title, and inlink anchor text retrieval runs; and Zhang, Lin, et al. (Zhang, Lin, Liu, Zhao, Ma, & Ma, 2003) employed an entry page location algorithm based on URL characteristics to boost the score of entry pages as well as utilizing site-compression techniques and the anchor text index.

System tuning based on query classification was one of the new approaches that emerged in the home/named-page finding tasks, where participants recognized the disadvantage of the "one size fits all" approach for the mixed search types. Using past Web track results as training data, Craswell, Hawking, McLean, Wilkinson, and Wu (2003) tuned anchor text weight and Okapi BM25 parameters to produce system optimizations for each home-page and named-page finding task and used a simple query classifier based on keyword occurrence to select an appropriate system to deploy for each given query. Amitay, Carmel, Darlow, Herscovici, et al.'s (2003) "Query-Sensitive Tuner" is another example of retrieval strategy adjustment by query typing. Query-Sensitive Tuner, which considers query length as well as the expected number of documents containing all query terms to classify queries, tunes the weighting of content, anchor text, and indegree contributions to the combined retrieval score based on query classification. Plachouras, Ounis, van Rijsbergen, and Cacheda (2003, p. 248) determined which combinations of evidence to use in a fusion formula based on the concept of "query scope," which is computed from statistical evidence obtained from the set of retrieved documents.

It is not clear which query classification approach is most effective, nor is it obvious whether a high-powered query classifier alone is sufficient to deal with a mixed search environment such as home/named-page task. However, the need to explore a diverse set of retrieval settings that reflects the true conditions of the Web search environment is apparent. By rigorously engaging in such endeavors, TREC has played an indispensable role in Web IR research. In addition to producing high-quality test collections for Web IR, and much associated experimental data from numerous researchers both in and outside the TREC community, TREC has led the way in establishing novel performance measures designed specifically for Web IR tasks. Furthermore, TREC has been instrumental in validating traditional IR approaches in the Web setting, as well as nurturing and fine tuning new ideas. For instance, TREC research validated term weighting and query expansion methods of traditional IR for the Web; established anchor text, URL, and document structure as important sources of Web evidence; and suggested such techniques as site compression, query classification, and fusion as promising areas of investigation for future Web IR research.

Research in Information Organization

Knowledge organization, the process of organizing concepts into an ordered group of categories (taxonomy), is the way humans understand the world. When we encounter a phenomenon for the first time, we try to understand it by comparing it with what we already know, thus "transforming isolated and incoherent sense impressions into recognizable objects and recurring patterns" (Langridge, 1992, p.3). We learn about the world by looking for differences and similarities in things and finding relationships among them. The process of grouping together similar things, perceptions, experiences according to some common quality or characteristic is an integral part of daily life. Infants begin by partitioning the world into those elements that give comfort and those that cause discomfort. As we grow and accumulate more information, we refine our understanding of the world by updating our knowledge organization system with more complex categories and relationships.

According to John Dewey (1925, p. 30), "knowledge is classification." Langridge (1992) states that without classification there could be no human thought, action, or organization. In other words, information organization (IO) is essential to human learning and knowledge. Not surprisingly, researchers in artificial intelligence, machine learning in particular, have been studying IO as a way to emulate human intelligence. IO has also been explored in information and library science as a mechanism with which to bridge the user's information need and the information collection. For centuries librarians have organized library collections in various ways (e.g., Library of Congress Classification, Dewey Decimal Classification) to help library patrons find information. Web directories, though more ad hoc and less standardized than traditional library systems, are another application of IO that guides users through the information discovery process.

IO approaches in IR reflect ideas from both machine learning and library science. Based on the notion that IO facilitates learning and is one of the basic building blocks of knowledge, some researchers have experimented with IO as a tool to develop intelligent IR systems that go beyond term matching and others have investigated how it could be used to help users better grasp the content of a collection. Although system- and user-oriented IO approaches differ in their focus, both hope to har- ness intelligence and knowledge to enhance retrieval by first organizing information in some manner.

Two types of information organization are studied in IR: classifica- tion[31] and clustering. Classification organizes entities by pigeonholing them into predefined categories, whereas clustering organizes informa- tion by grouping similar or related entities together. In classification, categories are first determined and objects are assigned to them accord- ing to characteristics of interest. In clustering, categories are revealed as a result of grouping objects based on some common characteristics. Although both clustering and classification have been active areas of research for some time, studies that investigate their application on the Web have been scarce. Yet, information organization is a concept reflected across the Web landscape, from Web directories and metadata initiatives to the Semantic Web and the digital library movement. Research that explores the adaptation of clustering and classification approaches for the Web may not only prove useful for ongoing Web IO activities, but also provide valuable insights into leveraging knowledge for information discovery on the Web.

Organizing Web Documents

Traditional text-based IR research uses homogeneous corpora with coherent vocabulary, high-quality content, and congruous authorship. The Web, however, introduces the challenges of diverse authorship, vocabulary, and quality. Furthermore, some Web documents are inten- tionally fragmented to facilitate navigation and hyperlinking, making it difficult to determine their topics from local content alone. Leveraging hyperlinks in the Web is also more problematic than in traditional hypertext IR research due to the number of link types that are hard to classify automatically.

Information organization on the Web inherits all the problems and challenges generally associated with Web IR. For instance, it is difficult to cluster or classify the whole Web because of its massive size and diver- sity. Even if such a feat were possible, most clustering approaches, which are not incremental, and text categorization approaches, which are based on a static classification scheme, would not be able to deal with the dynamic nature of the Web. Methods of organizing Web documents need to be efficient, flexible, and dynamic. Moreover, post-retrieval orga- nization of retrieved documents may be both a more desirable and a more realistic approach than trying to organize the entire Web.

Clustering the Web

There are several approaches to clustering Web documents, such as cocitation analysis to identify topic clusters (Larson, 1996), the "trawling" process to find emerging Web communities (Kumar et al., 1999, p. 403), and topic management approaches that collect and organize Web pages related to a particular topic (Modha & Spangler, 2000; Mukherjea, 2000a, 2000b). The post-retrieval clustering approach of Scatter/Gather (Cutting, Karger, Pedersen, & Tukey, 1992; Hearst & Pederson, 1996) has also been applied to the Web to produce topic-coherent clusters of retrieved documents dynamically (Sahami, Yusufali, & Baldonaldo, 1998).

Zamir and Etzioni (1998) conducted one of the few studies that tailored a clustering algorithm specifically for the Web. To satisfy the rigorous requirements of the Web environment, which demands fast, effective, and dynamic algorithms that produce concise and accurate descriptions, they proposed an incremental, linear-time clustering algorithm called Suffix Tree Clustering (STC), which clusters documents based on shared phrases. In a related study, Zamir and Etzioni (1999) used STC to group metasearch results into clusters dynamically by using snippets returned from search engines rather than the full text of documents in order to avoid fetching result pages, usually the major bottleneck in Web IR.[32]

Site clustering, sometimes called site compression, is the grouping of Web pages by authorship rather than by content similarity. Site compression in its simplest form, which uses the domain name portion of URLs to cluster pages by common authorship, was first used in link analysis to differentiate self-citational links from true recommendation links (Bharat & Henzinger, 1998; Kleinberg, 1998). Largely motivated by topic distillation and entry page finding tasks in TREC, researchers explored more sophisticated site compression methods, which combine multiple sources of Web evidence, such as page content, hyperlinks, and URL, not only to cluster pages by site but also to identify the site entry page (Amitay, Carmel, Darlow, Herscovici, et al., 2003; Amitay, Carmel, Darlow, Lempel, et al., 2003b; Wen et al., 2003; Zhang, Lin, et al., 2003; Zhang, Song, et al., 2003).

Clustering is also used to aggregate multiple Web pages that make up a coherent body of material on a single topic. Although link analysis approaches that produce topical clusters (Chakrabarti, Dom, Gibson, et al., 1998; Chakrabarti, Dom, Raghavan, et al., 1998; Kleinberg, 1998; Kumar et al., 1999) leverage link structure to achieve document clustering by topic, Eiron and McCurley (2003) propose an evidence fusion approach that combines analysis of anchor text similarity and hierarchical structure of hyperlinks to aggregate compound Web documents into a single information unit.

Classifying the Web

Although post-retrieval clustering of Web documents offers a viable and effective alternative to traditional, ranked-list retrieval, it lacks the clarity and purpose of a structured organizational hierarchy. Classification of

Web documents not only produces a useful organization of information that can be browsed or searched, but also offers a standardized way— similar to a thesaurus—to describe or refer to the content of Web pages, which can be leveraged to enhance retrieval performance. Chakrabarti, Dom, and Indyk (1998) explored the application of a term- and link-based classifier to hypertext documents and proposed a method that uses link analysis for better classification.

Geffner, Agrawal, Abbadi, and Smith (1999) suggest a method of exploiting the classification hierarchy. By representing a classification hierarchy as a data cube, where leaves become singleton ranges and nodes become the union of all ranges of their children, the data cube approach summarizes and encapsulates organizational structure in a multidimensional database of attributes. Yahoo!'s display of its categories with associated sizes and subcategories, for example, can be thought of as an application of the data cube idea to present a compact summary of the categories.

Jenkins, Jackson, Burden, and Wallis (1998) describe a method for using the Dewey Decimal Classification (DDC) to classify Web pages for WWLib.[33] WWLib is organized by an Automated Classification Engine (ACE), which uses class representatives constructed from DDC to classify documents hierarchically. The class representatives, consisting of DDC classmark and accompanying header text, as well as a manually selected set of keywords and synonyms, are compared to documents using the Dice similarity coefficient. ACE is similar to TAPER (Chakrabarti, Dom, Agrawal, & Raghavan, 1997) in that it employs the hierarchical classification approach using customized class representatives at each node of the DDC hierarchy. The ACE approach has the advantage of using a universal classification scheme that covers all subject areas at high levels of granularity, but it requires that class representatives be constructed manually.

Chekuri, Goldwasser, Raghavan, and Upfal (1996) describe a method of leveraging Yahoo! categories with the objective of increasing the precision of the Web search. Their main idea is to use a Yahoo!-trained classifier in conjunction with the traditional keyword search in an integrated search interface. In their proposed system, where users can specify both keywords and category terms, the classifier categorizes the keyword search results and presents filtered and organized documents to the user. In keeping with Larson's (1992) idea of automatic classification as a tool for assisting manual classification, Chekuri et al. suggest presenting the top k matching categories to the user and contend that a reasonably accurate classifier is sufficient in such a setting. To validate this contention, they trained a Rocchio classifier (Rocchio, 1971) using a sample of 2,000 Web pages from 20 high-level Yahoo! categories and classified 500 random documents from Yahoo!. More than 50 percent of test documents were classified correctly at the top-ranking category, more than 80 percent at top 3, and more than 90 percent at top 5, from which they concluded that displaying the top-10 categories for each of the search results would be sufficient for an interactive search interface.

A slightly different method of leveraging Yahoo! categories is described by Grobelnik and Mladenic (1998). Instead of using the actual Web documents associated with Yahoo! categories, they use Yahoo!'s descriptions of the categorized pages to train a hierarchical Naive Bayes classifier (Koller & Sahami, 1997). Because the feature space induced by such descriptions can be sparse, Grobelnik and Mladenic propagate the description terms upwards in the hierarchy with weights proportional to the node size where they appear. In classifying documents, they use a pruning strategy where categories with less than the required number of features in common with the target document are pruned using an inverted index of features by categories. They tested their method by training the classifier on three sets of training documents selected from the top 14 Yahoo! categories and classifying 100 to 300 randomly selected Web pages categorized by Yahoo!. The results showed that the correct category assignment probability was over 0.99 using only a small number of features.[34] They also found that pruning more than half the categories resulted in misclassification of only 10 to 15 percent of the documents.

Labrou and Finin (1999) also leveraged Yahoo! terms to train a classifier called Teltale. Their approach, which combined category labels, summaries, and link titles in a Yahoo! page and content of the Web pages referenced by the links to describe categories, revealed that the best results would occur when using very brief descriptions of category entries (i.e., Yahoo! summaries and titles).

The research reviewed so far leverages existing taxonomies to organize Web documents. A recent study by Glover et al. (2002) examined the use of anchor text for automatic classification and labeling of Web pages. They found that the fusion approach, which combines text-based classification with link-based classification by considering features in inlink anchor text as well as content text to classify a Web page, worked better than either approach by itself.

Based on the assumption that a site's functionality creates a site signature, Amitay, Carmel, Darlow, Lempel, and Soffer (2003a) examined the structural and connectivity-based properties of eight Web site[35] categories by function (e.g., Web hierarchies/directories, virtual hosting service, corporate site) to identify patterns for site-classifier feature determination. They then selected 16 features based on internal and external linkage patterns to and from the site (e.g., outdegree, outdegree per leaf page) as well as page distribution pattern in the site hierarchy (e.g., average page level, percentage of pages in most populated level). A decision-rule classifier with one or more of these features was then constructed using training data. The classifier operates with a series of decision rules, one of which may be of the kind "if feature #5 (outdegree) is greater than 5,000 AND feature #8 (outdegree per leaf page) is greater than 10, then assign 0.7 to category class #3 (Web Directory)." Amitay et al.'s approach to site classification addresses the principal weakness of link-based classifiers, the fact that they can be unduly influenced by the absolute link density. Effective classification of site by function could be useful for filtering or clustering search results as well as modifying link-weight propagation in link analysis.

The study by Liu, Chin, and Ng (2003) takes topic distillation a step further by infusing data mining and hierarchical classification into their topic mining algorithm. They argue that current term matching and link analysis approaches are inadequate for satisfying the information need associated with learning a new topic, which requires the retrieval of the topic hierarchy populated by pages that include definitions and descriptions of topic and subtopics. Their proposed topic mining algorithm, which is designed to find and organize pages containing definitions and descriptions of topic and subtopics, constructs and populates a topic hierarchy by recursively applying the following process until the topic hierarchy is complete (i.e., there are no more subtopics to be identified). First, obtain a set of relevant pages about a given topic by retrieving 100 top-ranked pages from a search engine with a topical query. Next, discover the salient concepts using HTML tag-emphasized phrases (e.g., <b>, <h>) to mine subtopic concepts from those pages. Then locate "informative pages" (Liu et al., 2003, p. 253) by analyzing HTML structure (e.g., header tag frequency, anchor text similarity) and pattern matching of definition cues (e.g., "known as," "defined as") to identify pages containing definitions of the topic and subtopics. If another level of topic hierarchy is desired, repeat the process by querying a search engine with subtopic concepts. In an experiment where the results of Liu et al.'s topic mining search were compared with the top-10 results of Google and AskJeeves, the topic mining method showed higher precision at rank 10 than other systems in 26 out of 28 topics tested.

The application of classification methods to Web IR is not limited to content, nor is its purpose limited to information organization. Query or task classification, as commonly practiced in Web IR research (Amitay et al., 2003a; Craswell, Hawking, McLean, et al., 2003; Plachouras et al., 2003), aims to categorize queries by task type and is typically a precursor to dynamic system tuning that will optimize system parameters for the identified query type. Recently, two new approaches that apply classification for the purpose of retrieval have been proposed. One approach exploits existing categories of Web pages to directly influence PageRank computations (Jeh & Widom, 2003; Haveliwala, 2002) and the other automatically identifies effective query modifications from iterations of the classification and feature selection process (Flake, Glover, Lawrence, & Giles, 2002).

Haveliwala's (2002, p. 517) research extends PageRank capability with a set of "topic sensitive" PageRank vectors, each of which is used to compute a Web page's importance score with respect to a given topic. A PageRank vector can be thought of as a vector of preference weights for all the pages, where a weight reflects the degree of preference for a given page. If we express the PageRank vector as a function, $PV(p)$ that returns the preference weight of page p, we can rewrite the PageRank formula in the following way:

$$R(p) = c \cdot PV(p) + (1 - c) \cdot \sum_{i=1}^{k} \frac{R(p_i)}{C(p_i)} \qquad (11)$$

where $C(p)$ is the outdegree of p, and p_i denotes the inlinks of p and c is a "teleportation" probability that a Web surfer breaks the normal traversal of hyperlinks and "teleports" to another page. In the case of global PageRank where page preference is not considered, $PV(p)$ returns a uniform weight of $1/T$ (T is the number of pages in the Web). In the case of topic sensitive PageRank, a PageRank vector biased toward a topic is constructed by assigning preference weights to pages about that topic. Such a PageRank vector introduces a computational bias toward its topic because the "teleportation" to a preferred page (e.g., topic pages with preference weights, pages linking to those pages, and so on) will produce a larger PageRank score than jumping to a nontopical page.

Haveliwala constructed a set of 16 PageRank vectors from pages classified in 16 Open Directory (http://dmoz.org) categories by assigning a uniform preference weight of $1/N$ to pages in the Open Directory category and zero to all other pages, where N is the total number of pages in a given category. He then computed 16 sets of topic-sensitive PageRank scores, which are combined at retrieval time to generate a final query-sensitive PageRank score. Topic-sensitive PageRank scores are combined by a weighted sum formula, where the weight of each topic-sensitive PageRank score is the class probability of the query belonging to a given class, computed by a probabilistic classifier.

Jeh and Widom (2003) generalize the notion of topic-sensitive PageRank vectors as Personalized PageRank Vectors (PPV), which are PageRank vectors that can reflect any number of personal preferences. Haveliwala's approach to computing preference-biased PageRank scores is resource intensive and can accommodate only a limited number of PageRank vectors. To address this problem, Jeh and Widom (2003, p. 272) propose a solution that represents PPV as a linear combination of "basis vectors," which are encoded as shared components that can derive PPV at query time in order to allow scalable PageRank computation. PPV is an innovative approach that fuses automatic classification and link analysis at a method level to dynamically optimize the retrieval strategy according to user preferences.

Flake et al.'s (2002) research can be viewed as tackling a similar problem in a quite different fashion. They address the problem of personalized retrieval by modifying a user-specified query with automatically generated query modifications that will effectively filter retrieval results by category. Query modifications (QM) are extracted from nonlinear SVMs using training data of positive and negative category examples in an iterative process. Both QM and PPV aim to accommodate a facet of information need not explicitly stated in a query, but approach the problem space from different directions. Although both approaches make use of prior evidence (e.g., personal preference data) and employ classification as an integral component of the overall strategy, the QM solution focuses on query optimization and the PPV solution focuses on dynamic tuning of the scoring function.

The idea of tackling the implicit query facet warrants further consideration. The facet in question, which might be a topic category, document type, or simply a set of preference features not explicitly stated in a query, could be mined from an existing taxonomy, link topology, or usage data among other things. Harnessing multiple sources of evidence and combining different approaches to satisfy multifaceted information needs may be the type of approach required to bring Web IR research to the next level.

Concluding Remarks

In this chapter, we examined the complexities of the Web search environment and reviewed strategies for finding information on the Web from two different perspectives. In both the retrieval and organizational approaches to information discovery on the Web, leveraging rich sources of evidence has been a consistent theme. Utilizing link information is a central approach in Web IR. The assumptions underlying the link-based methods (Craswell et al., 2001)—the recommendation assumption (page author is recommending the page that he/she is linking to), topic locality assumption (pages connected by links are more likely to be about the same topic), and anchor description assumption (anchor text of a link describes its target)—not only make intuitive sense but also have been shown to be valid in numerous studies (Craswell & Hawking, 2003; Davison, 2000; Dean & Henzinger, 1999; Kleinbeg, 1998; Page et al., 1998).

It is therefore curious to note that sophisticated link analysis techniques, such as PageRank and HITS, have yet to prove their effectiveness in the TREC arena, where other Web IR techniques, such as anchor text use and URL-based scoring, have been validated repeatedly. Whether this phenomenon is due to the small Web effect (Xue et al., 2003) or is an artifact of some other characteristics of the TREC environment, such as query, relevance judgments, or retrieval task, remains to be seen.

The Web search environment can reveal weaknesses in retrieval approaches based on single sources of evidence. For example, content-based IR approaches have difficulty dealing with the variability in vocabulary and quality of Web documents, and link-based approaches can suffer from incomplete or noisy link topology. The inadequacies of singular Web IR approaches coupled with the fusion hypothesis ("fusion is good for IR") make a strong argument for combining multiple sources of evidence as a retrieval strategy for Web IR.

The fusion strategies in early Web IR experiments achieved only moderate success, primarily by combining content-based results with anchor-text results (Hawking & Craswell, 2002). Fusion, however, has been widely adopted by today's Web IR researchers. Combining multiple sources of Web evidence such as document content, structure, hyperlinks, and URL information has become standard practice in TREC at the time of writing (Craswell, Hawking, Wilkinson, et al., 2003). Several innovative approaches that integrate both retrieval and information organization approaches have also been explored in recent studies (Liu et al., 2003; Flake et al., 2002; Haveliwala, 2002; Jeh & Widom, 2003).

Finding information on the Web is a complex and challenging task that requires innovative solutions. Research in Web IR has produced some approaches that effectively capitalize on the characteristics of the Web search environment and has also suggested the potential of fusion that combines both methods and sources of evidences. As for the future of Web IR, we may see a move toward information-rich areas such as user data mining and knowledge-based retrieval that leverage stored human knowledge (e.g., Web directories), where fusion and dynamic tuning are standard approaches to bring together multiple sources of evidence and multiple methods for personalized information discovery on the Web.

Endnotes

1. For the remainder of this chapter, IR will imply text-based retrieval unless otherwise stated.
2. DirectHit, a "popularity" search technology that powers other search services, no longer maintains its own site. See http://www.searchenginewatch.com/sereport/article.php/2164521 for details.
3. Web directories are manually constructed topical taxonomies of Web documents (e.g., Yahoo!).
4. Inktomi, which began as a University of California at Berkeley research project in 1995, is currently a Yahoo! subsidiary that provides Web search products and services. See http://www.inktomi.com for more details about the company.
5. Indegree of a document p denotes the number of inlinks of p, where an inlink of p denotes a document q that points (i.e., links) to p. Conversely, outdegree of a document p denotes the number of outlinks of p, where an outlink of p denotes a document q that is pointed to (i.e., linked) by p.
6. Bray employed a number of ad hoc rules to define a site, which was based on parsing of URLs with an intent to identify the logical location of documents.
7. In this chapter, the terms "Web documents" and "Web pages" are loosely interchangeable, although the former mostly implies the text content and the latter refers to both textual and nontextual content.
8. "The publicly indexable" Web includes only pages normally indexed by Web search engines, thus excluding pages with access restricted by the robot exclusion protocol, firewalls, or password protection, and hidden pages (e.g., cgi, dynamic pages).

9. Data as of February, 2003, published in Search Engine Watch (http://www.searchengine watch.com/reports/article.php/2156461).

10. The data, which were a random subset of Excite searches on March 10, 1997, consisted of 51,473 queries from 18,113 users.

11. The survey data consisted of 3,190 Web submissions and the search log data consisted of 400 random queries by AltaVista users in 2001.

12. The (i,j)th entry of the link matrix corresponds to the link from page i to page j.

13. The (i,j)th entry of A is 1 if there exists a link from page i to page j, and is 0 otherwise. In AT, the transpose of the link matrix A, the (i,j)th entry of A corresponds to the link from page j to page i. The (i,j)th entry of AAT gives the number of pages pointed to by both page i and page j (bibliographic coupling), whereas the (i,j)th entry of ATA gives the number of pages that point to both page i and page j (cocitation).

14. Incorporation of anchor weights into HITS can be better demonstrated by using matrix notations. If equation (4) and (5) are rewritten as a = ATh and h = Aa, where a and h are authority and hub vectors and A is a link matrix whose (i,j)th entry is 1 if a link exists from pi to pj and 0 otherwise, then ARC formulas can be written as a = ZTh and h = Za, where Z is a weighted link matrix whose (i,j)th entry is the anchor text similarity weight $w(i,j)$.

15. Average score ratio is the score ratio of search engine a to search engine b averaged over topic.

16. Because I could not find the details of the link weighting strategy by CLEVER, this statement about CLEVER is my inference only and may not be correct. Chakrabarti, Dom, Gibson, et al. (1998, p. 15) do not mention mutually reinforcing relationships, but simply say that CLEVER "varies the weight of links between pages based on the domain of their endpoints" to prevent pages by the same author from conferring authority upon one another.

17. Henceforth, BHITS will denote Bharat and Henzinger's fusion method of content and connectivity analysis.

18. Because the inverse document frequency of a term is impossible to obtain for the whole Web, idf is based on term frequencies measured in a crawl of 400,000 Yahoo! documents in January 1997.

19. In the ARC experiment, 28 queries were originally constructed but only 27 were used in the analysis.

20. Advertisement links are identified by certain characters in URLs, such as "=" and "?".

21. http://www.nzdl.org/html/prescript.html

22. http://www.research.compaq.com/SRC/virtualpaper/pstotext.html

23. Headers referred to here are bibliographic in character, usually consisting of author's name, title, and publication year.

24. http://home.netscape.com/escapes/related/faq.html

25. Actually, the scoring scale was 0, 1, and >-= for inaccessible, but >-= was considered the same as 0 in precision computations.

26. See the Search Engine Review section of Search Engine Watch for articles on Web search engine technology (http://www.searchenginewatch.com/resources/article.php/2156581). A summary of search engines based on Search Engine Watch data in 2000 can be seen at http://ella.slis.indiana.edu./~kiyang/wse/search/WSE.pdf (Appendix B).

27. Miller et al.'s approach is identical to the ARC formula except that the (i,j)th entry of the weighted link matrix is the frequency of link traversal from page i to j in the usage data instead of the anchor text similarity weight $w(i,j)$.

28. Singhal and Kaszkiel used 100 queries that sought a certain page/site in 17.8 million Web pages (217.5 gigabyte crawl in October, 2000) and counted the number of queries that retrieved the correct page in top ranks to compare the retrieval performances.

29. The current test collection of the Web track consists of 1.25 million Web pages (19 gigabytes) from .gov domain, which is larger, less diverse, and likely to be of higher quality than the previous collection, which was a 10 gigabyte subset of the Web crawl from Internet Archive.

30. Tomlinson classified Web pages into categories of root, subroot, path, and file, based on URL endings (e.g., index.htm, home.htm) and the number of slashes in the URL.

31. In this chapter, "classification" is used in the narrow sense, primarily in the context of text categorization.

32. A study comparing the clusters using snippets to those using full texts showed that the snippet approach produced only a moderate degradation of 15 percent average precision loss (Zamir & Etzioni, 1998).

33. WWLib (http://www.scit.wlv.ac.uk/wwlib), a virtual library at the University of Wolverhampton, is a searchable, classified catalog of Web pages in the United Kingdom.

34. Using n-grams instead of single words as features, they found the category feature vector size of three to four to be effective in most cases.

35. A site refers to the set of Web pages belonging to the same virtual entity.

References

Allan, J. (1996). Automatic hypertext link typing. *Proceedings of the 7th ACM Conference on Hypertext*, 42–52.

Amento, B., Terveen, L., & Hill, W. (2000). Does authority mean quality? *Proceedings of the 23rd ACM SIGIR Conference on Research and Development in Information Retrieval*, 296–303.

Amitay, E., Carmel, D., Darlow, A., Herscovici, M., Lempel, R., Soffer, A., et al. (2003). Juru at TREC 2003: Topic distillation using query-sensitive tuning and cohesiveness filtering. *The 12th Text Retrieval Conference (TREC 2003) Notebook*, 255–261.

Amitay, E., Carmel, D., Darlow, A., Lempel, R., & Soffer, A. (2003a). The connectivity sonar: Detecting site functionality by structural patterns. *Proceedings of the 14th ACM Conference on Hypertext and Hypermedia*, 38–47.

Amitay, E., Carmel, D., Darlow, A., Lempel, R., & Soffer, A. (2003b). Topic distillation with knowledge agents. *Proceedings of the 11th Text Retrieval Conference (TREC 2002)*, 263–272.

Arocena, G. O., Mendelzon, A. O., & Mihaila, G. A. (1997). Applications of a Web query language. *Proceedings of the 6th International WWW Conference*, 587–595.

Bar-Ilan, J. (2004). The use of Web search engines in information science research. *Annual Review of Information Science and Technology*, 38, 231–288.

Barabási, A. (2003). *Linked: How everything is connected to everything else and what it means for business, science, and everyday life.* New York: Plume.

Baron, L., Tague-Sutcliffe, J., & Kinnucan, M. T. (1996). Labeled, typed links as cues when reading hypertext documents. *Journal of the American Society for Information Science*, 47(12), 896–908.

Bernstein, M. (1998). Patterns of hypertext. *Proceedings of the 9th ACM Conference on Hypertext*, 21–29.

Bharat, K., & Henzinger, M. R. (1998). Improved algorithms for topic distillation in hyperlinked environments. *Proceedings of the 21st ACM SIGIR Annual International Conference on Research and Development in Information Retrieval*, 104–111.

Borgman, C., & Furner, J. (2002). Scholarly communication and bibliometrics. *Annual Review of Information Science and Technology*, 36, 3–72.

Botafogo, R. A., Rivlin, E., & Shneiderman, B. (1992). Structural analysis of hypertexts: Identifying hierarchies and useful metrics. *ACM Transactions on Information Systems, 10*(2), 142–180.

Bray, T. (1996). Measuring the Web. *Proceedings of the 5th International WWW Conference,* 994–1005.

Brin, S., & Page, L. (1998). The anatomy of a large-scale hypertextual Web search engine. *Proceedings of the 7th International WWW Conference,* 107–117.

Broder, A. (2002). A taxonomy of Web search. *ACM SIGIR Forum, 36*(2), 3–10.

Broder, A. Z., Kumar, S. R., Maghoul, F., Raghavan, P., Rajagopalan, S., Stata, R., et al. (2000). Graph structure in the Web: Experiments and models. *Proceedings of the 9th WWW Conference,* 309–320.

Brown, E. W., Callan, J. P., & Croft, W. B. (1994). Fast incremental indexing for full-text information retrieval. *Proceedings of the 20th VLDB Conference,* 192–202.

Carriere, J., & Kazman, R. (1997). WebQuery: Searching and visualizing the Web through connectivity. *Proceedings of the 6th WWW Conference,* 701–711.

Chakrabarti, S., Dom, B., Agrawal, R., & Raghavan, P. (1997). Using taxonomy, discriminants, and signatures for navigating in text databases. *Proceedings of the 23rd VLDB Conference,* 446–455.

Chakrabarti, S., Dom, B., Gibson, D., Kumar, S. R., Raghavan, P., Rajagopalan, S., et al. (1998). Experiments in topic distillation. *ACM SIGIR Workshop on Hypertext Information Retrieval on the Web,* 13–21.

Chakrabarti, S., Dom, B., & Indyk, P. (1998). Enhanced hypertext categorization using hyperlinks. *Proceedings of ACM SIGMOD Conference on Management of Data,* 307–318.

Chakrabarti, S., Dom, B., Raghavan, P., Rajagopalan, S., Gibson, D., & Kleinberg, J. (1998). Automatic resource list compilation by analyzing hyperlink structure and associated text. *Proceedings of the 7th International WWW Conference,* 65–74.

Chakrabarti, S., van der Berg, M., & Dom, B. (1999). Focused crawling: A new approach to topic-specific Web resource discovery. *Proceedings of the 8th WWW Conference,* 1623–1640.

Chekuri, C., Goldwasser, M., Raghavan, P., & Upfal, E. (1996). Web search using automatic classification. *Proceedings of the 6th WWW Conference,* Retrieved March 20, 2004, from http://www.cs.luc.edu/~mhg/publications/WWW1997

Cho, J., Garcia-Molina, H., & Page, L. (1998). Efficient crawling through URL ordering. *Proceedings of the 7th WWW Conference,* 161–172.

Craswell, N., & Hawking, D. (2003). Overview of the TREC-2002 Web track. *Proceedings of the 11th Text Retrieval Conference (TREC 2002),* 86–95.

Craswell, N., Hawking, D., McLean, A., Wilkinson, R., & Wu, M. (2003). TREC 12 Web Track at CSIRO. *The 12th Text Retrieval Conference (TREC 2003) Notebook,* 237–247.

Craswell, N., Hawking, D., & Robertson, S. (2001). Effective site finding using link anchor information. *Proceedings of the 24th ACM SIGIR Annual International Conference on Research and Development in Information Retrieval,* 250–257.

Craswell, N., Hawking, D., Wilkinson, R., & Wu, M. (2003). Overview of the TREC 2003 Web Track. *The 12th Text Retrieval Conference (TREC 2003) Notebook,* 220–236.

Croft, W. B. (1993). Retrieval strategies for hypertext. *Information Processing & Management, 29,* 313–324.

Cronin, B., Snyder, H., & Atkins, H. (1997). Comparative citation ranking of authors in monographic and journal literature: A study of sociology. *Journal of Documentation, 53*(3), 263–273.

Cui, H., Wen, J., Nie, J., & Ma, W. (2002). Probabilistic query expansion using query logs. *Proceedings of the 11th International WWW Conference,* 325–332.

Cutting, D. R., Karger, D. R., Pedersen, J. O., & Tukey, J. W. (1992). Scatter/gather: A cluster-based approach to browsing large document collections. *Proceedings of the ACM*

SIGIR Annual International Conference on Research and Development in Information Retrieval, 318–329.

Davison, B. (2000). Topical locality in the Web. *Proceedings of the 23rd ACM SIGIR Annual International Conference on Research and Development in Information Retrieval*, 272–279.

Davison, B. (2002). Predicting Web actions from HTML content. *Proceedings of the 13th ACM Conference on Hypertext and Hypermedia*, 159–168.

Dean, J., & Henzinger, M. R. (1999). Finding related pages in the World Wide Web. *Proceedings of the 8th International WWW Conference*, 389–401.

Dewey, J. (1925). *Experience and nature*. Chicago: Open Court.

Eiron, N., & McCurley, K. (2003). Untangling compound documents on the Web. *Proceedings of the 14th ACM Conference on Hypertext and Hypermedia*, 85–94.

Flake, G. W., Glover, E. J., Lawrence, S., & Giles, C. L. (2002). Extracting query modifications from nonlinear SVMs. *Proceedings of the 11th International WWW Conference*, 317–324.

Florescu, D., Levy, A., & Mendelzon, A. (1998). Database techniques for the World-Wide Web: A survey. *SIGMOD Record, 27*(3), 59–74.

Frei, H. P., & Stieger, D. (1995). The use of semantic links in hypertext information retrieval. *Information Processing & Management, 31*(1), 1–13.

Geffner, S., Agrawal, D., Abbadi, A. E., & Smith, T. (1999). Browsing large digital library collections using classification hierarchies. *Proceedings of the 8th ACM International Conference on Information and Knowledge Management*, 195–201.

Glover, E. J., Tsioutsiouliklis, K., Lawrence, S., Pennock, D. M., & Flake, G. W. (2002). Using Web structure for classifying and describing Web pages. *Proceedings of the 11th International WWW Conference*, 562–569

Grobelnik, M., & Mladenic, D. (1998). Efficient text categorization. *Proceedings of Text Mining Workshop on ECML-98*, 1–10.

Gurrin, C., & Smeaton, A. F. (2001). Dublin City University experiments in connectivity analysis for TREC-9. *Proceedings of the 9th Text Retrieval Conference (TREC-9)*, 179–188.

Harman, D. (1994). Overview of the second Text Retrieval Conference. *Proceedings of the 2nd Text Retrieval Conference (TREC-2)*, 1–20.

Haveliwala, T. (2002). Topic-sensitive PageRank. *Proceedings of the 11th WWW Conference*, 517–526.

Hawking, D. (2001). Overview of the TREC-9 Web track. *Proceedings of the 9th Text Retrieval Conference (TREC-9)*, 87–102.

Hawking, D., & Craswell, N. (2002). Overview of the TREC-2001 Web track. *Proceedings of the 10th Text Retrieval Conference (TREC 2001)*, 25–31.

Hawking, D., Craswell, N., Thistlewaite, P., & Harman, D. (1999). Results and challenges in Web search evaluation. *Proceedings of the 8th WWW Conference*, 243–252.

Hawking, D., Voorhees, E., Craswell, N., & Bailey, P. (2000). Overview of the TREC-8 Web track. *Proceedings of the 8th Text Retrieval Conference (TREC-8)*, 131–148.

Hearst, M., & Pedersen, J. O. (1996). Reexamining the cluster hypothesis: Scatter/gather on retrieval results. *Proceedings of the 19th ACM SIGIR Annual International Conference on Research and Development in Information Retrieval*, 76–84.

Hölscher, C., & Strube, G. (2000). Web search behavior of internet experts and newbies. *Proceedings of the 9th International WWW Conference*, 337–346.

Jansen, B. J., Spink, A., Bateman, J., & Saracevic, T. (1998). Real life information retrieval: Study of user queries on the Web. *SIGIR Forum, 32*(1), 5–17.

Jansen, B. J., Spink, A., & Saracevic, T. (1998). Failure analysis in query construction: Data and analysis from a large sample of Web queries. *Proceedings of the 3rd ACM International Conference on Digital Libraries*, 289–290.

Jansen, B. J., Spink, A., & Saracevic, T. (2000). Real life, real users and real needs: A study and analysis of users' queries on the Web. *Information Processing & Management, 36*(2): 207–227.

Jeh, G., & Widom, J. (2003). Scaling personalized Web search. *Proceedings of the 12th International WWW Conference,* 271–279.

Jenkins, C., Jackson, M., Burden, P., & Wallis, J. (1998). Automatic classification of Web resources using Java and Dewey Decimal Classification. *Computer Networks and ISDN Systems, 30,* 646–648.

Jijkoun, V., Kamps, J., Mishne, G., Monz, C., de Rijke, M., Schlobach, S., et al. (2003). The University of Amsterdam at TREC 2003. *The 12th Text REtrieval Conference (TREC 2003) Notebook,* 560–573.

Kahle, B. (1997, March). Preserving the Internet. *Scientific American, 276*(3), 82–84.

Katzer, J., McGill, M. J., Tessier, J. A., Frakes, W., & Das-Gupta, P. (1982). A study of the overlap among document representations. *Information Technology: Research and Development, 1,* 261–274.

Kehoe, C., Pitkow, J., Sutton, K., Aggarwal, G., & Rogers, J. D. (1999). *Results of GVU's tenth WWW user survey.* Retrieved November 11, 2003, from http://www.gvu.gatech.edu/user_surveys/survey-1998-10/tenthreport.html

Kessler, M. M. (1963). Bibliographic coupling between scientific papers. *American Documentation, 14,* 10–25.

Kim, H. J. (2000). Motivation for hyperlinking in scholarly electronic articles: A qualitative study. *Journal of the American Society for Information Science, 51*(10), 887–899.

Kleinberg, J. (1998). Authoritative sources in a hyperlinked environment. *Proceedings of the 9th ACM-SIAM Symposium on Discrete Algorithms,* 668–677.

Koller, D., & Sahami, M. (1997). Hierarchically classifying documents using very few words. *Proceedings of the 14th International Conference on Machine Learning,* 170–178.

Kopak, R. W. (1999). Functional link typing in hypertext. *ACM Computing Surveys,* 31(4es).

Kraaij, W., Westerveld, T., & Hiemstra, D. (2002). The importance of prior probabilities for entry page search. *Proceedings of the 25th ACM SIGIR Annual International Conference on Research and Development in Information Retrieval,* 27–34.

Kumar, S. R., Raghavan, P., Rajagopalan, S., & Tomkins, A. (1999). Trawling the Web for emerging cyber-communities. *Proceedings of the 8th WWW Conference,* 403–415.

Labrou, Y., & Finin, T. (1999). Yahoo! as an ontology: Using Yahoo! categories to describe documents. *Proceedings of the 8th ACM International Conference on Information and Knowledge Management,* 180–187.

Langridge, D. W. (1992). *Classification: Its kinds, elements, systems, and applications.* London: Bowker Saur.

Larson, R. (1992). Experiment in automatic Library of Congress Classification. *Journal of the American Society for Information Science, 43*(2), 130–148.

Larson, R. R. (1996). Bibliometrics of the World Wide Web: An exploratory analysis of the intellectual structure of cyberspace. *Proceedings of the 59th Annual Meeting of the American Society for Information Science,* 71–78.

Lawrence, S., & Giles, C. L. (1998). Searching the World Wide Web. *Science, 280*(4), 98–100.

Lawrence, S., & Giles, C. L. (1999a). Accessibility of information on the Web. *Nature, 400* (6740), 107–110.

Lawrence, S., & Giles, C. L. (1999b). Searching the Web: General and scientific information access. *IEEE Communications, 37*(1), 116–122.

Lawrence, S., Giles, C. L., & Bollacker, K. (1999). Digital libraries and autonomous citation indexing. *IEEE Computer, 32*(6), 67–71.

Lempel, R., & Moran, S. (2001). SALSA: The stochastic approach for link-structure analysis. *ACM Transactions on Information Systems, 19*(2), 131–160.

Li, L., Shang, Y., & Zhang, W. (2002). Improvement of HITS-based algorithms on Web documents. *Proceedings of the 11th International WWW Conference*, 527–535.

Liu, B., Chin, C., Ng, H. (2003). Mining topic-specific concepts and definitions on the Web. *Proceedings of the 12th International WWW Conference*, 251–260.

MacFarlane, A. (2003). Pliers at TREC 2002. *Proceedings of the 11th Text Retrieval Conference (TREC 2002)*, 152–155.

Marchionini, G. (1992). Interfaces for end-user information seeking. *Journal of the American Society for Information Science*, 43(2), 156–163.

Marchiori, M. (1997). The quest for correct information on the Web: Hyper search engines. *Proceedings of the 6th International WWW Conference*, 265–274.

McBryan, O. A. (1994). GENVL and WWW: Tools for taming the Web. *Proceedings of the 1st International WWW Conference*, 58–67.

Mendelzon, A., Mihaila, G., & Milo, T. (1996). Querying the World Wide Web. *Proceedings of the 1st International Conference on Parallel and Distributed Information Systems (PDIS'96)*, 80–91.

Miller, J. C., Rae, G., Schaefer, F., Ward, L. A., LoFaro, T., & Farahat, A. (2001). Modifications of Kleinberg's HITS algorithm using matrix exponentiation and Web log records. *Proceedings of the 24th ACM SIGIR Annual International Conference on Research and Development in Information Retrieval*, 444–445.

Modha, D., & Spangler, W. S. (2000). Clustering hypertext with applications to Web searching. *Proceedings of the 11th ACM Conference on Hypertext*, 143–152.

Mukherjea, S. (2000a). Organizing topic-specific Web information. *Proceedings of the 11th ACM Conference on Hypertext*, 133–141.

Mukherjea, S. (2000b). WTMS: A system for collecting and analyzing topic-specific Web information. *Proceedings of the 9th International WWW Conference*, 457–471.

Mukherjea, S., & Foley, J. (1995). Visualizing the World-Wide Web with the navigational view builder. *Proceedings of the 3rd WWW Conference*, 1075–1087.

Page, L., Brin, S., Motwani, R., & Winograd, T. (1998). *The PageRank citation ranking: Bringing order to the Web*. Retrieved November 30, 2003, from http://dbpubs.stanford.edu/pub/showDoc.Fulltext?lang=en&doc=1999-66&format=pdf

Pirolli, P., Pitkow, J., Rao, R. (1996). Silk from a sow's ear: Extracting usable structures from the Web. *Proceedings of ACM SIGCHI Conference on Human Factors in Computing Systems*, 118–125.

Plachouras, V., Ounis, I., van Rijsbergen, C. J., & Cacheda, F. (2003). University of Glasgow at the Web Track: Dynamic application of hyperlink analysis using the query scope. *The 12th Text Retrieval Conference (TREC 2003) Notebook*, 248–254.

Pollock, A., & Hockley, A. (1997, March). What's wrong with Internet searching? *D-Lib Magazine*. Retrieved November 30, 2003, from http://www.dlib.org/dlib/march97/bt/03pollock.html

Rasmussen, E. (2003). Indexing and retrieval from the Web. *Annual Review of Information Science and Technology*, 37, 91–124.

Resnik, P., & Varian, H. R. (1997). Introduction (to the special section on recommender systems). *Communications of the ACM*, 40(3), 56–59.

Rivlin, E., Botafogo, R., & Shneiderman, B. (1994). Navigating in hyperspace: Designing a structure-based toolbox. *Communications of the ACM*, 37(2), 87–96.

Rocchio, J. J., Jr. (1971). Relevance feedback in information retrieval. In G. Salton (Ed.), *The SMART retrieval system: Experiments in automatic document processing* (pp. 313–323). Englewood Cliffs, NJ: Prentice-Hall.

Sahami, M., Yusufali, S., & Baldonaldo, M. (1998). SONIA: A service for organizing networked information autonomously. *Proceedings of the 3rd ACM International Conference on Digital Libraries*, 200–209.

Salton, G., & Buckley, C. (1988). Term weighting approaches in automatic text retrieval. *Information Processing & Management, 24*, 513–523.

Salton, G., & Buckley, C. (1991, August 30). Global text matching for information retrieval. *Science, 253*, 1012–1015.

Salton, G., Buckley, C., Allan, J. (1994). Automatic structuring and retrieval of large text files. *Communications of the ACM, 37*(11), 97–108.

Savoy, J., & Picard, J. (1998). Report on the TREC-8 experiment: Searching on the Web and in distributed collections. *Proceedings of the 8th Text Retrieval Conference (TREC-8)*, 229–240.

Savoy, J., & Rasolofo, Y. (2001). Report on the TREC-9 experiment: Link-based retrieval and distributed collections. *Proceedings of the 9th Text Retrieval Conference (TREC-9)*, 579–516.

Schapira, A. (1999). *Collaboratively searching the Web: An initial study*. Retrieved November 30, 2003, from http://none.cs.umass.edu/~schapira/thesis/report

Shardanand, U., & Maes, P. (1995). Social information filtering: Algorithms for automating "word of mouth." *Proceedings of the ACM Conference on Human Factors in Computing Systems: Mosaic of Creativity*, 210–217.

Shaw, W. M., Jr. (1991a). Subject and citation indexing. Part I: The clustering structure of composite representations in the cystic fibrosis document collection. *Journal of the American Society for Information Science, 42*, 669–675.

Shaw, W. M., Jr. (1991b). Subject and citation indexing. Part II: The optimal, cluster-based retrieval performance of composite representations. *Journal of the American Society for Information Science, 42*, 676–684.

Shum, S. B. (1996). The missing link: Hypermedia usability research & the Web. *ACM SIGCHI Bulletin, 28*(4), 68–75.

Silverstein, C., Henzinger, M., Marais, H., & Moricz, M. (1998). *Analysis of a very large AltaVista query log* (Technical Report 1998-014). Palo Alto, CA: COMPAQ System Research Center.

Singhal, A., & Kaszkiel, M. (2001). A case study in Web search using TREC algorithms. *Proceedings of the 11th International WWW Conference*, 708–716.

Small, H. (1973). Co-citation in the scientific literature: A new measure of the relationship between two documents. *Journal of the American Society for Information Science, 24*(4), 265–269.

Sparck Jones, K. (1971). *Automatic keyword classification for information retrieval*. London: Butterworth.

Spertus, E. (1997). ParaSite: Mining structural information on the Web. *Proceedings of the 6th International WWW Conference*, 587–595.

Spink, A., Wolfram, D., Jansen, B. J., & Saracevic, T. (2001). Searching the Web: The public and their queries. *Journal of the American Society for Information Science, 53*(2), 226–234.

Tomlinson, S. (2003). Robust, Web and genomic retrieval with Hummingbird SearchServer at TREC 2003. *The 12th Text REtrieval Conference (TREC 2003) Notebook*, 372–385.

Trigg, R., & Weiser, M. (1983). TEXTNET: A network-based approach to text handling. *ACM Transactions on Office Information Systems, 4*(1), 1–23.

Voorhees, E. (2003). Overview of TREC 2002. *Proceedings of the 11th Text Retrieval Conference (TREC 2002)*, 1–16.

Voorhees, E., & Harman, D. (2000). Overview of the eighth Text Retrieval Conference. *Proceedings of the 8th Text Retrieval Conference (TREC-8)*, 1–24.

Weiss, R., Velez, B., Sheldon, M. A., Nemprempre, C., Szilagyi, P., Duda, A., et al. (1996). Hypursuit: A hierarchical network search engine that exploits content-link hypertext clustering. *Proceedings of the 7th ACM Conference on Hypertext*, 180–193.

Wen, J. R., Song, R., Cai, D., Zhu, K., Yu, S., Ye, S., et al. (2003). Microsoft Research Asia at the Web track of TREC 2003. *The 12th Text Retrieval Conference (TREC 2003) Notebook*, 262–272.

White, H. D., & McCain, K. W. (1989). Bibliometrics. *Annual Review of Information Science and Technology, 24*, 119–165.

Willett, P. (1988). Recent trends in hierarchic document clustering: A critical review. *Information Processing & Management, 24*, 577–597.

Wolfram, D., Spink, A., Jansen, B. J., & Saracevic, T. (2001). Vox populi: The public searching of the Web. *Journal of the American Society for Information Science and Technology, 53*(12), 1073–1074.

Woodruff, A., Aoki, P. M., Brewer, E., Gauthier, P., & Rowe, L. A. (1996). An investigation of documents from the World Wide Web. *Proceedings of the 5th International WWW Conference*, 963–980.

Xu, J., & Croft, W. B. (1996). Query expansion using local and global analysis. *Proceedings of the 20th ACM SIGIR Annual International Conference on Research and Development in Information Retrieval*, 4–11.

Xue, G., Zeng, H., Chen, Z., Ma, W., Zhang, H., & Lu, C. (2003). Implicit link analysis for small Web search. *Proceedings of the 26th ACM SIGIR Annual International Conference on Research and Development in Information Retrieval*, 56–63.

Zamir, O., & Etzioni, O. (1998). Web document clustering: A feasibility demonstration. *Proceedings of the 22nd ACM SIGIR Annual International Conference on Research and Development in Information Retrieval*, 46–54.

Zamir, O., & Etzioni, O. (1999). Grouper: A dynamic clustering interface to Web search results. *Proceedings of the 8th International WWW Conference*, 1361–1374.

Zhang, M., Lin, C., Liu, Y., Zhao, L., Ma, L., & Ma, S. (2003). THUIR at TREC 2003: Novelty, Robust, Web and HARD. *The 12th Text Retrieval Conference (TREC 2003) Notebook*, 137–148.

Zhang, M., Song, R., Lin, C., Ma, S., Jiang, Z., Jin, Y., et al. (2003). THU TREC 2002: Web track experiments. *Proceedings of the 11th Text Retrieval Conference (TREC 2002)*, 591–594.

Zhou, B., Chen, J., Shi, J., Zhang, H., & Wu, Q. (2001). Website link structure evaluation and improvement based on user visiting patterns. *Proceedings of the 12th ACM Conference on Hypertext and Hypermedia*, 241–244.

CHAPTER 3

Webometrics

Mike Thelwall
University of Wolverhampton

Liwen Vaughan
University of Western Ontario

Lennart Björneborn
Royal School of Library and Information Science

Introduction

Webometrics, the quantitative study of Web-related phenomena, emerged from the realization that methods originally designed for bibliometric analysis of scientific journal article citation patterns could be applied to the Web, with commercial search engines providing the raw data. Almind and Ingwersen (1997) defined the field and gave it its name. Other pioneers included Rodríguez Gairín (1997) and Aguillo (1998). Larson (1996) undertook exploratory link structure analysis, as did Rousseau (1997). Webometrics encompasses research from fields beyond information science such as communication studies, statistical physics, and computer science. In this review we concentrate on link analysis, but also cover other aspects of webometrics, including Web log file analysis.

One theme that runs through this chapter is the messiness of Web data and the need for data cleansing heuristics. The uncontrolled Web creates numerous problems in the interpretation of results, for instance, from the automatic creation or replication of links. The loose connection between top-level domain specifications (e.g., .com, .edu, and .org) and their actual content is also a frustrating problem. For example, many .com sites contain noncommercial content, although .com is ostensibly the main commercial top-level domain. Indeed, a skeptical researcher could claim that obstacles of this kind are so great that all Web analyses lack value. As will be seen, one response to this view, a view shared by critics of evaluative bibliometrics, is to demonstrate that Web data correlate significantly with some non-Web data in order to prove that the Web data are not wholly random. A practical response has been to develop increasingly sophisticated data cleansing techniques and multiple data analysis methods.

This review is split into four parts: basic concepts and methods, scholarly communication on the Web, general and commercial Web use, and topological modeling and mining of the Web. Webometrics is a new field based on analysis of new forms of data; therefore, methods of data collection and processing have been prominent in many studies. The second part, scholarly communication on the Web, is concerned with using link analysis to identify patterns in academic or scholarly Web spaces. Almost all of these studies have direct analogies in traditional bibliometrics and have sought to develop effective methods and validation techniques, the latter being an issue of particular concern on the Web. The general and commercial Web use section reviews link analysis studies that have used techniques similar to those applied to academic Web spaces. Some have their origins in social network analysis rather than information science, producing an interesting complementary perspective. The section also includes quantitative studies of the size of the "whole" Web and Web server log analysis. The final section, topological modeling and mining of the Web, covers mathematical approaches to modeling the growth of the Web or its internal link structure, mostly the product of computer science and statistical physics research. It concludes with detailed interpretations of "small-world" linking phenomena, an information science contribution to topological modeling of the Web.

Webometrics, Bibliometrics, and Informetrics

Being a global document network initially developed for scholarly use (Berners-Lee & Cailliau, 1990), the Web constitutes an obvious research topic for bibliometrics, scientometrics, and informetrics. A range of new terms for the emerging research field have been proposed since the mid-1990s, for instance, *netometrics* (Bossy, 1995), *webometry* (Abraham, 1996), *internetometrics* (Almind & Ingwersen, 1996), *webometrics* (Almind & Ingwersen, 1997), *cybermetrics* (also the name of the journal started in 1997 by Isidro Aguillo), *Web bibliometry* (Chakrabarti, Joshi, Punera, & Pennock, 2002), and *Web metrics* (the term used in computer science, for example by Dhyani, Keong & Bhowmick, 2002). Webometrics and cybermetrics are currently the two most widely adopted terms in information science, often used as synonyms.

Björneborn and Ingwersen (in press) have proposed a differentiated terminology, distinguishing between studies of the Web and studies of *all* Internet applications. They used an information science-related definition of webometrics as "the study of the quantitative aspects of the construction and use of information resources, structures and technologies on the WWW drawing on bibliometric and informetric approaches" (Björneborn & Ingwersen, in press). This definition thus covers quantitative aspects of both the construction and usage sides of the Web, embracing the four main areas of webometric research: 1) Web page content analysis, 2) Web link structure analysis, 3) Web usage analysis (e.g., exploiting log files of users' searching and browsing behavior), and

4) Web technology analysis (including search engine performance). The definition encompasses hybrid forms. For example, Pirolli, Pitkow, and Rao (1996) explored Web analysis techniques for automatic categorization utilizing link graph topology, text content, and metadata similarity, as well as usage data. All four main research areas include longitudinal studies of changes on the dynamic Web, for example, of page contents, link structures, and usage patterns. So-called *Web archaeology* (Björneborn & Ingwersen, 2001) could be important for recovering historical Web developments, for instance, by means of the Internet Archive (http://www.archive.org), an approach already used in webometrics (Björneborn, 2003; Thelwall & Vaughan, 2004; Vaughan & Thelwall, 2003).

Furthermore, Björneborn and Ingwersen (in press) have proposed cybermetrics as a generic term for "the study of the quantitative aspects of the construction and use of information resources, structures and technologies on the whole Internet, drawing on bibliometric and informetric approaches." Cybermetrics thus encompasses statistical studies of discussion groups, mailing lists, and other computer-mediated communication on the Internet (e.g., Bar-Ilan, 1997; Hernández-Borges, Pareras, & Jiménez, 1997; S. C. Herring, 2002; Matzat, 1998) including the Web. In addition to covering all computer-mediated communication using Internet applications, this definition of cybermetrics also covers quantitative measures of the Internet backbone technology, topology, and traffic (Molyneux & Williams, 1999). The breadth of coverage of cybermetrics and webometrics implies considerable overlap with proliferating computer science-based approaches in analyses of Web contents, link structures, Web usage, and Web technologies. A range of such approaches has emerged since the mid-1990s with names like *Cyber geography / Cyber cartography* (e.g., Dodge, 1999; Dodge & Kitchin, 2001; Girardin, 1996), *Web ecology* (e.g., Chi, Pitkow, Mackinlay, Pirolli, Gossweiler, & Card, 1998; Huberman, 2001), *Web mining* (e.g., Chen & Chau, 2004; Etzioni, 1996; Kosala & Blockeel, 2000), *Web graph analysis* (e.g., Broder, Kumar, Maghoul, Raghavan, Rajagopalan, Stata, et al., 2000; Chakrabarti, Dom, Kumar, Raghavan, Rajagopalan, Tomkins, et al., 1999; Kleinberg, Kumar, Raghavan, Rajagopalan, & Tomkins, 1999), and *Web intelligence* (e.g., Yao, Zhong, Liu, & Ohsuga, 2001). One reason for using the term *webometrics* is to show the connection with bibliometrics and informetrics and to stress an information science perspective.

There are different conceptions of informetrics, bibliometrics, and scientometrics. The diagram in Figure 3.1 (Björneborn & Ingwersen, in press) shows the field of *informetrics* embracing the overlapping fields of *bibliometrics* and *scientometrics* following widely adopted definitions by, among others, Brookes (1990), Egghe and Rousseau (1990), and Tague-Sutcliffe (1992). According to Tague-Sutcliffe (1992, p. 1), informetrics is "the study of the quantitative aspects of information in any form, not just records or bibliographies, and in any social group, not just scientists." Bibliometrics is defined as "the study of the quantitative aspects of the production, dissemination and use of recorded information" and scientometrics as "the study of the quantitative aspects of science as a discipline or economic activity" (Tague-Sutcliffe, 1992, p. 1). In the figure, politico-economic aspects of scientometrics are covered by the part of the scientometric ellipse lying outside the bibliometric one.

Webometrics may be seen as entirely encompassed by bibliometrics, because Web documents, whether text or multimedia, are recorded information (cf. Tague-Sutcliffe's definition of bibliometrics) stored on Web servers. This recording may be temporary only, just as not all paper documents are properly archived. In the diagram, webometrics is partially covered by scientometrics, because many scholarly activities today are Web-based. Furthermore, webometrics is contained within the field of cybermetrics as defined here.

In Figure 3.1, the field of cybermetrics exceeds the boundaries of bibliometrics because some activities in cyberspace are not normally recorded, but communicated synchronously, as in chat rooms. Cybermetric studies of such activities still fit in the generic field of informetrics as the study of the quantitative aspects of information "in any form" and "in any social group," as Tague-Sutcliffe (1992) stated.

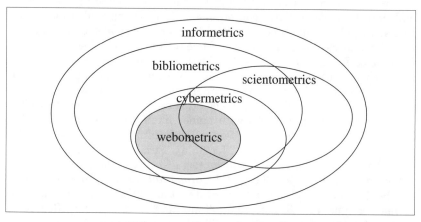

Figure 3.1   Infor-, biblio-, sciento-, cyber-, and webometrics (Björneborn & Ingwersen, in press). The sizes of the overlapping ellipses are made for sake of clarity only.

The inclusion of webometrics expands the field of bibliometrics because webometrics will inevitably contribute further methodological developments. As ideas rooted in bibliometrics, scientometrics, and informetrics have contributed to the emergence of webometrics, insights from webometrics will likely contribute to the development of these more established fields.

Scope of Coverage and Related Reviews

Although webometrics is still a young field, it has gained recognition: The Association for Library and Information Science Education lists webometrics in its Research Areas Classification and the *Encyclopedia of Social Measurement* includes an article on the topic (Vaughan, in press). Already many review articles offer partial coverage, including two *ARIST* chapters. The first is by Molyneux and Williams (1999), "Measuring the Internet." So much has changed since that chapter was written that it is now dated. The bibliometrics chapter by Borgman and Furner (2002) explicitly cast Web links as structurally equivalent to journal citations, developing a terminology in which both were referred to as links. Web research within and beyond information science was covered and there will be many connections from that chapter to this—especially the scholarly communication section—despite a relatively small overlap in articles reviewed. This similarity is important because many issues from standard bibliometrics are mirrored in Web behaviors and phenomena. A special issue of the *Journal of the American Society for Information Science and Technology* on webometrics is set to appear in 2004. It will contain a wide variety of different approaches to the quantitative study of the Web, some of which are reviewed in this chapter.

Bar-Ilan and Peritz (2002, p. 371) have published a review of "informetric theories and methods for exploring the Internet," which is similar in scope to ours but includes non-Web, Internet research, is more focused on general informetric techniques, and does not reflect recent significant advances in hyperlink analysis. Web metrics are surveyed from a computer science perspective by Dhyani, Keong, and Bhowmick (2002), covering mathematical techniques for measuring a wide variety of online phenomena. Rasmussen's (2003) *ARIST* chapter, "Indexing and Retrieval for the Web," provides useful additional information on Web crawlers and search engines. Our review differs from these in another regard: Its primary focus is on different types of Web link analysis.

Several general webometrics articles were published in 2003, reflecting a widening recognition of the importance of the topic and the existence of a critical mass of research. Park and Thelwall (2003) compared information science approaches to studying the Web to those from social network analysis. They found that information scientists emphasized data validation and the study of methodological issues, whereas social network analysts experimented with transferring existing theory to the Web. Wilkinson, Thelwall, and Li (2003) introduced Web-based quantitative methods to social scientists for use in their Web research. This coincided with the release of a special tool for social science link structure analysis, SocSciBot (http://cybermetrics.wlv.ac.uk/socscibot) and included a review section as well as a more prescriptive "how to do it" conclusion. Ingwersen's (1998) Web Impact Factor calculation was singled out for detailed coverage by Li (2003). Peres Vanti (2002) presented a general review.

Other review articles cover closely related topics. Search engine research is covered by Bar-Ilan (2004), search engines themselves by Arasu, Cho, Garcia-Molina, Paepcke, and Raghavan (2001), and data collection techniques in general by Bar-Ilan (2001) and Thelwall (2002d). Henzinger (2001) reviewed link structure analysis from a computer science perspective, showing how links could be used in search engine ranking algorithms. Barabási (2002) and Huberman (2001) have written popular science books explaining current research into mathematical modeling of the growth of the Web. Web mining (Chen & Chau, 2004; Kosala & Blockeel, 2000) and Web intelligence (Yao et al., 2001) are also relevant.

Online communication studies can contribute to webometric research by providing useful context information and explanatory theories. S. C. Herring (2002) gave an overview of computer-mediated communication on the Internet, taking a social science perspective and focusing mostly on non-Web media such as e-mail and newsgroups. One of the key general findings was that the use of new technology was very context specific, determined by users' needs rather than the technology. Ellis, Oldridge, and Vasconcelos (2004) explored communication in electronic communities. Finholt (2002) reviewed electronic virtual laboratories for scientists to share equipment or data (collaboratories). Finally, Kling and Callahan (2003) discussed e-journals and scholarly communication, which was useful background for our e-journals subsection.

Our review not only covers more recent research (including some currently unpublished research) but also has two areas of special emphasis: link analysis in academic Web spaces and basic concepts and methods. Other areas are covered, but less comprehensively, and either draw from or feed into the two main themes.

Basic Concepts and Methods

This section contains much terminology and many technical details. It is intended to provide a coherent basis for future webometric research and background to the studies reported. Some readers may wish to skip to the next section, "Scholarly Communication on the Web" on a first reading and return to this one later.

Terminology

Basic Link Terminology

Many new terms are used in the webometrics literature. For instance, a link received by a Web node (the network term "node" here covers a unit of analysis, such as a Web page, directory, Web site, or an entire top level domain of a country) has been named, variously, *incoming link*, *inbound link*, *inward link*, *back link*, and *sitation*; the latter term (McKiernan, 1996; Rousseau, 1997) clearly derives from bibliometric/citation analysis. An example of more problematic terminology is the two opposing meanings of an *external link*: either as a link pointing out of a Web site or a link pointing into a site. We recommend the consistent basic webometric terminology of Björneborn and Ingwersen (in press) for link relations between Web nodes, as briefly outlined in Figure 3.2. The proposed terminology has origins in graph theory, social network analysis, and bibliometrics.

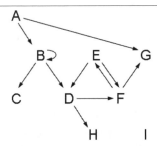

- B has an *inlink* from A
- B has an *outlink* to C
- B has a *selflink*
- E and F are *reciprocally* linked
- A has a *transversal* outlink to G: functioning as a shortcut
- H is *reachable* from A by a directed *link path*
- I has neither in- nor outlinks; I is *isolated*
- B and E are *co-linking* to D; B and E have *co-outlinks*
- C and D are *co-linked* from B; C and D have *co-inlinks*

Figure 3.2   **Basic webometric link terminology (see Björneborn & Ingwersen [in press] for a more detailed legend). The letters may represent different Web node levels, for example, Web pages, Web directories, Web sites, or top level domains of countries or generic sectors.**

The terms *outlink* and *inlink* are commonly used in computer science-based Web studies (e.g., Broder et al., 2000; Chen, Newman, Newman, & Rada, 1998; Pirolli et al., 1996). The term *outlink* implies that a directed link and its two adjacent nodes are viewed from the source node providing the link, analogous with the use of the term *reference* in bibliometrics. A corresponding analogy exists between the terms *inlink* and *citation*, with the target node as the spectator's viewpoint. In the conceptual framework of Björneborn and Ingwersen (in press), a link crossing a Web site border, like link *e* in Figure 3.3, is called a *site outlink* or a *site inlink* depending on the perspective of the spectator.

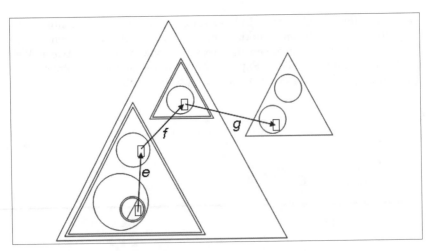

Figure 3.3   Simplified Web node diagram illustrating basic node levels
(Björneborn & Ingwersen, in press).

On the Web, *selflinks* are used for a wider range of purposes than self-citations in scientific literature. Page selflinks point from one section to another within the same page. Site selflinks (also known as *internal links*) are typically navigational pointers from one page to another within the same site. Most links on the Web connect Web pages containing cognate topics (Davison, 2000). However, some links may break a typical linkage pattern in a Web node neighborhood and connect dissimilar topical domains. Such (loosely defined) *transversal links* (Björneborn, 2001, 2003) function as cross-topic shortcuts and may affect so-called small-world phenomena on the Web (cf. the section on small worlds).

The two *co-linked* Web nodes C and D in Figure 3.2 with *co-inlinks* from the same source node are analogous to the bibliometric concept of *co-citation* (Small, 1973). Correspondingly, the two *co-linking* nodes B and E having *co-outlinks* to the same target node are analogous to *bibliographic coupling* (Kessler, 1963). *Co-links* is proposed as a generic term covering both concepts (Björneborn & Ingwersen, in press)

Basic Web Node Terminology and Diagrams

In webometric studies it may be useful to visualize relations between different units of analysis. Figure 3.2 illustrates some basic building blocks in a consistent Web node framework (Björneborn & Ingwersen, in press). In the diagram, four basic Web node levels are denoted with simple geometrical figures: *quadrangles* (Web pages), *diagonal lines* (Web directories), *circles* (Web sites) and *triangles* (country or generic top level domains, TLDs). Sublevels within each of the four basic node levels are

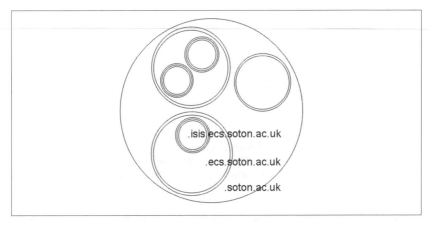

Figure 3.4   **Simplified Web node diagram of a Web site containing subsites and sub-subsites (Björneborn & Ingwersen, in press).**

denoted with additional borderlines in the corresponding geometrical figure. For example, a triangle with a double borderline denotes a generic second level domain (SLD), also known as a sub-TLD, assigned by many countries to educational, commercial, governmental, and other sectors, for instance: *.ac.uk, .co.uk, .ac.jp, .edu.au.*

The simplistic Web node diagram in Figure 3.2 shows a page *P* located in a directory of a subsite in a sub-TLD. The page has a site outlink, *e*, to a page at a site in the same sub-TLD. The outlinked page in turn is outlinking, *f*, to a page at a site in another sub-TLD in the same country. The link path *e-f-g* ends at a page at a site in another TLD.

Zooming in on a single Web site may reveal several subunits in the shape of subsites, sub-subsites, and so on, as indicated by hierarchically derivative domain names. For instance, as shown in Figure 3.4, the sub-subsite of The Image, Speech and Intelligent Systems Research Group (isis.ecs.soton.ac.uk) is located within the Department of Electronics and Computer Science (ecs.soton.ac.uk), one of the subsites at the University of Southampton, U.K. (soton.ac.uk). Subsites and sub-subsites are denoted as circles with double and triple borderlines, respectively. Subordinate sublevels would logically be denoted with additional number of borderlines. For the sake of simplicity, the diagram does not reflect actual numbers and sizes of elements.

Although some Web sites subdivide into derivative domain names, as noted, others locate the same type of subunits in folder directories. Obviously, such diverse allocation and naming practices complicate comparability in webometric studies. In Figure 3.5 directories, subdirectories, and so on, are denoted by one or more diagonal lines resembling URL slashes and reflecting the number of directory levels below the URL root level.

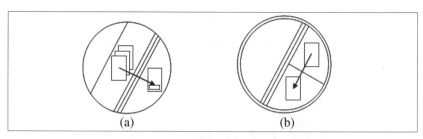

Figure 3.5   Simplified Web node diagrams of a Web site and a subsite with links between different directory levels including page subelements (Björneborn & Ingwersen, in press).

Web pages may also consist of subelements, such as text sections, and frames. Additional bands illustrate such page subelements as in the targets of the page selflink h and the page outlink i from the two sibling Web pages in the same directory in Figure 3.5a. More complex and numerous linkages within a site or subsite can be illustrated by combinations of elements in Figure 3.5, showing links between pages located either at different directory levels (Figure 3.5a) or in sibling directories at the same level (Figure 3.5b) in the Web site file hierarchies.

Naturally, any diagrammatic representation of large-scale hypertext structures will become too tangled to be of practical use, even less to be interpreted in any quantitative way. However, the proposed Web node diagrams with their simple and intuitive geometrical figures are intended to be used to emphasize and illustrate qualitative differences between investigated Web node levels. Moreover, the diagrams can illustrate important structural aspects of limited subgraphs of a given Web space.

Units of Analysis

The Definition of the Web

In this subsection we review different units of analysis in webometrics, after discussing the problematic issue of a precise definition of the Web itself. The Web is a loose concept (Berners-Lee, 1999) for which many definitions are possible (e.g., Boutell, 2003; Lavoie & Nielsen, 1999; webopedia.com, 2003) and the trend seems to be toward very inclusive definitions, including anything that can be accessed through a Web browser. It is important to be aware of this problem when devising a study and evaluating others' results. For example, non-HTML, automatically generated pages such as for search engine results, and password-protected pages could be reasonably included or excluded. Researchers should clearly state which definition they are using; but we do not recommend any one definition, only an awareness of the potential importance of differences. If the most generous definition were chosen, the size of the Web would be practically infinite due to the inclusion of dynamically created pages, pages that are created by Web servers in response to queries. The terms "Invisible Web" and "Deep Web" have been coined for the vast quantity of Web pages that are accessible with a Web browser but are difficult to find because they are of a type not indexed by search engines (Bergman, 2000). Invisible pages are a particular problem for large-scale studies of the Web, but researchers can explicitly state that they are excluded by a precise definition of the area studied. As an example of this, Lawrence and Giles (1999a) define their scope as pages on sites that can be reached by following links from the site home page. In practice, something similar to this often defines the scope of webometric studies.

Alternative Document Models

When counting objects on the Web, a decision needs to be made about the most appropriate unit of measure. Often this will be either the Web page or Web site: How many business Web sites are there in the U.S.? How many Web pages are there in the Stanford University Web site? But both of these have their problems and alternatives are possible.

There are technical difficulties in identifying Web pages, some associated with the definition of the Web itself, some to do with HTML and others concerned with a broader interpretation of what a document should be. An HTML-related problem is that a single page may be made up of multiple files if the frameset HTML tag is used. Counting files irrespective of whether they are in a frameset is a possible solution, but since individual frames in a frameset can change independently, there is no perfect answer (Thelwall, 2002d). A conceptual problem is that a Web author can choose various ways to divide a work. An online book, for example, could be presented as one huge Web page or thousands of section-based pages. An ideal solution would perhaps be to count pages or collections of pages by genre, so that in both of these cases what would be counted would be one instance of the online book genre. But genre identification on the Web is a highly complex, unsolved problem (Crowston & Williams, 2000; Rehm, 2002) and so heuristics have been developed as a crude substitute, notably the Alternative Document Models (ADMs) (Thelwall, 2002b). There are four main ADMs in use, the page, directory, domain, and site models.

The page ADM is the default unit, based upon individual files. With the directory ADM, all pages in the same (virtual) directory identified through the URL file name path are counted as one unit. In the online book example, if the Web page sections were stored in a single directory, then a thousand-page online book would be counted as a single entity, a Web directory. Clearly this is a heuristic, given that one might also store an entire library in a single directory. The domain ADM (similar to a subsite in Figure 3.4) aggregates all pages with the same domain name in their URL into a common entity. The site ADM (originally called the university ADM) allows multiple domain names in the same document by specifying only the domain name ending. The site ADM of the

University of Southampton in Figure 3.4 would include all pages from subsites, sub-subsites, and so on, with URLs that contain domain names ending in .soton.ac.uk.

Another unit of analysis has been proposed and extensively used by Aguillo (1998), the *seat* (named subsequently to its description in his article). This is similar to the site ADM and conventional understandings of the phrase "Web site" but contains elements of genre. A seat is a collection of Web pages created by a single person or organization for a specific purpose. Typically, seats will be larger than a single page and may be a collection of pages associated with an individual, however hosted on the Web. Seats will sometimes be the same as directory, domain, and site ADMs, but are more flexible. As a consequence, the concept is both more powerful and harder to apply automatically in practice.

When counting documents or links, the ADMs give the option of counting at different aggregation levels. For instance, the number of domains in a university site could be counted, as could the number of directories that contain a link to a given Web site. This additional flexibility can be used to bypass anomalies in the data and conceptual problems with counting pages, as reported in the subsections on university and departmental studies. The choice of ADM for any particular study depends upon the patterns of use of the Web in the area covered. Two techniques have been proposed. The first is to compare correlations between summary statistics calculated by different ADMs and an external data source (Thelwall & Harries, 2003; Thelwall & Wilkinson, 2003b). The second is a theoretical approach using assumptions about the process that causes document creation and introducing tests to assess the extent to which the assumptions fit the data (Thelwall, 2004, in press b).

Data Collection Methods

Personal Web Crawlers

When studying a small Web site, the researcher may be able to find all pages by browsing alone. On larger sites, an alternative is to use a commercial search engine or a Web crawler. A crawler, spider, or robot is software that can automatically and iteratively download pages and extract and download their links (Arasu et al., 2001; Heydon & Najork, 1999; Thelwall, 2001e). For example, a crawler could be given the URL of a site home page and fetch all pages on the site that can be found by following links from the home page. Crawler coverage will typically be less than 100 percent of a site, however. Crawlers find new pages by following links and, so, are likely to miss many isolated pages and collections of pages. Pages that are linked to can also be missed if the links are in a form that the crawler does not understand (for instance, JavaScript), if the pages are password protected, if the server is temporarily down, or if the owner requests that the page not be crawled (e.g., through the Robots Exclusion Protocol). In a large university Web site, the pages created by individual faculty or students would be missed unless they were integrated into the site by devices such as server-generated online lists of faculty/student pages, or links manually added to official pages, perhaps by Web administrators. Crawlers run by commercial search engines can partially offset this problem by adding to a crawl the URLs of all pages previously fetched from the site, URLs obtained by links from other sites, and any URLs submitted to the search engine by users (Thelwall, 2001b). All crawlers also have to employ heuristics such as a maximum crawl depth, maximum path depth, or maximum page fetches per site to avoid infinite loops. Commercial search engines may also have their crawling policies influenced by economic factors, with the most extreme example being pay-for-inclusion schemes. Cothey (in press) has shown how different crawling policies can give very different results on the same site, emphasizing the importance of full reporting of crawling parameters research. He also differentiated between content crawling (to find all unique documents) and link crawling (to map the link relationships between pages, irrespective of duplication), showing how the scope of a crawl will have to be tailored to its purpose.

It is possible to guarantee coverage of all (static) pages in a large Web site if privileged access is gained to the computer on which the server is operating. This is possible using a non-Web program that would not work by following links but instead by trawling all of the directories that were within the scope of the server. This approach would be difficult to apply to studies of more than one Web site due to problems in obtaining permission for special access to others' computers.

Commercial Search Engines

Commercial search engines have been used extensively for webometric research and are particularly useful for counts of pages over large areas of the Web. Some, including AltaVista, allow advanced Boolean queries to combine information requests about pages, links, and domains (Almind & Ingwersen, 1997), which can be used to return counts of pages or link pages that satisfy given criteria. This is the facility that originally spawned webometrics (advanced search help pages of search engines should be consulted for details of the full range of queries available for any particular engine).

Commercial search engines have a number of drawbacks. They do not cover all of the Web or even the entire publicly indexable Web; coverage was under 17 percent for the largest search engine in 1999 (Lawrence & Giles, 1999a). Their crawling and reporting algorithms are commercial secrets, which is a problem for their use in scientific research. Search engine results have been found to be unreliable and to fluctuate, even over short periods of time (Bar-Ilan, 1999; Ingwersen, 1998; Mettrop & Nieuwenhuysen, 2001; Rousseau, 1999; Snyder & Rosenbaum, 1999). AltaVista has been found to be more reliable recently (Thelwall, 2001c; Vaughan & Thelwall, 2003) and to give good coverage of academic Web sites (Thelwall, 2001b). However, researchers should always take steps to assess the accuracy and coverage of results because the algorithms are subject to change without notice and the effects of the advanced commands are not always fully documented. Accuracy can be checked by testing a sample of the pages returned to ensure that they do contain the features specified in the query. Coverage is more difficult to check, but for any given study it may be possible to use AltaVista's "host:" command to check the number of pages indexed by AltaVista in a set of key sites. Rousseau (1999) has proposed statistical techniques to compensate for fluctuations. Björneborn and Ingwersen (2001) recommend the use of sampling and filtering to clean up data from search engines, acknowledging that this is likely to be difficult. An alternative strategy is to employ multiple search engines and average the results; but this will increase theoretical coverage at the expense of weighting more highly the pages indexed by multiple engines (since they will be counted more than once) and making more time-consuming the process of checking the operating methods, accuracy, and coverage of the engines used. Alternatively, different search engines can be used in parallel, comparing the results of one against the other to see if internal differences substantially influence the results. This is an approach that deserves to be taken seriously. Of course, similarity in results does not rule out the possibility that both engines have similar flaws, perhaps for underlying technical reasons.

A future issue for the validity of data from search engines may be evolving commercial models, especially if they include pay-for-inclusion and pay-for-placement. There does not seem to be any logical reason why commercialization should not extend to universities and research groups paying for improved positioning in major engines. Clearly this would create problems for those using data from search engines to evaluate online impact. Additionally, the continued provision of advanced search interfaces should not be taken as a given: They may be discontinued, if judged not to be profitable. Both of these considerations are incentives for researchers to develop data collection techniques that do not rely upon AltaVista or any other search engine.

Commercial search engines display national biases in site coverage. One study of four countries showed that U.S. sites were much better covered by three major search engines than China, Taiwan, and Singapore (Vaughan & Thelwall, in press b). Site language did not seem to be a factor; lower coverage was related to the smaller number of indexed links targeting sites in countries that adopted the Web more recently, something research designs may need to take into account. A similar pattern was found for the Internet Archive (http://www.archive.org; Thelwall & Vaughan, 2004).

Time series data can be obtained from some search engines and also the Internet Archive. For example, AltaVista allows searches for pages by their last modified date (e.g., Leydesdorff & Curran, 2000). The results are only partially accurate because the search engine overwrites older versions when a page is modified; it thus returns only the date of the last modification. The Internet Archive Wayback Machine gives genuine time series data because it records all the dates on which a page was found, keeping copies of every indexed version so that changes can be tracked over time (e.g., Vaughan & Thelwall, 2003). The Archive can be used for an earliest known date for any page but not for an accurate creation date. This is because the earliest recorded date is merely when it was first found by the Archive crawler, not when it was actually created. Björneborn (2003) obtained indicative ages of different topological components (cf. the Overall Web Topology section) of the U.K. academic Web space based on first indexing dates from the Archive. A disadvantage of the Wayback Machine, at the time of writing, is that it does not allow global Boolean queries beyond simple wildcard matches, unlike AltaVista. An attractive alternative is to employ a crawler to collect data over time for a preset list of Web pages or sites (e.g., Koehler, 2002), but this is not possible for retrospective studies.

Web Log Files

In addition to studying the Web itself, it can also be useful to collect data on how humans interact with it: on how it is used. Data collected could include behavioral studies of humans using the Web, although this is beyond the scope of our review. Another source of information is Web server log files. These record the requests sent to a Web server by users' browsers (and Web crawlers) and can be mined for information about how they are using the site. Typical applications of server log analysis would be to see which pages are most frequently viewed or to identify patterns of surfing with a view to improving site navigation. The principal drawback from a webometric perspective is that log files typically cover one site or all sites owned by the same server. However, access is often restricted, so these are generally not a good data source for studying Web site use.

Search engine log files are a special case because these can furnish information about what a user was interested in and where he went to find it. These files are also normally secret, but useful studies have been conducted by researchers who have been given access to sampled, individual log files (e.g., Silverstein, Henzinger, Marais, & Moricz, 1999; Spink, Ozmutlu, Ozmutlu, & Jansen, 2002; Spink, Wolfram, Jansen, & Saracevic, 2001).

There are many inherent difficulties in analyzing Web log files. The main problems are in identifying a user and determining a search session. In traditional information retrieval (IR) systems, users have to log in and out, so a user and a search session are explicitly identified. The Web is "stateless," which makes this identification extremely difficult. The main measures that have been used to determine a unique Web user are cookies (small quantities of information stored by the user's machine) and Internet Protocol (IP) addresses. Silverstein et al. (1999) used the former technique to study AltaVista users. For queries in which the user disabled the cookies, they used the pair "domain IP/Web browser used" as a substitute for the cookie. They acknowledge that the cookie method is imperfect because different people can use the same browser, for example, a public computer station in a library. Following a user's search behavior over time is even more difficult because a person can use different computers, for instance, one at work and one at home. User identification by IP address can be even less precise because many users can share a single IP address in large Internet service providers and some organizations assign floating IP addresses to computers. More encouraging is the fact that 96 percent of the queries in Silverstein et al.'s (1999) study had cookie information. They analyzed their data in two ways, one for the full data set and the other for those with cookie information. The two results were very similar.

A common method for determining a search session is "timeout." Silverstein et al. (1999, p. 7) defined a session by a "series of queries by a single user made within a small range of time." The session is "timed out" after five minutes of user inactivity. The timeout method is commonly used by commercial Web database vendors to gather user statistics, but the length of time varies. Duy and Vaughan (2003) compared vendor data, based on IP addresses and timeouts, with local client-side data, based on login attempts, in a university library. They found that the two data sets had different numbers but similar patterns. Huang, Peng, An, Schuurmans, and Cercone (2003) employed a dynamic session identification method using n-gram language models to determine session boundaries. They found that this method performed better than the timeout method. Spink et al. (2001, p. 227) defined a session as the "entire set of queries by the same user over time." The user was identified by an anonymous user code assigned by the Excite server creating the logs. However, it is not clear how the code was assigned, possibly by IP addresses or by cookies, and how the session boundaries were determined. Another study using Excite data (Ross & Wolfram, 2000) revealed that Excite users were identified by cookies. The lack of standards in measurement makes the comparison of results from different studies very difficult. Jansen and Pooch (2001) pointed out this problem and proposed a framework for future studies.

There are other challenges in log file analysis. The log may not record the complete search process when either caching or a proxy server is used. Furthermore, it is difficult to distinguish search queries issued by humans from those of robots. Silverstein et al. (1999) found that robot-initiated queries in the AltaVista logs caused a number of seemingly bizarre results that should be removed prior to analysis. Additionally, it is difficult to determine user demographics such as geographical location. Although this kind of information can sometimes be obtained by an IP address reverse look-up, such data could be misleading because people may register their IP address in other countries for economic or other reasons. To minimize this problem, Nicholas, Huntington, and Williams (2003) proposed a micro-mining method where robust subgroups of users were analyzed. The processing of huge amounts of data contained in a server log file is another challenge. The traditional method of a master file and inverted indexes for text data storage and retrieval may not be sufficiently efficient and flexible. In response, Wang, Berry, and Yang

(2003) have implemented a relational database model that allows them to analyze longitudinal Web server log data with fine granularity. The study identified patterns of querying behavior that provide useful information for Web site design.

Selection and Sampling Methods

The decentralized and dynamic nature of Web pages and Web link creation means that it will not always be possible to find or analyze every page or link; thus sampling may be required. For a tightly defined area, for example, a study of the home pages (only) of the top 100 businesses, it may be possible to retrieve all relevant data. If, however, all business home pages were to be investigated, some kind of random sample would be needed. In any case, where the chosen set of pages would be a subset of the area studied, a potentially difficult problem is the design of an unbiased sample selection method.

Sampling Pages

Computer scientists have designed algorithms to sample pages from the Web randomly. A purely random selection procedure would be one where each page had an equal chance of being selected; but in practice this is not possible or necessarily desirable. A survey of sampling methods is provided by Rusmevichientong, Pennock, Lawrence, and Giles (2001). The general approach is to start with as large a collection of Web pages as possible and then to select from this. Pages are not selected completely at random, but by taking into consideration the link structure of the Web. The underlying model is of a surfer visiting pages partly at random and partly by following links—a "random walk" (Henzinger, Heydon, Mitzenmacher, & Najork, 2000). Algorithms based on this model claim to select popular pages more often than less popular ones, more closely reflecting the actual experiences of Web users. Some of the techniques require the creation of a large collection of pages, from which a random sample can be drawn. Others create the random sample at the same time they build the larger collection. Algorithms that use the second approach can save time and storage space by visiting pages only when necessary for the algorithm; they do not need to download every page in the larger collection. Nevertheless, the set of pages visited must still be much larger than the sample created; researchers wishing to apply this technique to the Web will need access to considerable computing resources.

A commercial search engine could be used in some circumstances. For example, if a sample of pages discussing "Neptune" was needed, a search for the word "Neptune" could be used to produce a list. If the list were too long, the search engine might display only, say, the first 200 matches; but this would be a biased subset, perhaps including those pages judged to be most relevant to the term. Bharat and Broder (1998) have proposed a different search-engine related algorithm for sampling pages. They use randomly sampled queries created from a Web-based lexicon to generate results pages with which to compare the coverage of different engines. The same basic approach could also be used to generate pages for other purposes. This approach should be less influenced by link structures than random walk methods, although, because search engines tend to use links in their ranking algorithms, there is still likely to be link bias.

Sampling Sites

If there are too many sites to be crawled, some technique must be employed to generate a random sample. If all home page URLs in the total set are known, this problem can be solved with a random number generator (Ju-Pak, 1999). Often, though, random sampling is much more difficult.

Lawrence and Giles (1999a) have used a site sampling method based on randomly sampling Internet Protocol addresses with the aim of generating a random selection of all Web sites. This is no longer recommended since the virtual server capability was introduced, allowing one IP address to host multiple domain names, but only one "main" name. This means that only the main sites run by each Web server could be identified (Fielding, Irvine, Gettys, Mogul, Frystyk, Masinter, et al., 1999; Thelwall, 2002d).

Suppose that the focus of a study is Mexican company Web sites. In order to obtain a representative sample, a complete list would be needed from which to sample with random numbers. Many companies will have domain names ending in .com.mx and administered by a central domain names registry in Mexico. But acquiring a complete list would mean requesting the information from the owning authority. Although direct online access to central domain names databases is a theoretical possibility (Albitz & Liu, 1992), this is not normally permitted. Domain name authorities may be reluctant to give out this kind of information due to fear of spam, although they have done so in the past (Weare & Lin, 2000). One way to circumvent this problem is to randomly generate legal domain names and crawl the associated sites, where they exist (Thelwall, 2002d). For example, a random generator could produce random domain names with all possible combinations between www.aaaa.com.mx and www.zzzz.com.mx (e.g., Thelwall, 2000a). This approach excludes longer domain names (e.g., www.aaaaa.com.mx) and therefore gives a biased sample. If newer Web sites tend to have longer domain names, this would result in age bias. In some circumstances, however, this is the only feasible approach.

This approach presents three other problems: Many Mexican companies probably use other domain names, including .com; some non-Mexican companies probably use .com.mx domains; and some company Web sites will not have their own domain names. All of these problems can be avoided by using an online directory of businesses, such as that in Yahoo!'s Mexican business directories, mx.dir.yahoo.com, but this introduces a different kind of bias, a human one related to how the sites were identified and selected for the directory. Probably the most reliable approach would be to sample the companies by name from an up-to-date, official directory and then find their Web sites by asking the companies directly. Alternatively, a search engine search could be used to find the named businesses, although this would introduce search engine coverage and indexing biases.

In summary, there is no fully effective method for sampling business Web sites within a single country. Investigators will have to choose the method that has the source of bias least likely to interfere with the research hypotheses and also record the nature of the bias in their publications.

Large Area Web Studies

Studies that require information about the whole Web currently have three options. The first is to use a crawler to fetch the pages themselves. This is an enormously expensive and time-consuming exercise. The other publicly available option is to access the crawls of more than 10 billion Web pages since 1996 available at the Internet Archive (http://www. archive.org), which will require software to process the data. Stanford University's WebBase project (http://www-diglib.stanford.edu/~testbed/doc2/WebBase) is similar, but less ambitious in terms of Web coverage. Researchers affiliated with commercial search engines have a third option: to use full search engine crawls. This approach has been used (Broder et al., 2000), but the crawls themselves have remained secret.

Studies that require information about large numbers of Web pages, but not necessarily the full contents of the pages themselves, have an additional option: to use advanced search engine queries to specify the pages to be counted, recording the totals reported. Counts of pages with links from one Web site to another are a common example of the kind of summary statistics that a commercial search engine can yield. However, such results should be treated with caution because of the limited coverage provided by search engines.

Scholarly Communication on the Web

The hope that Web links could be used to provide information similar to that extracted from journal citations has been a key factor in stimulating much webometrics research (Borgman & Furner, 2002; Cronin, 2001; Davenport & Cronin, 2000; Ingwersen, 1998; Larson, 1996; Rodríguez Gairín, 1997; Rousseau, 1997; Thelwall, 2002b). But can this hope be fulfilled? Although structurally very similar to Web links, journal citations are found in refereed documents and, therefore, their production is subject to quality control. They are part of the mainstream of academic endeavor but hyperlinks are not. This caused problems for the early hyperlink-citation analogies, as noted by, for instance, Meyer

(2000); Egghe (2000); van Raan (2001); Björneborn and Ingwersen (2001); and Prime, Bassecoulard, and Zitt (2002). In this section we summarize the results of a series of studies, organized by scale of units analyzed, before considering the underlying motivations for creating links. We conclude with a discussion of how far early hopes have been realized.

The goal underlying almost all of the research reported here is to validate links as a new information source. Such a task entails several different strategies, as reported by Oppenheim (2000) in the related context of validating patent citations. Two of these strategies are to correlate with other data sources of known value and to undertake qualitative studies to uncover creation motivations. With links between university Web sites, for instance, a positive correlation between link counts and research activity would provide some evidence that link creation was not completely random and could be useful for studying scholarly behavior. An important methodological issue is that nonparametric Spearman correlation tests are normally more appropriate than Pearson correlations, given the typically skewed nature of Web link data.

Journals and Scholars

Since much webometrics research has been stimulated by citation analysis, a natural step has been to see if the kinds of bibliometric techniques that are applied to journals and authors can also be applied to e-journals. This could potentially allow a wider range of journals to be studied, giving new insights into scientific publishing. The e-journal is an evolving phenomenon, one that is not yet well understood (Kling & Callahan, 2003).

Smith (1999) pioneered the application of citation analysis to e-journals with a selection of 22 Australasian, refereed e-journals, using AltaVista to count links to them from the rest of the Web. No significant relationship was claimed between these figures and impact factors calculated by the Institute for Scientific Information (ISI). One of the conclusions was that hyperlinks to e-journals are fundamentally different from citations because the former target the whole journal whereas the latter target individual articles. Harter and Ford (2000) reached similar conclusions for a heterogeneous set of 39 journals, also finding no significant correlation between link measures and ISI's impact factors. Vaughan and Hysen (2002) analyzed a discipline-specific set of journals, those from library and information science that were indexed by the ISI. The journals in their study were not e-journals in the sense of providing the full text of articles online, but rather traditional journals with associated Web sites. This time a significant correlation was found between inlink counts and impact factors. The better correlation perhaps reflected greater maturity in the Web. An additional factor may have been that earlier studies were cross-disciplinary, so that disciplinary variations in journal impact factors and Web use obscured underlying trends in the data. Online impact, as measured by total inlink counts, is therefore associated with offline impact, as measured by average citation counts per article. Why are significant correlations important? They suggest (but do not prove) that hyperlinks may be created for reasons similar to those for citations and therefore may have potential use in bibliometric-style calculations.

Factors not taken into account in the Vaughan and Hysen (2002) study were the content of the journal Web sites and their age. Vaughan and Thelwall (2003) incorporated these two factors for a larger exercise that included 88 law and 38 library and information science (LIS) journals (all ISI journals in the two disciplines that had Web sites). The findings confirmed that in both law and library and information science, counts of links to journal Web sites correlated with impact factors. Unsurprisingly, journals with more online content tended to attract more links, as did older journal Web sites. Evidence was also found that link counts for LIS journals tended to be higher relative to their impact factor than was the case in law. Vaughan and Shaw (2003) took a different approach, comparing citations to journal articles from the ISI's index with citations (not links) to them in the general Web, using Google to collect Web citation data (i.e., how often the journal articles were mentioned in Web pages). All papers published in 1997 in 46 LIS journals were used in this large-scale exercise, which produced predominantly significant correlations, suggesting that online and offline citation impact are in some way similar phenomena. A classification of 854 Web citations indicated that many "represented intellectual impact, coming from other papers posted on the Web (30 percent) or from class readings lists (12 percent)" (Vaughan & Shaw, 2003, p. 1313). Text based citations are more widespread than links to e-journals and are also more directly analogous to bibliometric citations because of the direct connection with individual articles. Data collection is, however, labor intensive. Thelwall (in press a) used this as an opportunity to compare the extent to which pure and applied science journals were invoked in commercial Web sites, but found that in fact businesses rarely mentioned either online. High technology companies were the exception, using published research to support product effectiveness claims.

Citations from the text of online articles rather than links on the general Web have also been objects of study. CiteSeer has been used for this because it maintains an automatically generated database of scholarly publications that are freely available on the Web (Lawrence & Giles, 1999b). CiteSeer incorporates a variety of file formats in addition to HTML. Goodrum, McCain, Lawrence, and Giles (2001) analyzed citation patterns in online PostScript and PDF-formatted computer science papers, finding significant differences when compared with ISI citation data. In particular, conference papers were more frequently cited online (15 percent), but only 3 percent of citations were from ISI computer science papers. This shows the different character of the two data sources. Note that in computer science, conference proceedings are often regarded as equivalent in quality to journals. Zhao and Logan (2002) conducted a similar study of Extensible Markup Language (XML) research, also using CiteSeer. They recommend that, where possible, online and offline citation analyses should be conducted in parallel, as the data sources did not have the same strengths and weaknesses. Strengths for the Web approach included greater coverage in terms of both number and variety of papers and the inclusion of more information from the full text of the papers. ISI data were more readily accessible, but CiteSeer had the potential for greater automation of data collection techniques, if programming expertise were available to a research team. Most significantly, Lawrence (2001) investigated online computer science research and found that papers publicly available online were more frequently cited (in the ISI database) than those that were not available. It seems likely that placing published articles freely online can generate more interest and facilitate easy access for other scholars. The converse is probably also true; some scholars may choose to publish only their most popular articles online.

The growth of many different types of digital library has created new opportunities for employing quantitative techniques to study aspects of journal article use. We have discussed some research based around the CiteSeer digital library, but there are also public digital libraries that are created by users submitting their articles directly. These are often called preprint archives. Posting research to a preprint archive offers two benefits: It is a public, time-stamped claim to have created research, one that is much faster than journal publication, and it is also a mechanism for dissemination of research. Harnad and Carr (2000) have shown how citations can be turned into links; and in fact the related e-print archive initiative has built in many log file-based tools for analyzing usage patterns. These tools have helped to show that readers in some fields prefer to download complete articles rather than abstracts (Brody, Carr, & Harnad, 2002). Most starkly, more than half of the downloads from the physics archive (http://arXiv.org) at any given time are attributable to articles that are less than a week old, and the average elapsed time between depositing a paper and its receiving a citation can be as short as one month.

Web server logs can also be used to give information about the usage of e-journals that are not in digital libraries. Rudner, Gellmann, and Miller-Whitehead (2002) tracked the readers of online education journals using an online survey, server access statistics, and content analysis. Their findings suggested that the readership for this type of journal was much wider and larger than that for print journals. They also found that the most current topics were the most widely read and provided a list of recommended actions for e-journal publishers.

Three studies have chosen scholars rather than journals or journal articles as the basic unit of analysis, collecting data based upon *invocations*: the mentioning of a scholar's name in any context in a Web page. Cronin, Snyder, Rosenbaum, Martinson, and Callahan (1998) found academics to be invoked online in a wide variety of contexts, including informal documents such as conference information pages and course reading lists. This was used to support a claim that "the Web should help give substance to modes of influence that have historically been backgrounded in the narratives of science" (Cronin et al., 1998, p. 1326) and that it could help to highlight "the diverse ways in which academic influence is exercised and acknowledged" (Cronin et al., 1998, p. 1326). Landes and Posner (2000) used Web invocations as a partial measure of public fame for intellectuals and Cronin and Shaw (2002) pursued this idea for information studies scholars; the population studied had low salience.

National University Systems

Analyzing the interlinking between universities within a single country offers the perfect scale for a study. The number of objects to analyze (one site per university in a country) is manageable and counting all links to a whole university site seems to give a sufficiently high level of aggregation to produce reliable results. In other words, the vagaries of link creation and data collection seem to average out at the level of entire universities, allowing trends to be identified.

Ingwersen (1998) proposed new measures of online impact. The most popular of these, the (External) Web Impact Factor (WIF or Web-IF), was designed to measure the average online impact of a set of Web pages by counting the inlinking pages outside the set in question, and then dividing by the number of pages inside the set. The WIF was later modified for universities by using another measure of university size: its full-time faculty (Thelwall, 2001d). This was justified by reasons that eventually led to the creation of the Alternative Document Models discussed earlier. For example, a university may post huge numbers of pages on its site, such as server log statistics, that are not intended for people outside the university. Server log statistics would be unlikely to attract links from other universities, but may run to tens of thousands of pages and, hence, reduce the average number of links per page. The modified (faculty denominator) WIF counts links per full-time faculty member, avoiding the problem of lower scores from posting large numbers of pages that are unlikely link targets. Most university studies have analyzed interuniversity links, but some have focused instead on connections between universities and other sectors of society, such as industry and government (Leydesdorff & Curran, 2000). If the total impact of university Web sites could be easily measured and compared through hyperlinks, this would clearly be useful information, given the importance of the Web for information seeking.

As was the case with journals, early studies centered on trying to show that link metrics correlated with research measures—a first step toward assessing their similarity to citations (e.g., Smith, 1999; Thelwall, 2000b). Early findings were negative (Smith, 1999; Thelwall, 2000b), and the reasons suggested included failures in the search engines used to obtain the link counts and the number of links created for reasons unrelated to research. Counts of links to a set of 25 U.K. universities were subsequently found to correlate significantly with average research productivity, (Thelwall, 2001b), although no claim was made that research activity caused link creation. Both AltaVista and a specialist information science Web crawler were used in this study, giving similar results. A comparable relationship was later found for Australia (Smith & Thelwall, 2002) and also Taiwan (Thelwall & Tang, 2003), using different national measures of research productivity. The correlations, although significant, were not very high and graphs of the data showed individual universities that did not fit the pattern. These anomalies occurred when individuals or automated processes had created huge numbers of links. For example, a biochemistry database at Warwick University contained tens of thousands of links to a similar online database at the University of Cambridge, dwarfing other link counts (Thelwall, 2002b).

The discovery of anomalies in link data led to the search for alternative methods that would be more robust, in particular to the development of the ADMs discussed previously. These were applied to the U.K. with an extended data set of 108 institutions, and produced much more significant results, particularly for the directory and domain versions (Thelwall, 2002b). In the experiment, results for the four ADMs were compared with estimated research productivity for the 108 U.K. institutions. Research productivity was calculated from a reliable, peer-review based official source (http://www.rae.ac.uk). The underlying hypothesis was that a better ADM would generate results that would correlate more strongly with research productivity figures. Although the Spearman correlation differences were small, the directory and domain graphs were much more linear, giving additional support for a closer relationship between links and research for the directory and domain ADMs. Hybrid ADM counting methods were later developed, where multiple links to a target document from the same university were removed, giving even more significant results (Thelwall & Wilkinson, 2003b). These "range" models used a counting approach based upon each of the four document models, but never counting the same link target more than once for the same source university. For example, if using the domain ADM and domains A and B in one university targeted domain C in another, this would score as only one link in the range model because the same domain was the target. The hypothesis behind the range models was that linking practices are frequently shared within a university, so that multiple links to the same target could cause discrepancies in counts. An example of this is when many pages in one university target a clickable list of U.K. university home pages. There are several of these link lists on the Web, but some universities have many links to just one. Presumably this is the result of imitation within the university.

The ADMs have also been combined with a manual classification scheme. By restricting inlink counts to target pages mainly connected with research, the Spearman correlations reached a value of 0.949 for the directory range model, compared to 0.940 for the same counting model applied to all target pages and 0.920 for the standard file model applied to all target pages (Thelwall & Harries, 2003). The effect of applying the directory and domain ADMs coincided with restricting the counts to classified pages, indicating that anomalies were predominantly found in pages not targeting academic content. This gives a high degree of confidence that links between university Web sites are connected with scholarly activity in some way, despite the number that are created for recreational reasons. Of course, it still does not establish a cause-and-effect relationship.

The significant correlations found between universities opened the door to attempts to mine deeper for patterns and to model the linking process. Metrics to measure the use of the Web by universities (via outlinks) and the connectivity of a university with its peers were proposed (Thelwall, 2003b). These were termed the Web Use Factor (WUF) and the Web Connectivity Factor (WCF), respectively. It was expected that the WUF would be less reliable than the WIF, being dependent upon the crawling of a single site to identify its outlinks, rather than upon multiple other sites to compile total inlinks. This was not borne out by a detailed examination of the data, however. The WCF was designed to restrict the impact of outliers on the data by assessing the link strength between a pair of universities to be the minimum of the count of links from the first to the second and the count of links from the second to the first. This should remove unidirectional anomalies. However, this more complex measure was not found to be significantly more robust, based upon the data set used (108 U.K. universities). An interesting development was the use of the measurements to provide baselines from which to compare the WIFs, WUFs, and WCFs of individual universities in order to identify those that were not well connected on the Web, perhaps indicating underlying problems in university Web usage or publishing policies.

In a different approach, the linking process has been modeled using link counts between all pairs of 86 U.K. universities as the raw data (excluding universities generating statistical anomalies), and regressing these against combinations of faculty size and average research productivity (Thelwall, 2002e). The university research productivity figures used were estimates derived by multiplying peer-reviewed research quality ratings of universities by total active researchers. This is a hybrid research measure (quality times quantity) but one that has some currency inside the U.K. because of its implicit use in newspaper ranking tables. The peer-review component, the Research Assessment Exercise, is used to direct U.K. government research funding and is very important to all universities. The best predictor of total links between a pair of institutions was the product of their two total research productivities. Combined with the previous study, this gave evidence that outlink creation is not fundamentally different from inlink reception. Both are equivalent in the model.

Four further investigations have given insights into the phenomenon of academic Web site interlinking. First, geographic factors for interlinking were investigated, revealing that the extent of interlinking between pairs of U.K. universities decreased with geographic distance (Thelwall, 2002c). In particular, neighboring institutions were very much more likely to interlink than average. This shows that despite the existence of collaboratories and other devices for virtual collaboration, and their undoubted use for international communication, the Web is not divorced from physical reality. Second, alternative sources of links were investigated, classified by top-level domain. It was found that links to U.K. universities from .edu sources produced results very similar to those between U.K. universities, suggesting that academic link attractiveness is a phenomenon that is numerically similar nationally and internationally (Thelwall, 2002a). Third, an investigation into personal home pages that linked to U.K. universities found that this source of links gave very similar quantitative results to inter-university links, even though almost a third of the links was for recreational purposes (Thelwall & Harries, 2004). It can be concluded that link attractiveness is a relatively robust construct for university Web sites, at least in the U.K., but that it is not exclusively dependent on the academic content of the pages linked to, even though link counts correlate highly with university research productivity. A fourth study compared different potential link attractors for Canadian universities, finding that faculty quality and language were the best predictors of inlink counts (Vaughan & Thelwall, in press a). French language universities in Canada attracted significantly fewer links than comparable English language universities.

The most significant quantitative result was that the voluume research produced by academics was the main reason for attracting links: Universities with better researchers attracted more links because the researchers produced more Web content, rather than because the content produced was of a higher "link attractiveness" (Thelwall & Harries, 2004b). This is in contrast to the case for formal scholarly publications, where "better" scholars tend to produce articles that attract more citations (Borgman & Furner, 2002), and is a critical finding suggesting that link counts should not be seen as in any way signifying quality.

International University Comparisons

Studies comparing university Web sites internationally have revealed interesting but fairly predictable descriptive patterns. These studies differ from those described in the previous section by being relational: looking for evidence of international linking trends. Smith and Thelwall (2002) compared linking between British, Australian, and New Zealand universities and found that New Zealand was relatively isolated on the Web, in line with findings from a previous bibliometric study for journals

(Glänzel, 2001). A new measure was introduced here, the propensity to link. This measure was taken to be the total links from all universities in one country to all universities in the second, divided by the total number of faculty in the first country and also by the total number of faculty in the second country. A crawler was used to collect the raw data. A normalizing calculation allowed these figures to be compared with links between universities in the same country. Not surprisingly, international links were dwarfed by national links, even for historically related and traditionally collaborating countries.

A larger follow-up study used AltaVista data and simple network diagrams (Thelwall, 2001a) to map the interlinking between universities in the Asia-Pacific region (Thelwall & Smith, 2002). In the four types of diagrams, the width of arrows between countries was proportional to raw link counts, links divided by target system size (total number of Web pages in all universities in the target country), links divided by source system size, and links divided by source and target system size. All four diagrams were very different, but showed that Australia and Japan were central to the academic link structure of the region, with smaller countries attracting attention disproportionate to their size. The diagrams were most useful when comparing Web collections of approximately the same size.

There have been three published studies of European university Web sites. Polanco, Boudourides, Besagni, and Roche (2001) clustered universities throughout the European Union (E.U.) based upon links. A general survey of E.U. university Web site sizes revealed a huge disparity between the West and East, one that could hinder attempts to use the Internet to integrate European research (Thelwall, Binns, Harries, Page-Kennedy, Price, & Wilkinson, 2002). AltaVista's linguistic capability was employed to examine the languages used for European Web pages and link pages in particular (Thelwall, Tang, & Price, 2003). English was the major language for the whole of the E.U. with the exception of Greece, both for all pages and for internationally linking pages. Beyond Greece and the English speaking nations, English language pages formed approximately half of all international link pages, with the other accounted for by indigenous languages. Not surprisingly, links between countries with a shared language were also common, particularly in the shared language. Apart from English, Swedish was the only other case of a language used extensively for international linking, mainly within Scandinavia; it should be noted that Swedish is very close to other Scandinavian languages.

Departments within a Discipline

In parallel with investigations into interlinking between universities, there have been studies of interlinking between departments within a single discipline. Success with research having a disciplinary focus could potentially lead to the development of techniques to measure subject communication patterns on the Web, perhaps even giving early indications of emerging interdisciplinary trends. Thomas and Willett (2000) studied U.K. library and information science departments, but found no significant correlation between inlink counts and research ratings. Significant differences between inlink counts and usnews.com rankings were found later for U.S. library and information science schools (Chu, He, & Thelwall, 2002). Since then, significant research and inlink correlations have been found for U.K. computer science departments (Li, Thelwall, Musgrove, & Wilkinson, 2003), U.S. psychology and U.S. chemistry departments (Tang & Thelwall, 2003). Interlinking between U.S. history departments was too low for patterns to be extracted, and there were significant disciplinary differences in patterns of interlinking within psychology, chemistry, and history (Tang & Thelwall, 2003). No geographic trends were evident, perhaps either because geography is less important in the U.S. than the U.K. or because the phenomenon is less evident within a single subject. History may actually be an anomaly in the humanities, because other humanities subjects seem to publish more on the Web (Thelwall, Vaughan, Cothey, Li, & Smith, 2003).

There is an important issue of scale when moving from analyzing entire university sites to analyzing individual departments. The studies reported here indicate that interdepartmental link data are sufficiently strong in some disciplines to support an analysis, but the research relationships found have been much weaker than for whole universities. Presumably this is due to the averaging effect of the larger units of analysis—in other words, the much larger number of links between universities than between departments makes trends more apparent. It would be an attractive research question to analyze the interlinking of departments in terms of cluster analysis approaches (analogous to Small's [1999] science mapping), but it is not yet clear that there is enough link data to do this meaningfully. In addition, the initial task of accurately identifying all Web sites associated with a discipline is far from trivial (Li et al., 2003). An alternative approach may be for more text-based analyses to circumvent the problem of link sparseness, despite the greater technical difficulty with this mixed approach.

Scale is also a relevant factor when considering the kinds of institutional analyses common in scientometrics, for example, evaluating all departments within a single university. The relatively small number of inlinks per department, coupled with the inherently skewed nature of Web linking, means that inlink counts can be only very weak indicators of visibility or impact and should certainly be interpreted with great caution. However, the Web is important to research communication, and thus departments attracting a low number of inlinks compared with other departments from the same discipline could reasonably be asked at least to address the issue of whether their Web presence is effective (see Thelwall, 2002f).

Link Creation Motivations

Link creation motivation studies are vital for developing an understanding of how link counts should be interpreted. They have tended to be less popular than statistical studies, although both are needed to validate link counts. That said, most of the correlation-based studies discussed here have included some basic analysis of Web pages in order to investigate factors related to motivations or anomalies (e.g., Thelwall, 2001d, 2002b; Thelwall & Wilkinson, 2003b), some through the employment of typologies to categorize link types. In fact, most motivation studies have investigated the context of links, rather than attempt to directly ascertain author motivations. Motivation studies should be viewed in relation to what is known about Web use in general. It is important to understand that Web use is not determined by technology, it is context specific (Hine, 2000). In particular, academics use the Web in different ways and this is likely to continue (Kling & McKim, 2000).

Kim (2000) investigated authors' motivations for creating links in e-journal articles. Some motivations were similar to those previously discovered for making citations, but others were new. The new motivations related to the accessibility and richness of electronic resources. S. D. Herring (2002) conducted an exploratory citation analysis of e-journal articles and found significant use of electronic resources.

Early studies of links found that many were created for nonacademic reasons, such as to connect to a list of local restaurants (Smith, 1999; Thelwall, 2001d), which undermined the use of link counts to investigate research relationships. Later studies also found links of this type but they were less prevalent (Thelwall, 2001b; Thelwall & Harries, 2003). A study of the 100 academic pages in the U.K. that were the target of most links from other U.K. universities did not find recreational pages, but mainly general-purpose pages, such as university home pages and some departmental pages (Thelwall, 2002g). Two studies separated links that were closely related to research from others, finding that counts based upon the research-related links associated more significantly with research productivity (Thelwall, 2001b; Thelwall & Harries, 2003). The most comprehensive study so far took a random collection of 414 links between U.K. university Web sites (using crawler data), which were then classified by two independent researchers (Wilkinson, Harries, Thelwall, & Price, 2003). The classification was problematic, because of a low level of agreement on categories. Nevertheless, by combining similar categories more reliable ones were formed and it was shown that more than 90 percent of links were related to scholarly activity in some way, including teaching. However, less than 1 percent were judged to be equivalent to journal or conference article citations. Combined with the correlation results, this gives strong evidence that links between university Web sites can be used as evidence of patterns of informal scholarly communication. Of course, this is only one type of informal scholarly communication, and there is no guarantee that other types will follow a similar pattern. One study investigated links to university home pages

(Thelwall, 2003c), finding different reasons for their use even within this subgroup. Some links did not appear to fulfill a navigation function and others served no communication purpose at all. This exercise showed the difficulty in both classifying link motivations and interpreting link counts, a major problem for academic Web link studies.

Conclusions

From the link motivation findings, it seems that link counts (either inlink counts or inlinking page counts) reflect a wide range of informal scholarly communication. Universities that conduct more and better research tend to attract more links, but only because they publish more Web content not because it is of higher quality. University Web link analysis is complex and should not be the basis for a simple online indicator of research or other impact. One practical use of academic link analysis is as a health check on universities (e.g., Thelwall & Aguillo, 2003). For instance, lower than expected WIF scores for individual institutions can be indicators of potential problems with the university Web site or Web publishing policy. A second application is the provision of data to complement other indicators—triangulation. This is probably most suitable for general impact evaluation at present. It is also valid to use link counts between academic spaces as partial measures of online scholarly communication, remembering that link creation reflects a wide variety of informal scholarly communication. Results must be interpreted carefully, however, because there is no reason to believe that hyperlinks would necessarily be representative of Web use, and other kinds of online scholarly communication, such as e-mail, may well have different patterns.

Heimeriks, Hörlesberger, and van den Besselaar (2003) have proposed using links in conjunction with other sources to develop a broader picture of collaboration; this method holds promise for future link research. They incorporated article citations (bibliographic coupling) and project collaborations in order to map communication and collaboration between institutions in specific advanced scientific fields. The methodological framework used also enables the construction of rich pictures with which to explore the social dimensions of science and university-industry-government (triple helix) relationships.

General and Commercial Web Use

Measurements of the Web

Given the Web's dynamic nature, there is a real need for benchmark data on its size and characteristics. These data also have practical implications for information retrieval, management, cataloging, and indexing. Unfortunately, as we shall see, reliable Web size estimates are impossible.

The Web Characterization project conducted by the Office of Research of the Online Computer Library Center, Inc. (OCLC) collects an annual Web sample based upon random IP addresses to analyze trends in the size and content of the Web (OCLC, Office of Research, 2003). The project compiles various statistics such as the number of Web sites, the annual growth rate of the Web, country of origin, language, and the 50 most frequently linked-to sites. The project estimated that the Web contained 9 million sites in 2002. Other estimates are available, but give very different numbers due to differences in sampling methods and operating definitions (e.g., total Web vs. indexable Web). Lawrence and Giles (1999a) used a random sample of IP addresses to estimate the number of servers and the mean number of pages per server. They then estimated the size of the indexable Web based on these two numbers. This approach suffers from the problem that the number of pages per server follows an extremely skewed distribution. The mean is not an appropriate measure, which limits the accuracy of the estimate, a fact acknowledged by the authors. Their definition of the publicly indexable Web was essentially all pages reachable by a robot following links from the home page of a site, excluding pages banned by the Robots Exclusion Protocol (http://www.robotstxt.org/wc/exclusion.html) and excluding sites with authentication requirements. However, as mentioned earlier, the advent of the virtual server meant that it was not possible to equate one IP address with one domain name or Web site. Therefore, IP sampling is no longer capable of giving reasonable Web size statistics (Thelwall, 2002d).

Perhaps more relevant and practicable is coverage of the Web by the major search engines. The well-known Search Engine Showdown site (http://searchengineshowdown.com) has regular updates on search engine statistics, such as relative size and database overlaps. The statistics on search engine size are based on counting total hits for single word queries (Notess, 2003). Longitudinal studies analyze changes in the Web rather than total size. Koehler (2002), for example, investigated Web page changes and persistence. He followed 361 pages weekly for four years and recorded changes in content and links. He found the half-life of a Web page to be roughly two years and that different types of Web pages (e.g., commercial, educational) behave differently. The limitation of the study is that these pages were randomly sampled from a single search engine, WebCrawler, in December 1996. These pages may not represent Web pages in general.

Web Server Log Analysis

A long tradition in information science involves examining user online search behavior by monitoring or recording searches in order to improve user interfaces or user training. Numerous studies have analyzed how people use traditional information retrieval systems such as Thomson Dialog or library OPACs (Online Public Access Catalogs). Those studies are typically conducted in experimental settings with academic users.

The Web user has much more heterogeneous search interests and habits. The Web server log, which unobtrusively captures real people's natural search processes, is a rich source of insights on Web use. Such information is useful not only for Web design but also for understanding the social impact of the Web.

Despite the difficulties in server log analysis, a significant body of research findings has been built up concerning the use of commercial search engines. Analyses of server logs from two search engines, AltaVista and Excite, reveal similar user search patterns. Web users differ significantly from users discussed in standard IR literature. Most people use short queries, seldom modify the query, rarely use advanced search features, and look at only the first 10 to 20 results (Silverstein et al., 1999; Spink, Jansen, & Ozmutlu, 2000; Spink, Wolfram, Jansen, & Saracevic, 2001). There are large numbers of mistakes or incorrect uses (Jansen, Spink, & Saracevic, 2000). Longitudinal analysis of Excite server logs has shown that search topics have shifted, but there has been little change in user search behaviors (Spink, Jansen, Wolfram, & Saracevic, 2002) even though there was increased use of advanced search features such as Boolean operators and relevance feedback (Wolfram, Spink, Jansen, & Saracevic, 2001). Ross and Wolfram (2000) investigated the topic content of queries submitted to the Excite search engine using a content classification of term co-occurrence within unique queries. This analysis provided a better understanding of the types of topics users were searching for and their relationships. Because unique queries, not specific users, were of interest, the problems of user and session identification were not at issue.

In addition to search engines, studies of the logs of individual Web sites can give useful information about user interactions with electronic sources. This is particularly relevant for information providers, such as health sites. Collating data from their previous studies (e.g., Nicholas, Huntington, Lievesley, & Wasti, 2000; Nicholas, Huntington, & Williams, 2002) and triangulating with questionnaire data, Nicholas, Huntington, Williams, and Dobrowolski (2004) summarized and characterized users' searching behavior as seldom penetrating a site to any depth, tending to visit a number of sites for a given information need, and seldom returning to sites previously visited. The triangulation approach, using different sources of information for the same research goals, is good practice for incorporating user perspectives in the construction of convincing theories of user behavior, especially given the problems of interpreting log files.

Chen and Cooper (2001) used clustering techniques to detect usage patterns without any demographic information in a Web-based library catalog system. Based on the usage groups classified, they further developed a Markov model that can predict users' moves from one search state to the next (Chen & Cooper, 2002). Although these studies were carried out on a Web-based library catalog system, the methods and the models developed could be applied to other Web systems.

A connection between log files and hyperlinks can be made because log files can record the URL of the page previously visited by the user in addition to the URL of the current one. This normally implies that a link on the previous page has been traversed. One team of researchers has pursued a fruitful line of research in which this kind of implied link navigation is used as a measure of the importance of links, or the likelihood of their future use. This approach has allowed them to devise improved tools for the navigation of individual Web sites by clustering pages for navigation interfaces in a way that helps users retrieve information from a site (Zhu, Hong, & Hughes, in press), by predicting paths that users will traverse within a site (Zhu, Hong, & Hughes, 2002), and by adapting PageRank to take into account the frequency of use of links in a site (Zhu, Hong, & Hughes, 2001). Thelwall (2001f) used the same server log information to study site inlinks and found that many generate little or no traffic. This was a novel use of both the inlink data (usually obtained from search engines) and the log file data. The combination of the two data sources provided a fuller picture than either data source alone.

Commercial Web Site Studies

Most webometric studies to date have focused on academic or scholarly Web sites, following in the tradition of bibliometrics. This is ironic given that the Web is dominated by commercial sites. Knowledge of commercial Web sites has become increasingly important with the rapid growth of e-commerce.

Thelwall (2000c) randomly sampled .co.uk domain names to determine the kind of U.K. businesses that were most prominent in the early stages (1999) of business Web use. The study found that media-related businesses and professions (e.g., legal firms) had a strong presence on the Web, although the computing industry remained the dominant publisher. In general, commercial use of the Web at that time was in an experimental phase. This was confirmed by a related international study (Thelwall, 2000a), which found many businesses did not design their Web sites to be search engine friendly. As a result, 23 percent of the sites were not registered at all in the five major search engines tested. About two thirds of the sites did not use HTML meta tags for indexing. Thelwall (2001c) also studied hyperlinks in commercial sites and found that roughly 66 percent did carry links pointing to other related Web sites; most often links were directed to sites with business relationships.

Vaughan and Wu (2003) studied the Web sites of China's top 100 information technology companies and found significant correlations between counts of inlinks to a company's home page and the company's business performance measures, including revenue, profit, and research and development expenses. There is also a statistically significant correlation between the age of a Web site and its inlink counts, with older Web sites having more inlinks. Vaughan (2004) used the same approach to examine the top U.S. information technology companies. Again, the inlink counts correlate significantly with business measures of revenues and profits. Moreover, the two sets of correlation coefficients for the two countries are remarkably similar, given that the two groups of Web sites have very different characteristics.

Many studies of commercial Web sites aim to discover information for competitive intelligence purposes. There are two approaches for this. One analyzes Web site content and the other focuses on link structures. Examples of the former are Zanasi (1998) and Madnick and Siegel (2002). Liu, Ma, and Yu (2001) proposed and tested a number of auto-mated methods to help users find novel information from competitors' Web sites, for example, unexpected services or products that competitors offer. The authors targeted competitors' Web site content but stated that their future work would use metadata and links to extract additional information. Reid (2003) presented a framework for mining the Web's hyperlink structure to gather business intelligence and demonstrated the method through a case study of MicroStrategy, Inc., a business intel-ligence vendor. Step one is to identify the implicit business community through a search of links to the MicroStrategy Web site. The Web sites of this community are then content analyzed to detect early warning sig-nals of relevance to MicroStrategy. For example, the Web community may create directories or news stories that link to the MicroStrategy site. If they express their perceptions about the company in these news stories, this may give MicroStrategy signals about its business opportu-nities and challenges.

Social Network Analysis

Social network analysis, a research approach developed primarily in sociology, social psychology, and communication science, focuses on pat-terns of relations among people, and among groups such as organiza-tions and states. Computer networks, including the Web, can host social networks because they connect people and organizations. Garton, Haythornthwaite, and Wellman (1999) have argued for the usefulness of social network analysis in the study of computer networks in general and the Internet in particular. They explain some basic concepts of social network analysis, describe data collection methods, and demonstrate their application. Their article is an excellent introduction to social net-works analysis and a comprehensive review of the related literature. Garton et al. (1999) discussed the applicability of social network analy-sis to computer networks in general; Jackson (1997) focused specifically on Web hyperlinks as a communication tool and discussed how to inter-pret Web link structures through social network analysis. Otte and Rousseau (2002) also pointed out the applicability of social network analysis to information science, specifically to the study of the Internet.

Two successful applications of social network analysis to Web phe-nomena are by Garrido and Halavais (2003) and Park, Barnett, and Nam (2002), both using links as their data source. The former studied networks of support for the Zapatista movement in Mexico, in which the Internet played a central role. The authors collected data on links to the Zapatista site and mapped these into a Zapatista network on the Web. This hyperlinked network provided an insight into the character of the Zapatista's success, for example, the strong connection with women's rights. Park et al. (2002) examined South Korea's 152 most popular Web sites. They operationalized affiliation between Web sites with links and applied cluster analysis to reveal the underlying network. Financial Web sites were found to occupy the most central position in the network.

Adamic and Adar (2003) studied home page owners in two U.S. universities to explore how well home page texts and links, as well as e-mail list membership, could predict social connections. The degree of success varied with the precise nature of the data: for example, two students with *co-outlinks* (analogous to bibliographic coupling, cf. Figure 2) to Yahoo! would not be as good an indicator of a likely social connection as *co-inlinks* (co-citation) from another student's "list of friends" would be. The software developed in the study was able to suggest reasons why two individuals could be friends, such as membership of common e-mail lists or shared topics of interest. The main limitation of the study was the atypical nature of the communities studied: students in two elite universities.

Links have been used in conjunction with other data sources to explore online social behavior around specific topics. Hine's (2000) virtual ethnography uses links as one element in a broad analysis of phenomena with online components. Her insistence that online actions should be interpreted in conjuction with offline actions is particularly welcome: The Web is not divorced from the real world. Web Sphere Analysis is another sociological method that has been used to study Web publishing in relation to U.S. Presidential elections (Foot, Schneider, Dougherty, Xenos, & Larsen, 2003). It includes links as part of a wider exploration of online relationships and genres of publishing. Detailed descriptions of both techniques are beyond the scope of this review but both represent thoughtful and appropriate frameworks for studying online phenomena.

Topological Modeling and Mining of the Web

If the Web is viewed as a collection of pages connected by links and all information about page contents, locations and URLs is discarded, then this produces what is called a *directed* graph in mathematics if the direction of the links is kept, or an *undirected* graph if the direction of the links is ignored. Physicists, computer scientists, and information scientists have attempted to build models of the Web or Web growth through link analysis, as well as to extract abstract clustering patterns. These approaches can be termed topological because they treat the Web as a mathematical graph and ignore spatial relationships and page contents. The Web is the largest network for which topological information is currently available, and it has thus become a testing ground for many modeling efforts (Albert & Barabási, 2002). Graph theoretic approaches have been used in bibliometrics and scientometrics since the 1960s for analyzing citation networks and other information networks (e.g., Furner, Ellis, & Willett, 1996; Garner, 1967; Hummon & Doreian, 1989; Nance, Korfhage, & Bhat, 1972) so it is natural for information scientists to be interested in Web graphs.

Two key algorithms that have served to underline the potential of hyperlink structure analysis are Kleinberg's (1999) HITS (Hypertext-Induced Topic Search) and Google's PageRank (Brin & Page, 1998). A lesser known approach, Constrained Spreading Activation (Cohen & Kjeldsen, 1987), has also had an impact in computer science when applied to links. The purpose of HITS is to identify the most useful Web pages for any given topic. The algorithm identifies relevant pages by text matching but picks out the most relevant through an iterative link-based algorithm that attempts to find "authorities," pages with many inlinks from within the topic, and "hubs," pages with many outlinks to topic pages. Good authorities and hubs are both desirable. The approach has been shown to be adaptable for different purposes (Lempel & Moran, 2002). In a topological analysis of the U.K. academic Web space, Björneborn (2003) found an indication that Web sites that are Kleinbergian hubs and authorities also have high *betweenness centrality* in the investigated Web space. Betweenness centrality is defined as the probability that a node (e.g., a Web site) will occur on a shortest path between two arbitrary nodes in a network graph (Freeman, 1977).

Unlike HITS, PageRank is not topic-centered but ranks all pages using the two assumptions that pages with more inlinks are more likely to be useful and that inlinks from useful pages are more important than inlinks from average pages (Brin & Page, 1998). This iterative algorithm is particularly valuable because its mathematical implementation makes it feasible to implement on billions of Web pages. Logically, PageRank should be a better indicator of value than simple inlink counts, but this is difficult to demonstrate because of the confounding effect of internal site selflinks for navigational purposes (Thelwall, 2003a). Highly ranked pages have hub-like properties (i.e., they point to, rather than directly contain, useful content): in university Web sites they are likely to be the home pages of universities and departments (Thelwall, 2003a).

Network Models of Linking Behavior

Fundamental to an understanding of linking on the Web is the occurrence of many so-called, scale-free network features (Barabási & Albert, 1999; Barabási, Albert, & Jeong, 2000). In a scale-free network there is no "typical" node, that is, no characteristic "scale" to the degree of connectivity (Ball, 2000). Scale-free distributions of inlinks and outlinks show long, power-law tails, meaning that only a small number of Web nodes receives or provides many links, whereas the majority has few inlinks or outlinks. A range of power-law distributions has been identified on the Web in, for example, scattering of TLDs on a given topic (Rousseau, 1997), inlinks per site (Adamic & Huberman, 2001; Albert, Jeong, & Barabási, 1999), outlinks per site (Adamic & Huberman, 2001), in-neighbors and out-neighbors per site (Björneborn, 2003), pages per site (Adamic & Huberman, 2001; Huberman & Adamic, 1999), visits per site (Adamic & Huberman, 2001; Huberman, Pirolli, Pitkow, & Lukose, 1998; Pitkow, 1998), and visited pages within a site (Huberman, Pirolli, Pitkow, & Lukose, 1998; Pitkow, 1998).

Barabási and Albert (1999) argue that scale-free link distributions are rooted in two generic mechanisms of many real-world networks: *continuous growth* and *preferential attachment* (the "rich-get-richer" phenomenon). In this framework, the Web is an open, self-organizing system that grows by the continuous addition of new nodes and links where the likelihood of connecting to a node depends on the number of links already attached to the node. This significance of preferential attachment for power-law distributions is well known in bibliometrics as the Matthew effect, "unto every that hath shall be given" (Merton, 1968), and "cumulative advantage" (Price, 1976).

Barabási & Albert's (1999) original model for explaining the emergence of scale-free, power-law properties in networks has been augmented with additional factors, including initial attractiveness (Dorogovtsev, Mendes, & Samukhin, 2000), competition and fitness (Bianconi & Barabási, 2001), optimization (Valverde, Ferrer i Cancho, & Solé, 2002), uniform attachment (Pennock, Flake, Lawrence, Glover, & Giles, 2002), transitive linking (Ebel, Mielsch, & Bornholdt, 2002) and lexical distance (Menczer, 2002). Moreover, Amaral, Scala, Barthélémy, and Stanley (2000) explain deviations in scale-free distributions by factors such as aging, cost, and the capacity constraints of network nodes.

The model presented by Pennock et al. (2002) explains the highly skewed and hooked shape of power law diagrams for linking in terms of competing tendencies for new links to be allocated to already highly linked pages and for links to be allocated at random (in reality, probably to new high quality pages). In fact, the balance of competition varies by domain. When the distribution of links was compared for the same type of page, it was found to be more log-normal than like a power law, with additional variations by type. For example, university home pages attract links following a pattern that is less skewed than scientists' home pages (i.e., links to university home pages are more equally distributed than links to scientists' home pages). This implies that scale-free network growth models are not necessarily useful for explaining individual linking behavior.

Overall Web Topology

Various topological properties of the overall structure of the Web have been investigated, which helps us understand how search engines work in practice and how the Web has developed. The first important contribution was a conjecture that any pair of Web pages would be connected by a short chain of links, with an average of only 19 links (Albert, Jeong, & Barabási, 1999). This was later discovered to be incorrect (Broder et al., 2000) because many pairs of Web pages are not connected at all. However, the small-world phenomenon (Watts & Strogatz, 1998)—where pages tend to cluster in groups but, almost paradoxically, pairs of pages can be connected across cluster boundaries without a torturously long chain of links—persists in large graph components of the Web (Broder et al., 2000).

The groundbreaking Web topology study was that of Broder et al. (2000), which used two full AltaVista crawls from May and October of 1999 and specially constructed software to process 200 million pages and 1.5 billion links. More than 90 percent of the links formed a huge connected group, provided the direction of the links was ignored. This central component split into four roughly equal parts. The core was a strongly connected component (SCC) in which all pages traced a directed link path to all others in the SCC. The other parts comprised a set of pages (OUT) that could be reached from the SCC by following (directed) links, a set of pages (IN) that could reach the SCC by following (directed) links, and the rest (TENDRILS). The remaining pages that were not connected in any way from the main 90 percent were dubbed DISCONNECTED. A problem with this study is the difficulty in extrapolating from it to the whole Web. AltaVista finds pages partly from user submissions of URLs, but primarily by following links from previously visited pages. As a result, pages that are not well linked to are more likely to be missed by its crawler. Thus DISCONNECTED is likely to be far greater for the whole Web, but it is not possible to estimate its size because there is no practical way to automatically find pages that are not linked to. Nevertheless, the results are interesting because AltaVista presumably covers the most used and most visited sites, and its database presumably reflects the Web that users actually see.

Evidence has been found that similar graph component structures are present in parts of the Web, including national university systems (Björneborn, 2003; Thelwall & Wilkinson, 2003a) and individual countries (Baeza-Yates & Castillo, 2001). Baeza-Yates and Castillo extended Broder's model by finding that some of the components could be subdivided. As a final point, it would be misleading to think of any of the components as homogeneous or made up of coherent sets of pages. For example, OUT will contain many pages without outlinks that are on the same site as SCC pages (Thelwall & Wilkinson, 2003a).

Clustering in the Web

The Web's structure does not allow the identification and extraction of small groups of pages by finding disconnected components, as the preceding discussion reveals; the connected components of the Web graph, such as the SCC can be very large. However, clustering Web pages is desirable for several reasons. Displaying similar pages close to each other is a useful technique for improving user interfaces (Amento, Hill, Terveen, Hix, & Ju, 1999). Search results can also be improved by fitting pages into the context of their relationships to topics (Kleinberg, 1999) and for direct webometric applications complete sets of pages associated with a given subject are often needed, including for department-based studies, as already discussed. As a result, it is important to investigate how the Web can be decomposed into coherent clusters other than the directed and undirected components. We will cover only topological clustering here, excluding algorithms that include page content analyses, such as Kleinberg's work (1999). In fact Menczer (in press) has shown that pages about the same topic tend to cluster together topologically, and that similar pages tend to interlink.

Traditional approaches to topological clustering consume too much computing time to make them practical for large graphs. A breakthrough was made, however, in the design of a fast algorithm that was able to create a viable topological cluster around any given set of connected pages (Flake, Lawrence, Giles, & Coetzee, 2000, 2002). This design was then adapted and applied to national collections of university Web sites, revealing the existence of rich community structures (Thelwall & Wilkinson, 2003a). An alternative clustering approach has also been applied, using co-inlinks (Polanco et al., 2001), but at the site level. A recent investigation into the feasibility of using links, co-inlinks (co-citations), and co-outlinks (bibliographic couplings) to identify similar pages concluded that the link topology was such that combinations of link structures were not likely to be more successful than links alone at identifying similar sites, but that the combinations would give a greater range of potentially similar sites (Thelwall & Wilkinson, 2004).

It is tempting to propose the use of topological clustering and social network analysis techniques for extracting field or subject interlinking patterns from national academic Webs, but this does not seem to be a promising line of research because of the relative scarcity of links. Essentially the problem is that while two journal articles on very similar topics should have references in common, there is no reason why two researchers, research groups, or departments with similar interests should link to the same pages: Some do, but many do not. Effective field clustering is likely to require text analysis in addition to link analysis.

Small Worlds

An intriguing dimension of Web topologies is the small-world properties of the Web's strongly connected component (SCC) (Broder et al., 2000). It is noteworthy that the originators of the World Wide Web envisaged small-world properties: "Yet a small number of links is usually sufficient for getting from anywhere to anywhere else in a small number of hops" (Berners-Lee & Cailliau, 1990, online). Small-world Web topologies are concerned with core information science issues such as the navigability and accessibility of information across vast document networks. For instance, short distances along link paths affect the speed and exhaustivity with which Web crawlers can reach and retrieve pages.

Small-world theory stems from research in Social Network Analysis on short distances between two arbitrary persons through intermediate chains of acquaintances. In a famous study, Milgram (1967) asked a sample of individuals in the U.S. midwest to forward a letter to a person not known to them on the east coast. The letter was to be mailed to a personal acquaintance (on a first-name basis) likely to know the target person. For the completed chains, a median length of five intermediate acquaintances was needed to forward the letter to the target person. This result contributed to the popularized notion of "six degrees of separation" between any two persons in the world through intermediate chains of acquaintances. Manfred Kochen, a visionary information scientist, was a pioneer in small-world theorizing (Kochen, 1989; Pool & Kochen, 1978). Another innovative information scientist and bibliometric pioneer, Eugene Garfield (1979), envisioned that small-world approaches could be used to identify gatekeepers and "invisible colleges" in informal scholarly communication networks. On the Web, small-world topologies could reflect such things as cross-disciplinary contacts and research fronts in the evolving interconnectedness of science (Björneborn, 2001, 2003; Björneborn & Ingwersen, 2001). In detailed case studies of the Web links, pages, and sites that function as connectors across dissimilar topical domains in the U.K. academic Web space, Björneborn (2003) found that personal link creators such as researchers and students were frequent providers of transversal (cross-topic) links, thus creating small-world phenomena in this Web space. These indicative findings also suggest that people and Web sites related to computer science are important cross-topic connectors. The role of computer science for short link distances in an academic Web space most likely reflects the auxiliary function of computer science in many different scientific disciplines, combined with a more experienced and unconstrained Web presence by many persons and institutions in computer science (Björneborn, 2003).

In a seminal paper, Watts and Strogatz (1998) introduced a small-world network model characterized by highly clustered nodes (high average clustering coefficient) as in regular graphs, yet with short characteristic path lengths between pairs of nodes as in random graphs. Watts and Strogatz showed that in a small-world network it is sufficient for a very small percentage of long-range links to function as short cuts connecting distant nodes of the network. This revival of small-world theory triggered an avalanche of research in a wide range of scientific domains including the Web (Adamic, 1999; Albert, Jeong, & Barabási, 1999; Barabási, 2001). The emergence of small-world topologies on the Web and in other complex networks can be attributed to the *scale-free* network features previously discussed. According to Albert & Barabási (2002), a heterogeneous scale-free topology is very efficient in bringing network nodes close to each other. Small-world structures may thus arise from a scale-free organization in which a relatively small number of well-connected nodes serve as hubs (Steyvers & Tenenbaum, 2001). Well-connected Web nodes often have multiple memberships providing strong and weak ties (cf. Granovetter, 1973) across many topical clusters. Such far-reaching bridging ties function as transversal links, shrinking the distances in a network to form a small world.

To help paint a mental picture of a small-world "crumpled-up" Web space, every new outlink on the Web may be regarded as a hook able to reshape the existing Web network (Björneborn, 2003). By pulling distant Web nodes close to each other (and thereby drawing together the neighborhoods and clusters to which these nodes already belong), transversal links may "crumple up" Web topologies and thus reinforce the small-world nature of the Web.

The scale-free properties of the Web imply a fractal structure where cohesive subregions display the same characteristics as the Web at large (Dill, Kumar, McCurley, Rajagopalan, Sivakumar, & Tomkins, 2001). Such scale-free properties are often regarded as a fingerprint of self-organization (van Raan, 2000). Analyses of the Web show a remarkable degree of self-organization in the shape of clustered hyperlink structures that reflect topic-focused interest communities (Flake et al., 2002; Gibson, Kleinberg, & Raghavan, 1998; Kleinberg & Lawrence, 2001; Kumar, Raghavan, Rajagopalan, & Tomkins, 1999). The coincidence of high local clustering and short global separation (Watts, 1999) means that small-world networks simultaneously consist of small local and global distances, leading to high efficiency in propagating information on both a local and global scale (Marchiori & Latora, 2000). However, Web links do not *directly* channel information flows as social networks, neural networks, or computer networks do. Web links *indirectly* reflect information diffusion among link creators, because added or removed links may reflect changes in topical interests and the social preferences of link creators. On an aggregated, macro level, such dynamic link adaptations could reflect cultural and social currents and formations (Björneborn, 2003).

From a microlevel perspective, the emergence of small-world properties on the Web is associated with decentralized Web page and link creations resulting in topic drift (Bharat & Henzinger, 1998) and genre drift (Björneborn, 2003). Diverse topics and genres of Web pages may be separated by one or just a few shortcut links. Web genres and topics thus exist side by side on the Web. As noted earlier, most links connect Web pages containing cognate topics (Davison, 2000). This topical propensity (Björneborn, 2003) leads to the emergence of topic-focused Web clusters. Such topical clusters may consist of diverse page genres; for instance, in an academic Web space it could be institutional home pages, link lists, course pages, conference pages, personal home pages, link lists, and others, interconnected by links in a *web of genres* (Björneborn, 2003). Hence, topical clusters with genre diversity entail genre drift along *intra*-cluster links. At the same time, some Web page genres, for example institutional or personal link lists, are more diversity-prone and thus function as providers of topic drift along *inter*-cluster links. According to Björneborn (2003), small-world phenomena on the Web may thus be created by complementarities of topic drift and genre drift.

The self-organized small-world architecture of large regions on the Web may be an important, nonengineered, organizing principle for structuring a vast information space (Björneborn, 2003). Webometric approaches may reveal characteristics of how such complex, bottom-up-aggregated Web structures provide traversal options and access points to information. In this context, scholarly Web activities are of special interest when one considers science as a complex, largely self-organizing, sociocognitive system (e.g, Leydesdorff, 2001; Sandstrom, 2001; van Raan, 2000). The Web with its capability to host self-organizing activities in the shape of "collaborative weaving" may thus be seen as a natural environment for scholars (Björneborn, 2003, p. 205).

Summary and Concluding Remarks

We have reviewed many different quantitative studies of the Web, concentrating on those with an information science background. A number of general conclusions can now be reached. First, large amounts of data are required for reliable studies, so investigations normally utilize special software or programs such as Web log file analyzers or commercial search engines. This is in the tradition of bibliometrics, with its use of ISI software or specifically designed programs to analyze large databases. The need to process large data sets is not new, but perhaps represents a solidifying trend in information science research.

Another theme is that the quantitative techniques require complementary qualitative methods (user studies, link creation motivation studies) to understand the results completely. A future trend may be to collaborate with other fields such as cultural studies, computer-mediated communication, social network analysis, and social informatics in order to gain the benefit of more qualitative techniques. The motivation and small world findings indicate that interpretations are likely to be complex and difficult to categorize, either on a small or large scale.

Given the many correlation studies reported here, it is important to bear in mind that significant correlations between link counts and research measures are not ends in themselves. The (partial) information they provide relates to the reliability and validity of using link data, but does not tell us how link structure information can be used for other purposes such as assessing the online impact of Web sites. Counts of links to academic Web sites can now be interpreted with some confidence as reflecting a range of informal scholarly communications. Inlink counts primarily indicate the productivity of the target institution, rather than the average quality or impact of the information presented. The most useful information to be gained from this relates to outliers: the identification of universities that appear to be making particularly good or poor use of the Web. One logical next step from correlations is to look for patterns in the data, perhaps clustering similar topics across different sites or employing social network analysis techniques. It may be that links alone will not be sufficient, however, since interuniversity links are too sparse to expect similar pages or even similar domains to be frequently interlinked. The future may therefore lie in combining link analysis with text analysis; it is likely that considerable additional computing power and expertise will be required for this development. As a result, a promising new direction for webometrics is to combine with computer science to build tools to harness our knowledge of Web phenomena and produce useful information: in other words, Web intelligence and Web mining. Applied to academic Web sites, techniques from these two fields, together with webometrics, form the emerging field of Scientific Web Intelligence (Thelwall, in press c).

Topological approaches to the Web are useful to give a big picture with which to set other link analyses in context. The future for information science research in this area seems to lie in building a more concrete understanding of linking phenomena by tracking individual links, as exemplified by the small-world approach. Links could then be utilized in social informatics for exploring networked knowledge creation and diffusion, including emerging cross-cultural formations; in Web mining for identifying fertile scientific areas for cross-disciplinary exploration; by search engine designers for more exhaustive Web traversal and harvesting, as well as improved visualization and navigation facilities for search results, perhaps stimulating serendipitous browsing.

The Web is continuously changing and developing, which will provide further research challenges and opportunities. The Semantic Web (Berners-Lee & Hendler, 2001) may eventually make the content of Web pages and the connections among them more explicit and transparent to automatic processes. This should open the door for more powerful and robust webometric analysis. It remains to be seen how far the Semantic Web will penetrate, however, and given the past history of the Web it will vary across user communities and its adoption will depend upon the ways in which the Web is able to provide valuable services for specific needs.

Other Web changes are social in nature. As the Web spreads to wider sections of the population of the richer countries and its use becomes more common in poorer nations, new uses can be expected. The Web is, therefore, likely to increase in importance for most aspects of human activity. A challenge and important future direction for webometrics will be to produce tools and techniques that are able to identify, monitor, and partially quantify these developments. These tools will be most useful if they can be made widely available to others studying the Web, including social scientists.

References

Abraham, R. H. (1996). *Webometry: Measuring the complexity of the World Wide Web.* Santa Cruz, CA: Visual Math Institute, University of California at Santa Cruz. Retrieved November 20, 2003, from http://www.ralph-abraham.org/vita/redwood/vienna.html

Adamic, L., & Adar, E. (2003). Friends and neighbors on the Web. *Social Networks, 25*(3), 211–230.

Adamic, L. A., & Huberman, B. A. (2001). The Web's hidden order. *Communications of the ACM, 44*(9), 55–59.

Aguillo, I. F. (1998). STM information on the Web and the development of new Internet R&D databases and indicators. *Online Information 98: Proceedings,* 239–243.

Albert, R., & Barabási, A. L. (2002). Statistical mechanics of complex networks. *Reviews of Modern Physics, 74*(1), 47–97.

Albert, R., Jeong, H., & Barabási, A. L. (1999). Diameter of the World-Wide Web. *Nature, 401,* 130–131.

Albitz, P., & Liu, C. (1992). *DNS and BIND.* Sebastopol, CA: O'Reilly.

Almind, T. C., & Ingwersen, P. (1996). *Informetric analysis on the World Wide Web: A methodological approach to "internetometrics"* (CIS Report 2). Copenhagen, Denmark: Centre for Informetric Studies, Royal School of Library and Information Science.

Almind, T. C., & Ingwersen, P. (1997). Informetric analyses on the World Wide Web: Methodological approaches to "webometrics." *Journal of Documentation, 53*(4), 404–426.

Amaral, L. A. N., Scala, A., Barthélémy, M., & Stanley, H. E. (2000). Classes of small-world networks. *Proceedings of the National Academy of Sciences, 97*(21), 11149–11152.

Amento, B., Hill, W., Terveen, L., Hix, D., & Ju, P. (1999). An empirical evaluation of user interfaces for topic management of Web sites. *Conference on Human Factors in Computing Systems,* 552–559.

Arasu, A., Cho, J., Garcia-Molina, H., Paepcke, A., & Raghavan, S. (2001). Searching the Web. *ACM Transactions on Internet Technology, 1*(1), 2–43.

Baeza-Yates, R., & Castillo, C. (2001). Relating Web characteristics with link based Web page raking. *Proceedings of SPIRE 2001,* 21–32.

Ball, P. (2000, Oct 25). The art of networking. *Nature Science Update.* Retrieved November 20, 2003, from http://www.nature.com/nsu/001026/001026-7.html

Bar-Ilan, J. (1997). The "mad cow disease," Usenet newsgroups and bibliometric laws. *Scientometrics, 39*(1), 29–55.

Bar-Ilan, J. (1999). Search engine results over time: A case study on search engine stability. *Cybermetrics, 2/3*(1). Retrieved November 20, 2003, from http://www.cindoc.csic.es/cybermetrics/articles/v2i1p1.html

Bar-Ilan, J. (2001). Data collection methods on the Web for informetric purposes: A review and analysis. *Scientometrics, 50*(1), 7–32.

Bar-Ilan, J. (2004). The use of Web search engines in information science research. *Annual Review of Information Science and Technology, 38,* 231–288.

Bar-Ilan, J., & Peritz, B. (2002). Informetric theories and methods for exploring the Internet: An analytical survey of recent research literature. *Library Trends, 50*(3), 371–392.

Barabási, A. L. (2001, July). The physics of the Web. *Physics World.* Retrieved November 20, 2003, from http://physicsweb.org/article/world/14/7/9

Barabási, A. L. (2002). *Linked: The new science of networks.* Cambridge, MA: Perseus.

Barabási, A. L., & Albert, R. (1999). Emergence of scaling in random networks. *Science, 286,* 509–512.

Barabási, A. L., Albert, R., & Jeong, H. (2000). Scale-free characteristics of random networks: The topology of the World-Wide Web. *Physica A, 281,* 69–77.

Bergman, M. K. (2000). *The deep Web: Surfacing hidden value.* Retrieved November 20, 2003, from http://128.121.227.57/download/deepwebwhitepaper.pdf

Berners-Lee, T. (1999). *Web architecture from 50,000 feet.* Retrieved November 20, 2003, from http://www.w3.org/DesignIssues/Architecture.html

Berners-Lee, T., & Cailliau, R. (1990). *WorldWideWeb: Proposal for a hypertext project.* Retrieved November 20, 2003, from http://www.w3.org/Proposal.html

Berners-Lee, T., & Hendler, J. (2001). Publishing on the semantic Web. *Nature, 410,* 1023–1024.

Bharat, K., & Broder, M. R. (1998). A technique for measuring the relative size and overlap of public Web search engines. *Computer Networks and ISDN Systems, 30,* 379–388.

Bharat, K., & Henzinger, M. R. (1998). Improved algorithms for topic distillation in a hyperlinked environment. *Proceedings of the ACM SIGIR 21st Annual International Conference on Research and Development in Information Retrieval,* 104–111.

Bianconi, G., & Barabási, A. L. (2001). Competition and multiscaling in evolving networks. *Europhysics Letters, 54,* 436–442.

Björneborn, L. (2001). Small-world linkage and co-linkage. *Proceedings of the 12th ACM Conference on Hypertext and Hypermedia,* 133–134.

Björneborn, L. (2003). *Small-world link structures across an academic Web space: A library and information science approach.* Unpublished doctoral dissertation, Royal School of Library and Information Science, Copenhagen, Denmark.

Björneborn, L., & Ingwersen, P. (2001). Perspectives of webometrics. *Scientometrics, 50*(1), 65–82.

Björneborn, L., & Ingwersen, P. (in press). Towards a basic framework of webometrics. *Journal of the American Society for Information Science and Technology.*

Borgman, C., & Furner, J. (2002). Scholarly communication and bibliometrics. *Annual Review of Information Science and Technology 36,* 3–72.

Bossy, M. J. (1995). *The last of the litter: "Netometrics."* Retrieved November 20, 2003, from http://biblio-fr.info.unicaen.fr/bnum/jelec/Solaris/d02/2bossy.html

Boutell, T. (2003). *What is the World Wide Web?* Retrieved November 20, 2003, from http://www.boutell.com/newfaq/basic/web.html

Brin, S., & Page, L. (1998). The anatomy of a large scale hypertextual Web search engine. *Computer Networks and ISDN Systems, 30*(1–7), 107–117.

Broder, A., Kumar, R., Maghoul, F., Raghavan, P., Rajagopalan, S., Stata, R., et al. (2000). Graph structure in the Web. *Computer Networks, 33*(1–6), 309–320.

Brody, T., Carr, L., & Harnad, S. (2002). Evidence of hypertext in the scholarly archive. *Proceedings of ACM Hypertext 2002,* 74–75.

Brookes, B. C. (1990). Biblio-, sciento-, infor-metrics??? What are we talking about? In L. Egghe & R. Rousseau (Eds.), *Informetrics 89/90: Selection of papers submitted for the Second International Conference on Bibliometrics, Scientometrics and Informetrics* (pp. 31–43). Amsterdam: Elsevier.

Chakrabarti, S., Dom, B. E., Kumar, S. R., Raghavan, P., Rajagopalan, S., Tomkins, A., et al. (1999). Mining the Web's link structure. *IEEE Computer, 32*(8), 60–67.

Chakrabarti, S., Joshi, M. M., Punera, K., & Pennock, D. M. (2002). The structure of broad topics on the Web. *Proceedings of the WWW2002 Conference.* Retrieved November 20, 2003, from http://www2002.org/CDROM/refereed/338

Chen, C., Newman, J., Newman, R., & Rada, R. (1998). How did university departments interweave the Web: A study of connectivity and underlying factors. *Interacting with Computers, 10*(4), 353–373.

Chen, H., & Chau, M. (2004). Web mining: Machine learning for Web applications. *Annual Review of Information Science and Technology, 38,* 289–329.

Chen, H.-M., & Cooper, M. D. (2001). Using clustering techniques to detect usage patterns in a Web-based information system. *Journal of the American Society for Information Science and Technology, 52*(11), 888–904.

Chen, H.-M., & Cooper, M. D. (2002). Stochastic modeling of usage patterns in a Web-based information system. *Journal of the American Society for Information Science and Technology, 53*(7), 536–548.

Chi, E. H., Pitkow, J., Mackinlay, J., Pirolli, P., Gossweiler, R., & Card, S. K. (1998). Visualizing the evolution of Web ecologies. *ACM SIGCHI Conference on Human Factors in Computing Systems,* 400–407.

Chu, H., He, S., & Thelwall, M. (2002). Library and information science schools in Canada and USA: A webometric perspective. *Journal of Education for Library and Information Science 43*(2), 110–125.

Cohen, P. R., & Kjeldsen, R. (1987). Information retrieval by constrained spreading activation in semantic networks. *Information Processing & Management, 23*(2), 255–268.

Cothey, V. (in press). Web-crawling reliability. *Journal of the American Society for Information Science and Technology.*

Cronin, B. (2001). Bibliometrics and beyond: Some thoughts on Web-based citation analysis. *Journal of Information Science, 27*(1), 1–7.

Cronin, B., & Shaw, D. (2002). Banking (on) different forms of symbolic capital. *Journal of the American Society for Information Science and Technology, 53*(14), 1267–1270.

Cronin, B., Snyder, H. W., Rosenbaum, H., Martinson, A., & Callahan, E. (1998). Invoked on the Web. *Journal of the American Society for Information Science, 49*(14), 1319–1328.

Crowston, K., & Williams, M. (2000). Reproduced and emergent genres of communication on the World Wide Web. *Information Society, 16*(3), 201–215.

Davenport, E., & Cronin, B. (2000). The citation network as a prototype for representing trust in virtual environments. In B. Cronin & H. B. Atkins (Eds.), *The web of knowledge: A Festschrift in honor of Eugene Garfield* (pp. 517–534). Medford, NJ: Information Today Inc.

Davison, B. D. (2000). Topical locality in the Web. *Proceedings of the ACM SIGIR 23rd Annual International Conference on Research and Development in Information Retrieval,* 272–279.

Dhyani, D., Keong W., & Bhowmick, S. S. (2002). A survey of Web metrics. *ACM Computing Surveys, 34*(4), 469–503.

Dill, S., Kumar, S. R., McCurley, K., Rajagopalan, S., Sivakumar, D., & Tomkins, A. (2001). Self-similarity in the Web. *Proceedings of the 27th International Conference on Very Large Data Bases,* 69–78.

Dodge, M. (1999). *The geography of Cyberspace.* (CASA Working Paper 8). London: Centre for Advanced Spatial Analysis, University College London. Retrieved November 20, 2003, from http://www.casa.ucl.ac.uk/cyberspace.pdf

Dodge, M., & Kitchin, R. (2001). *Mapping cyberspace.* London: Routledge.

Dorogovtsev, S. N., Mendes, J. F. F., & Samukhin, A. N. (2000) Structure of growing networks with preferential linking. *Physical Review Letters, 85,* 4633–4636.

Duy, J., & Vaughan, L. (2003). Usage data for electronic resources: A comparison between locally-collected and vendor-provided statistics. *Journal of Academic Librarianship, 29*(1), 16–22.

Ebel, H., Mielsch, L.-I., & Bornholdt, S. (2002). Scale-free topology of e-mail networks. *Physical Review E, 66,* 035103(R).

Egghe, L. (2000). New informetric aspects of the Internet: Some reflections—many problems. *Journal of Information Science, 26*(5), 329–335.

Egghe, L., & Rousseau, R. (1990). *Introduction to informetrics: Quantitative methods in library, documentation and information science.* Amsterdam: Elsevier.

Ellis, D., Oldridge, R., & Vasconcelos, A. (2004). Electronic communication and electronic communities. *Annual Review of Information Science and Technology, 38*, 145–186.

Etzioni, O. (1996). The World-Wide Web: Quagmire or gold mine? *Communications of the ACM, 39*(11), 65–68.

Fielding, R., Irvine, U. C., Gettys, J., Mogul, J., Frystyk, H., Masinter, L., et al. (1999). *Hypertext Transfer Protocol -- HTTP/1.1.* Retrieved December 12, 1999, from ftp://ftp.isi.edu/in-notes/rfc2616.txt

Finholt, T. (2002). Collaboratories. *Annual Review of Information Science and Technology, 36*, 73–107.

Flake, G. W., Lawrence, S., Giles, C. L., & Coetzee, F. M. (2000). Efficient identification of Web communities. *Proceedings of the 6th International Conference on Knowledge Discovery and Data Mining*, 150–160.

Flake, G. W., Lawrence, S., Giles, C. L., & Coetzee, F. M. (2002). Self-organization and identification of Web communities. *IEEE Computer, 35*, 66–71.

Foot, K., Schneider, S., Dougherty, M., Xenos, M., & Larsen, E. (2003). Analyzing linking practices: Candidate sites in the 2002 US electoral Web sphere. *Journal of Computer Mediated Communication, 8*(4). Retrieved December 9, 2003, from http://www.ascusc.org/jcmc/vol8/issue4/foot.html

Freeman, L. C. (1977). A set of measures of centrality based upon betweenness. *Sociometry, 40*(1), 35–41.

Furner, J., Ellis, D., & Willett, P. (1996). The representation and comparison of hypertext structures using graphs. In M. Agosti & A. F. Smeaton (Eds.), *Information retrieval and hypertext* (pp. 75–96). Boston: Kluwer.

Garfield, E. (1979). It's a small world after all. *Current contents, 43*, 5–10.

Garner, R. (1967). A computer oriented, graph theoretic analysis of citation index structures. In B. Flood (Ed.), *Three Drexel information science research studies* (pp. 3–46). Philadelphia: Drexel Press. Retrieved November 20, 2003, from http://www.garfield.library.upenn.edu/rgarner.pdf

Garrido, M., & Halavais, A. (2003). Mapping networks of support for the Zapatista movement: Applying Social Network Analysis to study contemporary social movements. In M. McCaughey & M. Ayers (Eds.), *Cyberactivism: Online activism in theory and practice* (pp. 165–184). New York: Routledge.

Garton, L., Haythornthwaite, C., & Wellman, B. (1999). Studying online social networks. In S. Jones (Ed.), *Doing Internet research: Critical issues and methods for examining the Net* (pp. 75–105). Thousand Oaks, CA: Sage Publications.

Gibson, D., Kleinberg, J., & Raghavan, P. (1998). Inferring Web communities from link topology. *Proceedings of the 9th ACM Conference on Hypertext and Hypermedia*, 225–234.

Girardin, L. (1996). Mapping the virtual geography of the World-Wide Web. *Proceedings of the 5th WWW Conference.* Retrieved November 20, 2003, from http://www.girardin.org/luc//cgv/www5/index.html

Glänzel, W. (2001). National characteristics in international scientific co-authorship relations. *Scientometrics, 51*(1), 69–115.

Goodrum, A. A., McCain, K. W., Lawrence, S., & Giles, C. L. (2001). Scholarly publishing in the Internet age: A citation analysis of computer science literature. *Information Processing & Management, 37*(5), 661–676.

Granovetter, M. S. (1973). The strength of weak ties. *American Journal of Sociology, 78*(6), 1360–1380.

Harnad, S., & Carr, L. (2000). Integrating, navigating, and analysing open eprint archives through open citation linking (the OpCit project). *Current Science, 79*(5), 629–638.

Harter, S. P., & Ford, C. E. (2000). Web-based analysis of E-journal impact: Approaches, problems, and issues. *Journal of the American Society for Information Science, 51*(13), 1159–1176.

Heimeriks, G., Hörlesberger, M., & van den Besselaar, P. (2003). Mapping communication and collaboration in heterogeneous research networks. *Scientometrics, 58*(2), 391–413.

Henzinger, M. (2001). Hyperlink analysis for the Web. *IEEE Internet Computing, 5*(1), 45–50.

Henzinger, M. R., Heydon, A., Mitzenmacher, M., & Najork, M. (2000). On near-uniform URL sampling. *Proceedings of the 9th International World Wide Web Conference, Computer Networks, 33*(1–6), 295–308. Retrieved November 20, 2003, from http://www9.org/w9cdrom/88/88.html

Hernández-Borges, A. A., Pareras, L. G., & Jiménez, A. (1997). Comparative analysis of pediatric mailing lists on the Internet. *Pediatrics, 100*(2), e8. Retrieved November 20, 2003, from http://www.pediatrics.org/cgi/content/full/100/2/e8

Herring, S. C. (2002). Computer-mediated communication on the Internet. *Annual Review of Information Science and Technology, 36*, 109–168.

Herring, S. D. (2002). Use of electronic resources in scholarly electronic journals: A citation analysis. *College & Research Libraries, 63*(4), 334–340.

Heydon, A., & Najork, M. (1999). Mercator: A scalable, extensible Web crawler. *World Wide Web, 2*, 219–229.

Hine, C. (2000). *Virtual ethnography.* London: Sage.

Huang, X., Peng, F., An, A., Schuurmans, D., & Cercone, N. (2003). Session boundary detection for association rule learning using n-gram language models. *Proceedings of 16th Canadian Conference on Artificial Intelligence,* 237–251.

Huberman, B. A. (2001). *The laws of the Web: Patterns in the ecology of information.* Cambridge, MA: MIT Press.

Huberman, B. A., & Adamic, L. A. (1999). Growth dynamics of the World-Wide Web. *Nature, 401*, 131.

Huberman, B. A., Pirolli, P. L. T., Pitkow, J. E., & Lukose, R. M. (1998). Strong regularities in World Wide Web surfing. *Science, 280*, 95–97.

Hummon, N. P., & Doreian, P. (1989). Connectivity in a citation network: The development of DNA theory. *Social Networks, 11*, 39–63. Retrieved November 20, 2003, from http://www.garfield.library.upenn.edu/papers/hummondoreian1989.pdf

Ingwersen, P. (1998). The calculation of Web Impact Factors. *Journal of Documentation, 54*(2), 236–243.

Jackson, M. H. (1997). Assessing the structure of communication on the World Wide Web. *Journal of Computer-Mediated Communication, 3*(1). Retrieved November 20, 2003, from http://www.ascusc.org/jcmc/vol3/issue1/jackson.html

Jansen, B. J., & Pooch, U. (2001). Web user studies: A review and framework for future work. *Journal of the American Society of Information Science and Technology, 52*(3), 235–246.

Jansen, B. J., Spink, A., & Saracevic, T. (2000). Real life, real users, and real needs: A study and analysis of user queries on the Web. *Information Processing & Management, 86*(2), 207–227.

Ju-Pak, K. H. (1999). Content dimensions of Web advertising: A cross-national comparison. *International Journal of Advertising, 18*(2), 207–231.

Kessler, M. M. (1963). Bibliographic coupling between scientific papers. *American Documentation, 14*(1), 10–25

Kim, H. J. (2000). Motivations for hyperlinking in scholarly electronic articles: A qualitative study. *Journal of the American Society for Information Science, 51*(10), 887–899.

Kleinberg, J. (1999). Authoritative sources in a hyperlinked environment. *Journal of the ACM, 46*(5), 604–632.

Kleinberg, J., Kumar, R., Raghavan, P., Rajogopalan, S., & Tomkins, A. (1999). The Web as a graph: Measurements, models, and methods. *Lecture Notes on Computer Science, 1627,* 1–18.

Kleinberg, J., & Lawrence, S. (2001). The structure of the Web. *Science, 294,* 1849–1850.

Kling, R., & Callahan, E. (2003). Electronic journals, the Internet, and scholarly communication. *Annual Review of Information Science and Technology, 37,* 127–177.

Kling, R., & McKim, G. (2000). Not just a matter of time: Field differences in the shaping of electronic media in supporting scientific communication. *Journal of the American Society for Information Science, 51*(14), 1306–1320.

Kochen, M. (Ed.). (1989). *The small world.* Norwood, NJ: Ablex.

Koehler, W. (2002). Web page change and persistence: A four-year longitudinal study. *Journal of the American Society for Information Science and Technology, 53*(2), 162–171.

Kosala, R., & Blockeel, H. (2000). Web mining research: A survey. *SIGKDD Explorations, 2*(1), 1–15.

Kumar, R., Raghavan, P., Rajagopalan, S., & Tomkins, A. (1999). Trawling the Web for emerging cyber-communities. *Proceedings of the 8th International World Wide Web Conference,* 403–415.

Landes, W. M., & Posner, R. A. (2000). Citations, age, fame, and the Web. *Journal of Legal Studies, 29*(1), 319–344.

Larson, R. (1996). Bibliometrics of the World Wide Web: An exploratory analysis of the intellectual structure of Cyberspace. *Proceedings of the 59th Annual Meeting of the American Society for Information Science,* 71–78. Retrieved November 20, 2003, from http://sherlock.berkeley.edu/asis96/asis96.html

Lavoie, B., & Nielsen, H. F. (1999). *Web characterization terminology & definitions sheet.* Retrieved November 20, 2003, from http://www.w3.org/1999/05/WCA-terms

Lawrence, S. (2001). Free online availability substantially increases a paper's impact. *Nature, 411*(6837), 521.

Lawrence, S., & Giles, C. L. (1999a). Accessibility and distribution of information on the Web. *Nature, 400,* 107–110.

Lawrence, S., & Giles, C. L. (1999b). Digital libraries and autonomous citation indexing. *IEEE Computer, 32*(6), 67–71.

Lempel, R., & Moran, S. (2002). Introducing regulated bias into co-citation ranking schemes on the Web. *Proceedings of the Annual Meeting of the American Society for Information Science and Technology,* 425–435.

Leydesdorff, L. (2001). *The challenge of scientometrics: The development, measurement, and self-organization of scientific communication* (2nd ed). Parkland, FL: Universal Publishers.

Leydesdorff, L., & Curran, M. (2000). Mapping university-industry-government relations on the Internet: The construction of indicators for a knowledge-based economy. *Cybermetrics, 4*(1). Retrieved November 20, 2003, from http://www.cindoc.csic.es/cybermetrics/articles/v4i1p2.html

Li, X. (2003). A review of the development and application of the Web Impact Factor. *Online Information Review, 27*(6), 407–417.

Li, X., Thelwall, M., Musgrove, P., & Wilkinson, D. (2003). The relationship between the links/Web Impact Factors of computer science departments in UK and their RAE (Research Assessment Exercise) ranking in 2001. *Scientometrics, 57*(2), 239–255.

Liu, B., Ma, Y., & Yu, P. S. (2001). Discovering unexpected information from your competitors' Web sites. *Proceedings of the 7th ACM SIGKDD International Conference on Knowledge Discovery and Data Mining.* Retrieved December 9, 2003, from http://www.comp.nus.edu.sg/~liub/publications/ kdd2001WebComp.ps

Madnick, S., & Siegel, M. (2002). Seizing the opportunity: Exploiting Web aggregation. *MIS Quarterly Executive, 1*(1), 35–46.

Marchiori, M., & Latora, V. (2000). Harmony in the small-world. *Physica A, 285*(3–4), 539–546.

Matzat, U. (1998). Informal academic communication and scientific usage of Internet discussion groups. *Proceedings IRISS '98 International Conference*. Retrieved November 20, 2003, from http://sosig.ac.uk/iriss/papers/paper19.htm

McKiernan, G. (1996). *CitedSites(sm): Citation indexing of Web resources*. Retrieved November 20, 2003, from http://www.public.iastate.edu/~CYBERSTACKS/Cited.htm

Menczer, F. (2002). Growing and navigating the small world Web by local content. *Proceedings of the National Academy of Sciences, 99*(22), 14014–14019.

Menczer, F. (in press). Lexical and semantic clustering by Web links. *Journal of the American Society for Information Science and Technology*.

Merton, R. K. (1968). The Matthew effect in science. *Science, 159*(3810, January 5), 56–63.

Mettrop, W., & Nieuwenhuysen, P. (2001). Internet search engines: Fluctuations in document accessibility. *Journal of Documentation, 57*(5), 623–651.

Meyer, E. K. (2000). Web metrics: Too much data, too little analysis. In D. Nicholas & I. Rowlands (Eds.), *The Internet: Its impact and evaluation: Proceedings of an international forum held at Cumberland Lodge, Windsor Park, 16–18 July 1999* (pp. 131–144). London: Aslib/IMI.

Milgram, S. (1967). The small-world problem. *Psychology Today, 1*(1), 60–67.

Molyneux, R. E., & Williams, R. V. (1999). Measuring the Internet. *Annual Review of Information Science and Technology, 34*, 287–339.

Nance, R. E., Korfhage, R. R., & Bhat, U. N. (1972). Information networks: Definitions and message transfer models. *Journal of the American Society for Information Science, 23*(4), 237–247.

Nicholas, D., Huntington, P., Lievesley, N., & Wasti, A. (2000). Evaluating consumer Website logs: A case study of the Times/The Sunday Times Website. *Journal of Information Science, 26*(6), 399–411.

Nicholas, D., Huntington, P., & Williams, P. (2002) Evaluating metrics for comparing the use of Web sites: Case study of two consumer health Web sites. *Journal of Information Science, 28*(1), 63–75.

Nicholas, D., Huntington, P., & Williams, P, (2003). Micro-mining log files: A method for enriching the data yield from Internet log files. *Journal of Information Science, 29*(5), 401–414.

Nicholas, D., Huntington, P., Williams, P. & Dobrowolski, T. (2004). Re-appraising information seeking behaviour in a digital environment: Bouncers, checkers, returnees and the like. *Journal of Documentation, 60*(1), 24–43.

Notess, G. (2003). *Search engine statistics: Relative size showdown*. Retrieved November 20, 2003, from http://searchengineshowdown.com/stats/size.shtml

OCLC. Office of Research. (2003). *Web characterization*. Retrieved November 20, 2003, from http://wcp.oclc.org

Oppenheim, C. (2000). Do patent citations count? In B. Cronin & H. B. Atkins (Eds.), *The web of knowledge: A Festschrift in honor of Eugene Garfield* (pp. 405–432). Medford, NJ: Information Today.

Otte, E., & Rousseau, R. (2002). Social network analysis: A powerful strategy, also for the information sciences. *Journal of Information Science, 28*(6), 441–454.

Park, H. W., Barnett, G. A., & Nam, I. (2002). Hyperlink-affiliation network structure of top Web sites: Examining affiliates with hyperlink in Korea. *Journal of the American Society for Information Science and Technology, 53*(7), 592–601.

Park, H. W., & Thelwall, M. (2003). Hyperlink analysis: Between networks and indicators. *Journal of Computer-Mediated Communication, 8*(4). Retrieved December 9, 2003, from http://www.ascusc.org/jcmc/vol8/issue4/park.html

Pennock, D. M., Flake, G. W., Lawrence, S., Glover, E. J., & Giles, C. L. (2002). Winners don't take all: Characterizing the competition for links on the Web. *Proceedings of the National Academy of Sciences, 99*(8), 5207–5211.

Peres Vanti, N. A. (2002). Da bibliometria à webometria: Uma exploração conceitual dos mecanismos utilizados para medir a informação e o conhecimento. *Ciência Da Informação, 31*(2), 152–162.

Pirolli, P., Pitkow, J., & Rao, R. (1996). Silk from a sow's ear: Extracting usable structures from the Web. *ACM SIGCHI Conference on Human Factors in Computing Systems.* Retrieved November 20, 2003, from http://www.acm.org/sigchi/chi96/proceedings/papers/Pirolli_2/pp2.html

Pitkow, J. (1998). Summary of WWW characterizations. *Computer Networks and ISDN Systems, 30*(1–7), 551–558.

Polanco, X., Boudourides, M. A., Besagni, D., & Roche, I. (2001). *Clustering and mapping Web sites for displaying implicit associations and visualising networks.* Patras, Greece: University of Patras. Retrieved November 20, 2003, from http://www.math.upatras.gr/~mboudour/articles/Web_clustering&mapping.pdf

Pool, I. d. S., & Kochen, M. (1978). Contacts and influence. *Social Networks, 1,* 5–51.

Price, D. J. D. (1976). A general theory of bibliometric and other cumulative advantage processes. *Journal of the American Society for Information Science, 27*(5), 292–306.

Prime, C., Bassecoulard, E., & Zitt, M. (2002). Co-citations and co-sitations: A cautionary view on an analogy. *Scientometrics, 54*(2), 291–308.

Rasmussen, E. (2003). Indexing and retrieval for the Web. *Annual Review of Information Science and Technology, 37,* 91–124.

Rehm, G. (2002). Towards automatic Web genre identification. *Proceedings of the 35th Hawaii International Conference on System Sciences.* Retrieved January 9, 2003, from http://www.computer.org (electronic restricted access document).

Reid, E. (2003). Using Web link analysis to detect and analyze hidden Web communities. In D. Vriens (Ed.), *Information and communications technology for competitive intelligence* (pp. 57–84). Hilliard, OH: Ideal Group.

Rodríguez Gairín, J. M. (1997). Valorando el impacto de la informacion en Internet: AltaVista, el "Citation Index" de la Red. *Revista Espanola de Documentacion Cientifica, 20,* 175–181.

Ross, N. C. M., & Wolfram, D. (2000). End user searching on the Internet: An analysis of term pair topics submitted to the Excite search engine. *Journal of the American Society for Information Science, 51*(10), 949–958.

Rousseau, R. (1997). Sitations: An exploratory study. *Cybermetrics, 1*(1). Retrieved November 20, 2003, from http://www.cindoc.csic.es/cybermetrics/articles/v2i1p2.html

Rousseau, R. (1999). Daily time series of common single word searches in AltaVista and NorthernLight. *Cybermetrics, 2/3*(1). Retrieved November 20, 2003, from http://www.cindoc.csic.es/cybermetrics/articles/v1i1p1.html

Rudner, L. M., Gellmann, J. S., & Miller-Whitehead, M. (2002). Who is reading on-line education journals? Why? And what are they reading? *D-Lib Magazine, 9*(12). Retrieved November 20, 2003, from http://www.dlib.org/dlib/december02/rudner/12rudner.html

Rusmevichientong, P., Pennock, D. M., Lawrence, S., & Giles, C. L. (2001). Methods for sampling pages uniformly from the Web. *Proceedings of the AAAI Fall Symposium on Using Uncertainty Within Computation,* 121–128.

Sandstrom, P. E. (2001). Scholarly communication as a socioecological system. *Scientometrics, 51*(3), 573–605.

Silverstein, C., Henzinger, M., Marais, H. & Moricz, M. (1999). Analysis of a very large Web search engine query log. *SIGIR Forum, 33*(1), 6–12.

Small, H. (1973). Co-citation in the scientific literature: A new measure of the relationship between two documents. *Journal of the American Society for Information Science, 24*(4), 265–269.

Small, H. (1999). Visualizing science through citation mapping. *Journal of the American Society for Information Science, 50*(9), 799–812.

Smith, A. G. (1999). A tale of two Webspaces: Comparing sites using Web impact factors. *Journal of Documentation, 55*(5), 577–592.

Smith, A. G., & Thelwall, M. (2002). Web impact factors for Australasian universities. *Scientometrics, 54*(3), 363–380.

Snyder, H., & Rosenbaum, H. (1999). Can search engines be used as tools for Web-link analysis? A critical view. *Journal of Documentation, 55*(4), 375–384.

Spink, A., Jansen, B. J., & Ozmutlu, H. C. (2000). Use of query reformulation and relevance feedback by Excite users. *Internet Research, 10*(4), 317–328.

Spink, A., Jansen, B. J., Wolfram, D., & Saracevic, T. (2002). From e-sex to e-commerce: Web search changes. *IEEE Computer, 35*(2), 107–109.

Spink, A., Ozmutlu, S., Ozmutlu, H. C., & Jansen, B. J. (2002). U.S. versus European Web searching trends. *SIGIR Forum, 32*(1), 30–37.

Spink, A., Wolfram, D., Jansen, B. J., & Saracevic, T. (2001). Searching the Web: The public and their queries. *Journal of the American Society for Information Science and Technology, 52*(3), 226–234.

Steyvers, M., & Tenenbaum, J. B. (2001). *The large-scale structure of semantic networks: Statistical analyses and a model for semantic growth.* Retrieved November 20, 2003, from http://arxiv.org/pdf/cond-mat/0110012

Tague-Sutcliffe, J. (1992). An introduction to informetrics. *Information Processing & Management, 28*(1), 1–3.

Tang, R., & Thelwall, M. (2003). Disciplinary differences in US academic departmental Web site interlinking. *Library & Information Science Research, 25*(4), 437–458.

Thelwall, M. (2000a). Commercial Web sites: Lost in cyberspace? *Internet Research, 10*(2), 150–159.

Thelwall, M. (2000b). Web impact factors and search engine coverage. *Journal of Documentation, 56*(2), 185–189.

Thelwall, M. (2000c). Who is using the .co.uk domain? Professional and media adoption of the Web. *International Journal of Information Management, 20*(6), 441–453.

Thelwall, M. (2001a). Exploring the link structure of the Web with network diagrams. *Journal of Information Science, 27*(6), 393–402.

Thelwall, M. (2001b). Extracting macroscopic information from Web links. *Journal of the American Society for Information Science and Technology, 52*(13), 1157–1168.

Thelwall, M. (2001c). The responsiveness of search engine indexes. *Cybermetrics, 5*(1). Retrieved December 9, 2003, from http://www.cindoc.csic.es/cybermetrics/articles/v5i1p1.html

Thelwall, M. (2001d). Results from a Web impact factor crawler. *Journal of Documentation, 57*(2), 177–191.

Thelwall, M. (2001e). A Web crawler design for data mining. *Journal of Information Science, 27*(5), 319–325.

Thelwall, M. (2001f). Web log file analysis: Backlinks and queries. *ASLIB Proceedings, 53*(6), 217–223.

Thelwall, M. (2002a). A comparison of sources of links for academic Web impact factor calculations. *Journal of Documentation, 58*(1), 60–72.

Thelwall, M. (2002b). Conceptualizing documentation on the Web: An evaluation of different heuristic-based models for counting links between university Web sites. *Journal of the American Society for Information Science and Technology, 53*(12), 995–1005.

Thelwall, M. (2002c). Evidence for the existence of geographic trends in university Web site interlinking. *Journal of Documentation, 58*(5), 563–574.

Thelwall, M. (2002d). Methodologies for crawler based Web surveys. *Internet Research, 12*(2), 124–138.

Thelwall, M. (2002e). A research and institutional size based model for national university Web site interlinking. *Journal of Documentation, 58*(6), 683–694.

Thelwall, M. (2002f). Research dissemination and invocation on the Web. *Online Information Review, 26*(6), 413–420.

Thelwall, M. (2002g). The top 100 linked pages on UK university Web sites: High inlink counts are not usually directly associated with quality scholarly content. *Journal of Information Science, 28*(6), 485–493.

Thelwall, M. (2003a). Can Google's PageRank be used to find the most important academic Web pages? *Journal of Documentation, 59*(2), 205–217.

Thelwall, M. (2003b). Web use and peer interconnectivity metrics for academic Web sites. *Journal of Information Science, 29*(1), 11–20.

Thelwall, M. (2003c). What is this link doing here? Beginning a fine-grained process of identifying reasons for academic hyperlink creation. *Information Research, 8*(3). Retrieved November 20, 2003, from http://informationr.net/ir/8-3/paper151.html

Thelwall, M. (2004). Methods for reporting on the targets of links from national systems of university Web sites. *Information Processing & Management, 40*(1), 125–144.

Thelwall, M. (in press a). Can the Web give useful information about commercial uses of scientific research? *Online Information Review, 28*(2).

Thelwall, M. (in press b). Data cleansing and validation for multiple site link structure analysis. In A. Scime (Ed.), *Web mining: Applications and techniques*. Hershey, PA: Idea Group.

Thelwall, M. (in press c). *Scientific Web intelligence: Finding relationships in university Webs*. Wolverhampton, UK: University of Wolverhampton.

Thelwall, M., & Aguillo, I. F. (2003). La salud de las Web universitarias españolas. *Revista Española de Documentación Científica, 26*(3). Retrieved January 29, 2004, from http://www.eicstes.org/EICSTES_PDF/PAPERS/La%20salud%20de%20las%20Web%20 universitarias%20espa%C3%B1olas%20(Thelwall-Aguillo).PDF

Thelwall, M., Binns, R. Harries, G. Page-Kennedy, T. Price, E., & Wilkinson, D. (2002). European Union associated university Websites. *Scientometrics, 53*(1), 95–111.

Thelwall, M., & Harries, G. (2003). The connection between the research of a university and counts of links to its Web pages: An investigation based upon a classification of the relationships of pages to the research of the host university. *Journal of the American Society for Information Science and Technology, 54*(7), 594–602.

Thelwall, M., & Harries, G. (2004a). Can personal Web pages that link to universities yield information about the wider dissemination of research? *Journal of Information Science, 30*(3), 243–256.

Thelwall, M., & Harries, G. (2004b). Do better scholars' Web publications have significantly higher online impact? *Journal of the American Society for Information Science and Technology, 55*(2), 149–159.

Thelwall, M., & Smith, A. (2002). A study of the interlinking between Asia-Pacific university Web sites. *Scientometrics, 55*(3), 363–376.

Thelwall, M., & Tang, R. (2003). Disciplinary and linguistic considerations for academic Web linking: An exploratory hyperlink mediated study with Mainland China and Taiwan. *Scientometrics, 58*(1), 153–179.

Thelwall, M., Tang, R., & Price, E. (2003). Linguistic patterns of academic Web use in Western Europe. *Scientometrics, 56*(3), 417–432.

Thelwall, M., & Vaughan, L. (2004). A fair history of the Web? Examining country balance in the Internet Archive. *Library & Information Science Research, 26*(2), 162–176.

Thelwall, M., Vaughan, L., Cothey, V., Li, X., & Smith, A. (2003). Which academic subjects have most online impact? A pilot study and a new classification process. *Online Information Review, 27*(5), 333–343.

Thelwall, M., & Wilkinson, D. (2003a). Graph structure in three national academic Webs: Power laws with anomalies. *Journal of the American Society for Information Science and Technology, 54*(8), 706–712.

Thelwall, M., & Wilkinson, D. (2003b). Three target document range metrics for university Web sites. *Journal of the American Society for Information Science and Technology, 54*(6), 489–496.

Thelwall, M., & Wilkinson, D. (2004). Finding similar academic Web sites with links, bibliometric couplings and colinks. *Information Processing & Management, 40*(3), 515–526.

Thomas, O., & Willett, P. (2000). Webometric analysis of departments of librarianship and information science. *Journal of Information Science, 26*(6), 421–428.

Valverde, S., Ferrer i Cancho, R., & Solé, R. V. (2002). *Scale-free networks from optimal design*. Retrieved November 20, 2003, from http://arxiv.org/abs/cond-mat/0204344

van Raan, A. F. J. (2000). On growth, ageing, and fractal differentiation of science. *Scientometrics, 47*(2), 347–362.

van Raan, A. F. J. (2001). Bibliometrics and Internet: Some observations and expectations. *Scientometrics, 50*(1), 59–63.

Vaughan, L. (in press). Web hyperlink analysis. In K. Kempf-Leonard (Ed.), *Encyclopedia of Social Measurement*. San Diego, CA: Academic Press.

Vaughan, L. (2004). Web hyperlinks reflect business performance: A study of U.S. and Chinese IT companies. *Canadian Journal of Information and Library Science, 28*(1), 17–31.

Vaughan, L., & Hysen, K. (2002). Relationship between links to journal Web sites and impact factors. *Aslib Proceedings, 54*(6), 356–361.

Vaughan, L., & Shaw, D. (2003). Bibliographic and Web citations: What is the difference? *Journal of the American Society for Information Science and Technology, 54*(14), 1313–1324.

Vaughan, L., & Thelwall, M. (2003). Scholarly use of the Web: What are the key inducers of links to journal Web sites? *Journal of the American Society for Information Science and Technology, 54*(1), 29–38.

Vaughan, L., & Thelwall, M. (in press a). A modeling approach to uncover hyperlink patterns: The case of Canadian universities. *Information Processing & Management*.

Vaughan, L. & Thelwall, M. (in press b). Search engine coverage bias: Evidence and possible causes. *Information Processing & Management*.

Vaughan, L., & Wu, G. (2003). Link counts to commercial Web sites as a source of company information. *Proceedings the 9th International Conference of Scientometrics and Informetrics*, 321–329.

Wang, P., Berry, M., & Yang, Y. (2003). Mining longitudinal Web queries: Trends and patterns. *Journal of the American Society for Information Science and Technology, 54*(8), 743–758.

Watts, D. J. (1999). Networks, dynamics, and the small-world phenomenon. *American Journal of Sociology, 105*(2), 493–527.

Watts, D. J., & Strogatz, S. H. (1998). Collective dynamics of "small-world" networks. *Nature, 393*, 440–442.

Weare, C., & Lin, W. Y. (2000). Content analysis of the World Wide Web: Opportunities and challenges. *Social Science Computer Review, 18*(3), 272–292.

webopedia.com (2003). *World Wide Web*. Retrieved November 20, 2003, from http://www.webopedia.com/TERM/W/World_Wide_Web.html

Wilkinson, D., Harries, G., Thelwall, M., & Price, E. (2003). Motivations for academic Web site interlinking: Evidence for the Web as a novel source of information on informal scholarly communication. *Journal of Information Science, 29*(1), 59–66.

Wilkinson, D., Thelwall, M., & Li, X. (2003). Exploiting hyperlinks to study academic Web use. *Social Science Computer Review, 21*(3), 340–351.

Wolfram, D., Spink, A., Jansen, B., & Saracevic, T. (2001). Vox populi: The public searching of the Web. *Journal of the American Society for Information Science and Technology, 52*(12), 1073–1074.

Yao, Y. Y., Zhong, N., Liu, J., & Ohsuga, S. (2001). Web intelligence (WI): Research challenges and trends in the new information age. *Lecture Notes in Artificial Intelligence, 2198,* 1–17.

Zanasi, A. (1998). Competitive intelligence through data mining public sources. *Competitive Intelligence Review, 9*(1), 44–54.

Zhao, D., & Logan, E. (2002). Citation analysis using scientific publications on the Web as data source: A case study in the XML research area. *Scientometrics, 54*(3), 449–472.

Zhu, J., Hong, J., & Hughes, J. G. (2001). PageRate: Counting Web users' votes. *Proceedings of the 12th ACM Conference on Hypertext and Hypermedia,* 131–132.

Zhu, J., Hong, J., & Hughes, J. G. (2002). Using Markov models for Web site link prediction. *Proceedings of the 13th ACM Conference on Hypertext and Hypermedia,* 169–170.

Zhu, J., Hong, J., & Hughes, J. G. (in press). PageCluster: Mining conceptual link hierarchies from Web log files for adaptive Web site navigation. *ACM Transactions on Internet Technology.*

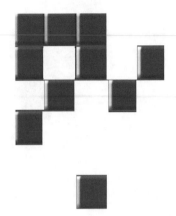

Technology
and Systems

CHAPTER 4

Information Visualization

Bin Zhu
Boston University

Hsinchun Chen
The University of Arizona

Introduction

Advanced technology has resulted in the generation of about one million terabytes of information every year. Ninety-nine percent of this is available in digital format (Keim, 2001). More information will be generated in the next three years than was created during all of previous human history (Keim, 2001). Collecting information is no longer a problem, but extracting value from information collections has become progressively more difficult. Various search engines have been developed to make it easier to locate information of interest, but these work well only for a person who has a specific goal and who understands what and how information is stored. This usually is not the case.

Visualization was commonly thought of in terms of representing human mental processes (MacEachren, 1991; Miller, 1984). The concept is now associated with the amplification of these mental processes (Card, Mackinlay, & Shneiderman, 1999). Human eyes can process visual cues rapidly, whereas advanced information analysis techniques transform the computer into a powerful means of managing digitized information. Visualization offers a link between these two potent systems, the human eye and the computer (Gershon, Eick, & Card, 1998), helping to identify patterns and to extract insights from large amounts of information. The identification of patterns is important because it may lead to a scientific discovery, an interpretation of clues to solve a crime, the prediction of catastrophic weather, a successful financial investment, or a better understanding of human behavior in a computer-mediated environment. Visualization technology shows considerable promise for increasing the value of large-scale collections of information, as evidenced by several commercial applications of TreeMap (e.g., http://www.smartmoney.com) and Hyperbolic tree (e.g., http://www.inxight.com) to visualize large-scale hierarchical structures.

Although the proliferation of visualization technologies dates from the 1990s when sophisticated hardware and software made increasingly faster generation of graphical objects possible, the role of visual aids in facilitating the construction of mental images has a long history. Visualization has been used to communicate ideas, to monitor trends implicit in data, and to explore large volumes of data for hypothesis generation. Imagine traveling to a strange place without a map, having to memorize physical and chemical properties of an element without Mendeleyev's periodic table, trying to understand the stock market without statistical diagrams, or browsing a collection of documents without interactive visual aids. A collection of information can lose its value simply because of the effort required for exhaustive exploration. Such frustrations can be overcome by visualization.

Visualization can be classified as scientific visualization, software visualization, or information visualization. Although the data differ, the underlying techniques have much in common. They use the same elements (visual cues) and follow the same rules of combining visual cues to deliver patterns. They all involve understanding human perception (Encarnacao, Foley, Bryson, & Feiner, 1994) and require domain knowledge (Tufte, 1990).

Because most decisions are based on unstructured information, such as text documents, Web pages, or e-mail messages, this chapter focuses on the visualization of unstructured textual documents. The chapter reviews information visualization techniques developed over the last decade and examines how they have been applied in different domains. The first section provides the background by describing visualization history and giving overviews of scientific, software, and information visualization as well as the perceptual aspects of visualization. The next section assesses important visualization techniques that convert abstract information into visual objects and facilitate navigation through displays on a computer screen. It also explores information analysis algorithms that can be applied to identify or extract salient visualizable structures from collections of information. Information visualization systems that integrate different types of technologies to address problems in different domains are then surveyed; and we move on to a survey and critique of visualization system evaluation studies. The chapter concludes with a summary and identification of future research directions.

Overview of Visualization

History and Background

Although (computer-based) visualization is a relatively new research area, visualization has a long history. For instance, the first known map was created in the 12th century (Tegarden, 1999), and multidimensional representations appeared in the 19th century (Tufte, 1983). Bertin (1967) identified basic elements of diagrams in 1967, and Tufte (1983) published his theory regarding maximizing the density of useful information in

1983. Both Bertin's and Tufte's theories have had substantial impact on subsequent information visualization. Most early visualization research focused on statistical graphics (Card et al., 1999) until the data explosion of the 1980s when supercomputers were able to run complex simulation models and advanced scientific sensors also generated huge quantities of data (Nielson, 1991). Researchers from earth science, physics, chemistry, biology, and computer science turn to visualization for help in analyzing copious data and identifying patterns. The National Science Foundation (NSF) launched its "scientific visualization" initiative in 1985 (McCormick, Defanti, & Brown, 1987) and the Institute of Electrical and Electronic Engineers (IEEE) held its first visualization conference in 1990.

At the same time, visualization technologies were being applied in many nonscientific contexts, including business, digital libraries, human behavior, and the Internet. As the application domains expanded and computer hardware and software became more powerful and affordable, visualization techniques continued to improve. Since 1990, a vast amount of nonscientific data has been generated as a consequence of easy information creation and the emergence of the Internet. The term "information visualization" was first used in Robertson, Card, and Mackinlay (1989) to denote the presentation of abstract information through a visual interface. Early information visualization systems emphasized interactivity and animation (Robertson, Card, & Mackinlay, 1993), interfaces to support dynamic queries (Shneiderman, 1994), and various layout algorithms on a computer screen (Lamping, Rao, & Pirolli, 1995). Later visualization systems presenting the subject hierarchy of the Internet (H. Chen, Houston, Sewell, & Schatz, 1998), summarizing the contents of a document (Hearst, 1995), describing online behaviors (Donath, 2002; Zhu & Chen, 2001), displaying Web site usage patterns (Eick, 2001), and visualizing the structures of a knowledge domain (C. Chen & Paul, 2001) have been stimulated by the networked and virtual nature of human society resulting from the adoption of advanced technologies.

Information visualization is unquestionably an interdisciplinary research field. It integrates the understanding of domain knowledge and human visual perception with computer graphics techniques. It also needs the support of information analysis algorithms (H. Chen et al., 1998). After a decade of focusing on system development, the lack of thorough, summative approaches to evaluating existing visualization systems has become increasingly apparent (C. Chen & Czerwinski, 2000). Special issues of *International Journal of Human-Computer Studies* have demonstrated the level of effort being extended to tackle this issue (C. Chen & Czerwinski, 2000). We believe more disciplines will contribute to visualization research as the technology moves forward and application domains expand.

A Theoretical Foundation for Visualization

Visualization research is important because the human eye can process many visual cues simultaneously. For example, humans can detect a single dark pixel in a 500 x 500 array of white pixels in less than a second. The display can be replaced every second by another, enabling a search of 15 million pixels in a minute (Ware, 2000). Also, people have a truly remarkable ability to recall pictorial images. In one study, Standing, Conezio, and Haber (1970) showed subjects 2,560 pictures, each for 10 seconds over seven hours, in a four-day period. Afterward, subjects were asked to classify pictures presented at a rate of 16 pictures per second and achieved better than 90 percent accuracy. People identify patterns through visual aids but may fail to do so when looking at tables and numbers. However, the human visual system identifies patterns according to its own rules. Because patterns will be invisible if they are not presented in certain ways, understanding visual perception can be helpful in the design of visualization systems. Ware (2000) surveyed perception studies related to visualization. Believing that the best visualization is one that can help problem solving, Ware (2000) sketched a model of human memory by synthesizing studies by Card, Moran, and Newell (1983); Anderson, Matessa, and Lebiere (1997); Kieras and Meyer (1997); and Strothotte and Strothotte (1997). According to Ware (2000), the human memory structure contains iconic, working, and long-term memories, each of which can be enhanced by visualization in a different way.

- *Iconic memory* is the memory buffer where pre-attentive processing operates. This involves a massive number of parallel processes that extract diverse visual cues for every visual point on the interface. Incoming visual information stays in iconic memory for less than a second before part of it is "read out" into working memory. Pre-attentive processing is important to visualization design because certain visual patterns can be detected at this stage without having to go through the cognition process. The theory of visual processing channels and their independent status is fundamental to understanding pre-attentive processing (Ware, 2000). Many studies work with this theory to help make an object visually appealing to viewers. Visual cues such as color and proximity are independent of each other because they are processed in different visual channels. As such, they can be employed independently to convey different attributes. This theory serves as the theoretical foundation for glyph representation (Chernoff, 1973). Other visual cues such as color and luminance can interfere with each other because their visual channels overlap. Gestalt laws (Koffka, 1935) suggest several ways to combine independent and related visual cues to deliver perceivable static patterns. Designing effective visualizations with computer animation also relies on understanding perception of motion patterns.

- *Working memory* integrates information extracted from iconic memory with information loaded from long-term memory for problem solving. Abstract visual patterns perceived by pre-attentive processing are mapped into patterns of the information space at this stage. Working memory holds information for pending tasks; people's attention decides the space allocated to a task. Similar to the RAM (random access memory) of a computer, input and intermediate results of an ongoing operation are stored, but discarded once the task is accomplished. Visualization can augment the working memory in two ways, *memory extension* and visual *cognition extension* (Ware, 2000). The high bandwidth of visual input enables working memory to load external information at the same speed as loading internal memory (Card et al., 1983; Kieras & Meyer, 1997). Visualization thus can serve as an external memory, saving space in the working memory. In addition to memory extension, visualization can facilitate internal computation. Because it makes solutions perceivable (Zhang, 1997), visualization reduces the cognitive load of mental reasoning and mental image construction that is necessary for certain tasks. Interaction with a visual interface can enhance such cognition extension. The best example is a computer aided design (CAD) system's helping an engineer design a product without having to build it.

- *Long-term memory* stores information associated with a lifetime's experiences. It is not just a repository of information; it is a network of linked concepts (Collins & Loftus, 1975; Yufik & Sheridan, 1996). The way this network is built determines whether certain ideas will be easier to recall than others. A sketch of links between concepts is believed to be an effective learning aid for students (Jonassen, Beissner, & Yacci, 1993). Using proximity to represent relationships among concepts in constructing a concept map has a long history in psychology (Shepard, 1962). Visualization systems such as Spatial Paradigm for Information Retrieval and Exploration (SPIRE) (Wise, Thomas, Pennock, Lantrip, Pottier, Schur, et al., 1995) and ET Map (H. Chen et al., 1998) also use proximity to indicate semantic relationships among concepts. Those systems generate from a large collection of text documents a concept map that can help users better understand the collection that is depicted.

In summary, visualization augments iconic memory, working memory, and long-term memory in different ways. Psychologists and neuroscientists have conducted many related studies; a complete survey is beyond the scope of this chapter. Interested readers are referred to Ware

(2000). Most perception studies can be helpful to the design of visualization systems, but converting their results to design principles that can be applied immediately remains a challenge.

Visualization Classification: Application Focus

Visualization is commonly classified based on application focus. Categories usually include scientific visualization, software visualization, and information visualization. These categories are not mutually exclusive and have fuzzy boundaries. For instance, scientific visualization often involves visualizing the multidimensional attribute space of a physical object; this overlaps with information visualization, which delivers patterns embedded in large-scale information collections. Seesoft (Eick, Steffen, & Sumner, 1992) is a system monitoring the change of software code. It has been discussed in books about both information visualization (Card et al., 1999) and software visualization (Stasko, Domingue, Brown, & Price, 1998). The abstract nature of input leads Card et al. (1999) to regard both software visualization and information visualization as information visualization.

Scientific Visualization

Scientific visualization helps scientists and engineers more efficiently understand physical phenomena embedded in large volumes of data (Nielson, 1991). The data may come from complex simulation models or from sensors such as satellites, medical scans, or telescopes. What distinguishes scientific visualization is the fact that it is always about physical objects. This condition provides natural counterparts such as the earth, the human body, the molecule, DNA, or an airplane to which the information can be mapped. Developing mathematical models to describe physical objects plays an essential role in mapping information. Colors or other visual cues are usually added to a physical object to describe different attributes. Isosurfaces, volume rendering, and glyphs are commonly used techniques for the description of attributes in scientific visualization. Isosurfaces depict the distribution of certain attributes. One example is the use of color contours to convey temperature distribution over a map. Volume rendering allows viewers to see the entire volume of 3-D data in a single image (Nielson, 1991). The 3-D data may come from medical magnetic resonance imaging (MRI), CAD, or remote sensing. Interaction between a visual display and its viewers directly affects the effectiveness of a volume-rendering visualization. Glyphs provide a way to display multiple attributes through combinations of various visual cues (Chernoff, 1973). Scientific visualization typically uses glyphs to describe flow information. A commonly used glyph is an arrow (Fayyad, Grinstein, & Wierse, 2002). A map with arrows representing magnitude and direction of wind at a place suggests the movement of air over a geographical area.

In addition to displaying distributions of attributes over a physical object, scientists and engineers also need visual aids to describe relationships among abstract attributes. Techniques used to visualize some of these attributes overlap with those used in information visualization and are discussed in the next subsection.

Software Visualization and Information Visualization

Unlike scientific visualization, software visualization and information visualization usually do not have inherent geometries by which to map information. They share approaches to representing abstract information on a computer screen. For instance, the TreeMap representation has been used to represent a hierarchical relationship in software (Jeffery, 1998), financial data (http://www.smartmoney.com/marketmap), and Usenet messages (Smith & Fiore, 2001). However, each visualization type has its own application focus.

Software visualization helps people understand and use computer software effectively (Stasko et al., 1998). Generally two types of software visualization are used, program visualization and algorithm animation. Program visualization, also positioned as a subfield of software engineering, helps programmers manage complex software (Baecker & Price, 1998). For instance, the Microsoft Windows 95 system has ten million lines of code, for which maintenance can be expensive. Program visualization tackles this problem by visualizing the source code (Baecker & Marcus, 1990), the data structure employed, changes made to the software (Eick et al., 1992), and run-time performance. Program visualization can be an effective tool for software maintenance, understanding, optimization, and debugging. Algorithm animation, on the other hand, is mainly used for education. Starting with the movie *Sorting Out Sorting* (Baecker, 1981), various algorithm animation systems have been developed to motivate and support the learning of computational algorithms.

Information visualization helps users identify patterns, correlations, or clusters. The information visualized can be structured or unstructured. Structured information, usually in numerical format, has well-defined variables. Examples include business transaction data, Internet traffic data, and Web usage data. Visualization of this type focuses on graphical representation to reveal patterns. Early on, standard, static graphics such as line graphs, scatter plots, bar charts, or pie charts were used to enhance understanding of stored data. Widely used commercial tools including Spotfire (http://www.spotfire.com), SAS/GRAPH (http://www.sas.com/technologies/bi/query_reporting/graph), SPSS (http://spss.com), ILOG (http://www.ilog.com), and Cognos (http://www. cognos.com) offer interactive visualizations to help users gain value from structured information.

The recent integration of this type of visualization with various data mining techniques has attracted attention, as huge volumes of data are routinely being generated and stored in databases. Computerized visualizations are vehicles for the delivery of patterns or structures identified by data mining algorithms. Without visualization, such patterns or structures might be too complex to be understandable (Fayyad et al., 2002). Interaction between the visualization system and the user also permits the inclusion of human expertise or feedback in data mining, leading to more effective data exploration. At the same time, data mining algorithms serve as preprocessors, finding appropriate perspectives and dimensions for visualization. Stronger interaction between visualization and data mining algorithms can be found in the systems of Thearling, Becker, and Decoste (2002) and Johnston (2002) where data mining models are visualized to help users understand back-end algorithms. Such interaction is usually employed to facilitate computational steering, a process defined as the ongoing intervention of users in the execution of an otherwise independent computational process (Parker, Johnson, & Beazley, 1997). Incorporating users' skill and expertise, the computational steering approach may improve the efficiency and performance of a data-mining tool.

Unstructured information, on the other hand, usually does not have well-defined variables. Examples of unstructured information include a collection of office documents, a collection of Web sites, or an e-mail archive. Unlike the visualization of structured information, this type of application often needs to identify variables (e.g., titles, locations, subject keywords) and to construct visualizable structures before the graphical representation. Several commercial visualization systems, including Vantage Point (http://www.thevantagepoint.com), SemioMap (http://www.entrieva.com/entrieva), and Knowledgist (http://www.inventionmachine.com), have applied different information analysis technologies to understand the semantics of unstructured information.

In summary, software visualization and information visualization transform data and map information into a visual space differently. But both use similar metaphors to represent abstract information and they adopt similar techniques for user–computer interactions. The next section provides detailed descriptions of those representation and interaction approaches.

A Framework for Information Visualization Technologies

Previous studies have constructed various taxonomies to categorize visualization research from different perspectives. Chuah and Roth (1996) list the tasks of information visualization, and Bertin (1967) and Mackinlay (1986) describe the characteristics of basic visual variables and their applications to different data types. Card and Mackinlay (1997) expand the research of Bertin (1967) and Mackinlay (1986) by constructing a data type-based taxonomy. Based on the features of data domains, the taxonomy divides the visualization field into several categories: scientific visualization, geographic information systems (GIS), multidimensional tables, information landscapes and spaces, nodes and links, trees, and text. Although a taxonomy based on data type may help the implementer select appropriate visualization technologies, the taxonomy Chi (2000) has proposed indicates how to apply these technologies. Chi (2000) breaks the visualization data pipeline into four distinct stages: value, analytic abstraction, visual abstraction, and view. Visualization techniques are thus classified based on the data stage at which they are applied. Chi (2000) contends that there are three types of techniques for transforming data from one stage to the next and four types of technology operating within each stage. The technologies applied at the early data stages extract or construct visualizable structures; those applied at later stages convert these structures into visual metaphors and provide appropriate user–interface interactions. Chi (2000) surveys thirty visualization systems and lists the technologies they apply at different data stages.

Because the development of a visualization system usually integrates several techniques, it may be helpful to provide a framework of visualization technologies based on their functionalities. Shneiderman (1996) identified two aspects of visualization technology that can be directly applied to a given structure. One focuses on mapping abstract information to a visual representation and the other provides user–interface interactions for effective navigation over displays on a screen. To fulfill users' requirements, visualization systems usually combine techniques from these two aspects. However, as indicated by C. Chen (1999), when it comes to visualizing unstructured or high-dimensional information, another set of technologies is needed to create structures that characterize the data set. Along with representation and user–interface interaction, information analysis technology also helps support a visualization system. It serves as a preprocessor, deciding what is to be displayed on a computer screen. Such automatic preprocessing becomes especially critical when manual preprocessing is not possible. The remainder of this section reviews the three research dimensions that support the development of an information visualization system: information representation, user–interface interaction, and information analysis. The framework described in this section is consistent with Chi's (2000) taxonomy, but focuses more on the characteristics of technologies available in each dimension.

Information Representation

Shneiderman (1996) proposed seven types of representation methods: the 1-D, 2-D, 3-D, multidimensional, tree, network, and temporal approaches. We use this framework to review related research.

- The 1-D approach represents abstract information as one-dimensional visual objects and displays them on the screen in a linear (Eick et al., 1992; Hearst, 1995) or a circular (Salton,

Allan, Buckley, & Singhal, 1995) manner. Representation in 1-D has been used to display either the contents of a single document (Hearst, 1995; Salton et al., 1995) or to provide an overview of a document collection (Eick et al., 1992). Colors usually represent some attributes of each visual object. For instance, colors indicate document type in the SeeSoft system (Eick et al., 1992) and depict the location in a document of search terms in TileBars (Hearst, 1995). A second axis may also play a role, presenting some characteristic of each visual object. One example is the SeeSoft system that piles up documents on the x-axis and uses the y-axis to visualize the number of lines in each document. Figure 4.1[1] displays an interface from the TileBars system that shows the occurrence of search terms in documents. The darkness of each tile indicates the frequency of a search term in a document.

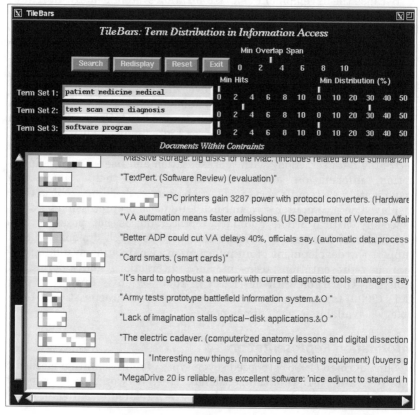

Figure 4.1   TileBars uses a 1-D approach to show term-document relevance (http://www.acm.org/sigchi/chi95/Electronic/documnts/papers/mah_fg4.gif, © 1995 ACM, Inc.). Figure available in color at http://www.asis.org/Publications/ARIST/vol39ZhuFigures.html

- A 2-D approach represents information as two-dimensional visual objects. Visualization systems based on 2-D output of a self-organizing map (SOM) (Kohonen, 1995) belong to this category. Such systems display categories created over a large collection of textual documents, with the layout of each category based on its location in the two-dimensional area of the SOM. Spatial proximity on the interface represents the semantic proximity of the categories created. The challenge in this approach is to help users deal with the large number of categories that will have been created for the mass textual data.

- A 3-D approach represents information as three-dimensional visual objects. One example is the WebBook system (Card, Robertson, & York, 1996) that folds Web pages into three-dimensional books. Realistic metaphors such as rooms (Card et al., 1996), bookshelves (Card et al., 1996), or buildings (Andrews, 1995) are employed to depict abstract information. Visualization systems using a 3-D version of a tree or network representation also belong to this category. One example is the 3-D hyperbolic tree created by Munzner (2000) to visualize large-scale hierarchical relationships. Figures 4.2 and 4.3 show screenshots of WebBook and WebForager, respectively, where the book metaphor is applied to organize Web pages from the same Web site and the WebForager provides a workspace to place books in use.

- The multidimensional approach represents information as multidimensional objects and projects them into a three-dimensional

Figure 4.2   The WebBook (http://acm.org/sigchi/chi96/proceedings/papers/ Card/skc1txt.html, © 1996 ACM, Inc.). Figure available in color at http://www.asis.org/Publications/ARIST/vol39ZhuFigures.html

Figure 4.3   The WebForager (http://www.acm.org/sigchi/chi96/proceedings/
papers/Card/skc1txt.html, © 1996 ACM, Inc.). Figure available in
color at http://www.asis.org/Publications/ARIST/vol39Zhu
Figures.html

or a two-dimensional space. This approach often represents textual documents as a set of key terms that identify the theme of a textual collection. A dimensionality reduction algorithm, such as multidimensional scaling (MDS), hierarchical clustering, k-means algorithms, or principle components analysis, is used to project document clusters or themes that have been sorted into a two-dimensional or three-dimensional space. The SPIRE system presented in Wise et al. (1995) and the VxInsight system (Boyack, Wylie, & Davidson, 2002) belong to this category. Figures 4.4 and 4.5 display two types of visualization developed for the SPIRE system. The Galaxy (Figure 4.4) clusters 567,437 abstracts of cancer literature based on the semantic similarity; the ThemeView (Figure 4.5) visualizes relationships among topics of a document collection. Glyph representation, another type of multidimensional representation, uses graphical objects or symbols to represent data through visual parameters that are spatial (positions x or y), retinal (color and size), or temporal (Chernoff, 1973). It has been used in various social visualization techniques (Donath, 2002) to describe human behavior during computer-mediated communication (CMC).

- The tree approach is often used to represent hierarchical relationships. The most common example is an indented text list. Other tree structure systems include the Tree-Map (Johnson & Shneiderman, 1991), the Cone Tree system (Robertson, Mackinlay, & Card, 1991), and the Hyperbolic Tree (Lamping et

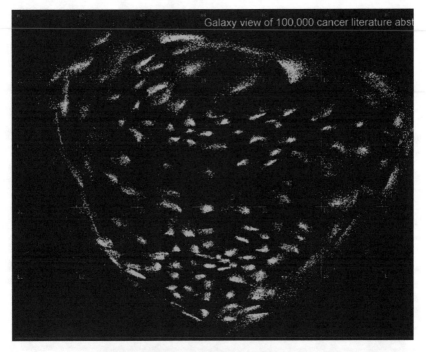

Galaxy view of 100,000 cancer literature abst

Figure 4.4   **Galaxy visualization of text documents (http://www.pnl.gov/ infoviz/gal_cancer800.gif, reprinted with permission). Figure available in color at http://www.asis.org/Publications/ARIST/ vol39ZhuFigures.html**

al., 1995). One crucial challenge to this approach is that the number of nodes grows exponentially as the number of tree levels increases. As a consequence, different layout algorithms have been applied. For instance, the Tree-Map (Johnson & Shneiderman, 1991) allocates space according to attributes of nodes, while the Cone Tree (Robertson et al., 1991) takes advantage of the 3-D visual structure to pack more nodes on the screen. Figure 4.6 displays the visual interface of the Cat-a-Cone system (Hearst & Karadi, 1997) that applies the 3-D Cone Tree to visualize hierarchies in Yahoo!. The Hyperbolic Tree (Lamping et al., 1995), on the other hand, projects subtrees on a hyperbolic plane and puts the plane into the range of display. A 3-D version of the hyperbolic tree has also been developed by Munzner (2000) to visualize large-scale hierarchies (Figure 4.7).

- The network representation method is often applied when a simple tree structure is insufficient for representing complex relationships. Complexity is evident, for example, in citations

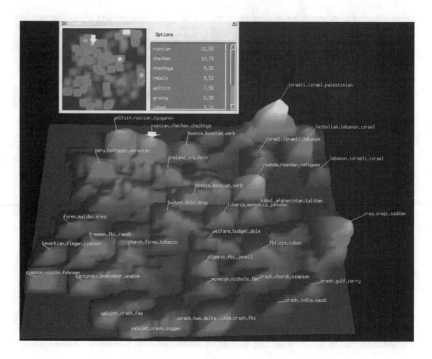

Figure 4.5   ThemeView. The height of a peak indicates the strength of a given topic in the collection of documents (http://www.pnl.gov/ inofviz/theme_cnn800.gif, reprinted with permission). Figure available in color at http://www.asis.org/Publications/ARIST/ vol39ZhuFigures.html

among academic papers (C. Chen & Paul, 2001; Mackinlay, Rao, & Card, 1995) or among textual documents that are distributed over, and linked by, the Internet (Andrews, 1995). Various network visualizations have been created to represent citation relationships (Mackinlay et al., 1995) or to display the World Wide Web (Andrews, 1995). The spring-embedder model, originally proposed by Eades (1984), along with its variants (Davidson & Harel, 1996; Fruchterman & Reingold, 1991), have become the most popular drawing algorithms for network relationships. Figure 4.8 presents the visualization of co-authorship among 555 scientists using a spring-embedder equivalent algorithm.

- The temporal approach visualizes information based on temporal order. Location and animation are two commonly used visual variables to reveal the temporal aspect of information. Visual objects are usually listed along one axis according to the

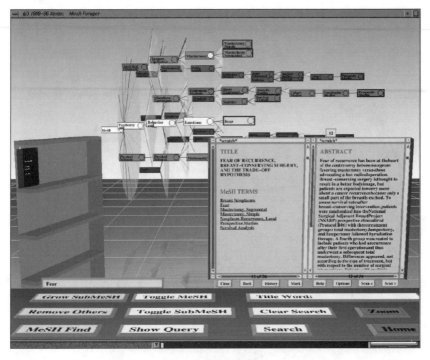

Figure 4.6   Cat-a-Cone tree that displays hierarchies in Yahoo!. The label of
a node can be brought to the foreground with a click (http://
www.sims.berkeley.edu/~hearst/cac-overview.html, © 1997
ACM, Inc.). Figure available in color at http://www.asis.org/
Publications/ARIST/vol39ZhuFigures.html

time when they occurred, while the other axis may be used to
display the attributes of each temporal object (Eick et al., 1992;
Robertson et al., 1993). For instance, the Perspective Wall
(Robertson et al., 1993) lists objects along the x-axis based on
time sequence and presents attributes along the y-axis. Using
animation is another way to display temporal information. In
the VxInsight system (Boyack et al., 2002), the landscape
changes its appearance as a user chooses a different point of
time on a time-slider.

The seven types of representation methods turn abstract textual documents into objects that can be displayed. A visualization system usually
applies several methods at the same time. For instance, there are 2-D
hyperbolic trees and 3-D hyperbolic trees. The multilevel ET map created by H. Chen et al. (1998) combines both 2-D and the tree structure,
where a large set of Web sites is partitioned into hierarchical categories
based on the sites' content. The entire hierarchy is organized in a tree

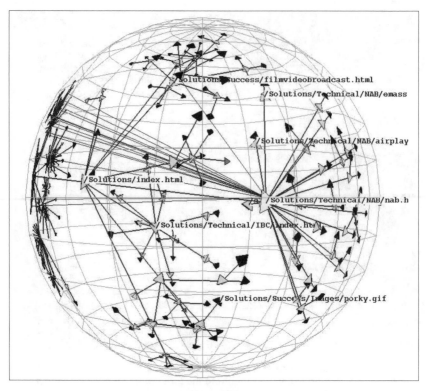

Figure 4.7   A 3-D hyberbolic space (http://graphics.stanford.edu/papers/ munzner_thesis/hyp-figs.html, reprinted with permission of Tamara Munzner). Figure available in color at http://www.asis. org/Publications/ARIST/vol39ZhuFigures.html

structure, and each node in the tree is a two-dimensional SOM on which the subcategories are displayed graphically.

Some representation methods also need to have a precise information analysis technique at the back end. For instance, the TileBar system (Hearst, 1995) employs a text-tiling analysis algorithm to segment a document; alternatively, the ThemeView and Galaxy (Wise et al., 1995) use multidimensional scaling to cluster and lay out documents on the screen.

The "small screen problem" (Robertson et al., 1993) is common to representation methods of any type. To be effective, a representation method needs to be integrated with the user interface. Recent advances in hardware and software allow rapid user–interface interaction, and various combinations of representation methods and user interface interactions have been employed. For instance, Cone Tree (Robertson et al., 1991) applies 3-D animation to provide direct manipulation of visual

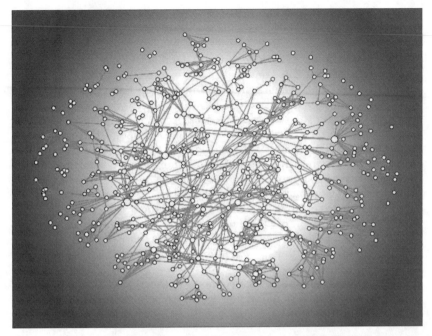

Figure 4.8   **Visualization of a large-scale co-authorship network (http://mpi-fg-koeln.mpg.de:80/~lk/netvis/Huge.html, reprinted with permission of Lothar Krempel). Figure available in color at http://www.asis.org/Publications/ARIST/vol39ZhuFigures.html**

objects, and Lamping et al. (1995) integrate hyperbolic projection with the fish-eye view technique to visualize a large hierarchy.

User–Interface Interaction

Immediate interaction between an interface and its users not only allows direct manipulation of the visual objects displayed, but also allows users to select what is to be displayed and what is not (Card et al., 1999). Shneiderman (1996) summarizes six types of interface functionality: overview, zoom, filtering, details on demand, relate, and history. Techniques have been developed to facilitate various types of interactions and this subsection briefly reviews the two most commonly used interaction approaches: overview + detail and focus + context (Card et al., 1999).

Overview + detail provides multiple views, with the first being an overview: providing overall patterns to users. Details about the part of interest to the user can be displayed. These views may be displayed at the same time or separately. When a detailed view is needed, two types of zooming are usually involved (Card et al., 1999): spatial zooming and semantic zooming. Spatial zooming refers to the process of enlarging selected visual objects to obtain a closer look, whereas semantic zooming provides additional information about a selected visual object by changing its appearance.

The focus + context technique provides detail (focus) and overview (context) dynamically on the same view. One example is the 3-D perception approach adopted by systems such as Information Landscape (Andrews, 1995) and Cone Tree (Robertson et al., 1991), where visual objects at the front appear larger than those at the back. Another commonly used focus + context technique is the fish-eye view (Furnas, 1986), a distortion technique that acts like a wide-angle lens to amplify part of the display. The objective is to simultaneously provide neighboring information in reduced detail and supply greater detail on the region of interest. In any focus + context approach, users can change the region of focus dynamically. A system that applies the fish-eye technique is the Hyperbolic Tree (Lamping et al., 1995), in which users can scrutinize the focus area and scan the surrounding nodes for the big picture. Other focus + context techniques include filtering, highlighting, and selective aggregation (Card et al., 1999).

Overview + detail and focus + context are the two types of interaction usually provided by a visualization system to help users deal with large volumes of information.

Information Analysis

Confronted with large quantities of unstructured information, an information visualization system needs to apply information analysis to reduce complexity and to extract salient structure. Such an application often consists of two stages, *indexing* and *analysis*.

The *indexing* stage aims to extract the semantics of information to represent its content. Different preprocessing algorithms are needed for different media types, including text (natural language processing), image (color, shape, and texture-based segmentation), audio (indexing by sound and pitch), and video (scene segmentation). This subsection briefly reviews selected approaches to textual document processing.

Automatic indexing (Salton, 1989) is a method commonly used to represent the content of each document as a vector of key terms. When implemented using multiword (or multiphrase) matching (Girardi & Ibrahim, 1993), a natural language processing noun-phrasing technique can capture a rich linguistic representation of document content (Anick & Vaithyanathan, 1997). Most noun phrasing techniques rely on a combination of part-of-speech-tagging (POST) and grammatical phrase-forming rules. This approach has the potential to improve precision over other document indexing techniques. Examples of noun-phrasing tools include the Massachusetts Institute of Technology's Chopper, Nptool (Voutilainen, 1997), and the Arizona Noun Phraser (Tolle & Chen, 2000).

Information extraction is another way to identify useful information from text documents automatically. It extracts names of entities of interest, such as persons (e.g., "John Doe"), locations (e.g., "Washington, D.C."), and organizations (e.g., "National Science Foundation") from textual documents. It also identifies other entities, such as dates, times, number expressions, dollar amounts, e-mail addresses, and Web addresses (URLs). Such information can be extracted based on either human-created rules or statistical patterns occurring in the text. Most existing *information extraction* approaches combine machine learning algorithms such as neural networks, decision trees (Baluja, Mittal, & Sukthankar, 1999), hidden Markov models (Miller, Leek, & Schwartz, 1999), and entropy maximization (Borthwick, Sterling, Agichtein, & Grishman, 1998) with a rule-based or a statistical approach. The best systems have been shown to achieve more than 90 percent accuracy in both precision and recall rates when extracting persons, locations, organizations, dates, times, currencies, and percentages from a collection of *New York Times* articles (Chinchor, 1998).

At the *analysis stage, classification,* and *clustering* are commonly used to identify embedded patterns. *Classification* assigns objects into predefined groups (using supervised learning), whereas clustering aggregates objects dynamically based on their similarities (unsupervised learning). Both methods generate groups by analyzing characteristics of objects extracted at the *indexing stage.* Widely used classification methods include the naïve Bayesian method (Koller & Sahami, 1997; Lewis & Ringuette, 1994; McCallum, Nigam, Rennie, & Seymore, 1999), *k*-nearest neighbor (Iwayama & Tokunaga, 1995; Masand, Linoff, & Waltz, 1992), and network models (Lam & Lee, 1999; Ng, Goh, & Low, 1997; Wiener, Pedersen, & Weigend, 1995).

Unlike classification, *clustering* determines groups dynamically. A commonly used clustering algorithm is Kohonen's self-organizing map, which produces a two-dimensional grid representation for *N*-dimensional features and has been widely applied in information retrieval (Kohonen, 1995; Lin, Soergel, & Marchionini, 1991; Orwig, Chen, & Nunamaker, 1997). Other popular clustering algorithms include multidimensional scaling, the *k*-nearest neighbor method, Ward's algorithm (Ward, 1963), and the *K*-means algorithm.

Information analysis represents each textual document with semantically rich phrases or entities (indexes) and identifies interesting patterns by using classification and clustering algorithms. Supporting a visualization system with these methods of analysis enables the system to deal with larger and more complex collections of information.

Emerging Information Visualization Applications

Information visualization can be applied to any domain where people need to extract insights from a vast amount of information. This is evidenced by the publication of several new books. Bederson and Shneiderman (2003) document various applications of visualization developed at the University of Maryland; Börner and Chen (2003) record different visualization applications in the development of digital libraries. In addition, C. Chen (1999) describes many visualization applications in virtual environments. This section explores various approaches to building visualization systems in the domains of digital libraries, the Web, and virtual communities, where large amounts of information are routinely generated.

Digital Library Visualization

Digital library research aims at enhancing information collection by facilitating access to, and the exploration of, stored information. A digital library may contain millions of objects including journal papers, books, maps, photographs, films, videos, and audio recordings. Because standard search engine techniques are no longer sufficient for accessing information in digital libraries, visualization can be applied to support both the browsing and the searching activities of users.

Browsing a Digital Library

Browsing is a way to retrieve information when a user does not have a specific goal (H. Chen et al., 1998; Marchionini, 1987). Visualization supports browsing by providing an effective overview that summarizes the contents of a collection. Interaction techniques are employed to lead a user to information of interest.

Providing a subject hierarchy is a conventional way to help browse information in a digital library. For example, MEDLINE, the largest and most widely used medical bibliographic database in the world, utilizes the vocabulary of the Medical Subject Headings (MeSH) to index its textual documents manually and organizes MeSH terms into 15 hierarchies called the MeSH tree structures (Lowe & Barnett, 1994). A user can traverse the MeSH tree to locate appropriate medical terms. Such a large-scale subject hierarchy can readily become unmanageable because users can easily become lost when scrolling through the headings (Lowe & Barnett, 1994).

Several visualization systems have been developed to display this large-scale hierarchy more effectively. The MeSHBROWSE system (Korn & Shneiderman, 1995) enables users to browse a subset of the MeSH tree interactively. Subcategories for a selected category are displayed, but the two-dimensional tree representation employed suffers from the problem of limited space. To utilize the space on a computer screen more effectively, Hearst and Karadi (1997) proposed using a three-dimensional Cone Tree and animation to display the MeSH tree.

However, being able to display a large-scale hierarchy is not enough. Both the MeSHBROWSE system and the 3-D Cone Tree rely on a MeSH tree that is manually generated. This approach cannot be adopted for other digital libraries unless there is an existing subject hierarchy. In addition, manual generation of a subject hierarchy is not only expensive but also too slow to catch emerging topics in a timely fashion.

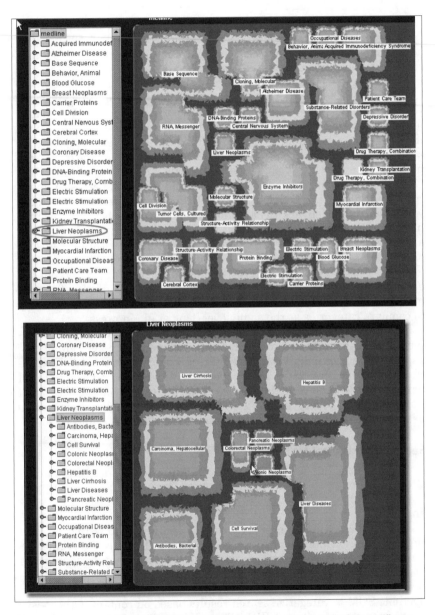

Figure 4.9   The Interface of CancerMap. Category "Liver Neoplasms" was
                    selected at the top level and the submap of "Liver Neoplasms"
                    was displayed. Figure available in color at http://www.asis.org/
                    Publications/ARIST/vol39ZhuFigures.html

The CancerMap system described by H. Chen, Lally, Zhu, and Chau (2003) adopted the SOM and Arizona Noun Phraser (Tolle & Chen, 2000) approaches to generate a subject hierarchy automatically. Figure 4.9 presents two consecutive screen shots, displaying the top-level categories and subcategories under the category of "Liver Neoplasms." The empirical study described by H. Chen et al. (2003) indicates that this approach generated a meaningful subject hierarchy to supplement or enhance human-generated hierarchies in digital libraries. The interface applies the overview + detail approach by combining the 2-D display of SOM with a 1-D text-based alphabetic display. Such a combination appears to be a promising approach to visualizing large-scale subject hierarchies (Ong, Chen, Sung, & Zhu, in press). Users can find a correct path systematically when using a 1-D display. The 2-D SOM map also provides more visual cues and delivers richer information about each node within a hierarchy by using spatial location to illustrate semantic relationships among categories (size for the number of documents within a category and color for the number of levels beneath a category). These features allow easy comparison of categories on the same level. It appears that the best strategy for using the interface is to use the 1-D display for path management when traversing a hierarchy and to use the 2-D SOM map to compare categories at the same level (Ong et al., in press).

In addition to subject hierarchy, other approaches to support browsing behavior have been proposed. If all documents are geo-referenced, users may browse a digital library by geographical locations. A clickable geographic map can serve as an overview (Cai, 2002). Christoffel and Schmitt (2002) built a virtual-reality interface to simulate a real-world library, aiming to provide an environment that would be familiar to users, who could navigate the interface as if walking in a library.

Searching a Digital Library

When a user has a specific goal, searching rather than browsing is often the preferred mode of interaction. Visualization can support searching behavior in two ways: query specification and search results analysis.

Providing a subject hierarchy not only facilitates browsing but also suggests appropriate query terms for searching. Users can combine the terms in the hierarchy to specify their queries. Visualization approaches to providing an overview may also be applied to help users organize search results. For instance, H. Chen, Chau, and Zeng (2002) used dynamic SOM to categorize search results based on content. Other visualization systems such as VIBE (Olsen, Korfhage, & Sochats, 1993) and TileBars (Hearst, 1995) provide visual cues to indicate the extent of match between a document returned and a query term. The VIBE system displays both documents and search terms, with the spatial distance between a document and a term indicating their semantic relationship— the shorter the distance the stronger the relationship. TileBars, on the other hand, uses grayscale colors to indicate the frequency of search terms in a document (Figure 4.1). Visualization can also help users maintain their search results. For example, in Hearst and Karadi (1997), the system organizes documents returned into a book, with the book cover showing the search terms, thereby helping store and manage search results (Figure 4.4). With the proliferation of digital library content and services, we believe that visualization can significantly enhance the value of a digital library by facilitating browsing and searching.

Web Visualization

The vast Web information space has probably become the most dominant information and communication resource for both academic researchers and the general public. Its rapid growth and constant changes also have posed a formidable challenge to visualization research.

Involving both academic and commercial efforts, Web visualization aims to provide a more effective way to access and maintain the Web. Two types of Web visualization, visualization of a single Web site and visualization of a collection of Web sites, will be discussed in the remainder of this subsection.

Visualization of a Single Web Site

The structure of a Web site can be visualized to provide "table of contents" information for effective Web site surfing and maintenance. Most sites have site maps for this purpose, but designing an effective graphical site map remains challenging, especially when a site may contain thousands of pages. A tree metaphor is commonly used to represent the hierarchical structure of a Web site. Visualizations such as the StarTree by InXight Software (http://www.inxight.com), the SiteBrain by Brain Technologies Corporation (http://mappa.mundi.net), and the Z-factor site map of Dynamic Diagrams (http://www.dynamicdiagrams.com) all employ a tree representation but differ in the type of tree used. Visual cues such as color, shape, or icon are applied to describe the attributes of a tree node in the hierarchy. The attributes may include the title of a page, the status of a page (the date of latest update), the type of page (text or image), or usage. A visualization system selects certain attributes for display based on the intended functionality. It may also link nodes with arrows in the tree to describe traffic direction within a Web site (Cugini & Scholtz, 1999). Eick (2001) describes several visual interfaces, all of which use the hyperbolic tree + fish-eye view approach. One interface assigns labels to nodes to construct a site map; another represents a node with a 3-D vertical line indicating the usage of the Web page. In addition, Chi, Pitkow, Mackinlay, Pirolli, Gossweiler, and Card (1998) used several Cone Trees along the x-axis in chronological order to depict the temporal evolution of a Web site. Each Cone Tree presents the usage pattern of the site over a four-week period, with colors to describe the usage of each Web page. Figure 4.10 shows an example of Web site visualization for a company (bestbuy.com) based on a hyperbolic tree.

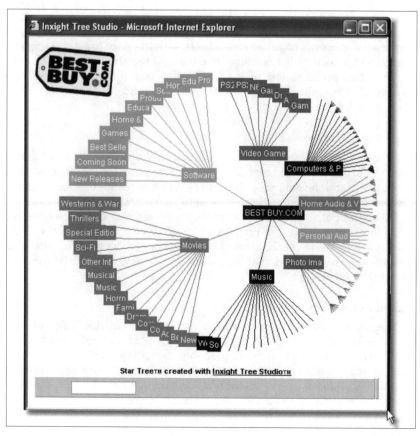

Figure 4.10   A graphical site map. StarTree (by InXight), which applies
hyperbolic tree and fish-eye view algorithms, was used to
visualize a Web site's structure. Colors were used to distinguish sub-trees (http://inxight.com/products/oem/star_tree/
demos.php). Figure available in color at http://www.asis.org/
Publications/ARIST/vol39ZhuFigures.html

Most existing visualization systems for a single Web site apply a tree
representation and use visual cues to describe each page, relying on
users to identify patterns. The challenge faced with this type of visualization is the same as that faced by the tree representation: how can a
very large-scale tree be displayed on a computer screen in an understandable way. Almost no information analysis technology is involved
because the tree metaphor appears to be a natural representation of the
hierarchical structure of a Web site. However, as Web log analysis
becomes more popular for understanding online behavior, visualization
of a single Web site may need to apply information analysis technology

to identify and display patterns embedded in the Web log data. Those patterns may include user demographics, browsing behaviors, and online purchases.

Visualization for a Collection of Web Sites (and Web Pages)

The common goal for visualizing collections of Web sites is to support information exploration over the Internet. Systems like ET map (H. Chen et al., 1998) organize Web pages based on content, applying the output of a self-organizing map to project categories. Other visualization systems organize cyberspace based on the link structure among Web pages (Andrews, 1995; Bray, 1996). Three-dimensional icons are presented on a two-dimensional map, where each icon represents a single Web page or a Web site. The mapping of 3-D icons is based on predefined hierarchical categories (Andrews, 1995) or on the strength of linkages among Web sites (Bray, 1996). Visual cues are supplied to each 3-D icon to represent attributes including size, type, number of incoming/outgoing links, and title of the Web page or Web site. Intensive computation is usually conducted to preprocess Web pages before visualization. ET Map (H. Chen et al., 1998) used automatic indexing to represent the content of a Web page and SOM to generate the subject hierarchy. Bray (1996) calculated links among Web sites to measure the "visibility" (number of links pointing to the site) and the "luminosity" (number of outgoing links) of each Web site.

With the ever-increasing quantity of Web sites and Web content, Web visualization promises to be a fertile ground for information visualization research.

Virtual Community Visualization

The Internet not only opens the door to information foraging but also offers new communication media such as e-mail, discussion groups, news groups, and chat rooms. These new media facilitate communication across geographical and time boundaries, stimulating the formation of virtual communities or new social networks centered on common interests and beliefs. The archives of communication contain rich information about discussion content and participant behavior, information that can be processed and displayed. The proliferation of computer-mediated communication and online communities inevitably poses challenges to people trying to locate a particular person or community, retrieve useful information from an archive, or manage their own communication archives. Many visualization systems have been developed to cope with these issues.

Visualization systems in this area generally belong to one of two categories: tools for communication management and tools for community analysis. ContactMap (Whittaker, Jones, & Terveen, 2002) and Chat Circles (Donath, Rarahalios, & Viega, 1999) belong to the first category. The ContactMap system acts like a visual address book with all contacts displayed on the computer screen as icons. An icon contains a picture and a name. A user can assign an icon to one or more predefined groups and that icon is mapped on the screen according to its groups. Interactions with a contact can be retrieved by a click on its icon. While ContactMap helps people manage their social networks, the Chat Circles system helps users form subgroups in a chat room. It assigns each user a colored circle enclosing text. The user needs to move his or her own circle closer to another circle in order to "speak to" and "hear" that person. Chat Circles 2 offers the capability of tracing the path of a circle in a chat room. Figure 4.11 presents a screen shot of Chat Circles 2 where the local user is "media lab." Hollow circles represent other people far away from the local user and semi-transparent, faded circles show the traces of people who have chatted on that spot before.

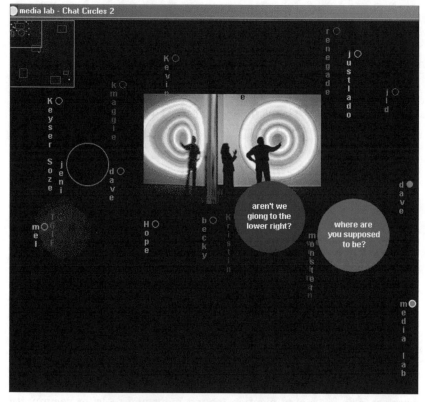

Figure 4.11   Interface of Chat Circles 2 (http://chatcircles.media.mit.edu, reprinted with permission of Judith Donath). Figure available in color at http://www.asis.org/Publications/ARIST/vol39Zhu Figures.html

Both ContactMap and Chat Circles facilitate communication within a community, but users may also need help to identify and to understand a community. Visualization systems such as the Loom (Donath et al., 1999), Conversation Map (Sack, 2000), Netscan Dashboard, and Netscan Treemap visualize the Usenet, the most popular discussion space on the Internet. Both the Loom and Conversation Map apply information analysis technology before visualization. The Loom system uses 2-D representation to describe the temporal patterns of postings in Usenet. Messages are mapped according to the sender and the time of posting. A rule-based algorithm is applied to classify messages into four categories: angry, peaceful, informational, and other. Conversation Map depicts a community by displaying its social and semantic relationships using the network metaphor. Information analysis techniques are applied to construct a semantic network. Message structure and quotation analysis are employed for constructing the social networks.

As part of the Netscan project in Microsoft Research, both Netscan Dashboard and Netscan Treemap use tree representation to describe different aspects of online discussion groups. Netscan Dashboard employs a conventional 2-D tree structure to display the hierarchical structure of a thread, while Netscan Treemap uses Treemap (Shneiderman, 1994) to present hierarchical relationships among Usenet newsgroups. These relationships can be inferred from the name of a newsgroup; the size of a node corresponds with the number of postings in a group.

PeopleGarden (Xiong & Donath, 1999) uses glyphs to summarize the social activity of a community. A flower metaphor is used to represent participants, with the number of petals representing the number of postings by the participant and the height of the flower conveying the length of time that the individual stays. As a community becomes a garden, the overall activity of this community can be seen at a glance. CommunicationGarden combines just such a floral representation with SOM to describe the liveliness of each subtopic within a community and to help locate the most active persons in a certain area. Active participants may not be the most knowledgeable, but will probably be the most helpful. Figures 4.12a, b, and c display the visualization components of the CommunicationGarden system: *Content Summary* (Figure 4.12a), *Interaction Summary* (Figure 4.12b), and *Expert Indicator* (Figure 4.12c). Each type displays a certain aspect of a computer-mediated communication process. In addition, *Content Summary, Interaction Summary,* and *Expert Indicator* divide their display panels into subgardens based on the output of SOM and the Arizona Noun Phraser. Thus each subgarden represents one subtopic.

Evaluation Research for Information Visualization

In spite of a decade of innovative visualization systems development, evaluation research for information visualization is still at an early stage (C. Chen & Yu, 2000). Our literature survey identified two types of

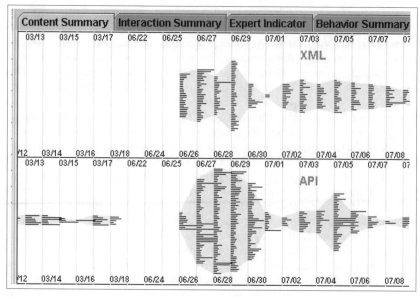

Figure 4.12a   Content Summary. The *x*-axis represents time; categories
generated by the SOM are laid vertically. Each dark line
represents one message. The vertical thickness of each
subtopic indicates its activity on a particular day. The length in
the *x*-dimension of each subtopic represents the time duration
of that subtopic. Figure available in color at http://www.asis.
org/Publications/ARIST/vol39ZhuFigures.html

empirical study in information visualization: 1) empirical usability studies that aim to understand the pros and cons of specific visualization
designs or systems, and 2) fundamental perception studies that try to
investigate basic perceptual effects of certain visualization factors or
stimuli. As the consequence of both the diversity of visualization systems and the relative novelty of computer-based visualization, stringent
metrics-based evaluations such as those adopted in TREC (Text
REtrieval Conference) or MUC (Message Understanding Conference)
(Chinchor, 1998) are nonexistent.

Empirical Usability Studies

Most empirical usability studies employ laboratory experiments to
validate the performance of visualization systems and designs, for example, comparing a glyph-based interface and a text-based interface (Zhu
& Chen, 2001), comparing different visualization techniques (Stasko,
Catrambone, Guzdial, & McDonald, 2000), or studying a visualization
system in a working environment (Graham, Kennedy, & Hand, 2000;
Pohl & Purgathofer, 2000).

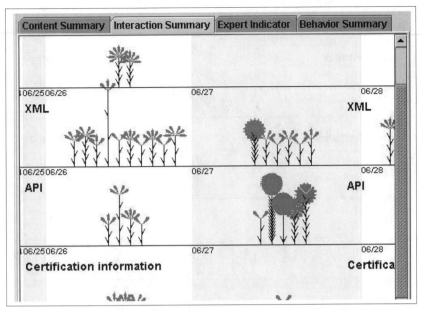

Figure 4.12b   Interaction Summary. The panel is divided into subgardens based on the SOM output. Each subgarden is a subtopic. Each flower represents one thread, where the number of petals represents the number of messages posted for the thread the number of leaves represents the number of participants in the thread and the height of the flowers represents the time duration of the thread. Figure available in color at http://www. asis.org/Publications/ARIST/vol39ZhuFigures.html

Studies such as Stasko et al. (2000), Graham et al. (2000), Morse and Lewis (2000), and Zhu and Chen (2001) use simple but basic visual operations for evaluation. Sometimes referred to as the "de-featuring approach," these studies examine generic operations such as searching objects with a given attribute value, specifying the attributes of an object, clustering objects based on similarity, counting objects, and visual object comparison. Accuracy of operation results and time to completion are two commonly used measures. Taking such an approach would make it easier to design an evaluation study and to attribute the task performance to differences in visualization designs. However, because of the complexity of real-life system interface tasks, the validity of such a design and the applicability of research conclusions are sometimes questioned by practitioners. For example, several studies have been conducted to evaluate popular tree representations such as Hyperbolic Tree (Pirolli, Card, & Van Der Wege, 2000), Treemap (Stasko et al., 2000), multilevel SOM (Ong et al., in press), and Microsoft Windows Explorer. These studies all involve simple visual operations of node searching and node comparison. Representations such as Treemap

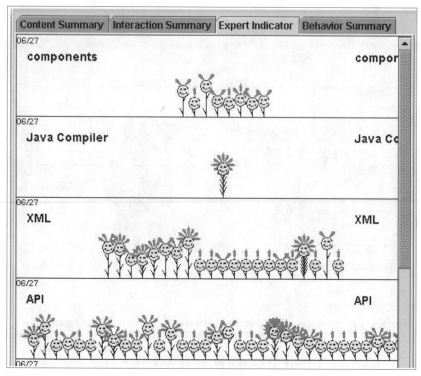

Figure 4.12c   Expert Indicator. The interface is divided into subgardens based on the SOM output. Each subgarden is a subtopic. Each flower represents one person, where the number of petals represents the number of messages posted by this person for this subtopic the number of leaves represents the number of threads participated by this person in the subtopic and the height of flowers represents how long this person has stayed in this subtopic. Figure available in color at http://www. asis.org/Publications/ARIST/vol39ZhuFigures.html

are multilevel SOM effective for node-comparison operations because they offer more visual cues for each node, while hyperbolic tree and Microsoft Windows Explorer, providing a global picture, are more effective in supporting node-searching operations. But how these basic node-searching and node-comparison operations are related to a user's real-life, complex searching or browsing tasks is unclear.

Complex, realistic, task-driven evaluation studies have been conducted frequently in visualization research, for example, Pohl and Purgathofer (2000); Risden, Czerwinski, Munsner, and Cook, (2000); and North and Shneiderman (2000). The experimental tasks are based on functionalities that the visualization system aims to provide. Subjects conduct tasks such as maintaining a hierarchy of subject categories

(Risden et al., 2000), writing a paper (Pohl & Purgathofer, 2000), or selecting appropriate visualization methods to display different information (North & Shneiderman, 2000). The usefulness of a given visualization system can be directly measured by this approach, but it is difficult to identify or isolate the visualization factors that contribute to user performance (partially due to the intertwining nature of the system, task, and user).

Although laboratory experimentation has been useful in information visualization research, we believe other well-grounded behavioral methods such as protocol analysis (to identify qualitative observations and comments), individual and focus group interviews (to solicit general feedback and group responses), ethnographic studies (to record behaviors and organizational cultures), and technology and system acceptance surveys (to understand group or organizational adoption process) also need to be considered. Instead of relying on a one-time, quantitative laboratory experiment, visualization researchers can triangulate and substantiate their findings using qualitative, long-term assessment methodologies.

Fundamental Perception Studies and Theory Building

Unlike empirical usability studies, fundamental perception studies are grounded in psychology and neuroscience. Theories from those disciplines are used to understand the perceptual impact of such visualization parameters as animation (Bederson & Boltman, 1999), information density (Pirolli et al., 2000), 3-D effect (Tavanti & Lind 2001), and combinations of visual cues (Nowell, Schulman, & Hix, 2002). What distinguishes this type of study from conventional perception studies is that it usually involves some form of computer-based visualization. For instance, Bederson and Boltman (1999) used the Pad++ program to study the impact of animation on users' learning of hierarchical relationships; a hyperbolic tree with fish-eye view was applied by Pirolli et al. (2000) to study the effect of information density. Hypotheses, tasks, and measures are developed under the guidance of theories from psychology. However, because of the unique system, task, and perception factor combinations, results may be applied only to the particular visualization system under study. A well-grounded visualization theory and research framework that can be used to guide visualization system development is urgently needed.

Summary and Future Directions

This chapter has reviewed information visualization research based on a framework of information representation, user–interface interaction, and information analysis. We have presented the field's history, theoretical foundations, and three important, emerging application domains: digital libraries, the Web, and virtual communities. We summarized the status of visualization system evaluation research and suggested areas for future research, in particular, long-term, qualitative, theory-grounded evaluation studies.

Although this chapter focuses on the visualization of textual information, many associated techniques can be applied to multimedia visualization. For example, the visualization system described in Christel, Cubilo, Gunaratne, Jerome, O, and Solanki (2002) applied the video indexing and segmentation techniques developed at the Carnegie Mellon University Informedia project (Wactlar, Christel, Gong, & Hauptmann, 1999) to help users browse video digital libraries.

In summary, information visualization can help people gain insights from large-scale collections of unstructured information. Developments in computer hardware and software will not only advance information visualization technology but also stimulate wider adoption. Even though more—and more innovative—visualization systems are expected to be developed soon, there is also a critical need for advancing visualization theories and evaluation.

Lastly, we suggest several promising research areas that could benefit from information visualization research: visual data mining, virtual reality-based visualization, and visualization for knowledge management.

Visual Data Mining

Visual data mining enables users to identify patterns that a data mining algorithm might find difficult to locate. Visualization could play two types of roles in a data mining tool. It could support interaction between users and data—the exploration of an unknown data set. Integrated with such user–interface interaction approaches as zooming or fish-eye view, representation methods such as scatter plot, parallel coordinates, glyphs, and self-organizing maps can be applied to project data (Simoff, 2002). Visualization can also support interaction with the analytical process and output of a data mining system. Such interaction can incorporate human expertise and judgment (Hinneburg, Keim, & Wawryniuk, 1999; Niwa, Fujikawa, Tanaka, & Oyama, 2001; Wong, 1999), which may be critical to the performance of a system but impossible to incorporate in computer code. How to integrate data mining algorithms and various visualization techniques seamlessly in an effective analytical process is still a pressing research challenge.

Virtual Reality-Based (Immersive) Visualization

Although most visualization tools rely on human visual perception to deliver patterns, virtual reality (or immersive) technology tries to take advantage of the entire range of human perceptions, including auditory and tactile sensations. However, in addition to the technological challenges such as input/output devices, virtual reality research still faces many human factors challenges, such as individual differences, input/sensor overload, and cyber-sickness (Kalawsky, 1993; Stanney,

1995). In spite of such challenges, we believe that the current generation, which has grown up with Internet surfing and video games, will be more ready to adopt future virtual reality-based visualization technologies.

Visualization for Knowledge Management

Visualization can support knowledge management by facilitating knowledge sharing and knowledge creation. Knowledge itself is difficult to visualize because it often exists only in someone's mind (referred to as tacit knowledge) (Nonaka, 1994). Visualization can accelerate internalization by presenting information in an appropriate format or structure or by helping users find, relate, and consolidate information (and thus helping to form knowledge) (C. Chen & Paul, 2001; Cohen, Maglio, & Barrett, 1998; Foner, 1997; Vivacqua, 1999). As knowledge management, data mining, and knowledge discovery research advances, we may begin to move from "information visualization" to "knowledge visualization."

Endnote

1. The figures in this chapter are available in color at http://www.asis.org/Publications/ARIST/vol39ZhuFigures.html

References

Anderson, J. R., Matessa, M., & Lebiere, C. (1997). ACT-R: A theory of higher-level cognition and its relation to visual attention. *Human–Computer Interaction, 12,* 439–462.

Andrews, K. (1995). Visualizing cyberspace: Information visualization in the Harmony Internet browser. *Proceedings of InfoVis'95, IEEE Symposium on Information Visualization,* 97–104.

Anick, P. G., & Vaithyanathan, S. (1997). Exploiting clustering and phrases for context-based information retrieval. *Proceedings of the ACM SIGIR Annual International Conference on Research and Development in Information Retrieval,* 314–323.

Baecker, R. (1981). *Sorting out sorting* [Film]. Toronto, Canada: Dynamic Graphics Project, University of Toronto.

Baecker, R., & Marcus, A. (1990). *Human factors and typography for more readable programs.* Reading, MA: Addison-Wesley.

Baecker, R., & Price, B. (1998). The early history of software visualization. In J. Stasko, J. Domingue, M. H. Brown, & B. A. Price (Eds.), *Software visualization: Programming as a multimedia experience* (pp. 29–34). Cambridge, MA: MIT Press.

Baluja, S., Mittal, V., & Sukthankar, R. (1999). Applying machine learning for high performance named-entity extraction. *Proceedings of the Conference of the Pacific Association for Computational Linguistics,* 365–378.

Bederson, B. B., & Boltman, A. (1999). Does animation help users build mental maps of spatial information? *Proceedings of the IEEE Symposium on Information Visualization,* 28–35.

Bederson, B. B., & Shneiderman, B. (2003). *The craft of information visualization: Readings and reflection.* San Francisco: Morgan Kaufmann.

Bertin, J. (1967). *Semiology of graphics: Diagrams, networks, maps.* Madison: University of Wisconsin Press.

Börner, K., & Chen, C. (2003). *Visual interfaces to digital libraries.* Berlin: Springer-Verlag.

Borthwick, A., Sterling, J., Agichtein, E., & Grishman, R. (1998). NYU: Description of the MENE named entity system as used in MUC-7. *Proceedings of the Seventh Message Understanding Conference (MUC-7)*. Retrieved January 3, 2004, from http://www.itl. nist.gov/iad/894.02/related_projects/muc/proceedings/muc_7_toc.html

Boyack, K. W., Wylie, B. N., & Davidson, G. S. (2002). Domain visualization using VxInsight for science and technology management. *Journal of the American Society for Information Science and Technology, 53*(9), 764–774.

Bray, T. (1996). Measuring the Web. *Computer Networks and ISDN systems, 28*(711), 992–1004.

Cai, G. (2002). GeoVIBE: A visual interface for geographic digital libraries. In K. Börner & C. Chen (Eds.), *Visual interfaces to digital libraries* (pp. 161–170). Berlin: Springer-Verlag.

Card, S. K., & Mackinlay, J. D. (1997). The structure of the information visualization design space. *Proceedings of IEEE Symposium on Information Visualization 1997 (InfoVis'97)*, 92–99.

Card, S. K., Mackinlay, J. D., & Shneiderman, B. (1999). *Readings in information visualization: Using vision to think*. San Francisco: Morgan Kaufmann.

Card, S. K., Moran, T. P., & Newell, A. (1983). *The psychology of human-computer interaction*. Hillsdale, NJ: L. Erlbaum.

Card, S. K., Robertson, G. G., & York, W. (1996). The WebBook and the WebForager: An information workspace for the World Wide Web. *Proceedings of the ACM SIGCHI Conference on Human Factors in Computing Systems (CHI'96)*, 111–117.

Chen, C. (1999). *Information visualization and virtual environments*. Berlin: Springer-Verlag.

Chen, C., & Czerwinski, M. P. (2000). Empirical evaluation of information visualizations: An introduction. *International Journal of Human–Computer Studies, 53*, 631–635.

Chen, C., & Paul, R. J. (2001). Visualizing a knowledge domain's intellectual structure. *IEEE Computer, 34*(3), 65–71.

Chen C., & Yu, Y. (2000). Empirical studies of information visualization: A meta-analysis. *International Journal of Human–Computer Studies, 53*, 851–866.

Chen, H., Chau, M., & Zeng, D. (2002). CI Spider: A tool for competitive intelligence on the Web. *Decision Support Systems, 34*(1), 1–17.

Chen, H., Houston, A. L., Sewell, R. R., & Schatz, B. R. (1998). Internet browsing and searching: User evaluation of category map and concept space techniques. *Journal of the American Society for Information Science, 49*(7), 582–603.

Chen, H., Lally, A., Zhu, B., & Chau, M. (2003). HelpfulMed: Intelligent searching for medical information over the Internet. *Journal of the American Society for Information Science and Technology, 54*(7), 683–694.

Chernoff, H. (1973). The use of faces to represent points in K-dimensional space graphically. *Journal of American Statistical Association, 68*(342), 361–368.

Chi, E. H. (2000). A taxonomy of visualization techniques using the data state reference model. *Proceedings of the IWWW Symposium on Information Visualization 2000 (InfoVis'00)*, 69–75.

Chi, E. H., Pitkow, J., Mackinlay, J., Pirolli, P., Gossweiler, R., & Card, S. K. (1998). Visualizing the evolution of Web ecologies. *Proceedings of ACM Conference on Human Factors in Computing Systems (CHI'98)*, 400–407.

Chinchor, N. A. (1998). Overview of MUC-7/MET-2. *Proceedings of the Seventh Message Understanding Conference (MUC-7)*. Retrieved January 3, 2004, from http://www.itl.nist.gov/iad/894.02/related_projects/muc/proceedings/muc_7_toc.html

Christel, M. G., Cubilo, P., Gunaratne, J., Jerome, W., O, E., & Solanki, S. (2002). Evaluating a digital video library Web interface. *Proceedings of the 2nd ACM-IEEE Joint Conference on Digital Libraries*, 389.

Christoffel, M., & Schmitt, B. (2002). Accessing libraries as easy as a game. In K. Börner & C. Chen (Eds.), *Visual interfaces to digital libraries* (pp. 25–38). Berlin: Springer-Verlag.

Chuah, M. C., & Roth, S. F. (1996). On the semantics of interactive visualizations. *Proceedings of IEEE Symposium on Information Visualization 1996 (INFOVIS '96)*, 29–36.

Cohen, A. L., Maglio, P. P., & Barrett, R. (1998). The expertise browser: How to leverage distributed organizational knowledge. *Workshop on Collaborative Information Seeking, Conference on Computer-Supported Cooperative Work (CSCW'98)*. Retrieved January 3, 2004, from http://domino.watson.ibm.com/cambridge/research.nsf/2b4f81291401771 785256976004a8d13/aa5a4c44fb619d3c852566f1006bf633?OpenDocument

Collins, A. M., & Loftus, E. F. (1975). A spreading activation theory of semantic processing. *Psychological Review, 82*, 407–428.

Cugini, J., & Scholtz J. (1999). VISVIP: 3D visualization of paths through Web sites. *Proceedings of the International Workshop on Web-based Information Visualization (WebVis'99)*, 259–263. Retrieved December 12, 2003, from http://www.itl.nist.gov/iaui/vvrg/cugini/webmet/visvip/webvis-paper.html

Davidson, R., & Harel, D. (1996). Drawing graph nicely using simulated annealing. *ACM Transactions on Graphics, 15*(4), 301–331.

Donath, J. (2002). Supporting community and building social capital: A semantic approach to visualizing online conversations. *Communications of the ACM, 45*(4), 45–49.

Donath, J., Rarahalios, K., & Viega, F. (1999). Visualizing conversation. *Journal of Computer-Mediated Communication, 4*. Retrieved December 12, 2003, from http://www.ascusc.org/jcmc/vol4/issue4/donath.html

Eades, P. (1984). A heuristic for graph drawing. *Congressus Numerantium, 42*, 149–209.

Eick, S. G. (2001). Visualizing online activity. *Communications of the ACM, 44*(8), 45–50.

Eick, S. G., Steffen, J. L., & Sumner, E. E. (1992). Seesoft: A tool for visualizing line-oriented software. *IEEE Transactions on Software Engineering, 18*(11), 11–18.

Encarnacao, J., Foley, J. D., Bryson, S., & Feiner, S. K. (1994). Research issues in perception and user interfaces. *IEEE Computer Graphics and Applications, 14*(2), 67–69.

Fayyad, U., Grinstein, G. G., & Wierse, A. (2002). *Information visualization in data mining and knowledge discovery*. San Francisco: Morgan Kaufmann.

Foner, L. N. (1997). Yeta: A multi-agency, referral-based matchmaking system. *Proceedings of International Conference on Autonomous Agents*, 301–307.

Fruchterman, T. M. J., & Reingold, E. M. (1991). Graph drawing by force-directed placement. *Software Practice and Experience, 21*, 1129–1192.

Furnas, G. W. (1986). Generalized fisheye views. *Proceedings of the ACM Conference on Human Factors in Computing Systems*, 16–23.

Gershon, N., Eick, S. G., & Card, S. (1998). Design: Information visualization. *ACM Interactions, 5*(2), 9–15.

Girardi, M. R., & Ibrahim, B. (1993). An approach to improving the effectiveness of software retrieval. *Proceedings of 3rd Annual Irvine Software Symposium*, 89–100.

Graham, M., Kennedy, J., & Hand, C. (2000). A comparison of set-based and graph-based visualizations. *International Journal of Human-Computer Studies, 53*, 789–807.

Hearst, M. (1995). TileBars: Visualization of term distribution information in full text information access. *Proceedings of the ACM SIGCHI Conference on Human Factors in Computing Systems*, 59–66.

Hearst, M. A., & Karadi, C. (1997). Cat-a-Cone: An interactive interface for specifying searches and viewing retrieval results using a large category hierarchy. *Proceedings of the ACM SIGIR Annual International Conference on Research and Development in Information Retrieval*, 246–255.

Hinneburg, A., Keim, D. A., & Wawryniuk, M. (1999, September/October). HD-Eye: Visual mining of high-dimensional data. *IEEE Computer Graphics and Application*, 22–31.

Iwayama, M., & Tokunaga, T. (1995). Cluster-based text categorization: A comparison of category search strategies. *Proceedings of the ACM SIGIR 18th Annual International Conference on Research and Development in Information Retrieval*, 273–281.

Jeffery, C. L. (1998). A menagerie of program visualization techniques. In J. Stasko, J. Domingue, M. H. Brown, & B. A. Price (Eds.), *Software visualization: Programming as a multimedia experience* (pp. 73–79). Cambridge, MA: MIT Press.

Johnson B., & Shneiderman, B. (1991). Tree-maps: A space-filling approach to the visualization of hierarchical information structures. *Proceedings of IEEE Visualization'91 Conference*, 284–291.

Johnston, W. (2002). Model visualization. In U. Fayyad, G. G. Grinstein, & A. Wierse (Eds.), *Information visualization in data mining and knowledge discovery* (pp. 223–228). San Francisco: Morgan Kaufmann.

Jonassen, D. H., Beissner, K., & Yacci, M. A. (1993). *Structural knowledge: Techniques for conveying, assessing, and acquiring structural knowledge*. Hillsdale, NJ: L. Erlbaum.

Kalawsky, R. S. (1993). *The science of virtual reality*. Workingham, UK: Addison-Wesley.

Keim, D. A. (2001). Visual exploration of large data sets. *Communications of the ACM, 44*(8), 39–44.

Kieras, D. E., & Meyer, D. E. (1997). An overview of the EPIC architecture for cognition and performance with application to human–computer interaction. *Human–Computer Interaction, 12*, 391–438.

Koffka, K. (1935). *Principles of Gestalt psychology*. New York: Harcourt-Brace.

Kohonen, T. (1995). *Self-organizing maps*. Berlin: Springer-Verlag.

Koller, D., & Sahami, M. (1997). Hierarchically classifying documents using very few words. *Proceedings of the 14th International Conference on Machine Learning (ICML'97)*, 170–178.

Korn, F., & Shneiderman, B. (1995). *Navigating terminology hierarchies to access a digital library of medical images* (Technical Report HCIL-TR-94-03). College Park, MD: University of Maryland.

Lam, S. L. Y., & Lee, D. L. (1999). Feature reduction for neural network based text categorization. *Proceedings of the International Conference on Database Systems for Advanced Applications (DASFAA '99)*, 195–202.

Lamping, J., Rao, R., & Pirolli, P. (1995). A focus + context technique based on hyperbolic geometry for visualizing large hierarchies. *Proceedings of the ACM SIGCHI Conference on Human Factors in Computing Systems*, 401–408.

Lewis, D. D., & Ringuette, M. (1994). Comparison of two learning algorithms for text categorization. *Proceedings of the Third Annual Symposium on Document Analysis and Information Retrieval (SDAIR'94)*, 81–93.

Lin, X., Soergel, D., & Marchionini, G. (1991). A self-organizing semantic map for information retrieval. *Proceedings of the ACM SIGIR 14th Annual International Conference on Research and Development in Information Retrieval*, 262–269.

Lowe, H. J., & Barnett, G. O. (1994). Understanding and using the Medical Subject Headings (MeSH) vocabulary to perform literature searches. *Journal of the American Medical Association, 271*, 1103–1108.

MacEachren, M. (1991). The role of maps in spatial knowledge acquisition. *The Cartographic Journal, 28*, 152–162.

Mackinlay, J. D. (1986). Automating the design of graphical presentations of relational information. *ACM Transactions on Graphics, 5*, 110–141.

Mackinlay, J. D., Rao, R., & Card, S. K. (1995). An organic user interface for searching citation links. *Proceedings of the ACM Conference on Human Factors in Computing Systems*, 67–73.

Marchionini, G. (1987). An invitation to browse: Designing full text systems for novice users. *Canadian Journal of Information Science, 12*(3), 69–79.

Masand, B., Linoff, G., & Waltz, D. (1992). Classifying news stories using memory based reasoning. *Proceedings of the ACM SIGIR 15th Annual International Conference on Research and Development in Information Retrieval,* 59–64.

McCallum, A., Nigam, K., Rennie, J., & Seymore, K. (1999). A machine learning approach to building domain-specific search engines. *Proceedings of the International Joint Conference on Artificial Intelligence (IJCAI'99),* 662–667.

McCormick, B. H., Defanti, T. A., & Brown, M. D. (1987). Visualization in scientific computing. *Computer Graphics, 21*(6), 1–14.

Miller, A. I. (1984). *Imagery in scientific thought: Creating 20th century physics.* Boston: Birkauser.

Miller, D. R. H., Leek, T., & Schwartz, R. M. (1999). A hidden Markov model information retrieval system. *Proceedings of the ACM SIGIR 22nd Annual International Conference on Research and Development in Information Retrieval (SIGIR '99),* 214–221.

Morse, E., & Lewis, M. (2000). Evaluating visualizations: Using a taxonomic guide. *International Journal of Human–Computer Studies, 53,* 637–662.

Munzner, T. (2000). *Interactive visualization of large graphs and networks.* Unpublished doctoral dissertation, Stanford University.

Ng, H. T., Goh, W. B., & Low, K. L. (1997). Feature selection, perception learning, and a usability case study for text categorization. *Proceedings of the ACM SIGIR 20th Annual International Conference on Research and Development in Information Retrieval,* 67–73.

Nielson, G. M. (1991). Visualization in science and engineering computation. *IEEE Computer, 6*(1), 15–23.

Niwa, T., Fujikawa, K., Tanaka, K., & Oyama, M. (2001). Visual data mining using a constellation graph. In S. J. Simoff, M. Noirhomme-Fraiture, & M. H. Böhlen (Eds.), *Proceedings of the International Workshop on Visual Data Mining.* Retrieved January 3, 2004, from http://www-staff.it.uts.edu.au/~simeon/vdm_pkdd2001/web_proceedings/03_niwa.pdf

Nonaka, I. (1994). A dynamic theory of organizational knowledge creation. *Organization Science, 5*(1), 14–37.

North, C., & Shneiderman, B. (2000). Snap-together visualization: Can users construct and operate coordinated visualizations? *International Journal of Human-Computer Studies, 53,* 715–739.

Nowell, L., Schulman, R., & Hix, D. (2002). Graphical encoding for information visualization: An empirical study. *Proceedings of the IEEE Symposium on Information Visualization (INFOVIS'02),* 43–50.

Olsen, K. A., Korfhage, R. R., & Sochats, K. M. (1993). Visualization of a document collection: The VIBE system. *Information Processing & Management, 29,* 69–81.

Ong, T. H., Chen, H., Sung, W. K., & Zhu, B. (in press). NewsMap: A knowledge map for online news. *Decision Support Systems.*

Orwig, R., Chen, H., & Nunamaker, J. F. (1997). A graphical self-organizing approach to classifying electronic meeting output. *Journal of the American Society for Information Science, 48*(2), 157–170.

Parker, S., Johnson, C., & Beazley, D. (1997). Computational steering software systems and strategies. *IEEE Computational Science & Engineering, 4*(4), 50–59.

Pirolli, P., Card, S. K., & Van Der Wege, K. M. (2000). The effect of information scent on searching information visualizations of large tree structures. *Proceedings of the Working Conference on Advanced Visual Interfaces (AVI 2000),* 161–172.

Pohl, M., & Purgathofer, P. (2000). Hypertext authoring and visualization. *International Journal of Human–Computer Studies, 53,* 809–825.

Risden, K., Czerwinski, M. P., Munsner, T., & Cook, D. D. (2000). An initial examination of ease of use for 2D and 3D information visualizations of Web content. *International Journal of Human–Computer Studies, 53,* 695–714.

Robertson, G. G., Card, S. K., & Mackinlay, J. D. (1989). The cognitive co-processor for interactive user interfaces. *Proceedings of UIST'89, ACM Symposium on User Interface Software and Technology*, 10–18.

Robertson, G. G., Card, S. K., & Mackinlay, J. D. (1993). Information visualization using 3D interactive animation. *Communications of the ACM, 36*(4), 56–71.

Robertson, G. G., Mackinlay, J. D., & Card, S. K. (1991). Cone Trees: Animated 3D visualizations of hierarchical information. *Proceedings of the ACM SIGCHI Conference on Human Factors in Computing Systems*, 189–194.

Sack, W. (2000). Conversation Map: A content-based Usenet newsgroup browser. *Proceedings of the 5th International Conference on Intelligent User Interfaces*, 233–240.

Salton, G. (1989). *Automatic text processing*. Reading, MA: Addison-Wesley.

Salton, G., Allan, J., Buckley, C., & Singhal, A. (1995). Automatic analysis, theme generation, and summarization of machine-readable text. *Science, 264*(3), 1421–1426.

Shepard, R. N. (1962). The analysis of proximities: Multidimensional scaling with unknown distance function: Part I. *Psychometrika, 27*(2), 125–140.

Shneiderman, B. (1994). Dynamic queries for visual information seeking. *IEEE Software, 11*(6), 70–77.

Shneiderman, B. (1996). The eyes have it: A task by data type taxonomy for information visualization. *Proceedings of IEEE Workshop on Visual Languages*, 336–343.

Simoff, S. J. (2002). VDM@ECML/PKDD2001: The International Workshop on Visual Data Mining at ECML/PKDD 2001. *ACM SIGKDD Explorations Newsletter, 3*(2), 78–81.

Smith, M., & Fiore A. T. (2001). Visualization components for persistent conversation. *Proceedings of the ACM SIGCHI Conference on Human Factors in Computing Systems (CHI'01)*, 136–143.

Standing, L., Conezio, I., & Haber, R. N. (1970). Perception and memory for pictures: Single trial learning of 2500 visual stimuli. *Psychonomic Science, 19*(2), 73–74.

Stanney, K. (1995). Realizing the full potential of virtual reality: Human factors issues that could stand in the way. *Proceedings of IEEE Virtual Reality Annual International Symposium (VRAIS'95)*, 28–34.

Stasko, J., Catrambone, R., Guzdial, M., & McDonald, K. (2000). An evaluation of space-filling information visualizations for depicting hierarchical structures. *International Journal of Human–Computer Studies, 53*, 663–695.

Stasko, J., Domingue, J., Brown, M. H., & Price, B. A. (1998). *Software visualization: Programming as a multimedia experience*. Cambridge, MA: MIT Press.

Strothotte, C., & Strothotte, T. (1997). *Seeing between the pixels*. Berlin: Springer-Verlag.

Tavanti, M., & Lind, M. (2001). 2D vs. 3D: Implications on spatial memory. *Proceedings of the IEEE Symposium on Information Visualization (INFOVIS'01)*, 139–148.

Tegarden, D. P. (1999). Business information visualization. *Communications of the Association for Information Systems, 1*, Article 4. Retrieved January 29, 2004, from http://cais.isworld.org/articles/default.asp?vol=1&art=4

Thearling K., Becker B., & Decoste, D. (2002). Visualizing data mining models. In U. Fayyad, G. G. Grinstein, & A. Wierse (Eds.), *Information visualization in data mining and knowledge discovery* (pp. 205–222). San Francisco: Morgan Kaufmann.

Tolle, K. M., & Chen, H. (2000). Comparing noun phrasing techniques for use with medical digital library tools. *Journal of the American Society for Information Science, 51*(4), 352–370.

Tufte, E. R. (1983). *The visual display of quantitative information*. Cheshire, CT: Graphics Press.

Tufte, E. R. (1990). *Envisioning information*. Cheshire, CT: Graphics Press.

Vivacqua, A. S. (1999). Agents for expertise location. *Proceedings of the AAAI Spring Symposium on Intelligent Agents in Cyberspace*, 9–13.

Voutilainen, A. (1997). *A short introduction to Nptool*. Retrieved December 12, 2003, from http://www.lingsoft.fi/doc/nttool/intro

Wactlar, H. D., Christel, M. G., Gong, Y., & Hauptmann, A. G. (1999). Lessons learned from the creation and deployment of a terabyte digital video library. *IEEE Computer, 32*(2), 66–73.

Ward, J. (1963). Hierarchical grouping to optimize an objection function. *Journal of the American Statistical Association, 58*, 236–244.

Ware, C. (2000). *Information visualization perception for design*. San Francisco: Morgan Kaufmann.

Whittaker, S., Jones, Q., & Terveen, L. (2002). Managing long term communications: Conversation and contact management. *Proceedings of the 35th Annual Hawaii International Conference on System Sciences*, 115b.

Wiener, E., Pedersen, J. O., & Weigend, A. S. (1995). A neural network approach to topic spotting. *Proceedings of the 4th Annual Symposium on Document Analysis and Information Retrieval (SDAIR'95)*, 23–34.

Wise, J. A., Thomas, J. J., Pennock, K., Lantrip, D., Pottier, M., Schur, A., & Crow, V. (1995). Visualizing the non-visual: Spatial analysis and interaction with information from text documents. *Proceedings of InfoVis'95, IEEE Symposium on Information Visualization*, 51–58.

Wong, P. C. (1999, September/October). Visual data mining. *IEEE Computer Graphics and Application*, 2–3.

Xiong, R., & Donath, J. (1999). Creating data portraits for users. *Proceedings of the 12th Annual ACM Symposium on User Interface Software and Technology*, 37–44.

Yufik, Y. M., & Sheridan, T. B. (1996). Virtual networks: New framework for operator modeling and interface optimization in complex supervisory control systems. *Annual Reviews in Control, 20*, 179–195.

Zhang, J. (1997). The nature of external representations in problem solving. *Cognitive Science, 21*(2), 179–217.

Zhu, B., & Chen, H. (2001). Social visualization for computer-mediated communication: A knowledge management perspective. *Proceedings of the Eleventh Workshop on Information Technologies and Systems (WITS '01)*, 23–28.

CHAPTER 5

Bioinformatics

Gerald Benoît
Simmons College

Introduction

This is the first review of bioinformatics to appear in *ARIST*; it is written primarily for computer and information scientists interested in exploring how information retrieval (IR), visualization, data modeling, algorithms, and Web-based resources might contribute to biology. It is written secondarily for biologists curious about the computer-based technologies and data resources that have been created to support their research.

The first goal of this review is to foster an appreciation of the very large amounts and specific properties of detailed data about humans and other species that exist. It also presents a range of bioinformatics applications relevant to molecular biology, clinical medicine, pharmacology, biotechnology, and associated disciplines. The review also suggests how a knowledge of computer science and information science techniques as applied to Internet-based databases can further research in many areas of biology.

For computer and information scientists, the field of molecular biology may be puzzling, but the range of products that it has created to address the needs of biology is impressive. While simultaneously conducting increasingly specialized research, biologists, in concert with computer scientists, are developing tools to integrate a wide range of data types. For example, data models are highly developed; programming applications and tools for Perl, Corba, and Java have evolved very quickly; specialized visualization techniques, too, are available for a range of data processing needs, from interactive information retrieval, information visualization, and data mining to sequence data and protein folds. The impetus for collaboration between biology and computer and information science resides in managing the large quantities of data necessary to reveal biological relationships and in using innovative techniques to locate, aggregate, manipulate, and present such data through user-friendly, cross-platform applications.

Wide-ranging though these efforts are, they can be categorized into two groups: data management and biological analysis. The first area involves manipulating files and strings of data, specifically strings representing

DNA (deoxyribonucleic acid) or proteins, making bioinformatics seem very much like an IR and data mining activity. There are also IR systems to aggregate and provide annotations for chemical sequences of DNA and to present the results through browsers. Biological analysis seeks to make data representing the structure and function of genomic sequences more comprehensible to research biologists. There are many tools already available for data management and manipulation, and these form the body of this review. But the challenge for biology, and the opportunity for computer and information science, lies in exploiting data to provide biologists with the tools for exploring increasingly sophisticated research questions.

This review follows the recommendation (Altman & Koza, 1996; Bayat, 2002; Denn & MacMullen, 2002) that the desired fusion of biology and computing—bioinformatics—is best served by providing computer and information scientists with the fundamentals of biology and simultaneously offering biologists enough technical knowledge to understand how computer technology can facilitate and augment their work. However, the field evolves daily, making defining the field and demonstrating the work of bioinformatics a challenge. Nevertheless, the benefits that bioinformatics offers society through a deeper understanding of genes, disease, and drug treatments warrant the involvement of as many researchers as possible from both domains. The intersection of skills creates a common foundation: a facility with using the Internet; knowledge of databases of sequence and structure data; specific biological knowledge of sequence analysis, sequence alignment, and phylogenetic analysis; and an understanding of data models, information retrieval, and predictive methods (based on Lesk, 2002). A good starting point is a data-centric synopsis of how records are created and manipulated by biologists.

The Basics of Biology and the Computer Files Produced Through Research

In brief, bioinformatics is the gathering of data about DNA; protein sequences and structures; genomes and proteomes; and the storage of these data in local, commercial, and freely accessible, Internet-enabled databases. Molecular biology generates tremendous amounts of data from DNA and protein sequences, macromolecular structures, and the results of functional genomics, all of which need to be made useful—scientifically sound and able to be interpreted by subject specialists—in a computationally efficient manner. Researchers need to know the nature of individual genomes, or the genetic material in the chromosomes of an organism, and their relationships. Early research efforts resulted in data that were stored in flat files and queried using tools such as Fasta and PSI-Blast for comparing protein sequences (Altshul, Madden, Schaffer, Zhang, & Zhang 1997; Korf, Yandell, & Bedell, 2003). Knowledge of how and why the data are gathered and modeled as they are, and how the research literature is being integrated into the biologist's toolkit, will suggest opportunities for computer and information scientists to further meet the information and processing needs of the field.

DNA Structure and Sequencing

The genetic material of all living organisms, the substance of heredity, is DNA. In 1953, British scientists James Watson and Francis Crick used X-ray crystallography to show that DNA took the form of a double helix of molecules in pairs of chemical bases held together by weak bonds (Watson & Crick, 1953). The four chemical bases are purines (adenine, abbreviated [A], and guanine [G]) and pyrimidines (cytosine [C] and thymine [T]). The bases form pairs only between A and T and between G and C, so one can deduce the base sequence of each single strand from its partner. The combination of oxygen, carbon, nitrogen, and hydrogen differs for each base. Each base is attached to a "deoxyribose," or sugar molecule, and to a phosphate molecule, creating a nucleotide. The nucleotides are linked in a certain order, or sequence, through the phosphate group. The precise order and linking of the bases within the DNA (i.e., the genotype) determines what proteins that gene produces and ultimately the phenotype of the organism.

In the simplest sense, a gene is a linear sequence of nucleotides that encode information for the corresponding linear sequence of amino acids that form a protein. The information in DNA is first transcribed to messenger RNA (ribonucleic acid) (mRNA), which is then decoded by ribosomes and other factors, translating the nucleotide code into amino acid code. The linear amino acid sequence folds into a three-dimensional conformation for a functional protein.

Colinearity of DNA and protein code is not exact because the nucleotide code is often interrupted by introns, segments that are removed from the mRNA. Thus, in most eukaryotic genes, the final protein code is created by the juxtaposition of exons (expressed segments). (Eukaryotic refers to nonbacterial, nonviral organisms; introns are DNA sequences that interrupt the protein-coding sequence of a gene and are transcribed into RNA but cut out of the message before it is translated into protein; exons are the protein-coding DNA sequence of a gene.)

The transcriptome refers to all the mRNA present in a cell (genes that are turned on), whereas the proteome comprises all the proteins that are made. There may not be a correspondence because gene expression can be regulated at various levels.

Biologists may look for functional clustering (e.g., based on metabolic pathways or sequence segments), or relationships between proteins, such as homologous (structurally and sequentially similar) or analogous proteins (related folds). The volume and heterogeneity of the data require the creation of algorithms for basic analysis, such as protein sequence analysis (Miller, Gurd, & Bass, 1999), and the uncovering of introns and exons (Boguski, 1999).

Some genome projects try to identify the small regions of DNA that vary between individuals. These differences may underlie disease susceptibility and drug responsiveness, particularly the most common variations that are called SNPs (single nucleotide polymorphism) (Human Genome Project Information, 2002). Other genome projects focus on nonhuman DNA sequences, such as plant genetics (Lim, 2002) and biotechnology (Chawla, 2002).

To explain how biologists generate data, Table 5.1 (National Center for Biotechnical Information, 2003) details the sequencing process:

Table 5.1 Sequencing process

Mapping	Identify set of clones that span the region of the genome to be sequenced
Library creation	Make sets of smaller clones from mapped clones
Template preparation	Purify the DNA and perform sequence chemistries
Gel electrophoresis	Determine sequences from smaller clones
Pre-finishing and finishing	Apply special techniques to produce high quality sequences
Data editing/annotation	Quality control, verification, biological annotation, submission to public databases

Sequencing and Analysis

First, chromosomes (as many as a couple of hundred million bases) are divided into smaller pieces, or "subcloned." A template is created from the shorter pieces to generate fragments, each differing only by a single base; this changing of the base is the mutagenesis. That single base is used as an identifier during template preparation of sequence reaction. Using florescent dyes, the fragments can be identified by color when they are separated by a process called gel electrophoresis. The base at the end of each fragment is now identified ("base-calling") to help re-create the original sequences of A, T, C, and G for each subcloned piece. A four-color histogram (chromatograph) is created to show the presence and location for each of the bases. Finally, the short sequences in blocks of about 500 bases (called the "read length") are assembled by computer into long, continuous stretches for analysis of errors, gene-coding regions, and other distinctions.

Microarrays

"Sequencing" is the determining of the order of nucleotides (the base sequences) in DNA or RNA or the order of amino acids in a protein. Analyzing these sequences provides the functional identification of genes. The analysis of whole genomes can be performed through DNA microarrays. This technique, described in Kohane, Kho, and Butte (2003) and in Bowtell and Sambrook (2003), is applied to many tasks, such as high-throughput genotyping, comparative genomic hybridization, monitoring of gene expression, and detection of single nucleotide polymorphisms. Such procedures can help determine the effect of gene expression, map disease loci, demonstrate chromosomal aberrations, and categorize tumor expression patterns.

Protein Structure and Sequencing

Proteins are made up of sequences of amino acids; to date approximately 400,000 protein sequences are known. There are 20 different amino acids, which, depending on their arrangement, can create larger macromolecular structures, such as the 51 amino acids that form insulin. Various techniques, such as X-ray crystallography and nuclear magnetic resonance (NMR), generate three-dimensional coordinate data (x-y-z), which are stored in an appropriately themed database, such as the Protein Data Bank (PDB) (Berman et al., 2000; Bernstein et al., 1977), for manipulation by programs (e.g., Orengo, 1999; Orengo & Taylor 1996).

Examples of Data Generation and Publication Projects

The Human Genome Project (HGP) is an example of a well known molecular biology project that generates data stored in publicly accessible, Internet-enabled databases. This project's original goal was to reveal all human genes, once estimated to be as many as 100,000. The human genome is built from almost 3 billion bases (Lander et al., 2001; Venter et al., 2001). With the completion of the sequencing of the human genome (i.e., determining the exact order of the genes in the chromosomes, described by the total number of base pairs), it turns out humans have approximately 21,000 genes (Goodman, 2003). However, the number of possible sequences of DNA pairs (about 3 billion) and 20 amino acids translates into a huge number of possible combinations in DNA and proteins, which helps explain why the data sets are so large.

Other data-generating projects include research into specific organisms' or cell types' gene expression (measuring mRNA produced in cells under different conditions) (Cheung, Dalrymple, Narasiman, Watts, Schuler, & Raap, 1999; Duggan, 1999; Eisen & Brown, 1999; Lipshutz, Fodor, Gingeras, & Lockhard, 1999). Still others concentrate on systems, such as metabolic pathways, regulatory networks, and protein-protein interaction data from 2-hybrid experiments.

The results of sequencing and expression research are released in publicly accessible databases, such as the National Center for Biotechnology Information's (NCBI) GenBank, which holds over 12 billion bases in 11.5 million entries (Benson, Karsch-Mizrachi, Lipman, Ostell, Rapp, & Wheeler, 2000). NCBI, a part of the National Library of Medicine (NLM) and National Institutes of Health (NIH), hosts PubMed, GenBank, molecular sequencing databases, and research literature databases. Researchers may also store the sequence data in other very large, multidimensional databases, such as the Wisconsin Package or Genetics Computer Group (GCG) sequencers. Ultimately, results appear in professional communication forums, intended to teach bioinformatics and to keep scientists up to date. Printed and online journals and conferences are accessible online (e.g., through Medline or the Science Citation Index).

The variety of data uses and formats, coupled with the desire for integrating resources through a single interface, raises the question of data integration (Gerstein, 2000; Gerstein & Jansen, 2000). Molecular biology's information needs (e.g., as described in Jiang, Xu, & Zhang, 2002) can be classified into three groups: raw data, data integrated with bibliographic systems and other data reflecting biological processes, and data used predictively to solve specific questions (Wilson, Kreychman, & Gerstein, 2000).

Successfully describing, retrieving, and presenting data from different sources is one aspect of bioinformatics; another is the manipulation of data in biologically meaningful ways. The problems facing scientists in both fields include how to address data redundancy and multiplicity in very large databases and how to integrate multiple data sources where different nomenclature and file formats are common. To understand how some of the differences arose, we examine some of the many definitions of bioinformatics. The review then considers specific genome projects, the myriad databases, and the techniques to retrieve and present these specialized data.

Bioinformatics Defined

Bayat (2002, p. 1019) broadly defines the discipline as "the application of tools of computation and analysis to the capture and interpretation of biological data." Operationally, "the main tools of a bioinformatician are computer software programs and the Internet. A fundamental activity is sequence analysis of DNA and proteins using various programs and databases available on the World Wide Web."

The National Center for Biotechnical Information (2003, online) also defines bioinformatics as a single broad discipline, but with three important subdisciplines:

- The development of new algorithms and statistics with which to assess relationships among members of large data sets

- The analysis and interpretation of various types of data including nucleotide and amino acid sequences, protein domains, and protein structures

- The development and implementation of tools that enable efficient access to and management of different types of information

Ellis (2003a) notes 40 published, operational definitions between 2000 and 2001 and another 37 (Ellis, 2003b) in 2003, suggesting the definitions vary by subdiscipline. It appears that several specialties were working out for themselves their roles in molecular biology (Ibba, 2002) and the influence of computational biology on their work. For example, bioinformatics is described in terms of: education (Brass, 2000; Ouzounis, 2002; Pearson, 2001; Sander, 2002), employment and retooling (Brass, 2000; Gardiner, 2001; Henry, 2002; Zauhar, 2001), computing (Sansom & Smith, 2000; Watkins, 2001), and genomics and proteomics (Zimmerman, 2003); others present subdomains or overviews of the field (Adler & Conklin, 2000; Bernstein, 2001; Cottle, 2001; McDonald, 2001). Some of the reviewers are biologists communicating across domain barriers to computer and information scientists, expressing concerns about how to manage the data (Attwood, 2000; Butte, 2001; Roos, 2001;Watkins, 2001).

A second reason for the many definitions is that gene research is expanding from its initial focus on gene sequences to encompass function prediction (Tsigelny, 2002), among other topics. This raises new areas for research, and collaboration with other fields, specifically computer science, information science, statistics (Ewens & Grant, 2001), and mathematics.

Additional questions concern how to make retrieval results more intelligible to biologists (for example, using visualization and data mining). As bioinformatics has matured, two distinct work practices (derived from biology and from information technology) have emerged to support collaboration on specific biological questions, such as drug discovery (Dougherty & Projan, 2003; Gatto, 2003; Hillisch & Hilgenfeld, 2003), pharmaceutics (Fagan & Swindells, 2000), pharmacogenomics (Jain, 2003; Kalow, Meyer & Tyndale, 2001), neurosurgery (Taylor, Mainprize, & Rutka, 2003), and medical practice (Brzeski, 2002; Grant, Moshyk, Kuskniruk, & Moehr, 2003).

Bioinformatics reviews cover biology and the emerging bioinformatics industry (Benaim Jalfon, 2001); and interest in employment and grants is addressed as well (Basi, Clum, & Modi, 2003; Calandra, 2002; Henry, 2002; Jenson, 2002; Kolatkar, 2002; Schachter, 2002; Van Haren, 2002).

Information science (informatics) and biomedical fields work together in several areas. Some researchers have defined their fields in relationship to bioinformatics (Altman & Dugan, 2003; Bayat, 2002; Fuchs, 2001; National Center for Biotechnical Information, 2003; Ouzounis, 2002). Altman (2000, 2003) tries to define the parameters of medical informatics, which sometimes includes bioinformatics; Grant et al. (2003) consider health informatics; Andersson, Larsson, Larsson, and Jacob (1999) address connections with mathematics. Other points of connection include traditional genomics (Altman, 2003; Chicurel, 2002; Rost, Honig & Valencia, 2002; Valencia, 2002), and proteins (Zimmerman, 2003).

Most intriguing, perhaps, for readers of *ARIST* are the information metaphors in biology. Nishikawa (2002), for example, proposes an "island model" of biology, which considers how, given a set of inputs, the proteins will cluster based on similarities in amino acid sequences. His description of clustering uses identifiable fields of molecular processes (such as amino acid sequences), which map directly to concepts in the information retrieval and clustering literature. In this description the behavior of polypeptides under physiological conditions parallels the behavior of a query under different user cognitive conditions. The critical role, however, of "information" (in the library and information science sense) remains (Denn & MacMullen, 2002; Gywne, 2002; Luscombe, Greenbaum & Gerstein, 2001), as does the search for resolving data modeling related problems, such as applying Extensible Markup Language (XML).

Historically, some of the work associated today with "bioinformatics" was viewed as genomics and computational biology (Fogel & Corne, 2003; Gascuel & Sagot, 2000; Priami, 2003). Faced with more data than could be efficiently processed and with new research questions, biology turned, as many fields do, to statistics and technology to help model phenomena and to expose interesting patterns and deviations from patterns. With the advent of computer technology, a greater variety of patterns could be examined more quickly and without the introduction of human error, making a picture of the invisible world of the molecule possible. Expanded modeling of these processes made biology more approachable and enabled research simultaneously on a broader descriptive level and on more highly focused questions. For instance, some computational molecular biologists (e.g., Leszczynski, 1999) moved to numerical simulation as a complement to traditional theoretical and experimental approaches; they probed *in silico* theories that could not otherwise be examined, such as phenomena at the atomic level. However, pursuing this path leads to quantum mechanical complications and the various simplifications (e.g., the Born-Oppenheimer approximation) employed, which are beyond the scope of this review. Nevertheless, it underscores the fact that computational molecular biology focuses on computerized and mathematical answers to biological questions.

Genomics "is operationally defined as investigations into the structure and function of very large numbers of genes undertaken in a simultaneous fashion" (University of California, Davis, Genome Center, 2003, online). Key areas are comparative genomics and functional genomics. Functional genomics infers the function of gene expression, typically based on eukaryotic homologues (nonviral, nonbacterial organisms with the same origin and function) or other model organisms, not usually tested *in vivo*. Functional genomics also includes mutagenesis (the production of changes in DNA sequences that affect gene products), the study of genotypes (the specific changes in DNA sequences in a mutant), and the effect of these on phenotypes (the biological consequence of a mutagen's presence).

Functional genomic testing of phenotypes relies heavily upon technology for analysis; analytical chemistry, imagery, robotics, and process automation are employed. Boundaries between genomics and bioinformatics are porous, the literature suggesting that genomics, like computational molecular biology, focuses on physical biological processes. Bioinformatics emphasizes the storage and retrieval of biological data and the related research literature: It involves organizing very large heterogeneous structures and determining algorithms for clustering, retrieving, and displaying subsets in meaningful ways, relying heavily upon information visualization, statistics, and integration of appropriate supporting bibliographies. For example, genomics may emphasize research into the microbiology or genetics of a specific organism, such as the *E. coli* genome or the fruitfly *Drosophila melanogaster*; bioinformatics will concentrate on manipulation of the generated data, although some work, such as DNA microarrays, may belong to both fields (e.g., Tessier, Benoît, Rigby, Houghes, van het Hoog, Thomas, 2000).

Recently published monographs define the field and specific work within it, at times focusing on one or another of the subdisciplines. For general introductions and textbook-type treatments, see especially Bergeron (2003); Baxevanis and Ouellette (2001); Lacroix and Critchlow (2003); Krane and Raymer (2003); Krawetz and Womble (2003); Pevsner (2003); Lesk (2002); Misener and Krawetz (2000); Mount (2001); Orengo, Jones, and Thornton (2003); Westhead, Parish, and Twyman (2002); and Dwyer (2003). Barnes and Gray (2003); Gibson and Muse (2002); Mewes, Seidel, and Weiss (2003); Primrose (2003); Sensen (2002); Wang, Wu, and Wang (2003); and Winter, Hickey, and Fletcher (2002) speak to geneticists; Tözeren (2004) to engineers and computer scientists. There are also many article-length discussions of bioinformatics (e.g., Kossida, Tahri, & Daizadeh, 2002).

Luscombe, Greenbaum, and Gerstein (2001, p. 347) propose a reasoned definition of bioinformatics, which they submitted to the Oxford English Dictionary; it is adopted here, with the small addition of "information science": "bioinformatics is conceptualising biology in terms of molecules (in the sense of physical chemistry) and applying '*informatics techniques*' (derived from disciplines such as applied maths, computer [and information] science and statistics) to *understand* and *organise* the *information* associated with these molecules, on a *large scale*. In short, bioinformatics is a management information system for molecular biology and has many *practical applications*" [emphases in original].

As these definitions demonstrate, the emphasis in bioinformatics is on the manipulation of data sources, improving scientists' understanding of the data, and facilitating data discovery and extraction. The challenge for computer and information science is to help biologists "organize data in a way that allows researchers to access existing information and to submit new entries as they are produced ... to develop tools and resources that aid in the analysis of data ..., and to analyse the data and interpret the results in a biologically meaningful manner" (Luscombe, Greenbaum, & Gerstein, 2001, p. 385).

Professional Communication

Journals

Professional communication plays a vital role in bioinformatics and is heavily dependent upon the Internet. Another important avenue is serials. Both McCain (personal communication, November 30, 2003) and Garfield (2002) offer ranked lists of journals (Table 5.2; note their use of different selection criteria), which suggests that the field is evolving. A complete list of journals publishing 25 or more items annually on bioinformatics is provided in Appendix 5.2.

Very Large Databases

The other and most critical form of professional communication is through very large Internet-based databases. Table 5.3 outlines some of the databases by theme or function. Some software packages support particular functions of gene sequencing. For example, there are at least 150 free software applications (Gilbert, 2000, pp. 157–184) addressing all aspects of sequencing. Accelrys's "GCG Wisconsin Package" (http://www. accelrys.com/about/gcg.html) is a popular package; an integrated suite of

Table 5.2 Ranked lists of bioinformatics journals

McCain (2003, personal communication)	Garfield (2002)
Journal of Computational Biology	*Bioinformatics*
Pacific Symposium on Biocomputing	*Genetic Engineering News*
Proceedings of the International Conference on Intelligent Systems for Molecular Biology	*Abstracts of Papers of the American Chemical Society*
Bioinformatics	*Nature*
In Silico Biology	*Scientist*
Briefings in Bioinformatics	*Nature Biotechnology*
Journal of Molecular Graphics and Modeling	*Science*

more than 130 program tools for manipulating, analyzing, and comparing nucleotide and protein sequences, it includes a graphic user interface, SeqLab, to interact with color-coded graphic sequences. The package includes sequence comparison statistics (alignment of two sequences to indicate gaps and best fit and x/y plotting of sequence similarity), database searching tools, and tools for manipulating text files. Database tools include: LookUp, StringSearch (for biological literature), BLAST (Basic Local Alignment Search Tool), NetBLAST, FASTA (for sequence strings), PAUPSearch, GrowTree, Diverge (for phylogenetic relations), Fragment Assembly, gene finding and pattern recognition tools, and protein analysis (e.g., PeptideMap, ChopUp, Reformat). Similarly, DNASTAR's Lasergene Sequence Analysis Software is a suite of applications to trim and assemble sequence data; discover and annotate gene patterns; predict protein secondary structure; create Boolean queries from sequence similarity, consensus sequence, and text terms; sequence, hybridize, and transcribe fragment data; create maps; and import data from other sources (Burland, 2000, p. 71). It is not possible to review all such packages and their capabilities here, but the reader can see that these desktop software applications emphasize the graphic display of data, pattern detection and sequence prediction, and integration of the literature.

Examples of Data Models, Sources, and Uses

Bioinformatics encompasses all aspects of molecular biology research and has made spectacular advances in understanding and disseminating information about molecular processes, not the least of which is the structure of protein molecules themselves and the biological sequences from which they are derived. Applying statistical, mathematical, and computer techniques, however, has pushed bioinformatics to explain the invisible and unanticipated, using such techniques as energy equations to model the dynamic behavior of molecules, linear and 3-D protein functions, and probabilistic models of sequences (Durbin, Eddy, Krogh, & Mitchison, 1998), especially Hidden Markov Models (Koski, 2001). The computer files containing these data are now used to visualize known biological structures (e.g., using Cn3D available from NCBI) and to help predict macromolecule structure in 3-D. Furthermore, and of considerable interest to pharmaceutical companies, bioinformatics applies genomic data to study variation in host and pathogen DNA and disease, thus helping to design drug treatments (Lengauer, 2002). Of particular significance to information science are the databases and software used to store, retrieve, and display data.

The vast amount of data generated in sequencing and recorded in the literature has resulted in a great number of databases and the attendant difficulties of querying across heterogeneous sources. As a result, portals like PubMed and topic- or function-specific databases have been created. Zdobnov, Lopez, Apweiler, and Etzold (2002b) propose the following categorization:

- Bibliographic

- Taxonomic

- Nucleic acid

- Genomic

- Protein and specialized protein databases

- Protein families, domains, and functional sites

- Proteomics initiatives

- Enzyme/metabolic pathways

In addition, data models specific to molecular biology and genomics have appeared. One such model is the Sequence Retrieval System (SRS) (Etzold, Ulyanov, & Argos 1996; Zdobnov, Lopez, Apweiler, & Etzold, 2002a); another is the NCBI Data Model.

This section outlines some of the major databases and tools: general databases, protein sequence databases and tools, nucleotide sequences, classification schema, error reduction techniques, the NCBI data model, and XML. For convenience, Web sites are listed in Appendix 5.1; Table 5.3 will help the reader classify the many resources.

Bibliographic and Taxonomic Databases

A comprehensive list of bibliographic and taxonomic databases is available at http://www.expasy.ch/alinks.html and in each January issue of *Nucleic Acid Research*. The most commonly used, publicly available resource is Medline (or PubMed). Agricola (Agricultural Online Access), from the U.S. National Agricultural Library, is also an important resource. Commercial databases include Embase (biomedical and pharmacological abstracts) and Biosis (the former *Biological Abstracts*). The NCBI maintains the most important taxonomic databases, whose hierarchical taxonomy is used by Nucleotide Sequence Databases, Swiss-Prot, and TrEMBL (along with derivatives such as NiceProt). Another important source is the Chemical Abstracts database, which includes the bibliographic file (CAPlus) and a file of compounds, the Registry File. According to the STN Database Summary Sheet (2004), there are about 52 million records, of which almost 31 million were sequences for either proteins or nucleic acids. Another commercial database on the STN International system is DGENE, Derwent's Geneseq database covering sequences from patents published by 40 patent offices worldwide. The December 2003 brochure for the database stated that "more than

Table 5.3 Examples of publicly accessible bioinformatics databases (Bergen, 2003, pp. 45–46).

Type of database	Examples	Description
Nucleotide sequencing	GenBank	13 billion bases from > 100,000 species
	DDBJ	DNA Database of Japan
	EMBL	European Molecular Bio Lab
	MGDB	Mouse Genome Database
	GSX	Mouse Gene Expression Database
	NDB	Nucleic Acid Database
Protein sequences		Protein knowledgebase
	TrEMBL	Annotated Supplement to SwissProt
	TrEMBLnew	Weekly, pre-processed update to TrEMBL
	PIR	Protein Information Resource
3-D structure data	PDB	
	MMDB	
	Cambridge	Small molecule structural database
Enzymes and compounds	LIGAND	Chemical compounds and reactions
Sequence motifs (alignments)	PROSITE	Sequence motif
	BLOCKS	Derived from Prosite
	PRINTS	
	Pfam	Protein families database of alignments and hidden Markov models
	ProDom	Protein domains
Pathway and complexes	Pathway	Metabolic and regulatory pathway maps
Molecular disease	OMIM	
Biomedical literature	PubMed	
	MedLine	
Vectors	UniVec	Identification of vector contaminates
Protein mutation	PMD	Protein Mutant Database
Gene expression	GEO	Gene Expression Omnibus
Amino acid indices	Aaindex	Amino Acid Index
Protein/peptide literature	LITDB	
Gene catalog	GENES	KEGG Gene Database

half of the sequence data that appear in DGENE is not available in any other public sequence database" (http://www.derwent.com/geneseqweb).

Genome Databases

The popularity of the Human Genome Project has introduced to the public three important databases for human genes. The primary human genome database is the Genome Database (GDB). Related to this is the

Online Mendelian Inheritance in Man (OMIM) database, which catalogs all human genes and genetic disorders. The Sequence Variation Database, like OMIM and GDB, maps genetic variation, but it also has links to many sequence variance databases (EBI-Mutations) and, via the Sequence Retrieval System (SRS) interface, to other human mutation databases. Increasingly, portals, such as GeneCard, are being developed to harmonize searches.

Perhaps most well known are the protein sequence databases. Like the nucleotide databases, the protein sequence databases fall into two groups: those containing data for all species or only for specific organisms. Of interest to information science is the further division of these databases into "sequence data" or "annotated sequence data." Swiss-Prot (Bairoch & Apweiler, 2000) is an annotated universal protein sequence database; it strives for quality by offering annotations and integration with other biomolecular databases. Each entry is analyzed by biologists; as of May 2000, there were more than 85,000 annotated sequence entries from more than 6,000 different species. A sister product, called TrEMBL (Translation of EMBL nucleotide sequence database), was created from Swiss-Prot to speed new sequence information to the public. SP-TrEMBL focuses on entries intended to be incorporated into Swiss-Prot. REM-TrEMBL contains other data that will not be integrated because they may be redundant, are truncated, or are not proteins or fragments legitimately translated *in vivo*. SPTR (SWALL) is another protein sequence database that provides nonredundant sequence data by focusing on data currency in Swiss-Prot, ignoring REM-TrEMBL, and by performing sequence comparisons against a database of all known isoforms. Table 5.4 outlines some genomic applications.

Nucleotides

GenBank (NCBI), the European Bioinformatics Institute (EBI) (Apweiler et al., 2003), and the DNA Data Bank of Japan have joined forces to create the International Nucleotide Sequence Database Collaboration. The quality and currency of the data vary between databases. The quality of the data in the nucleotide sequence databases is the responsibility of the authors or submitters (the scientists themselves; there is no external enforcement of standards). With more than 10 billion nucleotides in more than 10 million individual entries, one can imagine the potential error rate (EBI-Stats). The reader is referred to Rodriguez-Tomé (2000) for a description of EMBL and examples of interfaces for submitting and searching the databases and the Genome Monitoring Table for updates on the progress of genome sequencing projects.

Protein Sequence Databases

Specialized protein sequence databases perform different functions with data. CluSTr pre-clusters Swiss-Prot records (Apweiler et al., 2001), MEROPS and PepCards catalog and create structure based peptidases,

Table 5.4 Genomic applications (after Bergeron, 2003, p. 62).

Function	Example resource (database or software application)
Sequence search	BLAST, BLASTN, CLUSTALW, Fasta, Motif, PBLast, TBLASTIN
Submission	AceDB, Audet, BankIT, Sakura, Sequin, WebIN
Information retrieval	Entrez, DBGET, IDEAS
Linkage	LocusLink
Portal	KEGG
Structure match	DC, DALI, SCOP, Searchlite, Structure Explorer, VAST
Visualization	CAD, Cn3D, Mage, RasMol/MolMol, Swiss-PDB Viewer, VRML, WebMol
Protein-protein interaction	BRITE
Microarray gene Expression profiles	Expression
Open reading frame locator	ORF Index

including classification and nomenclature data along with hyperlinks for each peptidase, and family (FamCards) or clan (ClanCards). The Yeast Protein Database (YPD for *Saccharomyces cerevisiae*) details about 6,000 yeast proteins. Protein classification schemas define the cellular role, function, pathway, and other information about the functional data in the YPD Protein Reports.

Finding relationships when an unknown protein cannot be matched to known structures calls for examining the "sequence signatures." PROSITE, PRINTS, Pfam, ProDom, and especially InterPro attempt to derive patterns from sequence databases using various sequencing and clustering algorithms (Benson & Page, 2003; Gascuel & Moret, 2001; Guerra & Istrail, 2000; Guigó & Gusfield, 2002). InterPro (Integrated Resource of Protein Families, Domains, and Functional Sites) is an integrated documentation resource for PROSITE, PRINTS, and Pfam, which helps address the question of ambiguous biological relevance when a pattern is detected (e.g., by ignoring family discriminators) by linking to known protein sequences in Swiss-Prot and TrEMBL. InterPro entries are available as XML-formatted files (EBI-InterPro).

Some work focuses on learning more about organisms at various levels. For instance, the Kyoto Encyclopedia of Genes and Genomes (KEGG) and Proteome Analysis Initiative (PAI) provide information about the gene, transcript, protein, and function level. In addition, the proteome set information for all completely sequenced organisms in Swiss-Prot and TrEMBL are available through InterPro and CluSTr. This includes the amino acid composition and links to the homology

(Homology-derived Secondary Structure of Proteins [HSSP]). As evidence of the need to understand better the function of proteins (Ashburner, Ball, Blake, Botstein, & Butler, 2000) and genes and how to associate the literature more successfully, there is growing interest in ontologies (e.g., Bard, 2003). The very traditional library focus on classification schema for precoordination is making its mark in bioinformatics in response to the need to gain conceptual control over the literature and to create meaningful subsets of associated data. Paris (2002, p. 453) notes that annotations of bioinformatics resources pose a multilayered problem "both from the LIS viewpoint (e.g., authority files, correct and comprehensive attribution, recognizing the value of editors and compilers, building a collection of all editions) and from the scientific viewpoint (e.g., there are categorical disparities that make data mining very difficult)." Some of these concerns will be addressed through consistent, scientifically sound ontologies, such as the Gene Ontology (GO), XML, and other attempts at standardized nomenclatures. The GO collaborators "are developing three structured, controlled vocabularies (ontologies) that describe gene products in terms of their associated biological processes, cellular components and molecular functions in a species-independent manner" (Gene Ontology, 2004). The HUGO Nomenclature Committee (1999) reports on how standard vocabularies—and with them bibliographic, information retrieval, and semantic-mapping techniques—can service information needs. For instance, Blake (1999, online) contends that existing methods of assigning protein-coding regions based on sequence features and standardizing gene names are insufficient and that standardized vocabularies are required to manage the data. She writes,

> Recent technological advances now permit rapid accumulation of genomic sequence information and subsequent quick identification of protein-coding regions based on sequence features. Many of the new coding sequences are known only as having the characteristics of a coding sequence with a similarity to better known genes. The burden has become how to name these sequences, and indeed all coding regions, when each object starts with a minimum set of information, but over time accumulates information about gene function, membership in gene families, shared genomic features such as structural domains, and multiple expression patterns. Not only is it difficult to incorporate this information in the gene name, it is difficult to keep names current with the knowledge base. While we recognize that human cognitive needs require something more than a numbering system or unique identifiers for genes, the process by which we can maintain the capacity for timely decisions for naming unique genetic entities has not been easy to develop.

Classification Schemas

Because so many bioinformatics computer applications rely on post-coordinate retrieval, there is a need for classification schema to help manage the volume of data generated in biology. One example of a protein schema is the Enzyme Commission Classification, which assigns an "E.C. number." Hamilton (2004) describes its application:

> The most useful definition of enzyme function is through biochemical reaction, i.e., the chemical reaction catalyzed by the enzyme. The biochemical reaction catalyzed by an enzyme is assigned an EC (Enzyme Commission Classification) number. If a new enzyme activity is discovered then a new EC number has to be created. The EC scheme is hierarchical with four levels hence the four elements to an EC number, e.g., alcohol dehydrogenase has an EC number of 1.1.1.1. The first number is the top level of the classification (in this case an oxidoreductase), the last number refers to the specific enzyme (the specific oxidoreductase alcohol dehydrogenase). The middle numbers give details about the reaction catalysed.

In addition, standards are being developed for model organism databases, such as Lincoln Stein's MOD project (Harris & Parkinson, 2003).

Error Correction

Other projects attempt to reduce potential errors by clustering and specialization. Reminiscent of latent semantic indexing (Gordon & Dumais, 1998), clustering of data to remove redundant records is performed by UniGene and STACK (Sequence Tag Alignment and Consensus Knowledgebase). The Ribosomal Database Project (RDP), HIV Sequence Database (HIV), IMGT database, Transfac (transcription factors and transcription factor binding sites), EPD (Eukaryotic Promoter Database), REBASE, and GoBase are all examples of specialty resources. For instance, specialty resources such as the IMGT (International ImMunoGeneTics) information system, emphasize "expertly annotated information and data standardization" for "high-quality integrated information systems" use; by annotating and controlling the IMGT-Ontology "scientists and clinicians ... use identical terms with the same meaning" (International Immunogenetics Information System, 2004, online) meaning redundant records are more easily identified and removed and matching improved.

NCBI Data Model

The NCBI Data Model is of particular interest to information scientists.

This new and more powerful model made possible the rapid development of software and the integration of databases that underlie the popular Entrez retrieval system and on which the GenBank data is now built (Schuler, Epstein, Ohkawa, & Kans, 1996). The advances of the model (e.g., the ability to move effortlessly from the published literature to DNA sequences to the proteins they encode, to chromosome maps of the genes, and to the three-dimensional structures of the proteins) have been apparent for years to biologists using Entrez. (Baxevanis & Ouellette, 2001, p., 20)

The Entrez database (Tatusova, Karsch-Mizrachi, & Ostell, 1999) has the complete genome of about 300 archaeal, bacterial, and eukaryotic organisms, including exon and intron data.

The NCBI Data Model is an implementation of Abstract Syntax Notation One (ASN.1, n.d.), an ISO standard (ISO/IEC 8824-1:2002) for reliable encoding of data that is data-centered (here the DNA), human interpretable, and presented in computer readable flat files. Based on this, NCBI's Web site is a portal offering access to PubMed (for the biomedical literature), Entrez, BLAST, OMIM, and other services based on ASN.1. Specialized access points include protein, nucleotide, structure, genome, PopSet (aligned sequences from population studies), and UniSTS (unified sequence tagged sites).

Although readers can explore these sites for themselves, it may be helpful to show here an example of the results of a search, to suggest the file structure and how one might want to parse the file. Figure 5.1 shows the result of a search using the Entrez retrieval system to locate a gene in humans. The reader will immediately recognize some of the access points in the NCBI Sequence Viewer and notice the sequence of bases. Hyperlinks (in this case, organism, MedLine, PubMed, exon) in the record link the biological data to the support literature. Notice the traditional access points (authors, title, journal), links to other literature entries (the Medline unique identifier [MUID] and PubMed identifier [PMID]), and among others, the base. This record also includes the sequence identifiers (Seq-id) because the NCBI integrates sequence data from multiple sources.

The heterogeneity of data in addition to the various sequence databases and related resources calls for greater cross-discipline mapping (similar to the Unified Medical Language System [UMLS]) or semantically independent modeling schema, such as XML.

XML

The many XML schemas for biology highlight the variety of specific needs for semantically useful descriptors. Guerrinia and Jackson (2000) offer an introduction to XML and document type definitions (DTDs) in biology; Maher (2001, §3) demonstrates the application of BioML2SVG

```
LOCUS       HSDDT1                   166 bp     DNA
linear   PRI 01-FEB-2000
DEFINITION  Homo sapiens D-dopachrome tautomerase (DDT)
gene, exon 1.
ACCESSION   AF012432
VERSION     AF012432.1  GI:2352911
KEYWORDS    .
SEGMENT     1 of 3
SOURCE      Homo sapiens (human)
  ORGANISM  Homo sapiens
            Eukaryota; Metazoa; Chordata; Craniata;
Vertebrata; Euteleostomi; Mammalia; Eutheria; Primates;
Catarrhini; Hominidae; Homo.
REFERENCE   1  (bases 1 to 166)
  AUTHORS   Esumi,N., Budarf,M., Ciccarelli,L.,
  Sellinger,B., Kozak,C.A. and Wistow,G.
  TITLE     Conserved gene structure and genomic linkage
for D-dopachrome tautomerase (DDT) and MIF
  JOURNAL   Mamm. Genome 9 (9), 753-757 (1998)
  MEDLINE   98384542
  PUBMED    9716662
REFERENCE   , 2  (bases 1 to 166)
  AUTHORS   Esumi,N. and Wistow,G.
  TITLE     Direct Submission
  JOURNAL   Submitted (07-JUL-1997) Molecular Structure and
Function, NEI, Building 6, Rm. 331, NIH, Bethesda, MD,
20892, USA
FEATURES             Location/Qualifiers
     source          1..166
                     /organism="Homo sapiens"
                     /db_xref="taxon:9606"
                     /chromosome="22"
     exon            1..146
                     /gene="DDT"
                     /number=1
BASE COUNT      , 24 a     61 c     50 g     31 t
ORIGIN
        1 cttcttccgc cagagctgtt tccgttcctc tgcccgccat
gccgttcctg gagctggaca
       61 cgaatttgcc cgccaaccga gtgcccgcgg ggctggagaa
acgactctgc gccgccgctg
      121 cctccatcct gggcaaacct gcggacgtaa gcgtgggccg
ggcagc
```

Figure 5.1   Example of an NCBI record

to gene sequences. The variety of XML DTDs available today suggests, too, opportunities for mapping across schemas. A list of over twenty distinctive schemas is available at http://www.xml.com/pub/rg/ Bioinformatics. Some are domain specific, such as Neuron Markup Language (NeuroML), Genome Annotation Markup Elements (GAME), Ribonucleic Acid Markup Language (RiboML); others emphasize integration (Architecture for Genomic Annotation, Visualization and Exchange [AGAVE], Genbank to XML conversion [GB2XML], Integrated Taxonomic Information System [ITIS]), and even an XML-based Ontology Exchange Language (XOL). Visualization of data is critical in bioinformatics and several products are available that combine XML records and display techniques (EBI-Vis); some work emphasizes fine-grained parsing of XML records with interactive information retrieval.

Data Mining and Visualization

Some activities in molecular biology focus on predicting sequences where there are missing values or on establishing patterns that otherwise would be impossible to detect. Data mining, an "exploration and analysis by automatic and semi-automatic means, of large quantities of data in order to discover meaningful patterns and rules," when focused on biological processes turns bioinformatics into a data mining activity (Benoît [2002, p. 265]; see Benoît [2002] for a detailed overview of data mining or Kantardzic [2003] for an excellent course in statistical foundations). The EBI (European Bioinformatics Institute, 1999, online) hosted a conference exploring the intersection of bioinformatics and data mining:

> During the last few years bioinformatics has been overwhelmed with increasing floods of data, both in terms of volume and in terms of new databases and new types of data. We are now entering the post-genomic age, where, in addition to complete genome sequences, we are learning about gene expression patterns and protein interactions on genomic scales. This poses new challenges. Old ways of dealing with data item by item are no longer sustainable and it is necessary to create new opportunities for discovering biological knowledge "in silico" by data mining.

Other conferences have discussed the union of data mining and bioinformatics. One example is the BioKDD, Workshop on Data Mining in Bioinformatics, held in conjunction with the International Conference on Knowledge Discovery and Data Mining sponsored by ACM's special interest group, SIGKDD. Topics from that 2001 workshop demonstrate the coupling of biology and information technology: gene expression, sequence modeling, and clustering (http://www.cs.rpi.edu/~zaki/BIOKIDD01). These topics were also addressed at the Data Mining for Bioinformatics–Towards in Silico Biology conference supported by the EBI-Wellcome Trust (http://industry.ebi.ac.uk/datamining99). Jenssen,

Öberg, Andersson, and Komorowski (2001) explore methods for mining networks of human genes.

Several monographs have dealt with the marriage of data mining and biology: Schlichting and Egner (2001); Perner (2002); Calvanese, Lenzerini, and Motwani (2003); and Ye (2003).

The NCBI portal provides access to a range of tools, including: BLAST, genomes, Clusters of Orthologous Groups (COGs), Map Viewer (integrated views of chromosome maps), LocusLink (descriptive and sequence information on genetic loci), UniGene, ORF (Open Reading Frame) finder (to identify all possible ORFs in a DNA sequence), Electronic PCR(Polymerase Chain Reaction) to search DNA sequences for sequenced tagged sites (STSs), VAST Search (structure-to-structure similarity searches), Cancer Chromosome Aberration Project, Human–Mouse Homology Maps, VecScreen (identifies segments of a nucleic acid sequence), dbMHC (for human Major Histocompatibility Complex), Spidey (aligns mRNA to a genomic sequence), and the Cancer Genome Anatomy Project.

Computer software has been written to display atomic and molecular structures using information visualization techniques, or what is sometimes called "alternative metaphors" (Bergeron, 2003). The various visualization tools differentiate between structures, allowing a variety of explorations; for example, where the protein backbones are the same, is it possible to articulate the structures by hiding or showing atoms?

Most imaging uses data from PDB or the Molecular Modeling Databases (MMDB). After searching for a structure by protein name or identification label, e.g., "Glutamine synthetase" or "1FPY", the data can be used by different applications for various presentations. For instance, wireframe diagrams can be generated by PyMol to emphasize the atomic bonds; Chimera can create ribbon diagrams to highlight the protein's secondary structure. Figures 5.2 through 5.4 are examples of information visualization. Figure 5.2 shows the secondary structure of Colicin Ia, from *E. coli*, PDB ID number, 1CII.

There is a range of options for visualization. The data can be modeled using general purpose software, such as Excel, Strata Vision 3D, Max3D, 3D-Studio, Ray Dream Studio, StatView, SAS/Insight, MatLab (Bergeron, 2003), or creating one's own user interface using Java, VRML (Virtual Reality Modeling Language), C++, Python or computer-aided design (CAD) applications. There are also specific tools in existence that visualize sequences (NCBI-MapViewer) or structures. To visualize nucleotide locations use MapViewer; for protein structures there are many products (Swiss-PDB Viewer, WebMol, RasMol, Protein Explorer, Cn3D, VMD, MolMol, MidalPlus, PyMol, Chime, Chimera).

In addition to this type of visualization, researchers are developing applications with greater interactivity and functionality. For instance, Böhringer, Gödde, Böhringer, Schulte, and Epplen (2002, p. 51) have developed a software package for visualizing genome-wide information

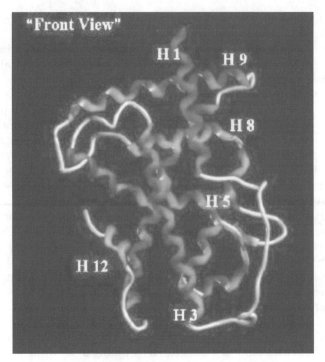

Figure 5.2   **Secondary structure of Colicin Ia**

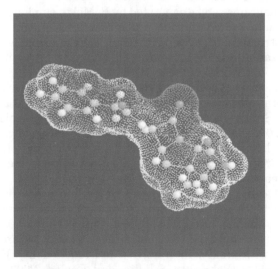

Figure 5.3   **Orthographic view of retinoic acid receptor, created using SYBYL 6.6 (Taylor, 2000)**

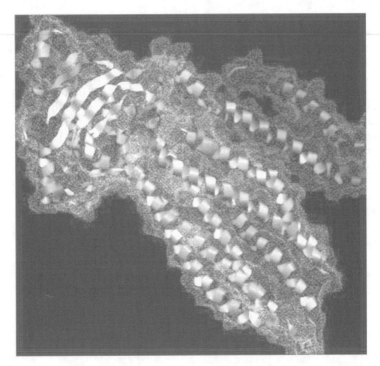

Figure 5.4   Surface potential of G-penicillin (RasTop, 2003)

and generating "arbitrary karyograms, banding patterns and chromosome groupings."

From an information and computer science perspective, many bioinformatics projects use familiar visualization techniques. One example is the TimeSearcher, which parses microarray data, sampling every few hours, and graphically represents specific cellular functions (http://www.cs.umd.edu/hcil/bioinfovis/timesearcher.shtml). Similarly, the concept map idea has been applied to gene ontology to generate directed acyclic graphs, according to biological functions (http://www.cs.umd.edu/hcil/bioinfovis/treemap.shtml). Indeed, most university campuses provide opportunities for molecular biologists to work with computer programmers and develop their skills. Monographs by C. Chen (1999); Ware (2000); and Card, Mackinlay, and Shneiderman (2002); online lectures by Keim (1997); and a series of articles (e.g., Lanzenberger, Miksch, Ohmann, & Popow, 2003; Rhodes, Bergeron, & Sparr, 2002; Schroeder, Gilbert, van Helden, & Noy, 2001) guide in selecting the types of interfaces that may be useful in identifying pitfalls.

There are also many workshops and conferences encouraging interdisciplinary visualization projects, such as the Bioinformatics Visualization Workshop sponsored by the University of Maryland's Human-Computer Interaction Lab, the IEEE Symposium on Information Visualization, and the various lectures organized by the EBI.

Visualization in bioinformatics is exceptionally valuable, and as the volume of data increases and becomes even more multidimensional, visualization techniques will integrate more with data mining techniques. Table 5.5 suggests how data can be subjected to visualization techniques and then combined with data mining approaches to be applied to biologically relevant tasks.

Table 5.5 Visualization applications (based on Bergeron, 2003).

Data (examples)	Microarray, genomic sequences, protein structures, gene ontologies, biological pathway data
Visualization techniques (examples)	2-D, 3-D scattergrams, heatmaps, hierarchical treemaps, temporal data searching, topographic displays, hierarchical presentations
Combined with data mining techniques	Supervised and unsupervised classification/categorization, principle component analysis, multidimensional scaling
Applied to biologically relevant tasks	Comparing samples, identifying similar and different genes, identifying targets, defining pathways, hypothesis generation

Collaborative Opportunities

Because bioinformatics aims to improve the organization and understanding of biological data, the future of collaboration between biology and the information and computer sciences is bright indeed. Denn and MacMullen (2002, p. 556) identify information science research areas ("domain analysis, information use, communication, and theories of information [basic]; systems analysis and design, data modeling, classification, storage and retrieval, and human–computer interaction [applied]") and map them "onto a generalized model of a molecular biology experimental cycle." They classify the "insertion points" of information science research into this experimental cycle as "the development of new tools and methods for managing, integrating, and visualizing data; the application of tools and methods for integration, inference and discovery; and theoretical approaches to biological information."

Denn and MacMullen (2002) identify several sources for information on each insertion point: (a) new tool development (Brazma, et al., 2001; Gene Ontology Consortium, 2001; Nucleic Acid Research, 2002; Paris, 1997; Searls, 2000), (b) applications (H. Chen, 1997; Corruble &

Ganascia, 1997; Juvan, 2001; Karp, Paley, & Zhu, 2001; Raychaudhuri, 2002; Wise, 2000), and (c) theory (de Jong, 2002; Poinçot, Lesteven, & Murtagh, 2000; Smalheiser & Swanson, 1996).

Information retrieval strives to improve the locating, retrieval, and ranking of full-text documents using a number of algorithms; this requires some knowledge of the collection (Baeza-Yates & Ribeiro-Neto, 1999). From an IR perspective bioinformatics data describing and representing both the biomedical literature and biological elements and functions can be viewed as decontextualized tokens. The manipulation of tokens from full-text literature and metadata can be lexical, based on the parsed texts or data files. Using these same tokens from a biological perspective suggests exploring the data to reflect manipulations of biological processes, which may expose unanticipated, interesting phenomena. For instance, in addition to mining biological data, it is possible to incorporate the literature of molecular biology to expose and visualize unanticipated relationships among the records. Stapley and Benoît (2000) describe a visualization technique and retrieval system based on co-occurrences of gene names in Medline abstracts.

Swanson has long been engaged in knowledge discovery across disciplines from noninteracting literatures (e.g., Swanson, 1991; Swanson & Smalheiser, 1997, 1999). Swanson's valuable literature investigations and Swanson and Smalheiser's (1997) work in finding complementary literatures are credited with stimulating the type of biological data/literature integration called for by Altman (2000) and others. Such work includes Cory (1997); Davies (1988); Lu, Janssen, Milios, and Japkowicz (2001); Jenssen, Öberg, Andersson, and Komorowski (2001); Valdés-Pérez (1999); and many others. Swanson and collaborators continue to further research in literature-based discovery of scientific knowledge; see, for example: Gordon and Lindsay (1996); Weeber (2001); Weeber, Vos, Klein, and de Jong-van den Berg (2001); Swanson, Smalheiser, and Bookstein (2001); and expressly in biology (Gardy & Brinkman, 2003).

Molecular biology's information resources have progressed from flat files through relational databases to Web pages and portals, and this trend hints at the need to construct robust digital library architectures to automate bibliographic searches and automatically integrate knowledge bases from the literature. Such automation necessarily incorporates improved string comparison methods and one-dimensional alignment algorithms, but also involves linking specific genes to different disease triads as represented in the structured data. This suggests data modeling projects such as the Gene Ontology and also XML schema for mapping between biological databases.

Naturally, computationally complicated proteome research requires greater data mining interaction to aid in prediction of nucleic acid structures (Tsigelny, 2002), motif and pattern identification, structural comparison, and 3-D matching. Computer and information science can contribute on several fronts:

- Machine learning (Baldi & Brunak, 2001)

- Clustering algorithms (e.g., Eisen 1998; Jajuga, Sokolowski & Bock, 2002; Laender & Oliveira, 2002; Toronen et al., 1999)

- Connectionist systems in bioinformatics (Kasabov, 2003)

- Artificial intelligence and heuristic methods in bioinformatics (Frasconi, 2003)

- Visualization of literature

- Metadata

- Graphic representations of surfaces and volumes in general and for disease-specific analysis

Some researchers have suggested that "systems biology" is the next wave in bioinformatics. For instance, the Institute for Systems Biology (2003, online) claims that the HGP "has catalyzed two paradigm changes in contemporary biology and medicine—systems biology and predictive, preventative and personalized medicine." Systems biology emphasizes causal relationships in cellular dynamics (e.g., Funahashi, Tanimura, Morohashi, & Kitano, 2003; Kitano, 2003) and even has its own Systems Biology Markup Language (Hucka et al., 2003).

Another collaborative opportunity lies in teaching computer programming skills. It is telling that biology feels a need to do more as far as data manipulation is concerned. Several recently published monographs focus on the string manipulating strength of the Perl programming language. For instance, *Genomic Perl: from Bioinformatics Basics to Working Code* (Dwyer, 2003), *Perl Programming for Biologists* (Jamison, 2003), *Beginning Perl for Bioinformatics* (Tisdall, 2001), and *Developing Bioinformatics Computer Skills* (Gibas & Jambeck, 2001) instruct biologists on how to parse full-text records. Dwyer (2003) cogently and effectively demonstrates the merging of computer technology and biological processes.

Using Perl emphasizes the string processing power needed for manipulating full-text files, although Java and C++ certainly offer equivalent file manipulation power along with greater computational efficiency and, at least in the case of Java, a large graphic library. Montgomery (2003) offers Java code to access Web-based databases GenBank, EMBL, and DDBJ. He describes several products to perform bioinformatic analysis: TIGR MultiExperimental Viewer, J-Express, BioJava, and Apollo, as well as Sockeye and WebMol for 3-D displays. Base4 hosts a large collection of Java applets, code, and "biowidgets," as well as references to Tk/Tcl and Corba projects.

Several free JavaBeans tools are available for manipulating BLAST and GSDB (e.g., BlastView and AnnotView, http://www.cbil.upenn.edu/

bioWidgets). Those with information storage and retrieval knowledge will see immediately the potential applications of parsing, matching, clustering, similarity measurements, and display methods from IR to bioinformatics records. The fact that biologists are learning the computing skills themselves suggests opportunities for information and computer science to assist.

Training

There are at least 45 undergraduate and 63 graduate programs (NIBIB) in bioinformatics (University of North Carolina, Chapel Hill. Bioinformatics Journal Club, 2003), which suggests that the need for integrated training is acute.

Another stimulus is the availability of grants aimed at cross training. For example, the National Science Foundation and the National Institutes of Health support training in bioengineering and bioinformatics (http://www.nibib1.nih.gov/training/trainingopps.html). These grants support several trainee programs throughout the country, along with post doctoral grants, e.g., the Institute for Pure and Applied Mathematics at UCLA (http://www.ipam.ucla.edu/programs/fg2000/fellowship_100301.html).

Integration with Clinical Informatics

Altman (2000, p. 442) describes the Stanford Medical Informatics program as the next step in a "post-genome age, [where] the interplay between basic biological data (sequences, structures, pathways, and genetic networks) and clinical information systems is, clearly, critical." Major concerns for the future (http://bits.stanford.edu) include robust computing for information acquisition, storage, retrieval, and management. By outlining six "affinity groups," Altman suggests greater integration (and hence opportunities for computer and information science) in

- Image acquisition and analysis (physical systems)

- Structural biology and genetics bioinformatics (physical systems)

- Biomechanical modeling for macroscopic systems (physical systems)

- Computer assisted interventions and robotics (physical systems)

- Data modeling, statistics, and informatics (informatics)

- Networked and computer enabled education (informatics)

Furthermore, Altman (1998, p. 53) predicts that "DNA sequence information and sequence annotations will appear in the medical chart with increasing frequency," which raises the ethical issue of making such data publicly accessible and also the technical questions of what data are stored and how to develop an expanded data model. Certainly advances in controlled vocabularies in clinical informatics can be applied to the representation of bioinformatic data.

Burge (2002, online) outlines several challenges:

- Precise, predictive model of transcription initiation and termination

- Precise, predictive model of RNA splicing/alternative splicing

- Precise, quantitative models of signal transduction pathways: ability to predict cellular responses to external stimuli

- Determining effective protein, DNA:protein, protein:RNA, and protein:protein recognition codes

- Accurate *ab initio* protein structure prediction

- Rational design of small molecule inhibitors of proteins

- Mechanistic understanding of protein evolution

- Mechanistic understanding of speciation: molecular details of how speciation occurs

- Continued development of effective gene ontologies: systematic ways to describe the functions of any gene or protein

- Education: development of appropriate curricula for secondary, undergraduate, and graduate education

The porous borders of clinical information systems, medical informatics, and bioinformatics—especially in light of Burge's and Altman's predictions of tighter integration of these fields—suggests great opportunities for data mining. Medical research and practice have generated tremendous amounts of data, beyond that created by pharmaceutical and biomedical research. For instance, electronic patient records and integrated medical information systems constitute a massive warehouse of clinical data online. By mining these data, bio- and other informaticians can detect trends and surprising events to support informed decision making by clinicians (e.g., evidence-based medicine) and even create "intelligent" systems that respond to the data (evidence-based adaptive medicine) to improve health care.

Conclusions

The trend in bioinformatics education is to train computer and information scientists in basic biology and genomics and to emphasize computerized database creation, searching, and parsing of files for particular research projects. This is supported by a growing number of undergraduate and graduate programs in bioinformatics throughout the world. In addition, conferences and monographs demonstrate the effectiveness of biologists working with computer and information scientists. Given the many grants that are available, the future seems bright for the training and hiring of new bioinformaticians.

The technology employed in the lab to create and process new genomic data forms a regular part of the molecular biologist's work. The equipment for general and specialized research advances rapidly, and the Internet is the primary means for keeping informed, along with conferences and annual reviews. Merely keeping abreast of current developments is a challenge, one which could be met by selective dissemination of information and the use of other "push" technologies—services that could be provided by centralized biotechnical digital libraries.

Scientists submit, and search for, records in a heterogeneous computing environment. Therefore, there is considerable opportunity for both computer scientists and information scientists to focus on their shared areas of expertise. The literature reviewed here demonstrates seemingly limitless opportunities for computer and information science to provide biology with expertise in machine centric work. Computer and information scientists could develop high-performance computing systems, compression algorithms, user interface design, refined similarity measures, metadata applications, interactive graphic user interfaces, and 3-D modeling in ways that will ultimately integrate sequence and expression data with the associated research literature. The results of shared efforts would naturally create a more robust computational environment and enable more flexible data manipulation, which should help answer increasingly sophisticated research questions.

Acknowledgments

The author wishes to thank Connie Chow, Harvard University School of Public Health, and the *ARIST* reviewers for their helpful comments.

Appendix 5.1. Resource Web Sites

Base4: http://www.cis.udel.edu/~vagrawal/bioinformatics/code; java/base4javabioresources

BIOML: http://bioperl.org/Projects/XML

bioXML: http://stateslab.bioinformatics.med.umich.edu

CluStr: http://www.ebi.ac.uk/clustr

EBI: http://www.ebi.ac.uk

EBI-InterPro: ftp://ftp.ebi.ac.uk/pub/databases/interpro

EBI-Mutations: http://www.ebi.ac.uk/mutations/index.html

EBI-Stats: http://www3.ebi.ac.uk/services/DBStats

EBI-Vis: http://industry.ebi.ac.uk/~alan/VisSupp/VisAware/index.html

EPD: ftp://ftp.ebi.ac.uk/pub/databases/epd

GDB: http://www.gdb.org

GeneCard: http://bioinfo.weizmann.ac.il/cards

GMT: http://www.ebi.ac.uk/~sterk/genome-MOT/MOTgraph.html

GO: http://www.geneontology.org

GoBASE: http://megasun.bch.umontreal.ca/gobase/gobase.html

HIV: http://hiv-web.lanl.gov

HGPI: http://www.ornl.gov/sci/techresources/Human_Genome/glossary

HSSP: http://www.sander.ebi.ac.uk/hssp

IMGT: http://imgt.cines.fr:8104/textes/IMGTindex/ontology.html

KEGG: http://www.genome.ad.jp/kegg

MEROPS: http://www.merops.co.uk

NCBI: http://www.ncbi.nlm.nih.gov

NIBIB: http://www.nibib1.nih.gov/training/coursetable.html

OMIM: http://www3.ncbi.nlm.nih.gov/Omim

PAI: http://www.ebi.ac.uk/swissprot/hbi/hpi.html

ProDom: http://www.toulouse.inra.fr/prodom.html

PubMed: http://www.ncbi.nlm.nih.gov/PubMed

RasTop: http://www.geneinfinity.org/rastop/gallery.htm

RDP: http://rdp.life.uiuc.edu/index.html

REBASE: http://rebase.neb.com/rebase

STACK: http://www.sanbi.ac.za/Dbases.html

Swiss-Prot: http://www.expasy.ch

Transfac: http://transfac.gbf.de/TRANSFAC/index.html

UniGene: http://www.ncbi.nlm.nih.gov/UniGene

YDP: http://www.proteome.com/databases

Appendix 5.2. Bioinformatics Journals

(supplemental to those listed by McCain [personal communication, November 30, 2003] and Garfield [2002])

Biochemie

Biochemistry

Bio-IT World

Biophysical Journal

Biotechniques

British Medical Journal

Cell

Cell Biology Education

Chemical and Engineering News

Drug Discovery Today

Engineering in Medicine and Biology

Genetic Epidemiology

Genetics

Genome Research

IEEE Transactions on Bio-Medical Engineering

Journal of Biological Chemistry

Journal of Molecular Biology

Journal of the American Medical Informatics Association

Journal of Theoretical Biology

Methods in Biochemical Analysis

Methods of Molecular Biology

Molecular Biology and Evolution

Molecular Medicine Today

Nature Biotechnology

Nature Genetics

Nature Structural Biology

New England Journal of Medicine

Nucleic Acid Research

Online Journal of Bioinformatics

Pharmacogenomics

Proceedings of the National Academy of Science

Proteins

Science

Science Next Wave

The Scientist

Trends in Biotechnology

References

ASN.1 Information Site (Abstract Syntax Notation One). (n.d.). Retrieved December 21, 2003, from http://asn1.elibel.tm.fr/en/index.htm

Adler, D. A., & Conklin, D. (2000). Bioinformatics. *Encyclopedia of Life Sciences.* New York: Nature Publishing Group.

Altman, R. B. (1998). Bioinformatics in support of molecular medicine. *Proceedings of the American Medical Informatics Association Annual Symposium*, 53–61.

Altman, R. B. (2000). The interactions between clinical informatics and bioinformatics: A case study. *Journal of the American Medical Informatics Association, 7*(5), 439–443.

Altman, R. B. (2003). The expanding scope of bioinformatics: Sequence analysis and beyond. *Heredity 90*(5), 345. Retrieved December 21, 2003, from http://www.nature.com/cgi-taf/DynaPage.taf?file=/hdy/journal/v90/n5/full/6800225a.html

Altman, R. B., & Dugan, J. M. (2003). Defining bioinformatics and structural bioinformatics. *Methods in Biochemical Analysis, 44*, 3–14.

Altman, R. B., & Koza, J. (1996). A programming course in bioinformatics for computer and information science students. In L. Klein & T. E. Hunter (Eds.), *Pacific Symposium in Biology* (pp. 73–84). Singapore: World Scientific.

Altshul, S. F., Madden, T. L., Schaffer, A. A., Zhang, J., & Zhang, Z. (1997). Gapped BLAST and PSI-BLAST: A new generation of protein database search programs. *Nucleic Acid Research, 25*(17), 3389–3402.

Andersson, S., Larsson, K., Larsson, M., & Jacob, M. (1999). *Biomathematics: Mathematics of biostructures and biodynamics.* Amsterdam: Elsevier.

Apweiler, R., Birney, E., Brazma, A., Brooksbank, C., Cameron, G., Camon, E., et al. (2003). The European Bioinformatics Institute's data resources. *Nucleic Acids Research, 31*, 43–50.

Apweiler, R., Biswas, M., Fleischmann, W., Kanapin, A., Karavidopoulou, Y., Kersey, P., et al. (2001). Proteome analysis database: Online application of InterPro and CluSTr for the functional classification of proteins in whole genomes. *Nucleic Acids Research, 29*(1), 44–48.

Ashburner, M., Ball, C. A., Blake, J. A., Botstein, D., & Butler, H. (2000). Gene ontology: Tool for the unification of biology. The Gene Ontology Consortium. *Nature Genetics, 25*, 25–29.

Attwood, T. (2000). The Babel of bioinformatics. *Science, 290*, 471–472.

Baeza-Yates, R. & Ribeiro-Neto, B. (1999). *Modern information retrieval.* Reading, MA: ACM Press and Addison-Wesley.

Bairoch, A., & Apweiler, R. (2000). The SWISS-PROT protein sequence database and its supplement TrEMBL. *Nucleic Acids Research, 28*, 45–48.

Baldi, P., & Brunak, S. (2001). *Bioinformatics: The machine learning approach.* Cambridge, MA: MIT Press.

Bard, J. (2003). Ontologies: Formalising biological knowledge for bioinformatics. *Bioessays, 25*(5), 501–506.

Barnes, M. R., & Gray, I. C. (Eds.). (2003). *Bioinformatics for geneticists.* Hoboken, NJ: Wiley.

Basi, G., Clum, R., & Modi, C. (2003, April 1). An array of opportunities. *Science Next Wave.* Retrieved December 21, 2003, from http://nextwave.sciencemag.org/cgi/content/f ull/2003/04/09/2

Baxevanis, A. D., & Ouellette, B. F. F. (2001). *Bioinformatics: A practical guide to the analysis of genes and proteins* (2nd ed.). New York: Wiley-Interscience.

Bayat, A. (2002). Science, medicine, and the future: Bioinformatics. *British Medical Journal 324,* 1018–1022. Retrieved June 24, 2003, from http://bmj.com/cgi/reprint/324/7344/1018

Benaim Jalfon, C. (2001). *Analysis of the bioinformatics industry.* Unpublished master's thesis, MIT Sloan School of Management, Cambridge, MA.

Benoît, G. (2002). Data mining. *Annual Review of Information Science and Technology, 36,* 265–310.

Benson, D. A., Karsch-Mizrachi, I., Lipman, D. J., Ostell, J., Rapp, B. A., & Wheeler, D. L. (2000). GenBank. *Nucleic Acids Research, 28*(1), 15–8.

Benson, G., & Page, R. (Eds.). (2003). *Algorithms in bioinformatics: Proceedings of the Third International Workshop, WABI, 2003.* Berlin: Springer.

Bergeron, B. P. (2003). *Bioinformatics computing.* Upper Saddle River, NJ: Prentice Hall.

Berman, H. M., Westbrook, J., Feng, Z., Gilliland, G., Bhat, T. N., Weissig, H., et al. (2000). The Protein Data Bank. *Nucleic Acids Research, 28*(1), 235–242.

Bernstein, F. C., Koetzle, T. F., Williams, G. J., Meyer, E. F., Brice, M. D., Rodger, J. R., et al. (1977). The Protein Data Bank: A computer-based archival file for macromolecular structures. *European Journal of Biochemistry, 80*(2), 319–324.

Bernstein, R. L. (2001). *What is bioinformatics?* Retrieved December 22, 2003, from http://www.swbic.org/education/bioinfo

Blake, J. (1999). From functional names to naming functions. *Second International Nomenclature Workshop INW2.* Retrieved January 25, 2004, from http://www.gene. ucl.ac.uk/nomenclature/workshop/INW2-abstract-txt.shtml

Boguski, M. S. (1999). Biosequence exegesis. *Science, 286*(5439), 453–455.

Böhringer, S., Gödde, R., Böhringer, D., Schulte, T., & Epplen, J. T. (2002). A software package for drawing ideograms automatically. *Online Journal of Bioinformatics 1,* 51–61.

Bowtell, D., & Sambrook, J. (2003). *DNA microarrays: A molecular cloning manual.* Cold Spring Harbor, NJ: Cold Spring Harbor Laboratory Press.

Brass, A. (2000). Bioinformatics education: A UK perspective. *Bioinformatics, 16*(2), 77–78.

Brazma, A., Hingamp, P., Quackenbush, J., Sherlock, G., Spellmann, P., Stoeckert, C., et al. (2001). Minimum information about a microarray experiment (MIAME): Towards standards for microarray data. *Nature Genetics, 29*(4), 365–371.

Brzeski H. (2002). An introduction to bioinformatics. *Methods of Molecular Biology, 187,* 193–208.

Burge, C. (2002). Top ten future challenges for bioinformatics. *Genome Technology, 17.* Retrieved June 24, 2003, from http://genes.mit.edu/burgelab/topten.htm

Burland, Timothy G. (2000). DNASTAR's Lasergene sequence analysis software. In S. Misener & S. A. Krawetz (Eds.), *Bioinformatics: Methods and protocols.* Totowa, NJ: Humana.

Butte, A. J. (2001). Challenges in bioinformatics: Infrastructure, models and analytics. *Trends in Biotechnology, 19,* 159–160.

Calandra, B. (2002, September, 2). Bioinformatics knowledge vital to careers. *The Scientist, 16*(17), 53–54.

Calvanese, D., Lenzerini, M., & Motwani, R. (Eds.). (2003). *Proceedings of the 9th International Conference on Database Theory, ICDT, 2003*. (Lecture Notes in Computer Science, 2572). New York: Springer.

Card, S. K., Mackinlay, J. D., & Shneiderman, B. (2002). *Readings in information visualization: Using vision to think*. San Francisco: Morgan Kaufmann.

Chawla, H. S. (2002). *Introduction to plant biotechnology*. Enfield, NJ: Science Publishers.

Chen, C. (1999). *Information visualization and virtual environments*. London: Springer.

Chen, H. (1997). A concept space approach to addressing the vocabulary problem in scientific information retrieval: An experiment on the Worm Community System. *Journal of the American Society for Information Science, 48*(1), 17–31.

Cheung, V. A., Dalrymple, H. L., Narasiman, S., Watts, J., Schuler, G., & Raap, A. K. (1999). Making and reading microarrays. *Nature Genetics, 21*(1 Supplement), 15–19.

Chicurel, M. (2002). Bioinformatics: Bringing it all together. *The Scientist, 419*, 751–757.

Corruble, V., & Ganascia, J.-G. (1997). Induction and the discovery of the causes of scurvy: A computational reconstruction. *Artificial Intelligence, 91*, 205–223.

Cory, K. A. (1997). Discovering hidden analogies in an online humanities database. *Computers and the Humanities 31*(1), 1–12.

Cottle, H. (2001, June 29). Bioinformatics for beginners. *Science Next Wave*. Retrieved December 23, 2003, from http://nextwave.sciencemag.org/cgi/content/full/2001/06/27

Davies, R. (1988). The creation of new knowledge by information retrieval and classification. *Journal of Documentation, 45*(4), 273–301.

de Jong, H. (2002). Modeling and simulation of genetic regulatory systems: A literature review. *Journal of Computational Biology, 9*(1), 67–103.

Denn, S. O., & MacMullen, W. J. (2002). The ambiguous bioinformatics domain: A conceptual map of information science applications to molecular biology. *Proceedings of the 65th Annual Meeting of the American Society for Information Science & Technology*, 556–558.

Dougherty, T. J., & Projan, S. J. (2003). *Microbial genomics and drug discovery*. New York: Marcel Dekker.

Duggan, D. J. (1999). Expression profiling using cDNA microarrays. *Nature Genetics, 21*(1 Supplement), 10–14.

Durbin, R., Eddy, S., Krogh, A., & Mitchison, G. (1998). *Biological sequence analysis: Probabilistic models of proteins and nucleic acids*. Cambridge, UK: Cambridge University Press.

Dwyer, R. A. (2003). *Genomic Perl: From bioinformatics basics to working code*. Cambridge, UK: Cambridge University Press.

Eisen, M. B. (1998). Cluster analysis and display of genome-wide expression patterns. *Proceedings of the National Academic of Science, 95*(25), 14863–14868.

Eisen, M. B., & Brown, P. O. (1999). DNA arrays for analysis of gene expression. *Methods of Enzymology, 303*, 179–205.

Ellis, L. (2003a). *What is bioinformatics? 2000–2001*. Retrieved June 24, 2003, from http://www.binf.umn.edu/whatsbinf2000.html

Ellis, L. (2003b). *What is bioinformatics? 2003*. Retrieved June 24, 2003, from http://www.binf.umn.edu/whatsbinf.html

Etzold, T., Ulyanov, A., & Argos, P. (1996). SRS: Information retrieval system for molecular biology data banks. *Methods of Enzymology, 266*, 114–128.

European Bioinformatics Institute. (1999). *Data mining for bioinformatics: Towards in silico biology*. Retrieved June 24, 2003, from http://industry.ebi.ac.uk/datamining99

Ewens, W. J., & Grant, G. R. (2001). *Statistical methods in bioinformatics: An introduction*. New York: Springer.

Fagan, R., & Swindells, M. (2000). Bioinformatics, target discovery and the pharmceutical/biotechnical industry. *Current Opinion in Molecular Therapeutics, 2*, 655–661.

Fogel, G. B., & Corne, D. W. (2003). *Evolutionary computation in bioinformatics*. San Francisco: Morgan Kaufman.

Frasconi, P. (Ed.). (2003). *Artificial intelligence and heuristic methods in bioinformatics*. Amsterdam: IOS Press.

Fuchs, R. (2001). From sequence to biology: The impact on bioinformatics. *Bioinformatics, 18*, 505–506.

Funahashi, A., Tanimura, N., Morohashi, M., & Kitano, H. (2003). CellDesigner: A process diagram editor for gene-regulatory and biochemical networks. *Biosilico 1*, 159–162.

Gardiner, K. (2001). Bioinformatics for biologists. *Trends in Genetics, 17*, 736–737.

Gardy, J., & Brinkman, F. (2003, January 17). The benefits of interdisciplinary research: Our experiences with pathogen bioinformatics. *Science Next Wave*. Retrieved June 24, 2003, from http://nextwave.sciencemag.org/cgi/content/full/2003/01/15/1

Garfield, E. (2002). *Bioinformatics*. Retrieved December 21, 2003, from http://www.garfield. library.upenn.edu/papers/bio/bioinformatics112002.html

Gascuel, O., & Moret, B. M. E. (Eds.). (2001). *Algorithms in bioinformatics: Proceedings of the First International Workshop, WABI 2001*. (Lecture Notes in Computer Science, 2149). New York: Springer.

Gascuel, O., & Sagot, M.-F. (Eds.). (2000). *Computation biology: First International Conference on Biology, Informatics, and Mathematics, JOBIM 2000*. (Lecture Notes in Computer Science, 2066). New York: Springer.

Gatto, J. G. (2003). The changing face of bioinformatics. *Drug Discovery Today 8*, 375–376.

Gene Ontology Consortium. (2001). Creating the gene ontology resource: Design and implementation. *Genome Research, 11*, 1425–1433.

Gene Ontology Consortium. (2004). What is GO? Retrieved January 25, 2004, from http://www.ebi.ac.uk/faq/cgi-bin/go?editCmds=hide&file=10&keywords=login&show Attributions=hide&showLastModified=hide&showModerator=hide

Gerstein, M. (2000). Integrative database analysis in structural genomics. *Nature Structural Biology, 7*(Supplement) 960–963.

Gerstein, M., & Jansen, R. (2000). The current excitement in bioinformatics, analysis of whole-genome expression data: How does it relate to protein structure and function. *Current Opinions in Structural Biology, 10*, 574–584.

Gibas, C., & Jambeck, P. (2001). *Developing bioinformatics computer skills*. Sebastopol, CA: O'Reilly.

Gibson, G., & Muse, S. V. (2002). *A primer of genomic science*. Sunderland, MA: Sinauer.

Gilbert, D. (2000). Free software in molecular biology for Macintosh and MS Windows computers. In S. Misener & S. A. Krawetz (Eds.), *Bioinformatics: Methods and protocols* (pp. 149–184). Totowa, NJ: Humana.

Goodman, L. (2003, July 15). Making a genesweep: It's official! *Bio-IT World News*, 12.

Gordon, M. D., & Dumais, S. (1998). Using latent semantic indexing for literature-based discovery. *Journal of the American Society for Information Science, 49*(8), 674–685.

Gordon, M. D., & Lindsay, R. K. (1996). Toward discovery support systems: A replication, reexamination, and extension of Swanson's work on literature-based discovery of a connection between Raynaud's disease and fish-oil. *Journal of the American Society for Information Science, 47*(2), 116–128.

Grant, A. M., Moshyk, A. M., Kushniruk, A., & Moehr, J. R. (2003). Reflections on an arranged marriage between bioinformatics and health informatics. *Methods of Information in Medicine, 42*(2), 116–120.

Guerra, C., & Istrail, S. (2000). *Mathematical models for protein structure analysis and design*. (Lecture Notes in Computer Science, 2666). New York: Springer.

Guerrinia, V. H., & Jackson, D. (2000). Bioinformatics and extended [sic] markup language (XML). *Online Journal of Bioinformatics, 1*, 1–13.

Guigó, R., & Gusfield, D. (Eds.). (2002). *Algorithms in bioinformatics: Proceedings of the Second International Workshop, WABI 2002* (Lecture Notes in Computer Science, 2452). New York: Springer.

Gywne, P. (2002, June 14). Informatics: Integrating information. *Science.* Retrieved June 24, 2003, from http://recruit.sciencemag.org/feature/advice/informatics.shtml

Hamilton, R. (2004). *Enzyme function prediction from multiple sequence alignments.* Edinburgh, UK: University of Edinburgh, Biocomputing Research Unit. Retrieved January 19, 2004, from http://www.bru.ed.ac.uk/~rhamilto/rsh_phd.html

Harris, M. A., & Parkinson, H. (2003). Conference report: Standards and ontologies for functional genomics: Towards unified ontologies for biology and biomedicine. *Comparative and Functional Genomics, 4,* 116–120.

Henry, C. M. (2002). Careers in bioinformatics: Field is not significantly affected by economic downturn; Qualified people are still hard to find. *Chemical and Engineering News, 79*(1), 47–55.

Hillisch, A., & Hilgenfeld, R. (2003). *Modern methods of drug discovery.* Boston: Birkhauser.

Hucka, M., Finney, A., Sauro, H. M., Bolouri, H., Doyle, J. C., Kitano, H., et al. (2003). The systems biology markup language (SBML): A medium for representation and exchange of biochemical network models. *Bioinformatics, 19,* 524–531.

HUGO Nomenclature Committee. (1999). *Second International Nomenclature Workshop INW2.* Retrieved January 25, 2004, from http://www.gene.ucl.ac.uk/nomenclature/workshop/INW2-abstract-txt.shtml

Human Genome Project Information. (2002). Retrieved June 24, 2003, from http://www.ornl.gov/hgmis/faq/seqfacts.html

Ibba, M. (2002). Biochemistry and bioinformatics: When worlds collide. *Trends in Biotechnology, 20,* 53–54.

International Immunogenetics Information System. (2004). *IMGT Marie-Paule page.* Retrieved January 25, 2004, http://imgt.cines.fr:8104/textes/IMGTindex/ontology.html

Institute for Systems Biology. (2003). *Predictive and preventive medicine.* Retrieved March 5, 2004, from http://www.systemsbiology.org/Default.aspx?pagename=predictiveandpreventive

Jain, E. (2003). Practical bioinformatics. *Pharmacogenomics, 4*(2), 119–121.

Jajuga, K., Sokolowski, A., & Bock, H.-H. (Eds.). (2002). *Classification, clustering and data analysis: Recent advances and applications.* New York: Springer.

Jamison, D. C. (2003). *Perl programming for biologists.* Hoboken, NJ: Wiley-Liss.

Jenson, D. (2002, October 18). Job market hype: Media misinformation puts a spin on biotech job market. *Science Next Wave.* Retrieved June 24, 2003, from http://nextwave.sciencemag.org/cgi/content/full/2002/10/17/2

Jenssen, T.-K., Öberg, L. M. J., Andersson, M. L., & Komorowski, J. (2001). Methods for large-scale mining of networks of human genes. In R. Grossman, V. Kumar, & J. Han (Eds.), *Proceedings SIAM Conference on Data Mining (SDM, 2001).* Retrieved December 21, 2003, from http://www.siam.org/meetings/sdm01/pdf/sdm01_10.pdf

Jiang, T., Xu, Y., & Zhang, M. Q. (2002). *Current topics in computational molecular biology.* Cambridge, MA: MIT Press.

Juvan, P. (2001). *Web-enabled knowledge-based analysis of genetic data* (Lecture Notes in Computer Science, 2119) (pp. 113–119). Berlin: Springer.

Kalow, W., Meyer, U. A., & Tyndale, R. F. (2001). *Pharmacogenomics.* New York: Marcel Dekker.

Kantardzic, M. (2003). *Data mining: Concepts, models, methods, and algorithms.* Piscataway, NJ: IEEE Press/Wiley-Interscience.

Karp, D. D., Paley, S., & Zhu, J. (2001). Database verification studies of Swiss-Prot and GenBank. *Bioinformatics, 17*(6), 526–532.

Kasabov, N. K. (2003). *Evolving connectionist systems: Methods and applications in bioinformatics, brain study and intelligent machines.* New York: Springer.

Keim, D. A. (1997). Visualization Techniques in exploring databases [Invited tutorial]. International Conference on Knowledge Discovery in Databases (KDD'97). Newport Beach, CA.

Kitano, H. (2003). A graphical notation for biochemical networks. *Biosilico, 1,* 169–176.

Kohane, I. S., Kho, A. T., & Butte, A. J. (2003). *Microarrays for an integrative genomics.* Cambridge, MA: MIT Press.

Kolatkar, P. (2002). Biocomputing at Singapore's top R&D institutes. *Science Next Wave.* Retrieved December 22, 2003, from http://nextwave.sciencemag.org/car.dtl

Korf, I., Yandell, M., & Bedell, J. (2003). *BLAST.* Farnham, UK: O'Reilly.

Koski, T. (2001). *Hidden Markov models for bioinformatics.* Dordrecht, The Netherlands: Kluwer Academic.

Kossida, S., Tahri, N., & Daizadeh, I. (2002, December). Bioinformatics by example: From sequence to target. *Journal of Chemical Education, 79,* 1480–1485.

Krane, D. E., & Raymer, M. L. (2003). *Fundamentals of bioinformatics.* San Francisco: Benjamin Cummins.

Krawetz, S. A., & Womble, D. D. (2003). *Introduction to bioinformatics: A theoretical and practical approach.* Totowa, NJ: Humana Press.

Lacroix, Z., & Critchlow, T. (2003). *Bioinformatics: Managing scientific data.* San Francisco: Morgan Kaufmann.

Laender, A. H. F., & Oliveira, A. L. (2002). (Eds). *String processing and information retrieval: Proceedings of the 9th International Symposium, SPIRE, 2002.* (Lecture Notes in Computer Science, 2476). New York: Springer.

Lander, E. S., Linton, L. M., Birren, B., Nusbaum, C., Zody, M. C., Baldwin, J., et al. (2001). Initial sequencing and analysis of the human genome. *Nature, 409,* 860–921.

Lanzenberger, M., Miksch, S., Ohmann, S., & Popow, C. (2003). Applying information visualization techniques to capture and explore the course of cognitive behavioral therapy. *Proceedings of the Symposium on Applied Computing,* 268–274.

Lengauer, T. (2002). *Bioinformatics: From genomes to drugs.* Weinheim, Germany: Wiley-VCH.

Lesk, A. M. (2002). *Introduction to bioinformatics.* New York: Oxford University Press.

Leszczynski, J. (Ed.). (1999). *Computational molecular biology.* Amsterdam: Elsevier.

Lim, H. A. (2002). *Genetically yours: Bioinforming, biopharming, biofarming.* River Edge, NJ: World Scientific.

Lipshutz, R. J., Fodor, S. P. A., Gingeras, T. R., & Lockhard, D. J. (1999). High density synthetic oligonucleotide arrays. *Nature Genetics, 21*(1), 20–24.

Lu, W., Janssen, J., Milios, E., & Japkowicz, N. (2001). *Node similarity in networked information spaces* (Technical Report CS-2001-03). Halifax, NS: Dalhousie University. Retrieved December 23, 2003, from http://www.cs.dal.ca/research/techreports/2001/CS-2001-03.html

Luscombe, N. M., Greenbaum, D., & Gerstein, M. (2001). What is bioinformatics? A proposed definition and overview of the field. *Methods of Information Medicine, 40,* 346–358.

Maher, R. (2001). BioML2SVG. In *Stirring XML: Visualisations in SVG.* Retrieved December 22, 2003, from http://www.svgopen.org/2003/papers/StirringXml-VisualisationsInSVG/#S3

McDonald, C. J. (2001). Hickham 2000: The maturation of, and linkages between, medical informatics and bioinformatics. *Journal of Laboratory and Clinical Medicine, 138,* 359–366.

Mewes, H.-W., Seidel, H., & Weiss, B. (Eds.). (2003). *Bioinformatics and genome analysis.* Berlin, Germany: Springer.

Miller, C., Gurd, J., & Bass, A. (1999). A RAPID algorithm for sequence database comparison: Application to the identification of vector contamination in the EMBL databases. *Bioinformatics, 15*(2), 111–121.

Misener, S. & Krawetz, S. A. (Eds.). (2000). *Bioinformatics: Methods and protocols* (Methods in Molecular Biology, 132). Totowa, NJ: Humana.

Montgomery, S. (2003). *Java for bioinformatics*. Retrieved December 23, 2003, from http://www.onjava.com/pub/a/onjava/2003/09/24/java_bioinformatics.html

Mount, D. W. (2001). *Bioinformatics: Sequence and genome analysis*. Cold Spring Harbor, NY: Cold Spring Harbor Laboratory Press.

National Center for Biotechnical Information. (2003, May 1). *What is bioinformatics?* Retrieved June 24, 2003, from http://www.ncbi.nlm.nih.gov/Education

Nishikawa, K. (2002). Information concept in biology. *Bioinformatics, 18*, 649–651.

Nucleic Acids Research. (2002). Annual database issue. Retrieved June 24, 2003, from http://nar.oupjournals.org/content/vol30/issue1

Orengo, C., Jones, D., & Thornton, J. (2003). *Bioinformatics: Genes, proteins, and computers*. Oxford, UK: BIOS Scientific.

Orengo, C. A. (1999). CORA: Topological fingerprints for protein structure families. *Protein Science, 8*(4), 699–715.

Orengo, C. A., & Taylor, W. R. (1996). SSAP: Sequential structure alignment program for protein structure comparison. *Methods of Enzymology, 266*, 617–635.

Ouzounis, C. (2002). Bioinformatics and the theoretical foundations of molecular biology. *Bioinformatics, 18*, 377–378.

Paris, C. G. (1997). Chemical structure handling by computer. *Annual Review of Information Science and Technology, 32*, 271–337.

Paris, G. (2002). Mining bioinformatics databases. *Annual Meeting of the American Society for Information Science and Technology, 39*, 452–453.

Pearson, W. R. (2001). Training for bioinformatics and computational biology. *Bioinformatics, 17*(6), 761–762.

Perner, P. (Ed.). (2002). *Advances in data mining: Applications in e-commerce, medicine, and knowledge management* (Lecture Notes in Computer Science, 2394). New York: Springer.

Pevsner, J. (2003). *Bioinformatics and functional genomics: A short course*. New York: Wiley-Liss.

Poinçot, P., Lesteven, S., & Murtagh, F. (2000). Maps of information spaces: Assessments from astronomy. *Journal of the American Society for Information Science, 51*(12), 1081–1089.

Priami, C. (Ed.). (2003). *Computational methods in systems biology: Proceedings of the First International Workshop, CMSB 2003*. (Lecture Notes in Computer Science, 2602). New York: Springer.

Primrose, S. B. (2003). *Principles of genome analysis and genomics*. Maldon, MA: Blackwell.

Raychaudhuri, S. (2002). Associating genes with gene ontology codes using a maximum entropy analysis of biomedical literature. *Genome Research, 12*, 203–214.

Rhodes, P. J., Bergeron, R. D., & Sparr, T. M. (2002). Database support for multisource multiresolution scientific data. *Proceedings of SOFSEM, 2002*, 94–114.

Rodriguez-Tomé, P. (2000). Resources at EBI. In S. Misener & S. A. Krawetz (Eds.), *Bioinformatics: Methods and protocols* (pp. 313–335). Totowa, NJ: Humana.

Roos, D. S. (2001). Bioinformatics: Trying to swim in a sea of data. *Science, 291*, 1260–1261.

Rost, B., Honig, B., & Valencia, A. (2002, July). Bioinformatics in structural genomics. *Bioinformatics, 18*, 897.

Sander, C. (2002). The journal *Bioinformatics*, key medium for computational biology. *Bioinformatics, 18*, 1–2.

Sansom, C. E., & Smith, C. A. (2000). Computer applications in biomolecular sciences. Part 2: bioinformatics and genome projects. *Biochemical Education, 28,* 127–131.

Schachter, B. (2002, June 12). Informatics moves to the head of the class. *Bio-IT World.* Retrieved June 24, 2003, from http://www.bio-itworld.com/archive/061202/class.html

Schlichting, I., & Egner, U. (Eds.). (2001). *Data mining in structural biology.* New York: Springer.

Schroeder, M., Gilbert, D., van Helden, J., & Noy, P. (2001). Approaches to visualization in bioinformatics: From dendrograms to Space Explorer. *Information Sciences, 139*(1–2), 19–57.

Schuler, G. D., Epstein, J. A., Ohkawa, H., & Kans, J. A. (1996). Entrez: Molecular biology database and retrieval system. *Methods of Enzymology, 266,* 141–162.

Searls, D. B. (2000). Bioinformatics tools for whole genomes. *Annual Review of Genomics and Human Genetics, 1,* 251–279.

Sensen, C. W. (Ed.). (2002). *Essentials of genomics and bioinformatics.* Weinheim, Germany: Wiley-VCH.

Smalheiser, N. R., & Swanson, D. R. (1996). Linking estrogen to Alzheimer's Disease. An informatics approach. *Neurology, 47*(3), 809–810.

Stapley, B. J., & Benoît, G. (2000). Biobibliometrics: Information retrieval and visualization from co-occurrences of gene names in Medline abstracts. *Pacific Symposium on Biocomputing, 5,* 538–549.

STN Database Summary Sheet (2004). Columbus, OH: Chemical Abstracts Service. Retrieved January 19, 2004, from http://info.cas.org/ONLINE/DBSS/registryss.html

Swanson, D. R. (1991). Complementary structures in disjoint science literatures. In A. Bookstein, Y. Chiaramella, G. Salton, & V. V. Raghavan (Eds.). *ACM SIGIR Annual International Conference on Research and Development in Information Retrieval,* 280–289.

Swanson, D. R. & Smalheiser, N. R. (1997). An interactive system for finding complementary literatures: A stimulus to scientific discovery. *Artificial Intelligence, 91*(2), 183–203.

Swanson, D. R. & Smalheiser, N. R. (1999). Implicit text linkages between Medline records: Using Arrowsmith as an aid to scientific discovery. *Library Trends, 48*(1), 48–59.

Swanson, D. R., Smalheiser N. R., & Bookstein A. (2001). Information discovery from complementary literatures: Categorizing viruses as potential weapons. *Journal of the American Society for Information Science, 52*(10), 797–812.

Tatusova, T. A., Karsch-Mizrachi, I., & Ostell, J. A. (1999). Complete genomes in WWW Entrez: Data representation and analysis. *Bioinformatics, 15*(7–8), 536–543.

Taylor, M. D., Mainprize, T. G., & Rutka, J. T. (2003). Bioinformatics in neurosurgery. *Neurosurgery, 52*(4), 723–731.

Tessier, D. C., Benoît, F., Rigby, T., Hogues, H., van het Hoog, M., Thomas, D. T., et al. (2000). A DNA microarray fabrication strategy for research laboratories. In C. Sensen (Ed.), *Essentials of Genomics and Bioinformatics* (pp. 209–220). Weinheim, Germany: Wiley-VCH.

Tisdall, J. D. (2001). *Beginning Perl for bioinformatics.* Sebastopol, CA: O'Reilly.

Toronen, P., Tamayo, P., Slonim, D., Mesirov, J., Zhu, Q., Kitareewan, S., et al. (1999). Analysis of gene expression data using self-organizing maps: Methods and application to hematopoietic differentiation. *Proceedings of the National Academy of Sciences, 96*(6), 2907–2912.

Tözeren. A. (2004). New biology for engineers and computer scientists. Upper Saddler River, NJ: Prentice Hall.

Tsigelny, I. F. (2002). *Protein structure prediction: Bioinformatics approach.* La Jolla, CA: International University Line.

University of California, Davis, Genome Center. (2003). *What is genomics?* Retrieved June 24, 2003, from http://genomics.ucdavis.edu/what.html

University of North Carolina, Chapel Hill. (2003). *Bioinformatics Journal Club*. Retrieved December 23, 2003, from http://ils.unc.edu/bmh/bioinfo/Bioinformatics_Programs_Brief_7-13-03.htm

Valdés-Peréz, R. E. (1999). Principles of human-computer collaboration for knowledge discovery in science. *Artificial Intelligence, 107*(2), 335–346.

Valencia, A. (2002, December). Bioinformatics: Biology by other means. *Bioinformatics, 18*, 1551–1552.

Van Haren, K. (2002 May, 24). Bioinformatics funding boost means more science training. *Science Next Wave*. Retrieved June 24, 2003, from http://nextwave.sciencemag.org/cgi/content/full/2002/05/23/1_

Venter, J. C., Adams, M. D., Myers, E. W., Li, P. W., Mural, R. J., Sutton, G. G., et al. (2001). The sequence of the human genome. *Science, 291*(5507), 1304–1351.

Wang, J. T. L., Wu, C. H., & Wang, P. (Eds.). (2003). *Computational biology and genomic informatics*. River Edge, NJ: World Scientific.

Ware, C. (2000). *Information visualization: Perception for design*. San Francisco: Morgan Kaufmann.

Watkins, K. J. (2001). Bioinformatics: Making sense of information mined from the human genome is a massive undertaking for the fledgling industry. *Chemical & Engineering News, 79*(8), 25–45.

Watson, J. D., & Crick, F. H. C. (1953). A structure for deoxyribose nucleic acid. *Nature, 171*, 737.

Weeber, M. (2001). *Literature-based discovery in biomedicine*. Unpublished doctoral dissertation, Rijksuniversiteit Groningen, The Netherlands.

Weeber, M., Vos R., Klein H., & de Jong-van den Berg, L. T. W. (2001). Using concepts in literature-based discovery: Simulating Swanson's Raynaud- fish-oil and migraine-magnesium discoveries. *Journal of the American Society for Information Science, 52*(7), 548–557.

Westhead, D. R., Parish, J. H., & Twyman, R. M. (2002). *Bioinformatics*. Oxford, UK: Bios Scientific.

Wilson, C. A., Kreychman, J., & Gerstein, M. (2000). Assessing annotation transfer for genomics: Quantifying the relations between protein sequence, structure and function through traditional and probabilistic scores. *Journal of Molecular Biology, 297*(1), 233–249.

Winter, P. S., Hickey, S. O., & Fletcher, H. L. (2002). *Genetics* (2nd ed.). Oxford, UK: Bios Scientific.

Wise, M. J. (2000). Protein annotator's assistant: A novel application of information retrieval techniques. *Journal of the American Society for Information Science, 51*(12), 1131–1136.

Ye, N. (2003). *Handbook of data mining*. Mahwah, NJ: Lawrence Erlbaum.

Zauhar, R. J. (2001). University bioinformatics programs on the rise. *Nature Biotechnology, 19*, 285–286.

Zdobnov, E. M., Lopez, R., Apweiler, R., & Etzold, T. (2002a). The EBI SRS server: Recent developments. *Bioinformatics, 18*(2), 368–373.

Zdobnov, E. M., Lopez, R., Apweiler, R., & Etzold, T. (2002b). Using the molecular biology data. In C. Sensen (Ed.), *Biotechnology: Genomics and bioinformatics* (vol. 5b, pp. 265–284). New York: Wiley-VCH.

Zimmerman, K.-H. (2003). *An introduction to protein informatics*. Boston: Kluwer Academic.

CHAPTER 6

Electronic Records Management

Anne Gilliland-Swetland
University of California, Los Angeles

Introduction

What is an electronic record, how should it best be preserved and made available, and to what extent do traditional, paradigmatic archival precepts such as provenance, original order, and archival custody hold when managing it? Over more than four decades of work in the area of electronic records (formerly known as machine-readable records), theorists and researchers have offered answers to these questions—or at least devised approaches for trying to answer them. However, a set of fundamental questions about the nature of the record and the applicability of traditional archival theory still confronts researchers seeking to advance knowledge and development in this increasingly active, but contested, area of research. For example, which characteristics differentiate a record from other types of information objects (such as publications or raw research data)? Are these characteristics consistently present regardless of the medium of the record? Does the record always have to have a tangible form? How does the record manifest itself within different technological and procedural contexts, and in particular, how do we determine the parameters of electronic records created in relational, distributed, or dynamic environments that bear little resemblance on the surface to traditional paper-based environments?

At the heart of electronic records research lies a dual concern with the nature of the record as a specific type of information object and the nature of legal and historical evidence in a digital world. Electronic records research is relevant to the agendas of many communities in addition to that of archivists. Its emphasis on accountability and on establishing trust in records, for example, addresses concerns that are central to both digital government and e-commerce. Research relating to electronic records is still relatively homogeneous in terms of scope, in that most major research initiatives have addressed various combinations of the following: theory building in terms of identifying the nature of the electronic record, developing alternative conceptual models, establishing the determinants of reliability and authenticity in active and preserved electronic records, identifying functional and metadata requirements for record keeping, developing and testing preservation strategies for archival records, and prototyping automated tools and techniques. Electronic records management, however, has also experienced difficulty in being accepted as an area of theoretical and applied research; this is due to a conspicuous absence of a clear articulation of electronic records management as an intellectual area, the constant need to advocate for and persuade a range of constituencies (including archivists themselves) of the importance of this research, and a lack of viable testbeds for implementing and evaluating technological solutions.

Although a recent *ARIST* chapter addressed the preservation of digital materials in general (Galloway, 2004), this is the first *ARIST* chapter dedicated to the topic of electronic records management. The chapter defines electronic records management as an area of research, identifying major research questions and conceptual and technological developments, and also discussing methodological issues that arise. In doing so, the chapter explicates key terms and briefly reviews the historical development of electronic records management as an area of research, with an emphasis on research activities dating from 1990. The chapter concludes with the identification of outstanding research questions and emerging areas of research and development.

The chapter is international in its coverage, but does not address materials published in languages other than English. Although there is some overlap between research activities relating to preserving digitized content and electronic records management (for example, research relating to the development of preservation metadata sets and migration of digital materials), the chapter does not directly address records that were originally created using traditional media and were subsequently digitized by archivists for access or, occasionally, for preservation purposes, albeit that these "digitized records" may present some of the same research issues regarding potential loss of evidential characteristics during conversion and preservation processes.

Definitions and Definitional Issues

Determining the nature of the record and reconceptualizing the role of the archive have been dominant foci of archival theory building for more than a decade, in large part driven by the challenges faced in electronic records research, but also as a result of changes in the nature of scholarship (Acland, 1992; Bailey, 1990; Bantin, 1998b; Cox, 1994b; Gavrel, 1990; McKemmish, 2001). Definitional concerns go far deeper than drawing distinctions between common terminological differences or apparent similarities between archivists and other information professionals, which are gradually being addressed through the adoption of the cross-domain terminologies promulgated by metadata standards and high-level models such as the Open Archival Information System Reference Model (OAIS) (Consultative Committee for Space Data Systems/International Organization for Standardization, 1999). As will be discussed, these definitional concerns are critical to the developing theoretical infrastructure within which electronic records research is located.

Problematizing the Record

The record, as an information construct and as an object and subject of research and development, has particular administrative, juridical, cultural, and historical dimensions and management needs that have tended to set it apart from research in information science and technology as more broadly conceived. Arguably, this separation, which has emphasized ways in which the record is different from, rather than similar to, other types of information objects, is attributable to several factors within the archival and records management communities that are most closely identified with record-related research and development. These communities focus their attention on the record; they consider other types of information only to the extent of determining that they are nonrecord. This is not only a theoretical, but also a pragmatic consideration, because legislation and organizational policy often mandate that distinctions be made between record and nonrecord for the purposes of implementing effective legal control over records created and maintained in bureaucratic contexts (Bearman, 1990; McClure & Sprehe, 1998). Historically, the needs and concerns of archivists and records managers were poorly articulated to other communities that might be able to provide additional expertise, such as information technology and policy research. In part, this was due to a lack of empirical and technological research skills and experience, but archivists and records managers were also concerned that their research issues might be submerged or compromised if they became part of larger information research agendas. The end results were not only separation, but also isolation from wider research communities.

Today, although much about this situation has changed, and a hallmark of ongoing electronic records research is its interdisciplinary nature, the record remains a problematic construct even within the archival community. Within the U.S., there is insufficient common understanding of the nature of the record and how the record as a construct might be operationalized in digital environments, such as distributed and multiprovenancial databases where there is often not a readily discernible physical information object that corresponds to paper notions of a record (Cox, 1994b, 1996; Gilliland-Swetland & Eppard, 2000; Roberts, 1994). Moreover, as both Bearman (1992a) and McKemmish (2001) have pointed out, definitions of common concepts, such as the record or even the archives, tend to be nationally and jurisdictionally contingent; this fact has not always been recognized at the outset of transnational archival research collaborations, but it inevitably needs to be addressed as those projects attempt to develop standards in areas such as terminology and metadata.

The standard American archival glossary produced by the Society of American Archivists (SAA) defines a record as a "document created or received and maintained by an agency, organization, or individual in pursuance of legal obligations or in the transaction of business" (Bellardo & Carlin, 1992, p. 28). Other definitions augment this statement with the notions that a record comprises content, context, and structure "sufficient to provide evidence of the activity regardless of the form or medium" (International Council on Archives, Committee on Electronic Records, 1997, p. 9); that a record may comprise one or more documents; and that a record cannot be changed (that is, it must have fixity) (European Commission, 2001). Records also embody record-keeping processes and transactions. Bearman (1996, p. 6) has argued that "records are at one and the same time the carriers, products, and evidence of business transactions. ... Business transactions must create records which logically are metadata encapsulated objects." In many organizational environments, the definition can be much simpler—a record is anything that an agency or legislation treats as a record (United Nations, Advisory Committee on Co-ordination of Information Systems, 1992). Although these definitions may appear to be fairly straightforward when it comes to determining what is and is not a record in the paper environment (that is, it was created in the course of practical activity and is used as a record, it encompasses more than content, it can be a collective information object, and it requires fixity), they do not provide the researcher with much assistance in identifying how this construct manifests itself in a digital world of dynamic and interactive, distributed databases, Web pages, electronic mail, and experiential systems. Not all information systems are record-keeping systems—as Bearman (1994, p. 35) has noted, "recordkeeping systems are a special kind of information system ... recordkeeping systems are distinguished from information systems within organizations by the role they play in providing organizations with evidence of business transactions (by which is meant actions taken in the course of conducting their business, rather than 'commercial' transactions." For example, when McClure and Sprehe (1998) investigated the practices of state and federal governments with regard to records management and preservation for digital materials on agency Web sites, they found a fundamental problem with the absence of a clear definition of what constitutes a record in the Web environment. Even if identifying a record were straightforward, these definitions provide few if any criteria for assessing the quality of an active or preserved record. (For example, an organization may treat an information object as a record even if it does not conform to all or even most of the characteristics identified here. Such an information object might indeed still be a record, by some definitions, but it is likely not a very good one.)

It has been a theoretical and a practical challenge, therefore, to operationalize such definitions for electronic records research and development purposes. As Bearman points out:

The essential difference between electronic and paper records is that the former are only logical things while paper records are usually thought of as only physical things. Physical things can be stored in only one place and in one observable order, logical things can be physically housed in many places but seen together. They can appear to have different arrangements depending upon the views accorded to their users. In other words, the properties of logical things are associated with them through formal, defined, logical relations while the properties of physical things are associated with them as material objects with concrete locations, attachments and marking. (Bearman, 1996, p. 1)

The SAA Glossary does not provide much guidance to either practitioners or researchers in its definition of electronic records: "records on electronic storage media" (Bellardo & Carlin, 1992, p. 12). Indeed, this definition today could be misleading, because current research findings such as those of the InterPARES Project indicate that medium is incidental to the status of recordness for electronic records (Gilliland-Swetland, 2002; MacNeil, 2002). Recent research has defined electronic records variously—a brief review of these definitions illustrates not only the debate over the nature of the records, but illuminates the conceptual bases of some of the research approaches that have been taken. For example, electronic records are "recorded information that is communicated and maintained by means of electronic equipment in the course of conducting a transaction" (Dollar, 1992, p. 85; Roberts, 1994). In this definition, the salient aspects are that a record is a type of information that is recorded, communicated, and is a result of a transaction. The communication must be between at least two agents, the creator and the receiver, and these agents may be human or computer. One benefit of working with such a definition using a systems design approach is that it can assist with identifying, and potentially capturing, a record through its association either with a computing event, such as a transaction, or when it passes across some communication boundary. It can also be used as the basis for research approaches such as those employed by the Pittsburgh and VERS projects that emphasized the embeddedness of records within their business and other procedural contexts, (Cox, 1994c; Heazlewood et al., 1999).

Another definition is that of the InterPARES Project, which states that an electronic record is "a record that is created (made or received, and set aside) in electronic/digital form," where a record is a "document made or received in the course of a practical activity as an instrument or by-product of such activity, and set aside for action or reference" (InterPARES Glossary, www.interpares.org). This deceptively simple definition harkens back to a more diplomatic conception of the record as a document that has inherent documentary characteristics and is either a probative or a dispositive instrument (that is, it either serves as proof of, or it effects an action); Duranti (1998) has expanded this conception to include the notion of supporting and narrative documents. This definition links the record to its association with an action, but does not directly link it to the concept of evidence. Key to this definition is that, to be a record, the document must somehow be "set aside" (MacNeil, 2000a, 2000b, 2002). However, the diplomatic emphasis on the "document" as the unit of analysis, even when the salient extrinsic and intrinsic elements of documentary form are translated into their nonpaper manifestations, can make this a problematic definition to use when attempting to identify a record within a multidimensional or relational system such as a database. A more complex definition that speaks to the technological, procedural and temporal complexity of electronic records is offered by Gilliland-Swetland and Eppard:

> Records are heterogeneous distributed objects comprising selected data elements that are pulled together by activity-related metadata such as audit trails, reports, and views through a process prescribed by the business function for a purpose that is juridically required. ... Records are temporally contingent—they take on different values and are subject to different uses at different points in time. Records are also time-bound in the sense that they are created for a specific purpose in relation to a specific time-bound action. (Gilliland-Swetland & Eppard, 2000, p. 2)

To sum up what can be derived about the nature of the record, whether electronic or not, from the varying definitions used in both archival research and practice, therefore, is the following: A record is always associated with some action or event, as an agent, product, or by-product; a record includes, at a minimum, a definable set of metadata that serves to provide evidence about that action or event. The scope of these metadata will be discussed later in this chapter.

Defining Electronic Records Management as an Area of Research

Any discussion of the problems with defining first a record, and then an electronic record, as an intellectual construct, a physical information object, and a unit of analysis for the purposes of research, must also cause us to reflect on the utility of the term "electronic records management" as it is applied to this area of research. As discussed later in this chapter, this area was originally referred to as "machine-readable records." The evolution of that term into "electronic records management" reflected a movement away from a data archives approach to one that was driven by the principles of managing records, both those that were archival and those that were created and actively used within their bureaucratic contexts. "Electronic records management" today is a blanket term that refers both to the practical management of electronic records, from birth to final disposition, and to theoretical and applied research relating to the nature, management, and use of those records. It is also distinct from another, less prevalent term, "archival informatics," that has tended to be used to refer to the design, development, and use of information systems containing description and digitized versions of archival holdings (although, with a developing focus on retrieval and use of archival electronic records, for example, through the development of Persistent Archives Technology, these two areas could merge). The use of the term "electronic records management" is indicative of a rapprochement that has taken place between the practice areas of records management and archivy because archivists have become, of necessity, more involved in the design of record-keeping systems and the management of the active record to ensure that it will be technologically possible to segregate and preserve the archival record. It is also, however, indicative of a bifurcation that has always been lurking within the archival community between traditional archival management and the management of electronic records. A solution to the problems of conceptualization, imprecision, and Balkanization engendered by these terminological issues is offered in recent work emerging from the Australian archival community, which has addressed these issues by fundamentally reconceptualizing the entire area of records and archival management and research under the rubrics of "record keeping" and "record-keeping research" (although even in Australia, there is a recognition that the record-keeping and archiving community is not "contiguous" with the records and archives community [McKemmish, 2001]). This reconceptualization is premised upon several notions including the evidentiary nature of records and archives, the workflow processes associated with registry systems, and the concept of multiple provenance (McKemmish, 1994, see also McKemmish, 2001; Reed, 1994). The Australian approach, however, although influential (particularly as it is reflected in the ISO 15489 Records Management Standard, which was based on the Australian standard), has not yet met with universal acceptance.

Reinventing Archives

Several other concepts integral to research in electronic records management have been undergoing redefinition, or have recently emerged. The term "archives" (in the plural), as used by the archival field, traditionally refers not only to records that are generated in the course of organizational activities that are no longer current but are still useful, but also to the repository that takes custody of those archival records and the program through which preservation and access is ensured (Bellardo & Carlin, 1992). The standard archival definition has been challenged in two significant ways as a result of research not only in electronic records, but in record keeping in general and also in broader areas of information science. First, as the worlds of record keeping, data management, and information systems design converge, electronic records researchers have had to work to differentiate between the "archive" and "archiving" as they are used with reference to backing up and storing data and the selective and distinct processes of appraisal and transfer of inactive records of continuing value into archival control. This distinction is becoming increasingly blurred by the adoption of the OAIS Reference Model (Consultative Committee for Space Data Systems/International Standards Organization, 1999), which is being used as the underlying information and process model by several current research projects that are investigating aspects of the preservation of and access to electronic records as well as other digital materials (InterPARES, 2002; OCLC/RLG Working Group on Preservation Metadata, 2002; Thibodeau, 2001).

The second way in which the traditional concept of archives is being defined has to do with how the role of the archives is conceptualized. In many national jurisdictions, the role of the archives is traditionally a custodial one involving the notion of records passing across an archival threshold. Under this approach, the archives takes physical custody of noncurrent records transferred into its control and thenceforth is responsible for preserving the physical and intellectual integrity of those records and making them available for secondary use. One of the most significant aspects to note about this approach is that the archivist takes on a unique role in providing for the physical and moral defense of the record, as advocated by eminent archivist Sir Hilary Jenkinson. Jenkinson argued in 1944 that "the Archivist has so to govern his own and other people's conduct in relation to the Archives in his charge as to preclude to the greatest possible extent, short of locking them up and refusing all access to them, any ... modification" (Daniels & Walch, 1984, p. 20). The reasoning behind this approach is that records creators have a compelling interest to maintain reliable records for as long as those records are actively used in daily business. Once the records become inactive, however, records creators may have a less compelling interest in maintaining the reliability of the records and may even have some reason to alter inactive records so that they reflect organizational activities in a more positive light. At the point when the records become inactive, therefore, the archivist must step in and ensure that the records are transferred into the physical and intellectual control of the archives, otherwise the continued reliability and authenticity of those records cannot be guaranteed.

In the U.S., this custodial approach is described within a life cycle model first developed more than fifty years ago within the National Archives and Records Administration. The life cycle is a simple, custodial model that addresses how records are created and used. It is premised on the assumption that records usage drops rapidly soon after they are created and continues to diminish until the records are either inactive and destroyed or are judged to have continuing value and are transferred to the archives and made available to secondary users such as historical scholars, journalists, and genealogists (Atherton, 1993).

Although some major research projects, notably the UBC (for University of British Columbia) and InterPARES Projects and the Persistent Archives Technology being developed by the San Diego Supercomputer Center in association with the U.S. National Archives and Records Administration (Duranti & MacNeil, 1996a; InterPARES, 2002; Moore et al., 2000a, 2000b), are still rooted in the life cycle approach, this model has been challenged increasingly by "post-custodial" or "noncustodial" approaches and the "continuum" model first developed in Australia and increasingly applied in Northern Europe. The life cycle and continuum models lie at the center of a major debate about how not only records but also the role of the archives as a physical and intellectual entity are conceptualized.

In the early 1990s, Dollar promoted the need to transform:

> the role of archival institutions from a custodian to a regulatory and access facilitative role. ... Archivists should define a centralized archives as an "archives of last resort" and take physical custody of electronic records only when their maintenance and migration across technologies can not be assured. Archivists should facilitate access to electronic records over time by helping to develop, promote, and implement international standards that minimize hardware and software dependence. ... Archivists should identify the functional requirements for the life cycle management of recorded information. (Dollar, 1992, pp. 75–76)

Likewise, Cook (1994, p. 300) has argued against the role of archivists as merely records custodians, calling for them to "shift [their] professional attention from archives to archiving." The postcustodial approach calls for the archivist to rise above being a mere custodian of records and take on more of a role as records and record-keeping consultants and access brokers within their organizations. The noncustodial approach reflects a growing reality for many archivists that there will never be sufficient technological, fiscal, or human resources to take physical custody of archival electronic records and that the records instead should remain within the record-keeping system and environment where they were created, but be subject to archival requirements and supervision. The approaches come together in advocating that archivists exercise an important intellectual role with relation to records rather than necessarily take all or any archival records into physical custody. Instead, archivists are to be involved from the inception of the record-keeping system in articulating functional and metadata requirements, monitoring compliance with these requirements by records creators, and brokering secondary user access to archival records that are held within the system (Cunningham, 1996). These approaches also incorporate the ideas that the record is more than what can be seen in a physical manifestation as paper in folders and boxes. The archives is viewed as much as a conceptual space as it is as a physical space, thus beginning to build an important bridge toward broader, record-keeping discourse and postmodern notions of "the archive" (in the singular) as both a place and a reflection of social and institutional authority and power (Cook & Schwartz, 2002; Harris, 2001; Ketelaar, 2002). Upward (1996, p. 5) presaged the considerable recent writing on this topic when he remarked on the "over-dependence upon the significance of archives as a physical space within which we hold society's most important legal, administrative and historical record." As already noted, postcustodial and noncustodial approaches have practical appeal, in that there is a growing realization that many records, especially those contained in databases, have the best chance of being preserved with their evidential value intact if they remain within the active record-keeping system being maintained by the creating unit, providing that both the system and unit continue to abide by technological and procedural requirements established by the archives. Moreover, archivists have more likelihood of raising their status within the organization, as well as being able to preserve archival records, if they come out of the archives and interact with and provide advice to those who are designing record-keeping systems and creating records. The immediate viability of this approach was tested in the New York State Department of Education's Building Partnerships Project (New York State Department of Education, 1994a, 1994b). There is, however, a need for future research to conduct a systematic evaluation of the effectiveness of noncustodial approaches on records preservation over longer periods of time, as bureaucratic regimes change.

The postcustodial approach also has theoretical appeal in that it promotes a continuum approach to record keeping. Upward (1996, 1997) delineates the Australian archivists' conceptualization of the records continuum as a more sophisticated way to think about the nature, role, use, and life of records moving through four dimensions: create, capture, organize, and pluralize. Upward (1996, p. 9) explains that "a records continuum is continuous and is a time/space construct not a life model ... no separate parts of a continuum are readily discernible, and its elements pass into each other ... it is built around 4 axes: identity, evidentiality, transactionality and recordkeeping entity. The axes encapsulate major themes in archival science, and each axis presents four coordinates which can be linked dimensionally." The continuum model today underpins Australian record-keeping practice and research, as is evidenced in both the *DIRKS Manual* distributed by the Australian National Archives (National Archives of Australia, n.d.) and the Recordkeeping Metadata Schema (RKMS) discussed later in this chapter. The continuum model has also been influential elsewhere, most notably in countries such as the Netherlands, but also in smaller, nongovernmental archives in the United States.

The debate between the custodial life-cycle approach and the continuum approach with its noncustodial option is a critical one in the electronic records research community, where research projects have tended to be premised upon either one approach or the other. In international research and standards development, in particular, the models often come into direct conflict. However, as Bantin (1998b, p. 18) has pointed out, "it is simply not a choice between one extreme or another, but a much more complicated and rich process or dialectic of combining and joining old and new into a modified theoretical construct." The outcome to date of such ferment, however, has been considerably expanded notions of both the role and expected activities of the archivist and an increasing level of discussion about the role and nature of the archives as a conceptual as well as physical space.

Development of Electronic Records Management as an Area of Research

Archivists and records managers have been identifying, preserving, and providing access to data sets and records generated or maintained using computers since the 1960s. The implementation in the 1990s, however, of high-profile agendas for electronic records research and development by national funding agencies, such as the U.S. National Historical Publications and Records Commission (NHPRC), and by government archival institutions in North America, Europe, and Australia—as well as the fallout from high profile litigation—have provided the impetus for the development of a strong evidence-based approach to the management of electronic records even while researchers and practicing archivists are grappling with the status of those records as technological and social constructs (Bearman, 1993, 1994; National Historical Publications and Records Commission, 1991). The development of electronic records management parallels developments in the record-keeping technology itself. Because of their charge to preserve the noncurrent, but still useful records of their organizations, archivists have found themselves in an unprecedented engagement, in some cases together with government and scientific agencies, in assessing the preservation implications of the new technologies and media on which those records will be created (National Academy of Public Administration, 1989; National Institute of Standards and Technology, 1989; National Research Council, 1995a, 1995b; U.S. House of Representatives, Committee on Government Operation, 1990); identifying specifications for future record-keeping software and systems, sometimes in collaboration with commercial software developers; and recommending strategies for active record keeping (Heazlewood et al., 1999; National Archives and Records Administration, 1990; National Archives of Canada, 1990; National Archives of Canada & The Canadian Workplace Automation Research Centre, 1991; National Archives of Canada & Department of Communication,1993; New York State Department of Education, 1994a, 1994b; Thibodeau & Prescott, 1996; United Nations, Advisory Committee for the Co-ordination of Information Systems, 1992), including analyzing and making recommendations about organizational workflow (Bantin & Bernbom, 1996).

In other words, as envisaged a decade ago by the postcustodialists, archivists are no longer the passive recipients of inactive records but instead are actively engaged with record keeping from the point of record-keeping system design and workflow development.

As mentioned earlier in this chapter, electronic records management was referred to until the late 1980s as the management of "machine-readable records," and these early stages of the development of the field can be characterized by data-centric approaches and a practice, rather than a research, orientation. Some of the earliest machine-generated materials with which archivists worked in the U.S. were a few data sets created using punch cards as part of World War II data processing applications such as firing tables, cryptology, aerodynamics, and meteorology. Even though the creation of records using computing technology was in its infancy, archivists as well as information scientists were challenged by the vision of Vannevar Bush, director of the Office of Scientific Research and Development (OSRD) in the 1940s. Bush anticipated research developments such as the work today of scientists at the San Diego Supercomputer Center, and demonstrated in his 1945 *Atlantic Monthly* article, "As We May Think," a keen awareness of the inter-relationships between record creation technologies and processes and the accumulation and exploitation of vast stores of knowledge. He wrote that:

> A record, if it is to be useful to science, must be continuously extended, it must be stored, and above all it must be consulted. Today we make the record conventionally by writing and photography, followed by printing; but we also record on film, on wax disks, and on magnetic wires. Even if utterly new recording procedures do not appear, these present ones are certainly in the process of modification and extension. (Bush, 1945, p. 104)

After World War II, the archival field very much took its lead from developments in the field of social science research, applying little traditional archival theory and practice in its work with machine-readable records. In 1946, the Elmo Roper Organization created one of the first social science data archives (SSDA), the Roper Public Opinion Research Center based at Williams College, to house machine-readable data from Roper surveys. For the next several decades, especially in the 1960s and 1970s, the SSDA community was at the center of a revolution in using quantitative methods in social science research. Unlike the social science data archives movement, however, in which universities played a leading role as repositories of research data, machine-readable records programs developed almost entirely in state and federal government settings. With no existing archival models to follow, archivists took their lead from those working with social science data such as statistical and survey files and applied these techniques to the mainframe-based, batch-processed materials generated for or by automated administrative functions such as accounts receivable and payroll (Cook & Frost, 1993). Although many of these materials had little evidential value because official or record copies were generally produced in paper form, they were retained for their value as statistically manipulable datasets. In one of the first archival articles on the subject, Morris Rieger of the National Historical Publications Commission noted the gradual shift in government to reliance upon the electronic versions of records:

> There has been a considerable growth in the special types of documentation (such as punch cards and magnetic tapes) associated with ADP [Automated Data Processing] procedures. Such documentation, when produced by governmental agencies, is necessarily of interest and concern to public archival institutions. For a long period, however, it was regarded by them as lacking in record character, as merely transitory work material linking the input and output records at the beginning and end of the ADP process. As it has become increasingly clear that creating agencies rely on parts of their ADP documentation for record purposes—preserving them for long periods or indefinitely and referring to them frequently in connection with official operations— archival attitudes are now changing, certainly on the national level. (Rieger, 1966, p. 109)

Archival consciousness as well as holdings began to grow from the 1960s onward. In 1963, Myron Lefcowitz and Robert O'Shea published a proposal in the *American Behavioral Scientist* calling for a National Archive of survey data (Lefcowitz & O'Shea, 1963), and in 1964, at its Fifth Congress, the International Council on Archives (ICA) began considering the implications of machine-readable records and the possibility of accessioning them. Encouragement from the United Nations Educational, Scientific, and Cultural Organization (UNESCO) and the newly formed United States Council on Social Science Data Archives (USCSSDA), encouraged an international effort to found social science data archives. In 1972, ICA established its Ad Hoc Working Party on the implications of ADP in archives. Also in 1972, national archival repositories in Canada, Sweden, the U.K., and the U.S. launched machine-readable records programs. The following year, the International Association for Social Science Information Service and Technology (IAS-SIST) was established and became an important cross-domain forum for those interested in machine-readable records (MRR), including archivists. IASSIST had three categories of membership: creators and disseminators of MRR, social science data archivists and librarians, and data users, especially social scientists.

Awareness of machine-readable records issues increased considerably in the 1980s with the rapid development of personal computing and computer networking. This decade saw the beginnings of several key state government machine-readable records programs, some of the most notable being those of Wisconsin, Kentucky, and New York, as well as increased activity on the subject by professional archival associations, the most influential of which were the Society of American Archivists' Committee on Automated Records and Techniques (CART), and the National Association of Government Archivists and Records Administrators' Information Technology (IT) Committee. Out of these conjunctions emerged a nascent research infrastructure in the form of programmatic bases, strategic collaborations, and intellectual forums through which to address the inevitable challenges that acquiring and preserving such records presented for archivists. In 1985 the State Archives of New York, which was to take an early lead in electronic records research, initiated the Special Media Records Project, in cooperation with the Governor's Office of Management and Productivity and nineteen state agencies. The project was to assess the adequacy of state government policies and procedures for the management of computer-generated, machine-readable records and to develop a program for the long-term preservation of selected machine-readable records at the state archives.

In 1987, the United States National Archives and Records Administration (NARA) contracted with the National Bureau of Standards (NBS), to investigate the role of standards in the creation, processing, storage, access, and preservation of electronic records. The resulting report led to NARA's strategy, adopted in 1990, for the development and implementation of standards for the creation, transfer, access, and long-term storage of electronic records to the federal government (National Institute of Standards and Technology, 1989). The strategies of most government archives during this period were still primarily data-centric in that they focused on rendering the records into software-independent form, maintaining accompanying documentation such as codebooks, and creating specialized indexes of selected materials to facilitate use. From a research perspective, these archives focused on determining the life expectancies of magnetic media used in recording the digital data, a topic that was also of interest to other communities such as electrical and sound engineers at the time, but that would gradually become less relevant as preservation became less media-dependent (Committee on Preservation of Historical Records, 1986; Cuddihy, 1980; Eaton, 1994; Geller, 1983).

During the same period, following much concern and early work by United Nations agencies such as the World Bank, the Advisory Committee for the Co-ordination of Information Systems (ACCIS) established a Technical Panel on Electronic Records Management (TP/REM). The charge to the Technical Panel was to develop guidelines for the implementation of electronic archives and records management programs for use in United Nations organizations, taking into account traditional archives and records management practice; to identify and describe standards that could facilitate effective utilization of the broad range of new technologies in U.N. organizations; and to facilitate coherent and integrated development of electronic archives and records management and electronic records transmission, so that the implementation and goals of these efforts could be jointly optimized wherever feasible (United Nations, Advisory Committee for the Co-ordination of Information Systems, 1990). What was significant about this initiative, as well as a study published by UNESCO and an article by Catherine Bailey, was that they began to frame electronic records issues within the context of traditional archival theory (Bailey, 1990; Gavrel, 1990; United Nations, Advisory Committee for the Co-ordination of Information Systems, 1990), marking the beginning of theory building around the electronic record.

Another critical component in developing electronic records awareness and seeding electronic records research initiatives was a series of Institutes on Advanced Archival Administration sponsored by NAGARA and held at the University of Pittsburgh from 1989 to 1996. These institutes not only educated government archivists, but also drew attention to the need for strengthened government management of information resources, especially records that needed to be preserved for long-term access (National Association of Government Archives and Records Administrators, 1991). In particular, government archivists recommended seeking a National Historical Publications and Records Commission program of challenge grants to develop electronic records programs; identify strategic issues such programs might encounter; implement a mechanism to establish a dialog between records administrators and information resource managers; and study applicable state laws (Olson, 1997).

In 1991, the NHPRC released a report, *Research Issues in Electronic Records*, which identified several applied research questions and called upon the archival community to undertake research and development activities to identify strategies and solutions to those questions:

1. What functions and data are required to manage electronic records in accord with archival requirements? Do data requirements and functions vary for different types of automated applications?

2. What are the technological, conceptual, and economic implications of capturing and retaining data, descriptive information, and contextual information in electronic form from a variety of applications?

3. How can software-dependent data objects be retained for future use?

4. How can data dictionaries, information resource directory systems, and other metadata systems be used to support electronic records management and archival requirements?

5. What archival requirements have been addressed in major systems development projects and why?

6. What policies best address archival concerns for the identification, retention, preservation, and research use of electronic records?

7. What functions and activities should be present in electronic records programs and how should they be evaluated?

8. What incentives can contribute to creator and user support for electronic records management concerns?

9. What barriers have prevented archivists from developing and implementing archival electronic records programs?

10. What do archivists need to know about electronic records?

The report was probably the single most important factor in developing an electronic records research front in North America, not only because it articulated research needs but also because it set the agenda for an NHPRC funding initiative devoted entirely to electronic records research and development, the first of its kind. Although today, electronic records research, with its increasingly empirical approach, emphasis on theory building, and growing convergence with the research interests of the digital libraries, digital preservation, and metadata development communities, has arguably outgrown this applied focus (which is currently being reevaluated in terms of directing it more toward translating research into practice through such activities as building model programs and education), much of the seminal research in the field for over a decade was conducted under the rubric of the NHPRC research agenda.

The NHPRC report not only set the stage for funded research related to electronic records management to develop, it also marked the end of the ascendancy of social science, data-driven approaches and the rise of a record- and evidence-driven approach informed by empirical study. A similar shift to a contextual, provenance-centered, evidential reorientation was also noted at the National Archives of Canada (Cook & Frost 1993). Both Cook and Cox argue that this shift from the so-called first generation of machine-readable records archivists to second-generation electronic records archivists indicated a new integration of electronic records management with the theoretical and practical concerns of traditional archivists (Cook, 1992; Cox, 1994a). Cox's study of whether the archival profession was prepared in the early 1990s to carry out its mission in the modern electronic information technology environment, however, concluded that the archival profession in the U.S. had not done well in structuring itself to manage electronic records, particularly in respect to educating electronic records practitioners and researchers (Cox 1994a). Cox, who also criticized the field for being reactive rather than proactive, noted the lack of consensus regarding the nature of the impact of electronic records upon archival theory and practice. He further noted that: State government archivist position descriptions did not reflect the skills and knowledge required to work with electronic records; almost no positions for electronic records specialists were advertised between 1976 and 1990; there was a very limited base of electronic records curriculum in graduate archival education programs; the advanced institute for state government archivists on electronic record and information policy offered by the University of Pittsburgh for 1989–1993 indicated that the archival profession still relied on continuing education to develop the practice base of electronic records management. Recognizing that education was a key component in creating this second generation of electronic records archivists, a CART curriculum, SAA workshops, and the first graduate school courses in electronic records management were all developed in the early 1990s (Walch, 1993a, 1993b).

The 1990s, therefore, was a critical decade for electronic records management. It saw a transition from a data-centric to a record-centric approach to electronic records management and a new emphasis on building an educational infrastructure to support the development of necessary archival expertise in the area. Most importantly for the topic of this chapter, it saw the beginnings of a robust research base, largely as a result of the funding agenda adopted by the NHPRC, which allowed archives other than very large governmental repositories to develop electronic record programs and research testbeds (notably, Indiana University, the City of Philadelphia, and states such as Michigan, Minnesota, and Mississippi). The support also allowed academic researchers to develop large-scale projects that would begin to generate a theoretical base for electronic records management and to experiment with technological requirements and tools.

The Shift from Information to Evidence

By the late 1980s, while archivists were concerned about systems obsolescence, deteriorating media, and the effect these would have on the integrity of records, they were increasingly realizing the fundamental importance of identifying what constitutes a record in the sense understood by the law (Newton, 1987). Roberts (1994) characterizes these definitional issues as especially relating to drawing distinctions between data management and administration, and the management of electronic records based upon their transactional and evidential nature. Acland writes that:

The pivot of archival science is evidence not information. Archivists do not deal with isolated and free-floating bits of information, but with their documentary expression, with what has been recently referred to in Australia as the archival document. ... A change in the traditionally perceived archival mindset is needed here to manage the records and their continuum, not the relics at the end stage in the record life cycle. ...

With the spotlight clearly on the record rather than the relic, the equilibrium can be adjusted to provide efficient, effective and innovative public record management with an intellectual control not custody axis, safeguarding and making accessible archival resources for good government, public accountability and future research needs. (Acland, 1992, pp. 58–59)

The emphasis on evidence by second-generation electronic records archivists has led to an increased research focus on the nature of the record, its legal requirements, its appraisal for legal and other values, and on preserving its evidence (Cook, 1995). However, it is important to note that the notion of evidence as applied in electronic records research is still tightly coupled with legal and business requirements, and there is an important research need to problematize evidence as a concept in order to understand the extent to which archival, historical, and cultural evidence and their requirements overlap with those of the law and of business. As previously mentioned, archivists were spurred on in this focus by the impact of high-profile, long-running litigation, such as Armstrong vs. the Executive Office of the President (a.k.a. the PROFS case), which revolved around the evidential status of electronic mail generated by the PROFS system in place in the Reagan-Bush White House and the role played by NARA in scheduling it for retention and disposition. The initial judgment, by Judge Richey, stated that electronic mail in its native state within the PROFS system was the official record because its electronic metadata resulted in it being a more complete record than a print version. The metadata, mostly the routing and header information, made it possible to identify who knew what, and when. The judgment also found that NARA had acted arbitrarily and capriciously by not promulgating adequate guidelines for the management of electronic mail (Armstrong v. Executive Office of the President, 1993; Bearman, 1993). This focus on evidence was also reflected in the NHPRC funding agenda; and it is hardly surprising, therefore, that the primary concerns of recent electronic records research activities have been with identifying functional requirements for creating reliable and preserving authentic electronic records, and the metadata and automated tools and techniques that will support those requirements.

Developing Functional Requirements for Electronic Records Management

As already discussed, one change in thinking that has occurred, in large part due to the challenges of working with electronic records, is to conceive of archives in functional rather than physical terms. Recent research has been dominated by attempts to identify unambiguously those functions as well as the functions of records creators and their records. Major projects have included the University of Pittsburgh Functional Requirements for Evidence in Electronic Recordkeeping Project (the Pittsburgh Project), the first, and probably the most influential major project funded by NHPRC, the Victorian Electronic Records Strategy (VERS) in Australia, and the Indiana University Electronic Records Project (Bantin, 1998a, 1999); the latter two based their work on the outcomes of the Pittsburgh Project, refining the functional requirements in the process. The National Archives of Canada's Information Management and Office Systems Advancement (IMOSA) Project (McDonald, 1993, 1995a, 1995b) and the United Kingdom Public Record Office's Electronic Records in Office Systems (EROS) Programme (Blake, 1998), are both examples of embedding functional requirements within electronic office systems. The Protection of the Integrity and Reliability of Electronic Records (UBC) Project resulted in a set of requirements that were subsequently built into the U.S. Department of Defense Design Criteria Standard for Electronic Records Management Software Applications (DOD 5015.2-STD) (Duranti, Eastwood, & MacNeil, 2002; U.S. Department of Defense, 1997) and have also been inputs into the European Commission's *Requirements for the Management of Electronic Records (MoReq Specification)* (European Commission, 2001) and the ongoing International research on Permanent Authentic Records in Electronic Systems (InterPARES) Project. The InterPARES Project, funded by government and private sector agencies in several countries, is further developing this work to identify conceptual requirements for reliability and authenticity not only in government, but also in science and the arts.

The Pittsburgh Project generated a set of functional requirements for good record keeping or "business acceptable communications" in different communities that were largely derived from an examination of literary warrant as well as case studies of record-keeping implementations in a range of settings. The use of literary warrant, essentially an analysis of laws, regulations, standards, guidelines, and best practices within those communities, was perhaps the most innovative aspect of this research and also resulted in the development of a methodology for identifying warrant in different settings (Cox & Duff, 1997; Duff, 1998). Based upon this analysis, the project identified three groups of attributes of evidentiality. The first of these groups addresses how a conscientious organization complies with meeting its legal and administrative accountability requirements; the second group specifies requirements for accountable record-keeping systems; and the third group specifies how the record itself needs to be created or captured, maintained, and is made available and usable. A third product of the Pittsburgh Project was a set of production rules that formally expressed each functional requirement as logical statements of simple, observable attributes that could be used by systems designers and metadata creators (Bearman, 1996; Hirtle, 2000).

The IMOSA Project, which ran from 1989 to 1992, was notable as a collaboration between several Canadian government agencies and the private sector and as an early example of integrating functional requirements into office automation systems (National Archives of Canada, 1990; National Archives of Canada & The Canadian Workplace Automation Research Centre, 1991). The resulting software, FORE-MOST (Formal Records Management for Office Systems Technologies), has been successfully applied by government agencies, and its utility in creating and maintaining reliable records was recently evaluated through case studies conducted by InterPARES (InterPARES Project, 2002). VERS, conducted by the Public Record Office Victoria in Australia working with the Department of Infrastructure, identified essential archiving requirements across the life of the record. These were identified by developing a testbed to prototype a potential system for document processing and record capture and then to test different techniques. The project concluded that it is possible to capture electronic records in a format suitable for long-term retention, with a large proportion of the contextual information automatically captured. The project delineated records capture requirements, archival system requirements, and records retrieval requirements, and also included process maps, metadata requirements, and technology cost analysis, thus laying the groundwork for future research in developing additional automated tools and techniques (Public Record Office Victoria, 1998).

Although there have been many sets of functional requirements from different theoretical stances generated since the mid-1990s, there is considerable agreement among them. Most, for example, require that an organization comply with existing warrant and ensure responsibility for record keeping. Records in the system should be able to be identified, fixed, segregated, and migrated to new software and hardware configurations. They should include an audit trail. It should also be possible to ensure that they are complete and that their physical and intellectual integrity has not been compromised in any way. The main criticism of these requirements by institutional systems staff and software vendors is that they remain too narrative and conceptual, although the Pittsburgh Project tried to obviate this through the generation of production rules, and the UBC and InterPARES Projects through the development of complex IDEF0 (Integration Definition for Function Modeling) models of the records and preservation appraisal processes. Some other key concerns with functional requirements research are that it is still struggling with fundamental definitional and conceptual issues and, without consensus on these issues, it is generating competing sets of functional requirements; few of the requirements sets have been implemented and tested iteratively and in a range of record-keeping domains due to a lack of real-life bureaucratic or archival testbeds; and perceptions on the part of institutions that the requirement sets are too complex and costly to implement, and may not reflect how people actually use software (Hirtle, 2000).

As alluded to previously, however, there have also been various methodological and theoretical points of contention between research approaches. For example, should research be deductive or inductive in its approaches? That is, should it start from theoretical first principles, as in the case of the UBC or InterPARES projects, or from observation, as has been the approach adopted by the Pittsburgh Project and most other electronic records research projects? The benefits of the first approach are that it is firmly rooted in archival principles and it underscores the continuities in role, characteristics, and use among records of all types, across time and space, regardless of media. The limitation is its restricted ability, as a consequence of being rooted only in the recognition of known characteristics of records and established principles of archival science, to discern if and when some truly new phenomenon is occurring in the electronic environment (Gilliland-Swetland, 2002). The strength of the second approach is that requirements are generated by analyzing actual electronic records and record-keeping applications. This approach also has a limitation, however, because almost all electronic records studied have, by definition, been created on systems that do not adhere to archival requirements and often, therefore, serve as poor examples of good records, this providing a weak basis for making recommendations about requirements. To counter these limitations and maximize the benefits, research projects are increasingly combining top-down and bottom-up approaches (Gilliland-Swetland, 2002, McKemmish, Acland, Ward, & Reed, 1999).

Another methodological question arises over the unit of analysis for electronic records research. Archival science as a discipline is still heavily material-centric, despite Australian work with continuum theory and record-keeping metadata that examines business, agents, business record-keeping entities and associated relationships, and mandates as well as records (McKemmish, Acland, & Reed, 1999). Diplomatics, one of the methodological approaches used by InterPARES, looks at individual, document-like objects and thus requires a close delineation of the physical and intellectual parameters of those objects, whereas archival science examines records in their aggregates and draws heavily upon different kinds of context to define both the scope of the record aggregate and its "recordness." However, in the process of electronic records research, the delineation of context as a concept has also been expanded to include technological context as well as the more customary juridical-administrative, procedural, and documentary contexts, thus making context a possible unit of analysis as well. Finally, there is, in recent years, a critical theoretical debate over the models on which functional requirements should be predicated. The InterPARES model is overtly predicated upon a life cycle archival model, whereas VERS and other Australian research activities are equally overtly predicated upon a continuum record-keeping model. This raises an important question as to whether functional, and indeed, metadata requirements can be devised that can be used regardless of the model being applied. Although such tensions may appear problematic, it is important to note that they are part of an intellectual ferment that is rapidly changing the face of archival research. Such ferment has led to significant and uniquely archival methodological advances, in particular the reconceptualization of the science of diplomatics as a tool for deriving requirements for establishing the authenticity of electronic records (Duranti, 1998; MacNeil, 2000a, 2000b), and the development of business process analysis as a tool for analyzing workflow and understanding the procedural context of record keeping.

Preserving Reliable and Authentic Electronic Records

The obsolescence of the technologies on which the records are created has been considered by many communities for some time to be more problematic than that of the media on which the records are recorded, so it is not difficult to discern overlap between archival researchers and these communities (Graham, 1994; Lesk, 1992). These obsolescence concerns are increasingly coupled with concerns over the ease with which the reliability and authenticity of an information object can be undermined due to the actions of its creators or preservers. Graham (1994, p. 1) addresses this when he introduces the concept of "intellectual preservation," which is concerned with the "integrity and authenticity of the information as originally recorded." He argues that "the ease with which an identical copy can quickly and flawlessly be made is paralleled by the ease with which a change may undetectably be made" (Graham, 1994, p. 1). Gilliland-Swetland and Eppard (2000) have argued that identifying the boundaries of such intellectually complex objects as records and then moving those objects forward through time and through migrations without compromising their authentic status is a significant research issue. In 2000, the Council on Library and Information Resources (2000, online) (CLIR) convened a group to ask "what is an authentic digital object," and to "create a common understanding of key concepts surrounding authenticity and of the terms various communities used to articulate them." As the authors of the report of that meeting note:

> "Authenticity" in recorded information connotes precise, yet disparate, things in different contexts and communities. It can mean being original but also being faithful to the original; it can mean uncorrupted but also of clear and known provenance, "corrupt" or not. … In each context, however, the

concept of authenticity has profound implications for the task of cataloguing and describing an item. It has equally profound ramifications for preservation by setting the parameters of what is preserved and, consequently, by what technique or series of techniques. (Cullen, Hirtle, Lynch, & Rothenberg, 2000, p. 4)

Among the questions asked by the report are "does the concept of an original have meaning in the digital environment?" and "what implications for authenticity, if any, are there in the fact that digital objects are contingent on software, hardware, network, and other dependencies?" (Cullen et al., 2000, p. vii). David Levy has responded that:

> One challenge comes from the fact that the digital realm produces copies on an unprecedented scale. It is a realm in which ... there are no originals (only copies—lots and lots of them) and no enduring objects (at least not yet). This makes assessing authenticity a challenge. (Levy, 2000, p. 1)

Because a recent *ARIST* chapter has reviewed research developments relating to the preservation of digital objects, this chapter focuses on the issues associated with understanding the nature of reliability and authenticity of records in the digital environment and how they need to be assured across the life of the record. Traditionally, in the life cycle model, the need for creators to rely upon their own active records, the fixity of those records, a documented unbroken chain of custody from the creators to the archivists, and the description of the archival record within a finding aid have been the perquisites of assuring authenticity of preserved records (Gilliland-Swetland, 2000; Hirtle, 2000). The UBC Project, 1994–1997, sought to identify and define the requirements for creating, handling, and preserving reliable and authentic electronic records (Duranti, 1995; Duranti & MacNeil, 1996a, 1996b; Duranti et al., 2002). The InterPARES Project, building on this work with an examination of the conceptual requirements for ensuring the continued authenticity of preserved records, found that the degree to which a record can be considered reliable depends upon the completeness of its form and the level of procedural and technical control exercised during its creation and management in its active life. Thus, reliability is the responsibility of the record creator. Authenticity, by contrast, is the responsibility of the preserver (which most commonly takes the form of archival management of inactive records) and is an absolute concept (Duranti et al., 2002; InterPARES Project, 2002). Yet again, the notion of what is reliable and authentic is heavily vested in ideas about evidence that are derived from legal, regulatory, and administrative warrant, and how that evidence is manifested in the records themselves and in record-keeping processes. The purpose of the authenticity requirements generated by InterPARES, together with the appraisal and preservation activity models demonstrating the application of those requirements, is to provide a risk management framework within which records preservers can assess the most appropriate preservation strategy or technique to use for a particular aggregation of records and to provide a blueprint for systems developers. Preservation strategies could potentially range from the familiar: computer output microform, rendering electronic records into software-independent form, and migration, to the emergent: emulation and Persistent Archives technology. Little research has examined whether the constructs of reliability and authenticity promulgated by archival requirements sets map onto the understandings of records creators and users (Park, 2001), although the second phase of InterPARES (InterPARES 2) is now examining conceptualizations of the constructs in scientific and artistic domains.

Metadata for Electronic Record Keeping

One of the first references to metadata in the archival literature was by David Wallace in 1993 (Wallace, 1993). Wallace's article raised the expectations of many in the archival community that metadata might provide the "magic bullet" to bring the problematic area of electronic records under control. In the period since he wrote, metadata has become a very specific area of research in electronic records management that encompasses much more than traditional archival description, with strong connections to metadata research agendas outside archival science (Gilliland-Swetland, 2003; Hedstrom, 2001). David Bearman explains why the need to address metadata is so pressing:

> Because the way that the records are organized on any storage device will not give evidence of their use or the business processes that employed them we must rely for such evidence on metadata (information about information systems and business processes) created contemporaneously with the record and its interaction over time with software functionality and user profiles. (Bearman, 1996, p. 1)

In record keeping, not only metadata about the record as an information object, but also event- and process-based metadata are required to document all the dimensions involved in the processes and technologies of record keeping. According to Duff and McKemmish (2000, p. 8), "a quality system requires three different types of documentation: records of business processes; business rules that control the business processes; and systems documentation." Metadata facilitates the management, continued use, and reuse of the records as they move forward through time, across space, and among users; and the responsibility for creating that metadata, through both automatic and manual means, is distributed across many different agents and domains of use (Gilliland-Swetland, 2003). It is through metadata that reliability and authenticity are documented, functional requirements are embedded in system design, and archived records and their components are described and made accessible in manipulable form to end users.

Many research projects such as the Pittsburgh Project have made recommendations about metadata, but two in particular deserve attention. The Australian Recordkeeping Metadata Project identified eight goals or purposes that metadata may serve: unique identification; authentication of records; persistence of records content, structure and context so that they can be re-presented with their meaning preserved for subsequent use; administration of terms and conditions of access and disposal; tracking and documenting use history, including record-keeping and archiving processes; enabling discovery, retrieval, and delivery for authorized users; restricting unauthorized use; and assuring interoperability in networked environments (Duff & McKemmish, 2000). These goals are embodied in the Recordkeeping Metadata Schema (RKMS), which employs a taxonomy of relationships between entity types—business, agent, records, and business record-keeping processes (McKemmish & Parer, 1998; McKemmish, Acland, & Reed, 1999; McKemmish, Acland, Ward & Reed, 1999). Arguing that "preservation metadata is the information infrastructure that supports the processes associated with digital preservation [and] more specifically ... is the information necessary to maintain the *viability, renderability* and *understandability* of digital resources over the long-term," the OCLC/RLG Working Group on Preservation Metadata (2002, p. 4) has developed an expanded conceptual structure for the OAIS information model and defined a set of metadata elements that were mapped to this conceptual structure to reflect the information concepts and requirements articulated in the OAIS Reference Model.

Ever mindful of Day's (2001, p. 8) comments that "more time and effort has [*sic*] been expended on developing conceptual metadata specifications than in testing them in meaningful applications. This is not intended as a criticism, but is just a reflection of how experimental the digital preservation area remains." The field is currently poised to move into several new areas of research. These include identifying how different types of metadata—process, event, and object-based—are going to interact in future record-keeping systems; identifying the requirements for metadata management, including more automatic ways in which metadata can be created, for example, through event triggers, inheritance, inference, or derivation, and managed by the various responsible agents (Baron, 1999; Gilliland-Swetland, 2002); and identifying techniques for long-term metadata management to ensure that metadata essential to identifying and authenticating records is preserved and that links between preserved records and associated metadata retain their referential integrity over time in the face of systems obsolescence, data migration, and evolution of metadata schema (Gilliland-Swetland, 2003).

Yet another area relates to the development of metadata-based tools and techniques that will help users working in a digital archives environment such as that delineated in the OAIS Reference Model to retrieve and manipulate electronic records and their components. Access and use have received scant attention from electronic records researchers in the past, who have been focused on identifying, acquiring, and preserving electronic records. However, as Hirtle states,

> We need self-conscious documentation by the creators and preservers of digital representations that details the methods employed in making and maintaining the representations. We also need to know what researchers need to know about the transformations from analog to digital format, as well as about any transformations that may occur as digital data are preserved. (Hirtle, 2000, p. 13)

One of the most promising developments addressing all of these potential areas of future research is the Persistent Archives technology being developed by the San Diego Supercomputer Center in collaboration with the U.S. National Archives and Records Administration. Based around the OAIS Reference Model, researchers are using computational power to ingest high volumes of records, identify commonalities in their structure, behaviors, and metadata attributes and create from these an XML (Extensible Markup Language) DTD (Document Type Definition) on the fly, and store the records as collections in infrastructure-independent form. At any later point, collections can be virtually recreated through the application of the DTD to the stored records. Moreover, the DTD can be used by researchers as a tool for querying and manipulating the records (Ludaescher, Marciano, & Moore, 2001; Moore et al., 2000a, 2000b). This work is also being factored into the ongoing research on metadata models and tools that is a part of InterPARES2.

Metadata seems likely to be a locus of considerable research and development for the foreseeable future. Although archival researchers will continue to work on questions such as "how much metadata is part of the record and how much resides outside but provides necessary context?" and "in what ways might functional requirements for record keeping be implemented in record-keeping systems through the use of metadata?" which itself begs the question of which kinds of metadata might be associated with each requirement, a whole new set of metadata questions seems to be emerging. For example, if metadata are essential to creating, managing, and preserving a reliable and authentic record, how do those metadata need to be managed? How do we ensure that a preserved record that contains a link to a metadata scheme continues over time to refer to the appropriate version of that scheme? If metadata continues to accrue around a preserved record as documentation of ongoing preservation and use processes, how do we ensure that only necessary metadata is preserved over time? Should we be building record-keeping systems for metadata? Another evolving area of research relates to use: for example, how to provide users of an OAIS-based archive with information packages they themselves specify and how to support increased demand for interoperability of systems containing preserved archival records with other information systems. In both cases, metadata will play an essential role.

Other Areas for Further Research

This chapter has described the movement in electronic records research away from concentration on the physical record to the record as an intellectual object embedded in a strong procedural and juridical-administrative context. This movement has been characterized as a change in emphasis from content—a data-centric perspective inherited from the data archives community—to context, with the subsequent expansion of the notion of context in archival theory. Certain contexts, however, have been privileged, thus largely excluding the social dimensions of electronic records. The social and cultural construction of the record is a subject of much intellectual activity in other areas of archival science at present, but this discussion has yet to be extended to the electronic record. Although information and computer scientists, preservationists, and digital library developers are all now interacting with electronic records researchers, no sociologists or anthropologists are involved. Conversely, the theory that is developing out of electronic records is only slowly being recognized as records theory and applied to records regardless of medium. There remains a strong focus on whether anything is qualitatively different about the electronic record. These two research directions should begin to inform each other, instead of progressing along separate trajectories.

One emerging, related area of research is digital archaeology, that is, the reconstruction of electronic records that have become unavailable as a result of damaged media or systems obsolescence. The records can be recovered through techniques such as baking, chemical treatments, searching the binary structures to identify recurring patterns, and support for the reverse engineering of the content. However, Ross and Gow (1999) warn that recovering binary patterns may not be sufficient for users to understand what those patterns represent, thus raising interesting questions about data intelligibility. This area may well become important simply because electronic records created since the advent of desktop computing, and in complex or Web-based environments, have no true paper counterparts, but have been created for almost two decades without archival requirements being factored into their design. There is a strong likelihood that if a need arises to review those records, digital archaeology may be the only viable approach. Closely linked to this, of course, is the rapidly developing area of electronic evidence forensics, in which electronic materials such as dump and backup tapes and computer hard drives are subjected to a barrage of technological processes in order to retrieve anything that might be relevant to a particular information need or, more likely, criminal investigation or litigation. Ironically, in this instance, electronic evidence that is thus retrieved has had a strong likelihood of being admitted in court, even though it does not meet the rigorous requirements for electronic records being established by the archival community.

From this review of electronic records research, three other areas emerge where there are important, under- or unaddressed research needs. Briefly, the first of these is in the area of electronic record-keeping policy, as well as associated areas such as privacy and digital rights management (Peterson, 2001). "Because of the speed of technological advances, the time frame in which we must consider archiving becomes much shorter. The time between manufacture and preservation is shrinking" (Hodge, 2000, p. 1). Hodge's study looked at "digital archiving," that is, "the long-term storage, preservation and access to information that is 'born digital' (created and disseminated primarily in electronic form) or for which the digital version is considered to be the primary archive" of scientific and technical information at the international level (Hodge, 2000, p. 2). The study identified intellectual property as a key concern relating to the acquisition of materials for archives. It points out that approaches vary from country to country because of variant national information policies or legal deposit laws. Identifying and addressing variances in the information policy infrastructure that affect electronic records management concerns has been an ongoing research focus within the InterPARES Projects. A study by Gilliland-Swetland and Kinney (1994) also identified rights management as a critical element in ensuring long-term access to preserved records relating to individuals communicating electronically in group settings such as electronic conferences. Hodge (2000, p. 12) identified several digital archiving access issues that relate to rights management: "What rights does the archive have? What rights do various user groups have? What rights has the owner retained? How will the access mechanism interact with the metadata created by archives to ensure that these rights are managed properly?" A further issue relates to the implications of acquiring and attempting to preserve electronic records that are encoded in software protected not only by copyright but also by patent restrictions.

A second closely related area is the need for economic metrics for assessing the costs of creating, preserving, making available, and using reliable and authentic electronic records over periods of time that may be longer than the lives of the creators and their institutions. This is emerging as an important area of research in the library community also, as it starts to address the financial implications of preserving digitized content and the transition from purchased to licensed electronic resources. Potentially, this is an area where libraries, archives, and digital library developers can come together to design standardized data-collection strategies and benchmarks from which metrics may be derived.

Finally, a third area of potential research would address the current and very noticeable absence of any effort to translate knowledge acquired in the process of working with electronic records to personal records and manuscripts that have been created and maintained in digital form (Cunningham, 1994). The development of electronic records management largely out of the government records community and the prevailing emphasis on legal evidence and bureaucratic record keeping has had the effect of excluding the concerns of archivists who work with personal papers and creative works that are now increasingly being born digital from this research area. Although a few research projects have addressed the preservation of Web pages, and InterPARES2 is currently investigating the creation and preservation of reliable and authentic records generated out of the scientific and the creative and performing arts communities, this entire area is ripe for study. Do approaches developed in bureaucratic environments transfer to more idiosyncratic and less controlled areas of digital records creation? Indeed, is it valid to think of materials such as Weblogs, personal electronic mail, word processed drafts of literary works, or digital photographs as records? This line of questioning brings us full circle to the need to define further our notions of what a record is, not only in the electronic environment, but also in terms of human experience. Examining records that are the products of human activities other than bureaucratic ones perhaps offers us a way to move beyond the juridically and technologically framed perspectives of electronic records research to date, perspectives that are increasingly being criticized as promoting a positivist and elitist paradigm for record keeping, toward a more inclusive and culturally based conceptualization of the human record as it is digitally inscribed.

References

Acland, G. (1992). Managing the record rather than the relic. *Archives and Manuscripts, 20*(1), 57–63.

Armstrong v. Executive Office of the President, 303 U.S. App. D.C. 107 (1993).

Atherton, J. (1993). From life cycle to continuum: Some thoughts on the records management-archives relationship. In T. Nesmith (Ed.), *Canadian archival studies and the rediscovery of provenance* (pp. 391–402). Metuchen, NJ: Scarecrow.

Bailey, C. (1990). Archival theory and electronic records. *Archivaria, 29*, 180–196.

Bantin, P. C. (1998a). Developing a strategy for managing electronic records: The findings of the Indiana University Electronic Records Project. *American Archivist, 61*, 328–364.

Bantin, P. C. (1998b). Strategies for managing electronic records: A new archival paradigm? An affirmation of our archival traditions? *Archival Issues, 23*(1), 17–34.

Bantin, P. C. (1999). The Indiana University Electronic Records Project revisited. *American Archivist, 62*, 153–163.

Bantin, P. C. & Bernbom, G. (1996). The Indiana University Electronic Records Project: Analyzing functions, identifying transactions, and evaluating recordkeeping systems: A report on methodology. *Archives and Museum Informatics: Cultural Informatics Quarterly, 10*, 246–266.

Baron, J. R. (1999). Recordkeeping in the 21st century. *Information Management Journal, 33*, 8–16.

Bearman, D. A. (1990). *Electronic records guidelines: A manual for policy development and implementation.* Pittsburgh, PA: Archives and Museum Informatics.

Bearman, D. A. (1992a). Diplomatics, Weberian bureaucracy, and the management of electronic records in Europe and America. *American Archivist, 55,* 168–181.

Bearman, D. A. (1993). The implications of Armstrong v. Executive Office of the President for the archival management of electronic records. *American Archivist, 56,* 674–689.

Bearman, D. A. (1994). *Electronic evidence: Strategies for managing records in contemporary organizations.* Pittsburgh, PA: Archives and Museum Informatics.

Bearman, D. A. (1996). *Item level control and electronic recordkeeping.* Paper presented at the Society of American Archivists 1996 Annual Meeting, San Diego, CA.

Bellardo, T., & Carlin, L. (1992). *Glossary for archivists, manuscript curators, and records managers.* Chicago: Society of American Archivists.

Blake, R. (1998). Overview of the Electronic Records in Office Systems (EROS) Programme. In *Electronic access: Archives in the new millennium* (pp. 52–58). London: Public Record Office.

Bush, V. (1945). As we may think. *Atlantic Monthly, 176,* 101–108.

Committee on Preservation of Historical Records (1986). *Preservation of historic records: Magnetic recording media.* Washington, DC: National Academy Press.

Consultative Committee for Space Data Systems/International Organization for Standardization. (1999). *Space data and information transfer system: Open archival information system: Reference model.* Geneva, Switzerland: International Organization for Standardization.

Cook, T. (1992). Easier to byte, harder to chew: The second generation of electronic records archives. *Archivaria, 33,* 202–216.

Cook, T. (1994). Electronic records, paper minds: The revolution in information management and archives in the post-custodial and post-modern era. *Archives and Manuscripts, 22,* 300–329.

Cook, T. (1995). It's 10 o'clock—Do you know where your data are? *Technology Review, 98,* 48–53.

Cook, T., & Frost, E. (1993). The electronic records archival programme at the National Archives of Canada: Evolution and critical factors of success. In M. Hedstrom (Ed.), *Electronic records management program strategies* (Archives and Museum Informatics Technical Report No. 18, pp. 38–47). Pittsburgh, PA: Archives and Museum Informatics.

Cook, T., & Schwartz, J. (2002). Archives, records, and power: From (postmodern) theory to (archival) performance. *Archival Science, 2,* 171–185.

Council on Library and Information Resources. (2000). *Authenticity in a digital environment.* Washington, DC: Council on Library and Information Resources. Retrieved July 23, 2003, from http://www.clir.org/pubs/abstract/pub92abst.html

Cox, R. J. (1994a). *The first generation of electronic records archivists in the United States: A study of professionalization.* New York: Haworth Press.

Cox, R. J. (1994b). The record: Is it evolving? *Records and Retrieval Report, 10,* 1–16.

Cox, R. J. (1994c). Re-discovering the archival mission: The Recordkeeping Functional Requirements Project at the University of Pittsburgh, a progress report. *Archives and Museum Informatics, 8,* 279–300.

Cox, R. J. (1996). The record in the information age: A progress report on reflection and research. *Records and Retrieval Report 12,* 1–16.

Cox, R. J., & Duff, W. (1997). Warrant and the definitions of electronic records: Questions arising from the Pittsburgh Project. *Archives and Museum Informatics, 11,* 223–231.

Cuddihy, E. F. (1980). Aging of magnetic recording tape. *IEEE Transactions on Magnetics, 16.*

Cullen, C. T., Hirtle, P. B., Lynch, C. A., & Rothenberg, J. (2000). *Authenticity in a digital environment.* Washington, DC: Council on Library and Information Resources.

Cunningham, A. (1994). The archival management of personal records in electronic form: Some suggestions. *Archives and Manuscripts, 22*(1), 94–105.

Cunningham, A. (1996). Journey to the end of the night: Custody and the dawning of a new era on the archival threshold. *Archives and Manuscripts, 24*(2), 312–321.

Daniels, M. F., & Walch, T. (1984). *A modern archives reader: Basic readings on archival theory and practice.* Washington, DC: National Archives and Records Service.

Day, M. (2001). Metadata for digital preservation: A review of recent developments. *Proceedings of the European Conference on Digital Libraries, ECDL 2001,* 161–172.

Dollar, C. M. (1992). *The impact of information technologies on archival principles and method.* Macerata, Italy: University of Macerata Press.

Duff, W. (1998). Harnessing the power of warrant. *American Archivist, 61,* 88–105.

Duff, W., & McKemmish, S. (2000). Metadata and ISO 9000 compliance. *Information Management Journal, 34.* Retrieved July 23, 2003, from http://rcrg.dstc.edu.au/publications/smckduff.html

Duranti, L. (1995). Reliability and authenticity: The concepts and their implications. *Archivaria, 39,* 5–10.

Duranti, L. (1998). *Diplomatics: New uses for an old science.* Lanham, MD: Society of American Archivists, Association of Canadian Archivists, and Scarecrow Press.

Duranti, L., Eastwood, T., & MacNeil, H. (2002). *Preservation of the integrity of electronic records.* Dordrecht, The Netherlands: Kluwer Academic.

Duranti, L., & MacNeil, H. (1996a). The protection of the integrity of electronic records: An overview of the UBC-MAS Research Project. *Archivaria, 42,* 46–67.

Duranti, L., & MacNeil, H. (1996b). Protecting electronic evidence: A third progress report on a research study and its methodology. *Archivi & Computer, 6*(5), 343–404.

Eaton, F. (1994). Electronic media and preservation. *IASSIST Quarterly, 181,* 14–17. Retrieved July 23, 2003, from datalib.library.ualberta.ca/iassist/publications/iq/iq18/iqvol181-2eaton.pdf

European Commission. (2001). *Requirements for the management of electronic records (MoReq Specification).* Brussels, Luxemburg: Cornwell Affiliates.

Galloway, P. (2004). Preservation of digital objects. *Annual Review of Information Science and Technology, 38,* 549–590.

Gavrel, K. (1990). *Conceptual problems posed by electronic records: A RAMP study.* Paris: UNESCO, International Council on Archives.

Geller, S. B. (1983). *Care and handling of computer magnetic storage media.* (NBS Special Publication 500–101). Washington, DC: Institute for Computer Sciences and Technology, National Bureau of Standards.

Gilliland-Swetland, A. J. (2000). *Enduring paradigm, new opportunities: The value of the archival perspective in the digital environment.* Washington, DC: Council on Library and Information Resources.

Gilliland-Swetland, A. J. (2002). Testing our truths: Delineating the parameters of the authentic archival electronic record. *American Archivist, 65*(2), 196–215.

Gilliland-Swetland, A. J. (2003). Metadata: Where are we going? In G. E. Gorman (Ed.), *International yearbook of library and information management 2003: Metadata applications and management* (pp. 17–33). London: Facet Publishing.

Gilliland-Swetland, A. J., & Eppard, P. B. (2000). Preserving the authenticity of contingent digital objects: The InterPARES Project. *D-Lib Magazine, 6.* Retrieved July 29, 2003, from http://www.dlib.org/dlib/july00/eppard/07eppard.html

Gilliland-Swetland, A. J., & Kinney, G. T. (1994). Uses of electronic communications to document an academic community: A research report. *Archivaria, 38,* 79–96.

Graham, P. S. (1994). *Intellectual preservation: Electronic preservation of the third kind.* Washington, DC: Commission on Preservation and Access. Retrieved July 23, 2003, from http://www.clir.org/pubs/reports/graham/intpres.html

Harris, V. (2001). Law, evidence and electronic records: A strategic perspective from the global periphery. *Comma, International Journal on Archives, 1–2*, 29–44.

Heazlewood, J., Dell'Oro, J., Harari, L., Hills, B., Leask, N., Sefton, A., et al. (1999). Electronic records: Problem solved? A report on the Public Record Office Victoria's Electronic Records Strategy. *Archives and Manuscripts, 27*(1). Retrieved July 23, 2003, from http://www.prov.vic.gov.au/vers/published/publcns.htm#VERSPapers

Hedstrom, M. L. (2001). Recordkeeping metadata: Presenting the results of a working meeting. *Archival Science, 1*, 243–251.

Hirtle, P. (2000). Archival authenticity in a digital age. In *Authenticity in a digital environment* (pp. 8–23). Washington, DC: Council on Library and Information Resources. Retrieved July 23, 2003 from http://www.clir.org/pubs/reports/pub92/hirtle.html

Hodge, G. (2000). Best practices for digital archiving: An information life cycle approach. *D-Lib Magazine, 6*. Retrieved July 23, 2003, from http://www.dlib.org/dlib/january00/01hodge.html

International Council on Archives. Committee on Electronic Records. (1997). *Guide for managing electronic records from an archival perspective.* Paris: International Council on Archives.

InterPARES Project (2002). *The long-term preservation of authentic electronic records: Findings of the InterPARES Project.* Retrieved July 23, 2003, from http://www.interpares.org/book/index.htm

Ketelaar, E. (2002). Archival temples, archival prisons: Modes of power and protection. *Archival Science, 2*, 221–228.

Lefcowitz, M. J., & O'Shea, R. M. (1963). A proposal to establish a national archives for social science survey data. *American Behavioral Scientist, 6*, 27.

Lesk, M. (1992). *Preservation of new technology: A report of the Technology Assessment Advisory Committee to the Commission on Preservation and Access.* Washington, DC: Commission on Preservation and Access. Retrieved July 23, 2003, from http://www.clir.org/pubs/reports/lesk/lesk2.html

Levy, D. (2000). Where's Waldo? Reflections on copies and authenticity in a digital environment. In *Authenticity in a digital environment* (pp. 24–31).Washington, DC: Council on Library and Information Resources.

Ludaescher, B., Marciano, R., & Moore, R. (2001). Towards self-validating knowledge-based archives. In *11th Workshop on Research Issues in Data Engineering (Rode).* Heidelberg, Germany: IEEE Computer Society. Retrieved July 29, 2003, from http://www.sdsc.edu/~ludaesch/Paper/ride01.html

MacNeil, H. (2000a). Providing grounds for trust: Developing conceptual requirements for the long-term preservation of authentic electronic records. *Archivaria, 50*, 52–78.

MacNeil, H. (2000b) *Trusting records: Legal, historical and diplomatic perspectives.* Dordrecht, The Netherlands: Kluwer Academic.

MacNeil, H. (2002). Providing grounds for trust II: The findings of the Authenticity Task Force of InterPARES. *Archivaria, 54*, 24–58.

McClure, C. R., & Sprehe, J. T. (1998). *Analysis and development of model quality guidelines for electronic records management on state and federal Websites: Final report.* Washington DC: National Historical Publications and Records Commission. Retrieved July 23, 2003, from http://istweb.syr.edu/~mcclure/nhprc/nhprc_title.html

McDonald, J. (1993). Information management and office systems advancement. In A. Menne-Hartiz (Ed.), *Information handling in offices and archives* (pp. 138–151). New York: K. G. Saur.

McDonald, J. (1995a). Managing records in the modern office: Taming the wild frontier. *Archivaria, 39*, 70–79.

McDonald, J. (1995b) Managing information in an office systems environment: The IMOSA Project. *American Archivist, 58*, 142–153.

McKemmish, S. (1994). Understanding electronic record keeping systems: Understanding ourselves. *Archives and Manuscripts, 22*, 150–162.

McKemmish, S. (2001). *"Constantly evolving, ever mutating:" An Australian contribution to the archival metatext.* Unpublished doctoral dissertation, Monash University, Melbourne, Australia.

McKemmish, S., Acland, G., & Reed, B. (1999). Towards a framework for standardising recordkeeping metadata: The Australian Recordkeeping Metadata Schema. *Records Management Journal, 9*, 177–202.

McKemmish, S., Acland, G., Ward, N., & Reed, B. (1999). Describing records in context in the continuum: The Australian Recordkeeping Metadata Schema. *Archivaria, 48*, 3–42.

McKemmish, S., & Parer, D. (1998). Towards frameworks for standardising recordkeeping metadata. *Archives and Manuscripts, 26*, 24–45.

Moore, R., Baru, C., Rajasekar, A., Ludaescher, B., Marciano, R., Wan, M., et al. (2000a). Collection-based persistent digital archives: Part 1. *D-Lib Magazine, 6.* Retrieved July 29, 2003, from http://www.dlib.org/dlib/march00/moore/03moore-pt1.html

Moore, R., Baru, C., Rajasekar, A., Ludaescher, B., Marciano, R., Wan, M., et al. (2000b). Collection-based persistent digital archives: Part 2. *D-Lib Magazine, 6,* Retrieved July 23, 2003, from http://www.dlib.org/dlib/april00/moore/04moore-pt2.html

National Academy of Public Administration. (1989). *The effects of electronic recordkeeping on the historical record of the U.S. Government: A report for the National Archives and Records Administration.* Washington, DC: National Academy of Public Administration.

National Archives and Records Administration. (1990). *A National Archives strategy for the development and implementation of standards for the creation, transfer, access, and long-term storage of electronic records of the federal government* (National Archives Technical Information Paper No. 8). Washington, DC: National Archives and Records Administration.

National Archives of Australia. (n.d.). *The DIRKS manual: A strategic approach to managing business information.* Retrieved July 23, 2003, from http://www.naa.gov.au/record keeping/dirks/dirksman/dirks.html

National Archives of Canada. (1990). *Managing information in office automation systems: Final report on the FOREMOST Project.* Ottawa, Canada: National Archives of Canada.

National Archives of Canada & The Canadian Workplace Automation Research Centre. (1991). *IMOSA Project: Functional requirements: Corporate Information Management Application (CIMA).* Ottawa, Canada: National Archives of Canada & The Canadian Workplace Automation Research Centre.

National Archives of Canada & Department of Communication. (1993). *The IMOSA Project: An initial analysis of document management and retrieval systems.* Ottawa, Canada: National Archives of Canada, Department of Communications.

National Association of Government Archives and Records Administrators. (1991). *A new age: Electronic information systems, state governments, and the preservation of the archival record.* Lexington, KY: NASIRE/The Council of State Governments.

National Historical Publications and Records Commission (1991). *Research issues in electronic records: Report of the working meeting.* St. Paul: Published for the National Historical Publications and Records Commission by the Minnesota Historical Society.

National Institute of Standards and Technology. (1989). *Framework and policy recommendations for the exchange and preservation of electronic records, prepared for the National Archives and Records Administration.* Washington, DC: National Institute of Standards and Technology.

National Research Council. (1995a). *Preserving scientific data on our physical universe: A new strategy for archiving the nation's scientific information resources.* Washington, DC: National Academy Press.

National Research Council. (1995b). *Study on the long-term retention of selected scientific and technical records of the federal government: Working papers.* Washington, DC: National Academy Press.

New York State Department of Education. (1994a). *Building partnerships: Developing new approaches to electronic records management and preservation: Final report.* Albany, NY: New York State Department of Education.

New York State Department of Education. (1994b). *Building partnerships for electronic recordkeeping: The New York State Information Management Policies and Practices Survey: Summary of findings.* Albany, NY: New York State Department of Education.

Newton, S. C. (1987). The nature and problems of computer-generated records. In M. Cook (Ed.), *Computer generated records: Proceedings of a seminar* (pp. 1–4). Liverpool, UK: University of Liverpool.

OCLC/RLG Working Group on Preservation Metadata. (2002). *Preservation metadata and the OAIS information model: A metadata framework to support the preservation of digital objects.* Retrieved June 9, 2003, from http://www.oclc.org/research/pmwg/pm_framework.pdf

Olson, D. (1997). "Camp Pitt" and the continuing education of government archivists, 1989–1996. *American Archivist, 60*(2), 202–214.

Park, E. G. (2001). Understanding "authenticity" in records and information management: Analyzing practitioner constructs. *American Archivist, 64,* 270–291.

Peterson, G. M. (2001). New technology and copyright: The impact on the archives. *Comma: International Journal on Archives, 1–2,* 69–76.

Public Record Office Victoria. (1998). *Victorian Electronic Records Strategy Final Report.* Melbourne, Australia: Public Records Office Victoria.

Reed, B. (1994). Electronic records management in transition. *Archives and Manuscripts, 22(1),* 164–171.

Rieger, M. (1966). Archives and automation. *American Archivist, 29*(1), 109–111.

Roberts, P. (1994). Defining electronic records, documents and data. *Archives and Manuscripts, 22*(1), 14–26.

Ross, S., & Gow, A. (1999). *Digital archaeology? Rescuing neglected or damaged data resources.* London: British Library and Joint Information Systems Committee. Retrieved July 29, 2003, from http://www.ukoln.ac.uk/services/elib/papers/supporting/pdf/p2.pdf

Thibodeau, K. (2001). Building the archives of the future: Advances in preserving electronic records at the National Archives and Records Administration. *D-Lib Magazine, 7.* Retrieved July 29, 2003, from http://www.dlib.org/dlib/february01/thibodeau/02thibodeau.html

Thibodeau, K., & Prescott, D. (1996). Reengineering records management: The U.S. Department of Defense, Records Management Task Force. *Archivi & Computer, 6*(1), 71–78.

United Nations. Advisory Committee for the Co-ordination of Information Systems. (1990). *Management of electronic records: Issues and guidelines.* New York: United Nations.

United Nations. Advisory Committee for the Co-ordination of Information Systems. (1992). *Strategic issues for electronic records management: Towards Open Systems Interconnection.* New York: United Nations.

Upward, F. (1996). Structuring the records continuum part one: Postcustodial principles and properties. *Archives and Manuscripts, 24,* 268–285.

Upward, F. (1997). Structuring the records continuum part two: Structuration theory and recordkeeping. *Archives and Manuscripts, 25,* 10–35.

U.S. Department of Defense. (1997). *DOD 5015.2-STD, Design criteria standard for electronic records management software applications.* Retrieved July 29, 2003, from http://www.dtic.mil/whs/directives/corres/html/50152std.htm

U.S. House of Representatives. Committee on Government Operations. (1990). *Taking a byte out of history: The archival preservation of federal computer records* (House report 101–978). Washington, DC: Committee on Government Operations, U.S. House of Representatives.

Walch, V. I. (1993a). Automated Records and Techniques Curriculum Development Project. Committee on Automated Records and Techniques. *American Archivist, 56*, 468–505.

Walch, V. I. (1993b). Innovation diffusion: Implications for the CART curriculum. *American Archivist, 56*, 506–512.

Wallace, D. A. (1993). Metadata and the archival management of electronic records. *Archivaria, 36*, 87–110.

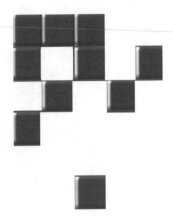

Social Informatics

CHAPTER 7

Interface Design and Culture

Ewa Callahan
Indiana University

Introduction

It is common knowledge that computer interfaces in different cultures vary. Interface designers present information in different languages, use different iconography to designate concepts, and employ different standards for dates, time, and numbers. These manifest differences beg the question of how easily an interface designed in one country can be used in and transferred to another country. Are the challenges involved in adaptation merely cosmetic or are they shaped by more profound forces? Do all cultures respond to interfaces in similar ways, or does culture itself shape user comprehension? If culture is a factor in explaining varied user reactions to comparable interfaces, what specific cultural dimensions are responsible for the divergences? Do differences reside mainly at the level of national cultures, or do they depend on other variables such as class, gender, age, education, and expertise with technology? In the face of a potentially large number of explanatory variables, how do we delimit a workable concept of culture and yet remain cognizant of other factors that might shape the results of culture and interface research?

Questions such as these have been asked in the ergonomics community since the early 1970s, when the industrialization of developing countries created a need for more research on cultural differences (Honold, 1999), resulting in an increased interest in the universal applicability of ergonomic principles. This trend continued after the reunification of Germany and the emergence of market economies in Eastern Europe (Nielsen, 1990). In the mid-1990s, as markets outside the U.S. rapidly expanded, it became necessary to develop appropriate user interfaces for non-Western cultures in order to facilitate international cooperation. This fresh impetus for research led to the development of practical guidelines and a body of case studies and examples of possible solutions. Most recently we have seen attempts to provide a theoretical foundation for cross-cultural usability engineering and experimental comparison studies (Honold, 1999).

The literature on international issues in human–computer interaction consists of three major types: 1) design guidelines, pointing out

257

issues of which designers should be aware when creating an interface for use in a foreign country, 2) theoretical works on cultural differences, which attempt to evaluate how cultural models from other fields can be applied to interface design, and 3) empirical research, dealing with various aspects of interface design in selected countries.

Design guidelines are the largest group and concentrate on culturally acceptable screen elements such as date and time formats, text flow, collating sequences, and numeric formats. These elements, if not used appropriately, may impair understanding of the interface features and functions. But design guidelines do not provide much understanding of how users from different cultures interact with computers, or any similarities and differences in communication behaviors. Theoretical works are relatively few. Research studies, also less numerous than design guidelines, typically include case studies, comparative examinations of interfaces, and/or users' preferences and performance.

The purpose of this literature review is to identify and integrate guidelines, theoretical approaches and research studies on the role of culture in interface design. The chapter will examine cultural influences in both the design process and in the resulting product, the computer interface. It will discuss features of international interfaces and the solutions and difficulties in crucial phases of creating or adapting a product to the international market.

The first part of this chapter introduces various definitions of culture and different cultural models and operationalizes the definition of culture used throughout. The next part introduces cultural interface design process solutions and discusses how human factors are involved in design. The subsequent parts discuss the linguistic and aesthetic aspects of interface design from a cultural perspective, as well as cultural differences in information structuring. The last part presents research and case studies on cultural dimensions of usability. Finally the conclusion highlights the various points outlined in the chapter, presents views on culturalization, and suggests future areas of research.

Culture

Definition of Culture

Before we attempt to answer the questions posed in the introduction, we need to establish a common understanding of the concept of *culture* and how it is understood in the human–computer interaction (HCI) literature. It seems that the term *culture* is widely understood. Operationalizing the construct of *culture*, however, results in multiple descriptions, some of which contradict one another. In their review, Kroeber and Kluckhohn (1952) collected more than 160 examples, dividing them into seven groups:

1. Descriptive definitions with an emphasis on enumeration of content

2. Historical definitions, emphasizing social heritage or tradition

3. Normative definitions, emphasizing rules, ideals, values, and behavior

4. Psychological definitions, describing culture as a problem-solving device, emphasizing the learning process or common habits

5. Structural definitions, emphasizing the patterns or organization of culture

6. Generic definitions, picturing culture as a product or artifact, emphasizing ideas and symbols

7. Incomplete definitions

Considering that their review was published in 1952, one can only imagine the much greater number of definitions that would be found today.

Straub and his colleagues (Straub, Loch, Evaristo, Karahanna, & Srite, 2002) categorized definitions of culture into three groups. The first type is *definitions based on shared values*. As the authors point out:

> Values are acquired early in life, mainly through the family and neighborhood and later through school. They provide us with fundamental values and assumptions about how things are. Once a value is learned, it becomes integrated into an organized system of values where each value has a relative priority. This value system is relatively stable in nature but can change over time, reflecting changes in culture as well as personal experience. (Straub et al., 2002, p.15)

The second type classifies *definitions based on problem solving*. As examples, the authors recall among others Kluckhohn and Leighton's (1946, p. 28 as cited in Straub et al., 2002, p. 17) definition according to which culture consists of "habitual and traditional ways of thinking, feeling and [reacting] ... that [are] characteristic of the ways a particular society meets its problems at a particular time."

The third type is *general all-encompassing definitions*, which includes definitions that are not based on shared values or outcomes. These definitions usually describe culture as a multilayered construct or as a concept characterized by various dimensions. Hoft (1996) presented an excellent review of these models and their applicability for interface design research:

- The objective culture and subjective culture model (Stewart & Bennett, 1991), where objective culture is described as the institutions and artifacts of a culture, such as its political structures, economic system, social customs, arts, crafts, and literary works. Subjective culture encompasses the psychological features of the culture, values, assumptions, and patterns of thinking.

• The iceberg model—this metaphor is used to suggest that only 10 percent of cultural characteristics are easily visible to the observer whereas 90 percent are hidden from view. Unspoken rules are hidden beneath the surface and are easier to research than unconscious rules, which lie even deeper.

• The pyramid model, introduced by the Dutch cultural anthropologist Hofstede (1997), positions culture (understood as learned, not inherited, characteristics common to a specific group or category of people) midway between personality specific to an individual (inherited and learned) and human nature (inherited) common to all human beings. The borderlines between personality and culture and culture and human nature are blurred. This model is especially useful in HCI research because it encompasses individual differences among users as well as similarities based on universal, inherited characteristics.

• The onion model presented by Trompenaars (1993) is similar to the subjective–objective culture model. The outer layer represents objective culture—the first things we encounter when introduced to a particular culture. The middle layer defines norms and values. The core of the cultural onion represents implicit characteristics of the culture and how people adapt to their environments.

Understanding these models can help us appreciate the problems designers and researchers encounter when trying to incorporate cultural dimensions in interface design. The models draw our attention to the fact that some cultural characteristics of an interface are easily visible— language first, followed by formats of date, time, and currency, then graphical representations and other aspects. Thus, these more obvious factors are most often discussed in the human–computer interaction literature and are objects of various guidelines on the cultural adaptation of interfaces. Other factors such as speech conventions, mental models, and metaphors are more difficult to grasp. These require more detailed research to understand their importance for interface users in a specific culture. Design guidelines point out the need for awareness—rather than offering ready-made solutions—and emphasize the need for consulting people native in the language and culture for which the interface is designed.

Considering the numerous definitions of culture—each emphasizing different aspects—it is surprising that most of the research studies described in this review do not operationalize the definition of culture used, possibly assuming that the concept is commonly understood. However, the following definitions do help us to understand how culture is understood in the HCI literature. Barber and Badre (1998) introduce culture as follows:

> The term "culture" is complex and a problematic one defined differently by various disciplines. We use the word "culture"—somewhat loosely—as a means of distinguishing among the different countries and their respective Web-sites. Our use of the term is not intended to be indicative of all the nuances and properties frequently implied by the term, but rather to permit discourse about the features that distinguish one country or region of the world from another in the electronic medium of the Web. (Barber & Badre, 1998, Introduction, ¶ 2)

Yeo (1996) combines several definitions:

> Culture is defined as behaviour typical of a group or class (of people) (Webster [Hypertext Interface], 1995). Culture is conceptualized as a "system of meaning that underlines routine and behaviour in everyday working life" (Bødker & Pedersen, 1991). Culture includes race and ethnicity as well as other variables and is manifested in customary behaviours, assumptions and values, patterns of thinking and communicative style (Borgman, 1992). (all citations in this paragraph are from Yeo, 1996, Cultures section, ¶ 1)

Many of the remaining studies leave it to the readers to create their own understanding of culture or introduce the research problem from the perspective of cultural differences, cultural effects, cultural factors, or cultural elements, based on the assumption that such dissimilarities among different cultural groups actually exist. Most of the studies have been conducted on an ad hoc basis without explicitly grounding the research in cultural theories. However, studies based on the work of Hofstede, which addresses theoretical concerns, have gained popularity in recent years. The Hofstede model of cultural dimensions will be discussed in a later section of this chapter.

For present purposes, culture can be understood as a complex construct encapsulating shared values, group behavioral patterns, mental models, and communication styles. However, each of these aspects will be given different levels of attention, depending on how extensively they become the focus in the interface design literature. Differences in values, for example, generally appear in the literature on graphical representations. Group behavioral patterns are discussed primarily in papers on usability testing and user interaction with the interface, although this latter category is underrepresented in the literature. Primary attention will be focused on the social and psychological differences among various cultures and their influence on design.

The concept of cultural group is also important in discussions on role of culture in interface design. In her article describing attitudes toward technology among different ethnic/linguistic groups in Switzerland, Rey (1998) defines culture as follows:

> I use the term in a pragmatic sense, and equate "culture" with the idea of "speaking the same language." This definition is practical in that it not only allows one to cite major differences—for instance between English and French nations—but it also covers more subtle variations such as vocabulary and accent or even finer difference such as those between the cultures of the upper classes and the ghettos. (Rey, 1998, ¶ 1)

Rey's definition allows us to define a *cultural group* as a congregation of people belonging to a specific culture. In interface design literature, a cultural group is usually considered a national group, although some studies do compare different linguistic groups that reside within a single country and others address issues of language group regardless of the countries in which they reside.

So, how important is the culture in interface design? Various studies have outlined the need for research on the role of culture in interface design. An analysis of the data collected during the 8th GVU (Graphics, Visualization, and Usability) WWW Survey suggested that people perceive cultural differences as important and that such differences could affect user performance and satisfaction while working with computers (Barber & Badre, 1998). Seventy-seven percent of respondents believed that in order to bring new people to the Web it will be necessary to provide sites in their native languages. An overwhelming number of respondents felt that organizations should provide multiple language translations of their Web sites, because it would be useful for more people (63 percent) or show respect for other cultures (36 percent). Forty-seven percent of respondents thought that culturally sensitive Web site design would lead to greater use of the Web (GVU's 8th WWW user survey, 1997). Some Middle Eastern and European respondents believed that American images (scenes, faces, architecture, and customs) make computers harder to learn (Barber & Badre, 1998). Because survey respondents were self-selected, the results may be biased, but they suggest that people do indeed perceive culture as important, and they indicate a need for more rigorous studies.

The degree of importance assigned to designing culturally sensitive user interfaces may vary from country to country. Nielsen (1993) introduced an unpublished study conducted by Andersen from the Technical University of Denmark, which analyzed software reviews from various computer magazines published in Denmark, the U.K., and the U.S. The American software reviews contained practically no discussion of international usability issues (0.06 comments per review). The numbers were higher for British reviews (0.41 comments per review) and Danish reviews (0.80 comments per review). Until 1990, most software was designed for the American market and did not need any modification; it could also be adopted easily in the U.K. with simple modifications. Denmark, however, has a different language and uses the Latin alphabet with additional characters, creating a greater need for localization. This may mean that the importance of culture in interface design is proportionally related to the extent to which the user's writing system differs from the designer's. However, the research on cultural perspectives on interface design emphasizes difficulties associated with differences in writing systems, especially those not based on the Latin alphabet. Most studies were conducted on interfaces designed for cultures (Asian, Middle-Eastern) considered "exotic" from a North American perspective.

Methodological Issues in Cultural Research

Researching culture and its role in the interface design process poses several difficulties. First, culture is not a homogenous construct. People from the same culture may behave similarly, but not in exactly the same ways. Moreover, the absolute boundaries of a culture cannot be determined. Chinese culture, for example, can refer to the inhabitants of Mainland China, Taiwanese, as well as Chinese-Americans, and among those groups there will be similarities as well as differences. Characterizing culture as a homogenous construct may lead to stereotyping (Fernandes, 1995).

Second, culture is a dynamic concept. Cultures interact with and influence each other and such interactions make culture susceptible to change (del Galdo, 1996). The Internet has created an unprecedented opportunity for cultural mixing, exposing many countries to Western values and the English language; at the same time as it has allowed many ethnic and linguistic groups to increase their visibility in the international arena (Nunberg, 2000).

Further, cultures are not mutually exclusive (Yeo, 1996). Users may belong simultaneously to multiple cultures, and each of them may influence preferences and behavior. Holstius (1995), for example, in presenting her model of international technology transfer, defines business culture as a cross-cut of a national culture and a set of corporate cultures. People who share a national culture may have a common language, religion, beliefs, attitudes, and ways of thinking and reacting. Business cultures have in common characteristics of business practice in different countries. And corporate culture is related to the strategy of a particular company or corporation. People from the same business culture will share some beliefs and attitudes, regardless of their nationality. This model could be adapted in the context of cultural overlap in

HCI. Computer culture connects computer users, who will share some characteristics regardless of their ethnic heritage. Internet users—due to the global nature of the medium—may be even more similar in the ways they interact with computers. Also within the Internet culture, Web sites that belong to a common genre, such as government or academia, may include similar genre-specific elements and/or be organized in similar fashion, regardless of the country/culture in which they where created (Barber & Badre, 1998; Sheppard & Scholtz, 1999).

This unifying nature of the computer/Internet culture introduces an essential and controversial issue in interface design for culturally (ethnically) diverse populations: Is the adaptation of interfaces for local markets (localization) necessary for improving understanding and showing cultural sensitivity or should one develop standardized solutions understood by an international audience? Both of these approaches have their pros and cons. Standardization facilitates communication and collaboration, but when standards are developed in one country and imposed on other regions, it may be perceived as a kind of cultural imperialism (Kurosu, 2001). Misapplied globalization efforts may result in solutions "understandable for all but appealing to none," and production of too many local variations may consume unnecessary resources (Marcus, Armitage, Volker, & Guttman, 1999, online). But lack of localization may result in the exclusion of some potential designers and users from the benefits of information technologies.

Complexity is inherent in cross-cultural research. Some of the obstacles described here may be difficult or impossible to overcome entirely; thus it is crucial that those conducting cultural research take account of this complexity in their study designs. It is important, for example, to provide the definition of culture used in a particular study, to delineate membership in the cultural group(s) under investigation, to provide some historical background on the culture studied, and to acknowledge the limitations of the study. The interpretation of the results should be restricted to the studied culture and cultural group(s), except where generalization is warranted on the basis of comparative evidence.

To be certain that cultures are measured in the same terms, researchers need to understand factors that may introduce bias in cross-cultural comparisons (Karahanna, Evaristo, & Srite, 2002). According to Van de Vijver and Leung (1997b), the three types of bias are:

- Construct bias, when a construct measured is not equivalent across cultures at a conceptual or operational level (for example, understanding of the concept of higher education varies across cultures)

- Method bias, when the research instrument or its administration is not adjusted across cultures, thereby not allowing subjects in the research to respond in the same manner (for example, sample nonequivalency in terms of user demographics or educational level)

- Item bias, which refers to bias in measuring instruments and scales

Various means are available to locate bias and eliminate it. In the case of construct bias, Van de Vijver and Leung (1997b) suggest two approaches to ensure that no single culture controls the design of the study. The first, *decentered,* allows for adoption of a culturally diverse perspective, for example, by having researchers from different cultures add culturally specific values to the research instrument. According to the second, *convergence,* each researcher designs his or her own instrument and all of the instruments are administered to all of the cultural groups being studied.

Method bias due to problems with the administration of instruments can be identified by repeated administration of the same instrument in various cultural groups. If subjects with similar scores on the first test show considerable discrepancies in later tests, one can assume that method bias is present. Another method of assessment is to vary the stimuli systematically across cultures—low consistency of responses points to method bias. Item bias may be avoided by back-translation and avoidance of words in the instrument that are not directly translatable into the language of the target culture.

Another problematic area in cross-cultural research concerns the sampling of the cultures studied (Karahanna et al., 2002). Many studies investigate cultures with which the researcher had contact or knows the languages spoken. Convenience sampling may result in bias.

Another form of bias, mirror imaging, can occur when the researcher, who is always a member of a specific culture, interprets or evaluates a different culture in terms of his or her own cultural assumptions (Encyclopædia Britannica, n.d.). To avoid this type of ethnocentric bias, some researchers invite colleagues from other cultures to conduct cooperative projects. Also, some studies seek to replicate analyses conducted in other countries. For example, Hofstede compared his results to the results of the Chinese Value Survey (Chinese Culture Connection, 1987; Hofstede, 2001).

Researching international users also poses many problems. For example, it may be difficult to judge which part of the users' behavior is due to their upbringing and which is due to their individual preferences (cf. the pyramid model of culture described earlier). A meaningful conclusion can be drawn only on the basis of similar results for other users from the same culture.

Experimental research on cultural differences in HCI is challenging because of the many potentially explanatory variables that need to be controlled. Bias can also be reflected in the sampling of the subjects, if other characteristics (such as gender, age, and education level) are not taken into consideration. For example, in usability testing differences in performance between male and female subjects may be more striking in cultures where male and female gender roles are more pronounced. One solution is to match the samples in terms of demographic characteristics, so that sample differences do not influence the results of the study (Van de Vijver & Leung, 1997a). More complex factorial designs may also show significant interactions of culture and other variables such as gender or education level.

Interface Design

Cultural Interface Design Solutions

An interface is commonly defined as a point of interaction or communication between a user and a computer system. In the case of international interfaces, the interface has to be translated and culturally adjusted to facilitate communication. Of the various approaches to international interface design and adaptations, the most common is the internationalization-localization process. Internationalization (often abbreviated *i18n,* where the number 18 represents the number of letters between the first and last letter) can be viewed as a process for enabling different national versions of the product (Karat & Karat, 1996); it is accomplished by removing all culture-specific content (language, culturally meaningful icons) from the interface or by developing a core product that can be easily modified for any culture (Aykin, 1999). It can also be accomplished by creating a product equipped with features supporting a variety of cultures. Internationalization usually occurs in the country where the product was originally developed. In most cases, developers focus only on the elements related to text, numbers, and dates (Russo & Boor, 1993). Localization enhances the product to fit the users of a particular culture and to accommodate local requirements such as language, country-related information, writing direction, and so forth. Karat and Karat (1996) compare localization (nationalization as they call it) to the process of customization, which in general terms refers to adaptation of the product developed for one group of users to the needs of another. Customization is for a specific set of "national" concerns. Interface designers have to make decisions about which elements or functions of the interface should be internationalized and which should be localized.

The process of cultural adaptation of the interface for use on the international market is guided mainly by the purpose the interface is intended to serve. In the literature on the subject, two different concepts are often mixed. Bourges-Waldegg and Scrivener (1998) point out differences between interfaces targeted for a particular culture and those intended to be shared by culturally diverse participants. In the first case, the interface is adapted (reproduced) for each culture separately, whereas in the latter, the interface must be adjusted taking into consideration all possible users. Additional sets of issues (e.g., machine translation support) need to be considered in interfaces that facilitate international computer-mediated communication (Connolly, 1996).

Implementing international standards (Blanchard, 1997; Dzida, 1996; Schrier, Williams, MacDonell, Peterson, Strijland, Wichansky, et al., 1992) is an example of internationalization. Dzida (1996) differentiates between standard user interfaces and standards for user interfaces. The first derives from consensus in an industry. Standard interfaces aim to provide a uniform presentation of information (so-called *look*) and an interaction (so-called *feel*) among different application programs. Style guides help design the interface according to commonly accepted standards. There are also international usability standards such as those established by the International Standards Organization or other standard-guiding parties. Blanchard (1997) has presented a summary of the work of such organizations and international standards for HCI.

Standardization is convenient for the computer industry because changes do not have to be made for diverse markets. It may be beneficial also for international users because it provides a uniform platform, allowing them to communicate and cooperate easily. Conversely, standards may present several obstacles. In some cases customs and habits may not coincide with the standards (Kurosu, 2001). As mentioned earlier, concerns about cultural imperialism and exclusion of users who do not possess good knowledge of English are also important. International standards may not coincide with national standards. For example, for some time, U.S. companies could not market personal computers (PCs) in Germany because German standards called for keyboards with certain ergonomic features that could not be accommodated in the original design (Karat & Karat, 1996). Poland provides another example: The Polish language uses additional letters, but Poland adopted the international QWERTY keyboard so the additional letters must be entered using the ALT key. Users in countries that started to use computers with American software before the idea of localization became widespread may also oppose localization. Switching to localized versions of software and hardware would require relearning.

There are alternative approaches to the internationalization-localization process, based on the idea of customization. Peterson suggests that the interface should enable local customization rather than provide the customizations themselves (Peterson, 1990). Marcus (2001) gives an example of the PlanetSABRE program with an interface that can be customized by users to better represent their cultural preferences. Similarly, Unified User Interface Development suggested by Stephanidis and his colleagues would require the design of only culture-free "core" elements, while platform-, interaction metaphor-, and user-specific interface properties could be handled by special-purpose tools (Stephanidis, Savidis, & Paramythis, 1997).

Yeo (1996) proposed a Cultural User Interface for each target culture, which would take advantage of the shared or common knowledge of the culture based on its various elements. The interface could work with a range of applications, reducing the need for localization of each product separately. Yeo suggested that this could translate into a personal user interface (PUI) based on additional, noncultural data about the user. Others propose an intelligent agent, which could learn from user behavior and modify the interfaces accordingly (O'Connell, 2000). Each of these approaches has its strengths and weaknesses, but all emphasize the importance of heeding the needs of culturally diverse users.

Human Factors in Interface Design

Schemata and Mental Models

The guidelines on the internationalization-localization process concentrate on screen elements, but the psychological aspects of human–computer interaction also receive some treatment. To understand the cultural aspects of interface design we have to consider how users from different cultures interact with an interface and how they understand and perceive the interaction.

In the field of human–computer interaction, user understanding of computer interfaces is commonly explained on the basis of schemata and mental models. Schemata are knowledge structures that represent the regularity of concepts and recognize them as typical. Schemata are based on information stored in long-term memory; they include mentally run scenarios to anticipate effects of actions. We call upon schemata to make sense of, or to help us interpret, information that comes to our senses and to enable us to be cognitively efficient in responding to our physical and social worlds. For example, in order to recognize a face as a face, we must first cognitively create a schema of a face as an object that is composed of eyes, lips, nose, cheeks, chin, and ears in a particular order and then recall the schema when we see it.

Schemata are not limited to objects. According to Taylor and Crocker (1981, as cited by Nishida, Hammer, & Wiseman, 1998) schemata can be classified into five groups: *person schemata*, which contain knowledge about different types of people; *self-schemata*, which contain knowledge about ourselves; *role schemata*, which characterize knowledge about social roles; *event schemata*, which relate to predictable sequences of events; and *content free schemata*, which contain information about processing rules.

Schemata are relatively stable, but as an individual acquires new knowledge, additional facts can be added to existing schemata (accretion), minor changes in schema structure can be made to account for a more refined view of the world (tuning), or major structural changes based on new information can occur (restructuring) (Satzinger & Olfman, 1998). When existing schemata do not explain observed phenomena, new schemata need to be created (schema induction).[1]

Because construction of the schema network occurs when users interact with the environment, users from different cultures may have different knowledge structures for similar concepts. Anderson, a schema theorist, noted:

> The knowledge a person already possesses is the principal determiner of what a person can come to know. Knowledge, in turn, is conditioned by culture. Therefore, a person's culture is a principal determiner of what he or she can come to know. (Anderson, 1984, p.8)

As an example, the author cites research he conducted with his colleagues on the effects of culture on reading comprehension (Steffensen, Joag-Dev, & Anderson, 1979). East Indian and American subjects were shown detailed descriptions of Indian and American wedding ceremonies and were asked to recall details. Subjects recalled more information about the ceremonies from their native countries but also tended to interpret the facts about the foreign ceremony in the light of knowledge about weddings from their own country. (For example, the heirloom dress of the American bride ["something old"] seemed old fashioned and inappropriate to Indian subjects; Indian gifts for the in-laws were interpreted as party favors, a diamond ring on the bride's hand was interpreted as gift from her family by Indian subjects, and as an engagement ring by Americans.) Culturally dependent schemata were labeled by D'Andrade (1981, 1989) as cultural schemata—socially shared ways in which cultural groups organize their behavior.

A classic example of how schemata (scripts) work in real-life situations was presented by Schank and Abelson (1977) in their discussion of the restaurant schema. The story: "John went to the restaurant. He ordered chicken. He left the large tip." is easily understood even though important parts of the story have been omitted, because we have a mental schema of behaviors associated with a restaurant. The reader, based on his or her schema for a restaurant, will assume that John sat at a table, looked at a menu, was served chicken, ate the chicken, paid the bill, and left the restaurant, and that the food and service were good because he left a big tip. However, the restaurant schema may need to be adjusted for a buffet, where the customer pays first and then chooses his or her own food. It may also be modified in some regions because different countries vary, for example, with regard to tipping practices (paid along with the check, left at the table, expected, or not customary at all). The restaurant schema example can be modified in the context of human–computer interaction. The sentences "John turned the computer on. He browsed the Web. He downloaded the application." will be understood by computer literate people, even if a lot of the information has been omitted, such as opening the Web browser, following links, typing a URL, or following commands needed to download an application.

Schemata play an important role in intercultural communication. People from different cultures may have different schemata for social interaction patterns. They may also have, however incomplete or distorted, schemata for behaviors of peoples from other cultures (Beamer, 1995). Mutual understanding of the cultural norms of social expectation is crucial for successful interaction. Schemata of social interaction may be mirrored in user interaction with computers. Reeves and Nass (1996), in their book *Media Equation,* argue that people do not treat computers as tools but rather as social actors and that they expect the interaction with the computer to follow the schemata of interaction with persons. The authors cite a number of studies in which people unconsciously treated media as human beings and expected them to react accordingly. When the computer violates social norms such as politeness (e.g., by displaying culturally offensive error messages), the computer itself is viewed as offensive. Because there are cross-cultural differences in politeness, international interfaces should be adjusted to the socially accepted schema of communication of their intended users. The authors point out, however, that although the rules for polite behavior vary, every culture "recognizes politeness, everyone tries to obey politeness rules, and everyone feels bad when they are broken" (Reeves & Nass, 1996, p. 36). Cooper (1997), inspired by *Media Equation,* formulated 14 rules for polite software—software that acts according to the schemata of interpersonal communication. Cooper's presentation of the politeness rules along with examples of social interaction and human–computer interaction provides useful, albeit general, guidelines for developers from the human rather than the computer perspective.

In user interaction with the computer, schemata are employed to form mental models of the system. Mental models can be defined as dynamically constructed mental representations of the world or its parts based on existing knowledge. From such models, the user can predict the behavior of the system in response to his or her commands, plan methods for accomplishing new tasks, and deal with error situations by evaluating the system's state according to the model and then choosing actions to leave that state (Card & Moran, 1986).

The difference between schemata and mental models is that mental models are dynamic, whereas schemata are relatively fixed. Mental models are dynamically constructed based on users' schemata (Eberts, 1995). Successful interaction depends strongly on the extent to which the user's mental model of the system matches the designer's conceptual model of the system. Because the communication between the system designer and the user takes place through the interface, if the interface does not make the design model clear and consistent, then the user will end up with an incorrect mental model (Norman, 1990). The designer uses a conceptual model to design a task. This model is conveyed to the user through the interface and the user has to build a mental model of the task that resembles the designer's model. If models employed by the designer and user are based on different schemata, user interaction with the computer may be impaired. In what follows, I analyze the available literature from this perspective to see if it provides evidence to support the assumption.

Perception and Reasoning in Interaction Process

Cultural differences can influence how people perceive and interpret their interaction with the computer. Ito and Nakakoji (1996) developed a model user–computer interaction and considered the way culture influences each step. According to the authors, interaction happens in two modes: the listening mode, when the user is presented with the computer's reaction and the speaking mode, in which the user gives instructions to the interface. In both modes the cultural background of the user plays a role in the way the information is comprehended.

In the listening mode, user cognitive activity will go through three phases—perception, semantic association and logical reasoning—to comprehend the information presented. The *perception phase* is least affected by culture because the physiology of color or shape depends on sensory monitors only.[2] In the *semantic association phase*, the user, based on previously created schemata, associates meaning with text and graphics. Metaphors are decoded on this level, so cultural differences play a larger role. Metaphors will be discussed later in this chapter. *The reasoning phase* is most affected by culture, because most cognitive reasoning depends on social norms and background culture.

In the speaking mode, users convey their intentions to the computer. In the *affordance perception phase*, users identify what they can do with presentations displayed by the system. The next phase, *applicability check,* requires users to validate the actions they are going to take. Regarding cultural impacts on this phase, the authors note differences in interface exploration: The trial-and-error method is considered tedious and time consuming in Japan, whereas in other cultures it is associated with freedom and exploration. During these two phases, the user builds a mental model of the system. In the third phase, *enactment with expectations,* the user will carry out the action. Ito and Nakakoji suggest that the action may depend on language. In languages where grammatical rules call for specifying the object before the action (as in Japanese), users will look for the object first and then perform the action. In languages with the opposite word order, the action may be chosen first. The last phase, *confirmation*, may also be culturally dependent, as it depends on different perceptions of time, hence how much value is placed on speed of the interaction and error avoidance. The issue of time perception will be discussed further in the section on "Cultural Dimensions of Usability."

The Use of Language at the Interface

Translation Issues

Because language is one of the most visible signs of culture, it is appropriate to discuss textual elements of the interface separately from graphical representations. For interfaces designed for specific cultures, translation of the interface into a target language is an obvious step.

With the newest technologies, translation may be particularly difficult because the specific words may not exist in a target culture. To help alleviate translation problems, Microsoft has published a dictionary of the terms used in their products translated into 14 European languages (The GUI guide, 1993). Even this did not cover all possible terms, as the number of software applications and the number of new functions they carried continued to grow. The translations suggested in this guide are not always the best possible, so software designers have tried to find terms that fit better the cultural schemata for the functions needed. For example, the exact Greek translation for a "header" (used in Word 6.0 for Windows) is a rather uncommon word (Lepouras & Weir, 1999). The Greek edition of the spreadsheet Lotus Ami Pro 3.0 uses a word in reverse translation meaning "supertitle." WordPerfect uses a phrase that literally means "page title." Similarly the word used for "paste" in the Chinese version of Word for Windows is a translation of the word "post," with the meaning that something was stuck on but can be removed (Leung & Cox, 1997). Even localized versions of certain Microsoft products tend to display some interface elements in both English and the local language at the same time (cf. Wang, Fu, Jin, & Li, 1999).

Direct translation is only one of several ways to assign appropriate terms (Lepouras & Weir, 1999). When translation is impossible or undesirable, designers' other options include: 1) create a new word (neologism), sometimes constructed from existing words, that would represent the concept; 2) retain the original word in its original spelling; 3) adjust the word to the spelling rules of the target language (*interfejs* for interface in Polish); and/or 4) adjust it to the target language morphology (*kliknij* for click in Polish); 5) assign a new meaning to the existing word, which represents a similar concept. As an example of the second option, Nielsen (1990) recounts that during a visit to Japan, he was shown a menu-based interface. All the commands were written in Kanji, but the "undo" command was in English. He was told that because there is no good translation of the concept, it was better to just leave it in Latin characters.

Assigning a new meaning to an existing word is common for new technological concepts, both in the language where the concept originated and in the language to which the concept is being translated. For example, "diskette," "disk drive," "zooming," and "panning" did not exist in Thai prior to being borrowed from English (Sukaviriya & Moran, 1990). These words (or their new meanings) did not exist in English either before the artifacts and concepts were invented. They were created on the basis of similar concepts; "zoom" was adapted from photography, for example. Thus, new technical words in other countries have to be created by adapting English words or creating new ones based on native concepts. For example, in Polish the word "browser" was translated as *przegl_darka* (a device to watch slides) because browsing through the pages of the Web is closely related to switching through slides, a mental model commonly understood in Polish culture.

Problems arise when different software developers assign different words to the same function. In a comparative study of two commonly used word processors, Lepouras and Weir (1999) found that the terminology was consistent in English versions, but that the number of common terms in their Greek counterparts was 30 percent lower. Consistency makes interfaces easier to learn (Nielsen, 2002), as users can build mental models of the system on pre-existing schemata. Satzinger and Olfman (1998) found that consistency in syntax and hierarchy design between different applications measurably improved user performance.

Translations need to be made with great care. Mistakes can cause confusion and also suggest incompetence on the part of the designer. Russo and Boor (1993) observed that when Sun Microsystems translated their Open Windows Developer's Guide into Cantonese, the word "menu" was translated as a "list of food items." But even proper translation does not prevent problems if other aspects of cultural discourse are not adjusted. For example, after rebooting the Macintosh computer by pulling the plug, the message "This computer was shut down by an unusual way [sic]. Please use the normal shut down method" would appear (Kurosu, 2001, p. 143). This message will feel offensive to Japanese speakers, because it blames the user. Similarly, requests in English may be expressed as interrogatives (e.g., "Why don't you?"), but in Polish these would be interpreted as direct questions (Connolly, 1996). The schema of interpersonal interaction is replayed during interaction with a computer. The issue of cross-cultural pragmatics is widely discussed in the linguistics literature (cf. Wierzbicka, 1991).

Another issue associated with style is the translator's choice of words. Translations are most effective when done by native speakers of the language who also specialize in technical translations; translations using words and expressions that are below the reading level of the user makes them sound puerile. In some cultures people are insulted by the use of dialects that are considered "low status" (Fernandes, 1995).

The localized interface should also be designed to fit the local writing style. Some languages (e.g., Arabic) are written right-to-left, so the reading of the document starts at the top right-hand corner. This obliges designers of graphical interfaces to help the user enter the text in the right-to-left direction. Other details such as cascading menus need to be right-to-left oriented (Amara & Portaneri, 1996), and scrolling bars must be appropriately adjusted as well. This last feature is even more complicated because full justification is done in English by adding spaces between words and in Arabic by stretching the last letter of the word in a line. Thus, designers of word processing programs for the Arabic market have to include these options in their products. Another example comes from the Thai language, where all words in a sentence are written with no spaces between them; justification by inserting spaces is impossible.

Problems can also arise with hyphenation because languages have different conventions, some of which involve spelling changes when a word is hyphenated. For example, Dutch words with diaereses, such as *beëindining*, lose the umlaut: *be-eindining*; and Hungarian words with double consonants, such as *asszony*, regain their root consonant: *asz-szony*. German words with *ck*, such as *backen,* were spelled with two *k*s when hyphenated—*bak-ken* (Sprung 1990), but grammar reform in 1998 changed the hyphenation rules and computer programs had to be adjusted accordingly (Woestenburg, 2003).

It is important to take into consideration not only the character set, but also the manner in which the words of a given language are written. In Thai (Sukaviriya & Moran, 1990), some vowels are placed on top of the preceding consonant. The remaining vowels are placed in the left-to-right character sequence. A tonal marker is always placed on top of the previous consonant. If the vowel has already been placed on top of the previous consonant, the tonal marker will be placed above that vowel. This means that a designer in Thai has to consider how the user might enter the characters in four different lines rather than one (not counting subscript and superscript), as in English.

Translation of the menus, boxes, and icon text can also be problematic because the length of words varies between languages. The length of a text can increase up to 200 percent after translation (Dray, 1996). Those who redesign interfaces for another country have to make tradeoffs between maintaining the consistency of the two interfaces and presenting the text in a form that best suits user needs. The problem is amplified when the language in question does not have the necessary words in its vocabulary, especially when a translation requires many descriptive terms. In the early 1990s, in the Polish version of dBASE III +, the word "browse" (*przeszukuj*) was translated as *buszuj* because of limited space. *Buszuj* roughly means to "rummage" and has rather pejorative meaning. Furthermore, the word "list" (*sporzadz liste*) was translated as *listuj* (English word with Polish ending), which does not exist in the Polish language. However, this word is similar to the word *lista* which means "a list." The translator apparently assumed that users will somehow understand or grasp the proper meaning.

Coding Issues

For text to be displayed correctly in the user language it must be properly encoded. The first standard code ASCII (American Standard Code for Information Interchange) character set includes 94 characters plus 43 functions, and works fully only for English, Indonesian, and Swahili (Character Encoding, 1996). Thus, to incorporate other writing systems, other character sets had to be developed. ASCII was extended into the International Organization for Standardization (ISO) 8855 series that now defines more than 15 language groups. Unicode, a 16-bit character set that encompasses 65,536 characters, can include practically all possible writing systems; Unicode seems promising, but it is still not used universally (Erickson, 1997).

But the ISO 8855 series are not the only character sets. Numerous character sets exist for each language and software, and Web site designers can choose among them. This can create problems especially in the Web environment. Because Web pages may use different encoding, in some cases users will have to adjust their browser options for a page to be displayed correctly. Today's browsers (e.g., Internet Explorer) offer an automatic option to download fonts needed to display correctly pages in languages other than English. Commercial search engines have also become more international. Several offer the interface in a choice of languages. This has allowed users to restrict retrieved sites by language or country domain, adjust the language of the interface depending on the user's Internet Protocol (IP) address, and sometimes even translate retrieved sites. The limitation to a country domain is especially useful when a query includes words with the same spelling (sometimes with different meaning) in several languages, which is often the case in languages from the same linguistic group (for example: *burza* means stock exchange [Czech], or thunderstorm [Polish]).

The presence of multiple characters sets for the same language also creates problems in performing Web searches. Search engines look for a string of characters, so the same phrase entered in two different encodings are actually two different sets of characters for the search engine, unless it has cross-encoding conversion. Designers who want to make their sites easy to find by search engines must be aware of the ways search engines look for pages in different encodings. For example, for the pages coded in ISO 8855-2 (Eastern European Languages, based on the Latin alphabet with diacritical marks), entering search terms without diacritical marks generally retrieves a numbers of hits, but including the diacritical marks should yield a greater number of hits, although it is difficult to generalize because search engines use different algorithms (Callahan, 2002). Among six Polish and polonized search engines researched by Marek Sroka (2000), only one retrieved the same number of results regardless of whether terms were entered with or without diacritics. Entering words with special characters may be difficult for users who do not reside in their country of origin and do not have the required hardware and software to enter proper fonts. Some Web designers, aware of this problem, include keywords without diacritical marks in their metadata.

Multiple Languages on the Web—an Overview

The problems reported in the previous section are becoming more urgent. Despite the current dominance of the English language on the Internet (Large & Moukdad, 2000), the number of non-English speaking users is steadily increasing as is the number of Web pages designed in languages other than English. This increase is so rapid that it is difficult to estimate. According to research conducted by Global Reach (Global Internet Statistics, 2003), 63.9 percent of the total online population speaks a language other than English. The other major groups of Internet users speak Chinese, Japanese, or Spanish. Accenture predicts that by the year 2007 Chinese will be the dominant language of the Internet (*So Now It Gets Interesting*, n.d.).

Several researchers have attempted to estimate the number of pages in different languages on the Web. Grefenstette and Nioche (2000) counted the frequency of words commonly occurring in a specific language. Their counts, conducted in 1996, 1999, and 2000 using AltaVista indexes, suggest that Web English has grown 800 percent over four years, while German and Spanish have grown 1,500 percent and 1,800 percent, respectively. Large and Moukdad (2000) used another technique. They limited the retrieved results to one language only (an option given by AltaVista) and entered a meaningless string of characters preceded by a minus sign. The number of pages retrieved indicated the number of pages in the selected language indexed by the search engine. According to their calculations in June 1999, AltaVista had indexed more than 198 million pages in English, more than 20 million in German, and more than 5 million in Japanese, followed by French and Spanish, each with more than 4 million pages. The results of these studies do not give full answers to how many Web pages exist in a specific language, because none of the search engines indexes the whole corpus of the World Wide Web. They can give us, however, an idea about the proportional relations of English and other languages and strongly suggest that the presence of other languages is increasing.

The dominance of English on the Internet is historically related to the global role played by English over the last few centuries. British colonial expansion established the preconditions for worldwide use of English in the 19th century. This was followed by the economic and political rise of the U.S. as a global leader, which spread the English language alongside U.S. economic, technological, and cultural influence (Graddol, 1997). In addition, English speakers in the U.S. developed the Internet. That English dominates is in part a legacy of the Internet's origins.

The number of speakers of other languages (especially in South America and Asia) is increasing at a faster rate than English language speakers because population growth in those countries is more dynamic. This may result in generations of new Internet users who will create content in their own languages, and for whom learning English will not be a prerequisite to enjoying new communication technologies.

The availability of Web content in specific languages may correlate with how people use the Web. A Statistics Canada (1998) survey showed that only 26.2 percent of Quebec residents communicate by computer in a typical month, as compared with 45.1 percent in Alberta (as cited in Clavet, 2000). One reason may be the lack of information and services in French. Another survey conducted by IDC Project Atlas II asked users if they preferred to visit Web sites in English or in their native language

(Josephson, 2002). Among Europeans, 52 percent preferred their native language over 23.3 percent in English (for the remaining 24.7 percent English was the native language). In Japan, the figures stand at 83.9 percent for Japanese and 7.7 percent for English, and the numbers for The People's Republic of China are 84.9 percent and 14.8 percent. Interestingly, in the case of Asia/Pacific (excluding Japan) the figures were only 39.2 percent for native language and 31.4 percent for English. In Latin America, 76.3 percent preferred native language sites.[3]

The exclusive use of English in international interfaces may be preferred in certain situations: The user works in an international corporation where English is the language of communication, the software or Web page was not translated into the user language, or the user feels more comfortable with English computer terminology. Bourges-Waldegg and Scrivener (1998) provide the example of a Chinese user who complained about some English expressions on the interface. Asked if he would prefer to have it in Chinese, his first language, he answered that in the case of computers English is his first language because he does not know the proper terminology in Chinese. Similarly, attempts to develop Japanese programming languages did not succeed because other Western programming languages were already commonly in use (Ito & Nakakoji, 1996).

However, learning to use English language interfaces may present problems for the user, even one relatively proficient with the language. As Agnes Kukulska-Hulme (2000) has pointed out, difficulties are caused primarily because the words in the interface are presented without context, making it difficult to guess the *actual* word meaning as opposed to *potential* (dictionary definition) meanings. She specifies ten areas of difficulty for non-native speakers:

- Words similar in form (users may have problems distinguishing the meaning of words like *box* and *border*; *refresh* and *restore*)

- Incorrect pronunciation (users may have problems distinguishing written forms of words like *previews* and *previous*, if they know only the pronunciation)

- Words related in meaning (users may have problems distinguishing between words such as *mistake, error, fault* or *delete* and *remove*)

- Ambiguous words (users may have problems with words with multiple meanings such as *draw* and *move* or with nouns that have the same form as verbs because the articles *a, the* and *to* are usually omitted)

- False friends (users may incorrectly identify the word by comparing it with similarly sounding words in their native language that have a different meaning)

- Culture-specific meanings

- Abbreviations

- Semi-technical terms (the user may have trouble understanding words not normally encountered in language learning, such as *cropping, merge*)

- Staked modifiers (users may have problems understanding expressions such as *the re-enter password box* as *the box for re-entering a password*, since those phrases do not follow proper grammatical rules)

- Idiomatic expressions and phrasal verbs

Metaphoric expressions also can cause potential problems for international users. For example, the White Pages metaphor is not internationally recognized as signifying the phone book, even if users understand the meaning of the individual words *white* and *pages* (Bourges-Waldegg, 1999).

Several proposals have been made to facilitate the use of English-language interfaces by non-native speakers. One of the most promising is the use of simplified English. Simplified English was developed by the Association Européenne des Constructeurs des Matérial Aerospatial (European Association of Aerospace Industries—ECMA) because of problems with English language manuals (Mills & Caldwell, 1997). Simplified English aims to reduce the number of synonyms: There is only one word for each idea—for example, the words *begin, commence, initiate, originate*, and so on are always expressed by the word *start*. Basic Simplified English is comprised of about 900 words. Simplified English is used in the aviation industry, but it shows promise also in the area of human–computer interaction. Research shows that the use of Simplified English makes text easier to read, reducing the number of errors especially among non-native speakers (Mills & Caldwell, 1997).

Even if English is the lingua franca of the Internet, many entities present their Web sites in several linguistic versions to accommodate various cultural groups. Reasons for multilingual sites may be cultural, economic (increasing customer base), or political (language preservation). Two French-language associations filed a suit against the Georgia Institute of Technology for presenting English-only information on its Web site about its educational programs in France, claiming that it violated a French law that outlaws the use of foreign words in advertising (Ladner, 1996).

Statistics from Forrester Research show that 80 percent of European-based corporate sites are multilingual (Dunlap, 1999). A multilingual Web site presents significant opportunities to expand business to new groups of customers. This can be seen in the example of the Howard Johnson International hotel chain Web site, which received double the number of hits after launching a Spanish version of its Web page (Whitford, 2000).

One of the problems is which version of the language should be used on the Web site. English exists in numerous variations—British, Australian, and North American, among others (Kachru, 1982). If the site is located in the country where English is one of the official languages, the version of the language spoken in that country is an obvious choice. If English is the second language of the site, the designers have to make a decision based on the target audience. Other languages spoken in several countries will present similar variations. For example, people from Taiwan and China use a different expression for the World Wide Web (Chu, 1999).

Steve W. Chu (1999) compared the problem of designing a bilingual Web site to table setting in Chinese restaurants outside China. Restaurant owners are faced with the choice of giving the guest chopsticks (because it is a Chinese restaurant) or a fork (because it is in a foreign country, guests may not feel comfortable eating with chopsticks). The owner can also offer both to all customers or discriminate and offer forks to Westerners and chopsticks to Asian guests.

Bi/multilingual solutions could be implemented in several ways from the perspective of site design. Cunliffe (2001) distinguished three methods: splash page, monolingual page, and bilingual page. In the splash page approach, the home page is preceded by a splash page that directs users to the home page in the language of their preference. In the monolingual home page approach, the user is presented with the home page designed in one language and has to follow a link to access a page in another language. Bilingual home pages are usually split horizontally or vertically to incorporate both languages. Bilingual home pages are rare, due to encoding problems.

However, sometimes the bilingual function may be implemented only partially. Among business Web sites in Russia studied by Travica and Cronin (1996), some provided only top-level pages in English and the rest of the site in Russian. The other problem was that English language pages were linked to Russian language pages without any warning. Additionally, the English on some pages was rather awkward.

Multilingual CMC Accessed Through the Web

Computer-mediated communication (CMC) and computer-supported cooperative work (CSCW) systems for culturally diverse users who work either in real time or asynchronously pose the biggest challenge for interface designers (Bourges-Waldegg, 1999):

> Cultural factors cannot only be addressed by providing users with multiple culturally-adapted interfaces, or with customizable views designed with the aid of culturalization guidelines. ... users may become confused by the existence of several languages and other culturally-specific options in their workspace, or may need to talk about the computer tools they are using, and if these tools have different names or appearance, breakdowns in collaboration or communication may occur. (Bourges-Waldegg 1999, p. 2)

Bourges-Waldegg recognizes that people using multicultural systems are usually already aware of some of the cultural differences. Her approach, Meaning in Mediated Action (MIMA), involves a cycle of observation, evaluation, analysis, and design. This process is similar to the usability lifecycle process, but concentrates on representations and how users understand them.

CMC interfaces are especially difficult when the script of a language cannot be entered easily. Alphabetical languages such as Greek may use transliteration, relating letters of their alphabet to Latin on a phonetic or orthographic basis (Tseligka, 2002). Transliteration and Romanization (for nonalphabetical languages) pose problems, however, when there are variant spellings of a word, for example Peking vs. Beijing, Mao Tse-tung vs. Mao Zedong, Tchaikovsky vs. Chaikovskii (Borgman, 1997).

Languages—Future Research Agenda

International users and designers face many challenges in attempting to use information technologies to their full potential. On the other hand, developments in computer technologies can have a profound influence on cultures. English may be the lingua franca of the Internet, but the increasing presence of other languages underscores the need for research on the influence of the Internet on linguistic changes and how these changes are, or should be, reflected in interface design. English has contributed greatly to the language of computing. It would be interesting to explore how speakers of other languages perceive these influences. Do they follow the French in rejecting foreign words or do they perceive them as cultural enrichment? Are discussions on this issue important in specific cultures or treated marginally? Can the culturalization solutions achieved in more developed countries be used in cultures that have just started to utilize information technologies? Which words are most appropriate for interfaces and what attempts are made to achieve consistency of word use? The literature on this subject addresses the issue from the perspective of morphology rather than from the viewpoint of actual speakers. Additionally, most of those studies are reported in native languages, so they are not available to an international audience.

Another issue that deserves attention is users' language preferences for Internet use and how this corresponds to the availability of Web content in their native languages. The next thing to look at would be how users and designers overcome problems created by multiple coding standards, for example, by studying the strategies they apply when searching for pages in languages other than English and also in CMC contexts.

Graphical Elements of the Interface

Cultural Aesthetics

The visual appearance of the interface may also be perceived differently across cultures. Tractinsky (1997) points out that in the HCI literature, aesthetic concerns, although valued, often take the blame for decreased usability of interfaces. Aesthetics tends to be portrayed in terms of facilitating information processing, not in terms of engaging the user in a pleasing experience. Tractinsky conducted a series of experiments where Israeli subjects rated interfaces used in automated teller machines (ATMs) for their usability and aesthetic values. The subjects viewed overheads of 26 design layouts (20 seconds for each screen) and were asked to rate their usability and beauty. The methodology was adapted from a study by Kurosu and Kashimura (1995), who used Japanese subjects. Tractinsky compared their results with findings from her study to examine whether culture played a role in perceived usability of the interface. The findings suggested that aesthetic perception and its relation to usability were culturally dependent. The author hypothesized that Japanese culture is more sensitive to aesthetics than Israeli culture and thus would emphasize more the role of aesthetics in interface design. The findings suggested the contrary; in Israel the correlations between perceived usability and aesthetics were considerably higher than in Japan. This suggests that our knowledge of the role of aesthetics in interface design in different cultures is still limited and in many cases based on cultural stereotypes.

Differences in aesthetic preferences can be explained by cultural differences in perception. In research studies conducted by Masuda and Nisbett (2001), participants were shown an animated aquatic scene in which one large fish swam among small fish and other underwater creatures. The participants were asked to describe what they saw. American subjects tended to concentrate on the big fish, whereas Japanese participants began by describing the background. Japanese subjects made about 70 percent more statements about the background than their American counterparts and twice as many statements about the relationships between animate and inanimate objects. In the second part of the study the same scene was put against a different background; the

Japanese subjects had trouble in identifying the big fish, indicating that their perception of the objects was linked to their perception of the background scene.

Differences in perception have been documented in anthropological research. Fussell and Haaland (1978) found that in Nepal a line drawing with shading and details produced more accurate identifications than black-and-white photographs. Cook (1980) discovered that in Papua New Guinea people would scan pictures in a circular pattern and not left-to-right, top-to-bottom (both citations from Sukaviriya & Moran, 1990).

Icons

The two visual characteristics discussed most commonly in HCI guidelines are icons (and associated metaphors) and colors, especially their symbolism. Icons are commonly used in graphical user interfaces because they can substitute for lengthy text and can be understood by people with various levels of reading ability. Iconic metaphors are designed to make the user experience more intuitive and playful and to make the idea of working with a computer less intimidating (Johnson, 1997). People use file folders to store paper documents in their offices, so it makes sense to store computer documents in computer-generated folders that look similar to actual file folders. But folders in the U.S. are flat and have tabs that can indicate the contents of the folder. In Europe and Japan documents are stored in cardboard-box-like containers (Duncker, 2002). Using metaphors in the interface may be helpful, but the designer has to choose the metaphor carefully.

When designing icons and symbols for foreign countries, the designer must consider three things: Will the symbol be understood? Will the symbol be appropriate? Will the symbol be culturally acceptable? Correct understanding of the icons' meanings is key to the users' interaction with a graphical interface. If the user cannot recognize the icon, and consequently cannot understand the meaning it represents, communication with the system is impaired. For example, as a cultural outsider, I was able to recognize an icon of a U.S. rural mailbox (because it had MAIL printed on its side) and understand its function, but failed to understand the significance of the red flag, which I interpreted as a decorative element.

The mailbox and trash can metaphors are most commonly cited as inappropriate in international interface design because their appearance and shape in the real world vary from country to country; in some countries mail is delivered door to door or picked up at the post office (Duncker, 2002). Del Galdo (1990) has suggested that an envelope would be more suitable to represent e-mail communication because this graphical representation is recognized internationally. But, as Mrazek and Baldacchini (1997) point out, over the years many international users have been exposed to those icons, so they will not only recognize them and understand the meaning, but may also expect them. Thus, substituting the icon with something else in a localization process may cause confusion.

However, the mailbox and trash can icons have been modified; some countries redesigned the graphical representations to more closely resemble actual objects (examples of mailbox icons in different countries can be found in Ito & Nakakoji [1996] and Prabhu & Harel [1999]). Some redesign suggestions may seem humorous to Western users. Sukaviriya and Moran (1990) noted that the trash can icon in the design presented by Macintosh has no meaning to Thai people because in Thailand the trash can is normally a basket with "some flies on top of it to be more realistic" and proposed a similar graphical representation for the Thai interface

As has been mentioned, some symbols and icons may not be acceptable in some countries. In Chinese culture, a green turtle is considered backward, incompetent, and cowardly. Moreover, this symbol can refer to a man who has been cuckolded. For this reason, including a picture of a crawling turtle may not be appropriate in interfaces designed for the Chinese market (Zhang Qibo, personal communication, 1997). The examples can be multiplied. In Taiwan, the elephant is a symbol of strength, but in Thailand it is a national emblem so its use in funny or disrespectful ways should be avoided.[4] Some symbols are meaningless in certain contexts. For example, a picture of snowflakes to mean "cold" is less meaningful for cultures that do not have snow (Sacharow, 1983).

The meaning of certain symbols, however, should not be overemphasized. Designers should not make judgments based on anecdotal evidence, but be aware of the fact that some graphical representations may not be appropriate in particular cultures. Thus, it is important that questions about the appropriateness and acceptability of visual elements be asked while gathering information about the target population. Even if users understand the meaning of the icon designed in another culture and the depiction is not culturally sensitive, they may prefer a graphical representation of the object and concept that is not culturally foreign.

Care should also be given to the presentation of pictures. Some cultures are very sensitive to how human features are represented. For example, when a picture developed in the U.S. was shown in an Egyptian female hygiene program, women protested that the woman presented on the pictures did not look like them (Russo & Boor, 1993). In many cases icons can work internationally and be understood even if the software—or in most cases the Web site—is in a foreign language. In the real world, pictorials are commonly used in public communication (Marcus, 1997; Marcus et al., 1999). But even icons considered international are not globally understood. In studies by Brugger (1990), only 13 percent of Japanese recognized a first-aid symbol based on the Red Cross, and most did not associate the symbol/letter i as a sign referring to information services. Lin (1999) also observed differences in recognition and design preference regarding public information symbols (fire alarm, information, lost property, etc.).

Recognition of the symbol does not happen automatically but requires learning (Marcus, 1997), even within the native culture. Learnability can be simplified when the picture closely resembles the actual object or concept (Ossner, 1990). Unfamiliar depictions will require a longer learning time for international users. Users in other cultures will eventually learn the meaning of the icons representing functions they use often, although they may not recognize other functions depicted by metaphoric buttons. The preference for icons over alphanumeric descriptions may also vary among cultures. Choong and Salvendy (1998) found that Americans performed tasks more quickly and with lower error rates when using an interface with textual elements, whereas Chinese people had better results using an interface with icons.

Some metaphors can introduce cultural conflicts. Nielsen (1996) suggests that the best visual symbol for language is a flag, but this raises profound issues. Is language an indication of ethnicity and is the flag a symbol of nationality? Which flag would be the best for indicating English on the Web: British, American, or Australian? What about other countries where English is spoken (India, several African countries)? What about the German language, which is spoken in Germany, Austria, and parts of Switzerland? Do people in multilingual countries have to click on the flag of a foreign country to retrieve a page in their own language (Pemberton, 1998)? The idea of using a flag to represent a language seems untenable and potentially politically explosive. And we are still left with the problem of the existence of thousands of languages that cannot be represented by a flag.[5]

Colors

Colors, and their possible meanings in various cultures, are also a subject of discussions in the interface design literature. For example, in North America a blue ribbon signifies the best or first in a given class and a red ribbon signifies second; in the U.K. it is the opposite (del Galdo, 1990); and in other countries, such as Poland, colored ribbons are not used as a sign of victory. Russo and Boor (1993) offer a good presentation of the meaning of colors: red means happiness in China, aristocracy in France, and death in Egypt; yellow means cowardice in the U.S. and success in India; green means safety in the U.S. and criminality in France. Designers should have knowledge of colors that have positive and negative connotations and also of how certain color combinations are perceived: Japanese people prefer the contrast of red and white and dislike the combination of yellow and black (*Packaging design in Japan*, 1986); in Poland purple with silver signifies death.

The subject of culturally meaningful colors is discussed in design guidelines as often as the subject of icons and their metaphorical meanings, but both issues have to be approached with caution. For example, in the case of colors, usually not all hues have a specific meaning, so it is difficult to specify the right one in written guidelines. Also, the meaning of some colors may be generally acknowledged in a specific culture, whereas the meaning of certain other colors or their combinations may be meaningful only to older generations or in specific regions. Even in the same culture, colors can have multiple meanings. Black is the color of mourning in the U.S. but it also signifies elegance (Bourges-Waldegg & Scrivener, 1998). The most important aspect is that the meaning of the color depends on the context. Bourges-Waldegg and Scrivener (1998) present the meanings of colors using a model adapted from Searle (Searle, 1995 as cited in Bourges-Waldegg & Scrivener, 1998)—*R* (Representation) means *M* (Meaning) in Context *C*. Thus, for example, "color purple (R) means God (M) in the context of Japanese religion (C)". For this reason a *specific* cultural context may influence user understanding and preferences. It is important that designers are aware that colors may have a specific meaning in different cultures. However, color choices should not be made on the basis of rigid guidelines. More helpful would be actual knowledge about color preferences in different cultures. Barber and Badre (1998) noticed that Israeli and Lebanese Web sites frequently included the color green, and French sites used red, white, and blue heavily. The color of the national flag was frequently used in governmental sites, with the exception of Brazil, where all sites tended to use multiple, bright colors regardless of the site genre.

Duncker, Theng, and Mohd-Nasir (2000) observed differences in color choices among Web sites developed by international students. English students used pastel colors with much gray and low contrast; Scandinavians preferred dark colors. Students with a Jamaican background chose bright colors with high contrasts and African students most often chose black as the base color and added some brighter colors. European and U.S. students tended toward bright backgrounds, black text, and some colorful objects.

These examples help us gain an understanding of color preferences in various cultures. One still needs to bear in mind that those studies reflect general preferences, which can be diverse even within the population. Other information on this topic can be found in the marketing literature, specifically product packaging and advertising (cf. Clarke & Honeycutt, 2000; Madden, Hewett & Roth, 2000).

Graphical Elements—Future Research Agenda

The articles discussed in this section have been written over the last 10 or more years, which naturally raises the question of what changes in understanding and acceptance of iconic metaphors and graphical representations have occurred since then. Some metaphoric icons—for example the trash can and mailbox—have become standard so that misunderstanding should not be an issue.

Interface guidelines still use these metaphors as anecdotal evidence that culturally specific metaphors may not be understood and for this reason should be avoided. However, the number of studies on the understanding of the culturally dependent meaning of icons is rather small. More important, research on the subject of graphic representation on the interface has to go beyond the meaning and acceptability of particular "culturally meaningful" icons and colors to explore how they influence effectiveness, efficiency, and international user satisfaction. Thus, the issue of the relationship between aesthetics and usability from a cultural perspective should be studied on the basis of actual user performance. Barber and Badre's (1998) study of the use of *cultural markers*, interface features and elements that are dominant and possibly preferred by a particular cultural group, sparked a series of studies on their effect on user interaction with the computer. Sheppard and Scholtz (1999) evaluated the performance of North American and Middle Eastern users on specially designed, culturally enhanced sites and compared their performance. As they worked on assigned tasks, the users provided more correct responses when interacting with the site designed for their own culture. North Americans also reported that they preferred the U.S. site. Sun (2001) suggested that cultural markers can increase user satisfaction and improve navigation, although when cultural markers conflict with usability, some users may prefer usability to cultural sensitivity. Both studies are considered preliminary because the research instrument was not fully developed, few subjects were studied, and there were other limitations. In a more elaborate study, Badre (2000) augmented interfaces with U.S., Italian, or Greek cultural markers and examined use by U.S. and Italian subjects. Although the U.S. subjects' performance was not affected, the culturally meaningful navigation icons improved efficiency for the Italians. The results of these studies, although limited, suggest that cultural aesthetics can play an important role in usability, and should be given serious consideration in future research.

The Structuring and Content of Information

Formats

Cultures also vary in how they present information. Different numerical schemes can serve as an example. Depending on location and convention, formats for displaying numbers, time, and dates vary and may cause considerable confusion. There are many standards for displaying a date in the Gregorian calendar: day-month-year, month-day-year, year-month-day. Because all of these formats are often displayed using numbers only, the designer should remember to suggest the date scheme.

Del Galdo (1990, p. 4) suggests that "If an application will have an audience of more than one nationality, for example information kiosks in airports or train stations, use only alphanumeric [*sic*] characters for the display of month to avoid confusion caused by the various conventions of date formats." Although practical, this guideline is limited by the assumption that in a multinational crowd everyone knows the English names of months and their abbreviations.

The format to present numbers can also cause confusion (del Galdo, 1990). In Europe, the period is used as a separator for units of thousands and the comma is used as a decimal point; the reverse of North America and the U.K. Further, the European standard includes zero in front of the decimal point (0.5), whereas in North American it can sometimes be omitted (.5). Currency formats vary from country to country as well. The currency symbol may be up to four characters long and placed before the amount, after the amount, and/or as a decimal separator.

Layout and Content

There are also differences in information organization. For example, television schedules differ in the U.S., France, and Russia (Donskoy, 1997). In the U.S., the channels are listed horizontally and show times are organized by columns. In Russia the channels are listed vertically and there is no show time structure. In France, the schedule is not organized as a table; the listings for each channel are organized in sequential fashion.

Differences in layout have also been observed on Web sites designed in different cultures. Schmid-Isler (2000) noted that Chinese home pages (especially of news sites) are frequently divided into many independent spaces. In contrast, Western style prefers a focused layout with pictures as a visual attractor and other elements arranged around the focal point in an orthogonal way. These layout solutions represent browse vs. focus preferences in information reception, and are based on culturally dependent schemata of information storage and display. Dormann and Chisalita (2002) distinguished "visual" (based on graphics) from "index" (link based) sites.

Knowledge representation (abstract or concrete) and the structure of information presented (functional or thematic) can also influence search performance. Choong and Salvendy (1999) tested such influences on a group of 80 subjects from the U.S. and mainland China. For Chinese participants, the thematic structure of information and concrete knowledge representation reduced performance time as well as the error rate. A functional interface reduced the number of errors in the American group. However, users may prefer different solutions depending on the cultural or linguistic context. For example, Chinese participants in the experiment conducted by Dong and Salvendy (1999) performed better using a horizontal menu structure when they interacted with the English version of the interface. Vertical menus yielded better results for the Chinese version. Presumably users built different mental models for the English and Chinese interfaces, based on previous interactions with other interfaces of both kinds.

The differences in information organization may also impair user interaction with the system. Walton, Vukovic', and Marsden (2002) questioned whether the Western hierarchical tree, present in file structures, databases, and Web design, is suitable for systems designed for South African users. Their study suggested that although South African students did not have problems navigating the tree, its hierarchical meaning caused difficulties. The visual conventions that express the tree structures through the layout also caused problems.

The Maori, indigenous people of New Zealand, had difficulty in comprehending the digital library metaphor, because their oral culture transfers knowledge by stories, songs, and art, rather than by written texts (Duncker, 2002). Although Maori students learned to navigate the digital library, the study showed that the Western digital library model was not efficient. They did not have a concept of publishing forms, nor were the subject headings representative of the way information would be organized in Maori culture. Ducker suggested that providing search and browsing tools that enable a user to zoom into documents that contain a particular phrase without requiring any knowledge of how the documents are organized or published would make the digital libraries more usable for Maori users.

Metaphors that represent whole concepts (desktop, online store, digital library) rather than singular functions of the interface (e.g., trash can) may be more difficult to grasp. Evers (2001) investigated cultural understanding of the virtual campus metaphor and discovered that it was not universally understood. Understanding differed on several levels: Real-world experiences shaped understanding of the design metaphor and individual functions, knowledge of language and speaking conventions influenced understanding of textual elements, and different information needs influenced navigation and the perceived usefulness of the Web site. Users' understanding of the virtual campus metaphor influenced the meanings they attributed to specific icons.

Several research studies have examined the similarities and differences in interface design in different cultures, concentrating on the interface as a whole rather than its constituent features (Marcus & Gould, 2000; Robbins & Stylianou, 2002; Dormann & Chisalita, 2002). What unites these studies is that they are based on the works of Dutch anthropologist Hofstede (1997, 2001) and used his framework of cultural dimensions. According to Hofstede, world cultures vary along five consistent dimensions: power distance, collectivism vs. individualism, femininity vs. masculinity, uncertainty avoidance, and long- vs. short-term orientation. Power distance refers to the degree to which unequal distribution of power is accepted in a culture. Countries with a high power distance index tend to have centralized power, hierarchies, and considerable differences in status. Low power distance countries tend to downplay differences among people in terms of power and wealth.

Individualistic societies value personal achievement, whereas collectivistic ones emphasize the benefits of working in a social group. High masculine cultures make a strong distinction between gender roles and place priorities on values such as assertiveness and competition, whereas feminine cultures tend to dissolve gender differences and put more emphasis on relationships and living environments. Uncertainty avoidance measures how willing people are to take risks. Long-term-oriented cultures focus on practice and patience in achieving goals, whereas short-term-oriented cultures desire immediate results. The dimensions are not distinct; they overlap, and sometimes neutralize one another.

Hofstede's survey research was conducted in international subsidiaries of IBM (50 countries and three regions) between 1967 and 1973, with the results entered into an IBM database. The primary goal was to understand the work attitudes of international employees. Although the survey was conducted in, and referred only to, organizational contexts, Hofstede's five-dimensional model of culture is often used to interpret a range of phenomena in cross-cultural research. According to the *Social Sciences Citation Index,* Hofstede's books, *Culture's Consequences* and *Cultures and Organizations,* have together been cited over 3,500 times since their publication. Hofstede's dimensions of culture have been used in a large variety of disciplines: cross-cultural and organizational psychology, sociology, management, and communication. The popularity of his approach may reside in the fact that his dimensions of culture, especially power distance and individualism/collectivism, seem intuitive and their effects can be directly observed in interpersonal, cross-cultural encounters.

Hofstede (2001, p. 73) responded to the criticism that a study of the subsidiaries of one company cannot provide information about entire national cultures and its results cannot be generalized beyond the scope of the organizational subculture: "What were measured were *differences between* national cultures. Any set of functionally equivalent samples from national populations can supply information about such differences. The IBM set consisted of unusually well matched samples of an unusually large number of countries." In *Culture's Consequences,* Hofstede (2001) analyzed more than 200 cross-cultural studies from different disciplines as a basis for comparison with his results and found many similarities.

The popularity of Hofstede's model in intercultural communication research has led a number of interface design researchers to use it to examine differences in Web sites designed in various cultures. Marcus and Gould (2000) suggest that Web sites in high power countries will have highly structured access to information; employ tall hierarchies of mental models; and place significant emphasis on social and national order and their symbols, on security and restrictions of access, and on the prominence given to leaders. Less structured access to information; shallow hierarchies; fewer access barriers; and only a slight focus on expertise, authority, logos, and official stamps will characterize Web sites in countries with low power distance. Furthermore, Marcus and Gould argue that interfaces in highly individualistic countries will include, inter alia, frequent images of success demonstrated through materialism and consumerism, content emphasis on change and progress, and personal information, whereas Web sites in collectivist countries will emphasize socio-political achievements, the prominence of age and experience, and tradition and history. Masculine interfaces, Marcus and Gould contend, will focus on tasks with quick results, navigation oriented to exploration and control, games and competitions, graphics, and sounds and animation for utilitarian purposes. In contrast, feminine interfaces will incorporate cooperation, exchange, and support, and will attract attention through poetry and visual aesthetics.

According to Marcus and Gould (2000), uncertainty avoidance also plays a large role in Web site structure. Interfaces in countries with high uncertainty avoidance will be simple, with limited choices and a high number of redundant clues to prevent users from getting lost. Low uncertainty avoidance Web sites will be more complex, with less control of navigation and help systems that are concept—rather than task—oriented. Similarly, Web sites for long-term-oriented countries would require patience, whereas those for short-term-oriented countries would be designed to achieve goals quickly. The authors do not conduct systematic studies but provide examples that support their assessment. However, their approach appears promising. More systematic study in this area would help to identify culture type-specific interface elements.

Robbins and Stylianou (2002) took a similar approach in their empirical study of commercial Web sites in several geographical regions. They conducted frequency counts of Web site elements operationalized from Hofstede's five dimensions. According to the researchers, the power distance would be indicated by the presence of organizational charts, biosketches of top leaders, and messages from chief executive officers (CEOs). Uncertainty avoidance would be represented, for example, by listings of job openings and cookie disclosures. Site registration requirements, security provisions, and privacy policy statements would relate to individualism/collectivism and annual reports and financial highlights would represent the site's masculinity/femininity. Long- or short-term orientation would be indicated by the presence or lack of features such as a search engine, site map, frequently asked questions (FAQs), and corporate history. Their results, based on those dimensions, showed significant differences among corporate Web sites based in different countries.

A third study using Hofstede's dimensions to evaluate Web sites was conducted by Dormann and Chisalita (2002). They examined university Web sites in five countries on the masculinity/femininity index. This study is especially interesting because the authors did not conduct the evaluation themselves but rather designed a questionnaire and asked participants from the Netherlands (feminine culture, low masculinity

[MAS] index) and Austria (masculine, high MAS index) to evaluate Web sites from countries with different values on the masculinity index. The subjects rated higher all the feminine values in feminine Web sites, but with regard to masculine values only authority and toughness were rated higher in home pages from masculine countries. Although there were only limited correlations between the MAS index and masculine and feminine values presented on the sites, this work is important because it adopts the perspective of people from different cultures rather than the Web site elements as such.

The differences noticed in these studies suggest that designers from different countries have diverse preferences in organizing Web site layout and content. But a more important question has to be asked: Do the cultural differences evident in graphic design influence Web site usability? Do people from countries at opposite ends of Hofstede's dimensions prefer Web sites with widely divergent designs, and more importantly, do they perform better using the interfaces designed specifically for their cultures? The influence of design on user performance should be the main focus for studies of international interfaces. Critical work evaluating the suitability of Hofstede's model for research in visual representation could also provide a basis for more empirical studies on differences in Web and interface design.

Interface Design Process

Interface Design Lifecycle

A poorly designed interface can have a significant effect on user performance and satisfaction. It may lower productivity, increase training costs, result in costly user errors, or makes one's product less marketable than those produced by companies that have invested more effort in designing user-friendly interfaces (Mayhew, 1999a). Usability engineering—a process that incorporates prospective user input at all stages of design—helps identify interface problems before a product is released. Because of rapid advances in information technology, most products will be redesigned for subsequent release so that users have more functionality.

The interface design process is cyclical. Detailed descriptions of various stages of interface design lifecycle can be found in the works of Jacob Nielsen (1993) and Deborah Mayhew (1999b), among others. Nielsen's and Mayhew's models of the interface design lifecycle vary in terms of the number and order of their steps and substeps. Nielsen emphasizes that the design process is not separate from other processes in the organization, including steps such as parallel design (when several developers work independently on solutions to find the best one) and financial impact analysis. Mayhew approaches the problem from the system's perspective, incorporating platform-capabilities constraints (display size, number of colors, input devices, system speed, etc.). What these models have in common is highlighting the importance of involving potential users in most of the design substeps to ensure that users' mental models of the interface are reflected in the design.

Localization of the interface repeats the creation cycle and redesigned elements and functions have to be evaluated by a new cultural group of users. As Nielsen (1990, p. 39) points out: "an interface which is used in another country than the one [for which] it was designed, is a new interface. One cannot trust the original usability work on the user interface to necessarily have produced a design which will be equally usable around the world." It follows that many of the interface design lifecycle steps could be affected by cultural differences. The first step in interface design is collecting detailed information about future users of the product. "Individual user characteristics and variability of the tasks are the two factors with the largest impact on usability, so they have to be studied carefully" (Nielsen 1993, p. 73). Prior to designing the interface for another country, engineers should gather data about the potential user population. An even more important step is to analyze the differences in goals and the ways tasks are performed in the target culture. Questionnaires for user information gathering may also be employed in this phase, but the observation of users at work is most beneficial. An outcome of task analysis is typically a list of the tasks users want to accomplish using the system, the steps needed to carry out those tasks, and a list of possible outcomes. The system should accommodate all possible exceptions from normal performance; if user requests cannot be met by the system, a meaningful message should be displayed to the user (Nielsen, 1993). Task analysis is also helpful in determining the level of acceptability of the results. (System acceptability is discussed further in the usability testing section.) International users should take part in all steps of the design process. The original interface may be used in a competitive analysis to determine differences in task performance. Financial analysis should also be considered, especially in deciding the methods of usability testing. Style guidelines should not only be based on the principles of good design, but also include information about cultural preferences and taboos.

Cultural Dimensions of Usability

The usability of interfaces is operationally defined as the effectiveness, efficiency, and satisfaction with which users can perform particular tasks in a given environment and express their attitudes toward the interface (Nielsen, 1993). Effectiveness can be defined as the accuracy and completeness with which users achieve their goals. It is usually measured by the number of completed tasks or their elements and the number of errors made during the interaction with the interface. Efficiency is defined as the relationship between the accuracy and completeness of tasks and the time needed to accomplish them. As a primary index of efficiency, a task completion time is measured; however, in some cases, learning time is also calculated. Satisfaction is measured through rating scales that record positive or negative attitudes about the system. Satisfaction is usually measured by standardized attitude questionnaires. These are not limited to assessing whether the user likes or dislikes the system, but also measure perceived efficiency and effectiveness. These measurements are commonly used to compare interfaces and are contextually defined in operational terms as targets to meet.

> Users should be able to perform specified tasks with new tool after W minutes training, with X percent effectiveness, at least Y percent efficiency, and Z percent greater satisfaction than the old interface. (Dillon & Morris, 1999, p. 1018)

Barber and Badre (1998) point out that in discussing design for different cultures, usability must be redefined, because what is "user friendly" for one culture can be vastly different for another. When comparing the usability of a localized version of an interface with the original, it might seem logical to assume that if they are rated similarly on usability measurements, their overall usability ratings will also be similar. But the specification of what is an acceptable score on each of the usability measurements has to be adjusted. As regards the usability of an interface developed or adapted for another culture, differences in time perception may affect the interpretation of usability measurements. Anthropologists describe two concepts of time: monochronic (M-time) and polychronic (P-time) (Hall, 1990). Monochronic cultures view time as a linear dimension, along which events take place one at a time, in ordered, calculated fashion. Monochronic time is based on a belief in the logical progression of past, present, and future. Monochronic cultures are committed to doing one task at a time; following schedules and meeting set deadlines is important. The U.S., Germany, Switzerland, and Scandinavian countries are cultures with a monochronic perception of time. Cultures with a polychronic conception (Mediterranean, Latin American, and Arab countries) tend to perform multiple tasks simultaneously. Punctuality is defined loosely and deadlines are flexible. The present matters most. Both types of time perception are context dependent, so cultures may be characterized in terms of which time perception is dominant. Because of these differences, simply measuring task completion time may not be the best way to determine the efficiency of a system. A quicker completion time may not have the same value for people from monochronic and polychronic cultures. Because polychronic cultures are more task-oriented than monochronic ones, there may be significant differences in interpreting effectiveness. Similarly, satisfaction may be greater with higher efficiency and lower effectiveness in one culture and with lower efficiency and higher effectiveness in others (or might require high values in both dimensions).

This is an important issue, one that has yet to be studied. The only research I could locate on the topic was an experiment conducted by Evers and Day (1997), who compared the attitudes of Indonesian, Chinese, and other Asian students with a control group of Australian students. The Chinese students found usefulness more desirable, whereas Indonesians found ease of use to be more important and closely correlated with satisfaction. Among the Chinese, satisfaction was equated with a combination of design feature preferences and system usefulness. Evers and Day's results are derived from participant questionnaires only, so the variables are based on perceptions and not actual measures of behavior.

Usability Measuring Techniques

Applying other usability measurement techniques should be done with caution. Noiwan and Norcio (2000) used a heuristic evaluation method modeled by Keevil and Associates (Keevil, 1998) to examine four Thai and U.S. academic Web sites. The authors saw differences in the number of violations of design guidelines but were not able to conclude whether those were due to differences in usability preferences between the two groups or to the limitations of the evaluation instrument. Combining heuristic evaluation with other usability assessment techniques may help answer this question.

Herman (1996) suggests that there may also be inconsistencies between objective usability evaluation results (performance measures) and subjective evaluation results (questionnaires and interviews), that are due to cultural/behavioral characteristics of the users. However, Yeo's (2000) studies, in which he compared positive and negative comments in think-out-loud protocols (objective measures) with a System Usability Scale—a questionnaire—and interviews (subjective measures), showed different results. There was a moderate positive correlation, but the correlation was strongest for users who had previous experience with similar software. This example shows the complexity of cultural research when multiple variables have to be taken into consideration.

Usability measurement techniques may yield more information in some research populations than in others. Evers (2002) observed users from the U.S., the U.K., the Netherlands, and Japan and discovered differences in the effectiveness of data gathering methods (questionnaire, task observation, interview) among subjects from the target groups. Questionnaires posed problems for North Americans (especially answering questions about ethnic background) and Japanese subjects (who treated it as a test and needed reassurance to start). Task observations were problematic for all groups except North Americans. Japanese and U.K. participants did not feel comfortable thinking aloud, and the Dutch used much sarcasm and humor, which was difficult to transcribe and translate. Interviews were problematic in the case of North Americans, because their responses were often inconsistent with the results of the task observation sessions.

Usability Apparatus

Material used for usability studies (such as questionnaires) has to be adjusted to account for cultural differences, and results should be interpreted with caution. For example, in studies by Yeo (2001), the comments on the questionnaires were generally very positive. Malaysia is a collectivistic society, where maintenance of harmony is important. Thus, users avoided negative evaluations given that this might disturb the harmony. Herman (1996) obtained similar results. In Japan, for example, questions about subjects' feelings had to be removed from the questionnaires because test participants did not feel comfortable answering them (Fernandes, 1995). Translated satisfaction questionnaires should be used with caution as they may not be understood or accepted in the same manner as the original questionnaires.

When transferring usability measurement instruments to another culture, one has to ensure that the translation and adaptation reflect corresponding concepts. To assess the quality of evaluation methods for comparing culture, the following criteria should be applied (Helfrich, 1993, as cited in Beu, Honold & Yuan 2000, p. 356):

- Equivalence of constructs or concepts to assure translations of questions back and forth between languages will help designers to come up with the vocabulary which best represents the concepts as understood in both cultures

- Equivalence of operators will have to be ensured for validity in different cultures

- Equivalence of the assessment process, hence, the way the indicators are measured must be identical between the cultural groups.

As Beu and his colleagues (Beu et al., 2000) point out, most standardized questionnaires were validated for the language and culture in which they were created. Developing a similar instrument for China, for example, poses a challenge because discrete rating scales are not common in that country.

Usability Procedure

Cultural differences can also become evident during interaction with usability session participants. In some countries, evaluations may be more effective when done by natives; in others, this may not be an issue (Dray, 1996). It is also important to be aware of gender relations. A female facilitator may be effective in one culture, but in others it may be wiser if she observes from another room. In some cultures, Western women are perceived differently from local women and may face considerably less discrimination. Female test participants may behave in different ways—Claris Corporation's researchers in Japan observed that women spoke very softly, so the microphones used for the study had to be of good quality. Women also talked very little when participating in co-discovery sessions with male partners. A similar situation was observed when people of different status were put in the room together (Fernandes, 1995).

Yeo (2001) observed that Malaysian usability subjects who ranked higher professionally than the session facilitator tended to be "harsher" in their negative comments than those who were lower in the hierarchy. Malaysia was ranked first among 50 countries on power distance in Hofstede's studies.

Beu et al. (2000) defined several barriers in focus groups conducted in cultures that can also influence other subjective methods of information gathering (such as interviews, focus groups, and questionnaires):

- Modesty with regard to one's achievements

- Expression of respect

- Caution in speech

- Fear of losing face

- Differentiation of action and language usage depending on hierarchies and situations

- Avoidance of direct criticism

These observations are consistent with Herman's (1996) case study of a usability session conducted with Far East users. The author suggested that unobtrusive observations might yield better results than testing in the presence of the observer.

Case studies are excellent sources for learning about conducting usability studies in other countries. During a comparative study conducted in the U.S., Germany, and France by a human factors team at Hewlett-Packard (Dray & Mrazek, 1996), the researchers investigated how families used computers and printers in their homes. The team observed 20 families in six different locations. Dray and Mrazek emphasized the benefits of including natives in the observation team, as local translators will not only translate the conversation but also interpret the nonverbal actions typical of the culture and report acts of appreciation from the subjects that may not have been noticed by the researchers. When preparing study materials they may also point out cultural issues of which researchers are unaware.

Russo and Boor (1993) provided an example of how using native experts in the design process may help usability researchers avoid serious mistakes. The authors cite a story from Hapgood about a product that was designed by Lotus Corporation for the Japanese market. Because the Japanese keep dates in the years of the Emperor, designers thought that it would be helpful to provide a feature that would allow the Emperor's name to be changed when needed. Native consultants explained that it would be inappropriate because the immortality of the Emperor would be questioned.

These case studies show examples of cultural differences in usability procedures uncovering a need for more rigorous research. The most important question that should be addressed in future studies is: Do all cultures define usability in the same manner? Answering this question is essential for understanding which functions of the interface should be emphasized in localized design. The second important aspect of usability requiring greater attention is whether and how users' tasks vary in different cultures and how they are executed.

Conclusions

Summary

The literature surveyed in this chapter provides evidence that culture plays a role in interface design on multiple levels. Cultural differences are manifested on the textual level through written language formats (character set, direction, etc.), vocabulary (use of pre-existing words to capture a computer function or invention of new words), and systems for keeping time and dates, which involve coding and translation. On the level of graphic design, differences include preferences for colors, layout, and culturally familiar icons. However, cultural differences go beyond system features and aesthetic preferences. To these differences should be added cognitive and cultural aspects such as conceptual models of culturally constructed schemata, differences in perception of time, and preference for efficiency or effectiveness. Last but not least, cultural differences may be evident not only during user interaction with the computer, but also in the user data-gathering phase of the interface design lifecycle and/or usability testing.

Implications

Differences in technical features, graphic design, and cognitive and cultural traits all potentially have an impact on whether users can operate with the interface and their level of satisfaction in their interaction with the computer. Therefore, the short and simple answer to the question of whether cultural differences matter in interface design is a resounding "yes." The more important and harder question is which cultural differences matter most, and under what circumstances. Differences that impair user interaction with a system will be of critical importance. Language seems to be the most obvious obstacle. Users who do not know the language of the interface will not be able to communicate with the system; and even those who are proficient in the language of the interface may not have been provided with the necessary resources (fonts, keyboard) to carry out the desired tasks.

Other important barriers potentially exist on the cognitive level. If users' understanding of interface functions is different from the system's conceptual model, they may not be able to interact efficiently until new mental models of the system are created. An example of such a barrier is the culturally specific interface metaphor. If users do not understand the metaphor, they cannot successfully exploit the interface's functionality. For this reason, use of culturally specific metaphors should be avoided.

Developers often use schemata from the physical world to design computer interfaces that work in analogous ways. But because the physical world is not common across all cultures, an interface designed to reflect reality in one country may be problematic for other cultural groups. For example, the schema/metaphor of the typewriter was used to design the first word processor program. This concept was foreign, for example, to Japanese users who did not use typewriters (Ito & Nakakoji, 1996), but rather a ruled writing pad with 20 x 20 grids—with the writing flowing from top to bottom and right to left. To be able to use the word processing program, Western users had to add some facts to their existing typewriter schema (schema accretion), whereas Japanese users had to create a whole new schema (schema induction).

Other observable cultural differences might be of little importance for accomplishing tasks, especially after users have become familiar with the interface. For example, a color that may not seem culturally appropriate will likely have little, if any, effect on the interaction. The same will be true of decorative elements in the interface that seem culturally foreign. User satisfaction, however, should not be underestimated because it shapes user perception of the system's usefulness and may, therefore, be a decisive factor in the choice of which system to use in a marketplace of competing products. These points are summarized in Table 7.1.

The importance of culturally and morally acceptable content touches on a concern that goes deeper than finding the right elements and design solutions to ensure usability by members of other cultural groups. It is rooted in the importance of a culture to its members.

Culture is important to self-perception and identity, although people vary in the degree to which they consider themselves to be representatives of a specific culture. The need to experience a sense of belonging is enduring and fundamental to human nature. Cultural identification determines which cultural groups individuals consider as in groups and which as out groups (Rogers & Steinfatt, 1999). It gives members a sense of security and it influences how people behave, what they believe, and how they interact with others.

Table 7.1 Importance of culturally variable interface elements to user interaction

	Textual elements	Graphical elements
Critical for interaction – interaction cannot otherwise occur or is severely hampered	• language of the interface is known to user • ability to enter proper fonts • ability to specify appropriate formats (numbers, date, time, etc.)	
Important for interaction – interaction cannot occur until user learns new information	• discourse style is understandable to the user • transparent relation between translated word and system function • understandable formats (date, time, etc.)	• culturally understandable graphical metaphors • transparent relation between culture specific icon and system function
Important for system acceptance – interaction can occur otherwise, but user may reject the system	• option to interact in native language • use of discourse style of native language	• culturally appealing / appropriate colors • culturally acceptable graphical representations • information display characteristic of user's own country • culturally/morally accepted content

Strong influence exercised by other cultures may be perceived as a threat to cultural sovereignty, which some cultures may go to great lengths to protect. Foreign cultures may be perceived in adversarial fashion. The desire to preserve one's own culture may permeate all aspects of life, including the design and use of information technologies. For example, Iceland requested a localized version of the Microsoft operating system on the grounds that in its absence, young Icelanders will lose proficiency in their native language. India, a country of many languages in which knowledge of English is necessary for success, started to develop a coding system, Indian Standard Code for Information Interchange (INSCII), to cover major Indian languages. The French, in alliance with French speaking Canadians and Francophonic Africa, make systematic efforts to ensure that French is preserved in information technologies—software imported to France and Web sites developed in that country must by law be in French. The Germans share similar concerns about the hegemony of English in the computer world—although as a senior German telecommunications official commented off the record, "We let the French do the talking for us" (all examples from Keniston, 1998).

These examples show the urge to maintain one's culture in the realm of information technology. Cultural differences in interface design elements should thus be taken into consideration not only because they may impair foreign user interaction with technology, but as a matter of cultural sensitivity. Even if the impact of culturally variable interface factors on human–computer interaction were negligible (which it is not), it would still be important to attend to cultural variables out of respect for other groups of users.

Research Questions Revisited

Although the literature on the role of culture in interface design is still limited—in that the research domain is relatively new—and geographically biased—in that most of the research has been produced by Western scholars—the works surveyed here can begin to provide some answers to the questions raised at the beginning of this chapter.

> 1. *How easily, after translation, can an interface designed in one country be used in / transferred to another country?*

Most of the research available concentrates on studying differences between Western and Chinese or other Asian cultures, which are more profound than differences among Western cultures. In such cases, transfer can be challenging, although it is still possible. For two cultures that have affinity, the process of adaptation should be less involved than in cultures that are dissimilar.

> 2. *Are the challenges involved in adaptation merely cosmetic or do more profound forces shape them?*

The iceberg model of culture can shed light on the nature of the challenges associated with interface adaptation. The tip of the iceberg is cosmetic modifications, such as adjustments of date, which can easily be made. However, designing interfaces that fit cultural schemata poses greater challenges for the designer who must have not only basic familiarity with the culture's standards, but in-depth knowledge about users' goals and tasks. Deeper yet, the social, economic, and political forces that privilege technologically advanced nations over less developed nations may shape users' attitudes toward, and acceptance of, information technologies in ways that are beyond the control of individual designers. These must be addressed at the level of policy.

> 3. *Do all cultures respond to interfaces in similar ways, or does culture itself shape user comprehension?*

User schemata of computer interaction and of social interaction jointly shape communication with computer interfaces. Users from different cultures may have similar expectations of the outcome of the interaction, but how they perceive the interaction itself may vary because their schemata of interaction are culturally dependent. Increased contact with computers and greater familiarity with their functions, should enable users to understand better the rules of interaction, but if the rules do not match their mental models this understanding will take time.

4. *If culture is a factor in explaining varied user reactions to comparable interfaces, what specific cultural dimensions are responsible for the differences?*

Because cultural dimensions are themselves part of the definition of culture, and the studies reviewed here used different definitions of culture, it is difficult to draw generalizable conclusions. Van de Vijver and Leung (1997a) recommend that culture be decomposed into a set of psychologically meaningful constructs in order to determine which aspects of a given culture are responsible for observed differences. Among the papers reviewed, only a few tried to explain observed cultural differences from this perspective. Hofstede's dimensions of culture constitute a promising model in this respect.

5. *Do usability differences reside mainly at the level of national cultures, or do they depend on other variables such as class, gender, age, education, and expertise with technology?*

Not all differences among people are rooted in culture. Within a given culture, individual differences such as physical characteristics, gender, age, social class, religion, multicultural exposure, education level, linguistic ability, expertise with technology, and task experience account for variation as well (Dillon & Watson, 1996). Some uneasiness with the system or user mistakes may, in fact, have nothing to do with cultural differences (Mrazek & Baldacchini, 1997).

6. *In the face of a potentially large number of explanatory variables, how do we delimit a workable concept of culture and yet remain cognizant of other factors that might shape the results of culture and interface research?*

It is important to recognize that the definition of culture one chooses will partially determine one's conclusions. In the literature, the definition of cultural group is usually narrow. It assumes that a (national) group of users will essentially share the same cultural characteristics. From a practical research perspective, this approach works, in as much as it eliminates the need to include multiple variables that are difficult to control. Given that research on cultural differences in interface design is still embryonic, this approach may be appropriate for now. However, as our knowledge of cultural differences increases, it will be essential to re-evaluate our operating definition of culture and to account for other variables through multifactor studies.

Future Research Agenda

Culture in the literature reviewed in this paper is defined in two predominant ways that are actually opposite sides of the same coin: as a unifying factor (people from the same culture have a lot in common) and as a distinguishing factor (cultures vary, thus people from different cultures will have different characteristics). One shortcoming is that most of the research studies provide comparisons of only two or three different cultures. These studies provide us with a better understanding of the research issues they cover, but suggest a need for studies that are more geographically extended. Also, the studies on international interface design are limited to specific geographic regions, comparing in most cases Western and Eastern cultures. Latin America and Africa are almost never mentioned. Examining more diverse cultures would show the differences in international interface perception between early and late information technology adopters, on both sides of the digital divide.

Research might also explore the degree to which designers in different cultures apply Western models or look for solutions rooted in their own traditions. Such studies would allow us to examine the relationship between technology and culture not only from the perspective of how culture influences interface design, but also how international users must adjust for information technologies to transform their ways of life.

There is also a need for comparative, multivariate studies to account for the differences between novice and experienced computer users. One might assume that cultural differences play a greater role in the case of novice users than individuals who can build mental models of the interface on the basis of other applications. Similarly, cultural differences will presumably have less impact on users who have had greater exposure to Western culture.

We should also bear in mind the need to acknowledge cultural differences among national and ethnic groups depending on their political situation, geographical location, and emotional attachment to their own culture. Catalan users may emphasize their cultural uniqueness, whereas minorities within Great Britain may feel that designing culturally sensitive interfaces will make them second-class citizens (Duncker, Theng, & Mohd-Nasir, 2000). In many cultural groups where computers and Internet access are perceived as status symbols, the user populations will be inclined to learn English and assimilate a Western style of thinking; other cultures may use Internet technology to promote their own heritages.

As our knowledge of cultural differences increases, it will be essential to evaluate and re-evaluate our operating definition of culture and expand it in flexible ways to account for other variables. In so doing, we will not only be able to offer more detailed and culturally aware prescriptions for cultural interface design, but we will also be better positioned to refine our answers as to what cultural differences matter, to what extent, when, for whom, and why.

Acknowledgments

The author would like to thank Blaise Cronin, Julia Fox, and especially Susan Herring for their guidance and helpful comments on earlier versions of this chapter. The input from three anonymous reviewers helped in structuring the chapter and illuminating major points.

Endnotes

1. Schema theory is not the only theory that attempts to explain how we learn to make sense of our environment and to act accordingly. Other cognitive, as well as social and behavioral, theories could also be used in making sense of interacting with computers. Schema theory has been used often in cultural research; it thus seems appropriate for discussing cultural aspects of interface design. A good review of learning theories is provided by Schunk (2000).

2. The Ito and Nakakoji definition of *perception* is rather narrow. In the physiological literature, *perception* is understood as mediated by different levels of representation: object representation (abstract objects), implicational representation (qualitative, holistic interpretation of stimuli), and prepositional representation (semantic, relational facts about perceived scene) (May, 2000); thus all the phases of the listening mode can be interpreted as perception. This broad definition of perception is the one I will use throughout the chapter.

3. The figures do not add up to 100, because the remaining percent are native English speakers.

4. This, of course, is dependent on the context, because governmental or other official Thai sites would be expected to display the national symbol.

5. There are more than 6,000 languages in the world (Ethnologue, 2000), not all of them present on the Internet, and only 193 countries, many of them with multiple official languages (U.S. Department of State, 2003).

References

Amara, F., & Portaneri, F. (1996). Arabization of graphical user interfaces. In E. del Galdo & J. Nielsen (Eds.), *International user interfaces* (pp. 127–150). New York: Wiley.

Anderson, R. C. (1984). Some reflections on the acquisition of knowledge. *Educational Researcher, 13*, 5–10.

Aykin, N. (1999). Internationalization and localization of the Web sites. *Proceedings of HCI International '99 - the 8th Conference on Human Computer Interaction*, 1218–1222.

Badre, A. N. (2000). *The effects of cross cultural interface design orientation on World Wide Web user performance* (Technical Report No. GIT-GVU-01-03). Atlanta: Graphics, Visualization and Usability Center, Georgia Institute of Technology.

Barber, W., & Badre, A. (1998). Culturability: The merging of culture and usability. *Proceedings of the 4th Conference on Human Factors & the Web*. Retrieved November 12, 2000, from http://www.research.att.com/conf/hfweb/proceedings/barber

Beamer, L. (1995). A schemata model for intercultural encounters and case study: The emperor and the envoy. *Journal of Business Communication, 32*, 141–162.

Beu, A., Honold, P., & Yuan, X. (2000). How to build up an infrastructure for intercultural usability engineering. *International Journal of Human–Computer Interaction, 12*, 347–358.

Blanchard, H. E. (1997). International standards on human–computer interaction: What is out there and how will it be implemented? In G. Salvendy, M. J. Smith, & R. J. Koubek (Eds.), *Design of computer systems: Cognitive considerations* (pp. 590–602). Amsterdam, The Netherlands: Elsevier.

Bødker, K., & Pedersen, J. S. (1991). Workplace cultures: Looking at artifacts, symbols and practices. In J. Grenbaum & M. Kyng (Eds.), *Design at work: Cooperative design of computer systems* (pp. 121–136). Hillsdale, NJ: L. Erlbaum.

Borgman, C. L. (1992). Cultural diversity in interface design. *SIGCHI Bulletin, 24,* 31.

Borgman, C. L. (1997). Multi-media, multi-cultural, and multi-lingual digital libraries, or how do we exchange data in 400 languages? *D-Lib Magazine, 6.* Retrieved January 6, 2003, from http://www.dlib.org/dlib/june97/06borgman.html

Bourges-Waldegg, P. (1999, April). *Dealing with cultural differences in computer-supported collaborative work.* Paper presented at Changing Places: A One-Day Workshop on Workspace Models for Collaboration, London. Retrieved January 12, 2003, from http://www.dcs.qmul.ac.uk/research/distrib/Mushroom/workshop/final-papers/bourges.pdf

Bourges-Waldegg, P., & Scrivener, S. A. R. (1998). Meaning, the central issue in cross-cultural HCI design. *Interacting with Computers, 9,* 287–309.

Brugger, C. (1990). Advances in the international standardization of public information symbols. *Information Design Journal, 6,* 79–88.

Callahan, E. (2002, October). *Web search engines: Information retrieval in less common languages.* Paper presented at the Association of Internet Researchers conference Internet Research 3.0: NET / WORK / THEORY, Maastricht, The Netherlands.

Card, S. K., & Moran, T. P. (1986). User technology: From pointing to pondering. *Proceedings of the ACM Conference on the History of Personal Workstations* (pp. 183–198). New York: ACM.

Cooper, A. (1997). *The inmates are running the asylum.* Indianapolis, IN: Macmillan Computer Publishing.

Character encoding. (1996). Retrieved August 20, 2002, from http://babel.alis.com:8080/codage/index.html

Chinese Culture Connection. (1987). Culture value and the search for culture free dimensions of culture. *Journal of Cross-Cultural Psychology, 18,* 143–174.

Choong, Y. Y., & Salvendy, G. (1998). Design of icons for use by Chinese in mainland China. *Interacting with Computers, 9,* 417–430.

Choong, Y. Y., & Salvendy, G. (1999). Implications for design of computer interfaces for Chinese users in mainland China. *International Journal of Human–Computer Interaction, 11,* 29–46.

Chu, S. W. (1999). Using chopsticks and fork together: Challenges and strategies of developing a Chinese/English bilingual Web site. *Technical Communication, 46,* 206–219.

Clarke, I., III, & Honeycutt, E. D., Jr. (2000). Color usage in international business-to-business print advertising. *Industrial Marketing Management, 29,* 255.

Clavet, A. (2000). The government of Canada and French on the Internet: Special study, August 1999. *Proceedings of INET 2000, The 10th Annual Internet Society Conference.* Retrieved June 7, 2001, from http://www.isoc.org/isoc/conferences/inet/00/cdproceedings/8k/8k_1.htm

Connolly, J. (1996). Problems in designing the user interface for systems supporting international human–human communication. In E. del Galdo & J. Nielsen (Eds.), *International user interfaces* (pp. 20–40). New York: Wiley.

Cook, B. L. (1980). Picture communication in Papua New Guinea. *Educational Broadcasting International, 13,* 78–83.

Cunliffe, D. (2001). *Guidelines for bilingual usability: Unitary authority Web sites in Wales.* (University of Glamorgan, School of Computing Technical Report). Retrieved February 8, 2003, from http://www.comp.glam.ac.uk/~Daniel.Cunliffe/bilingual/techreport.doc

D'Andrade, R. G. (1981). The cultural part of cognition. *Cognitive Science, 5,* 179–195.

D'Andrade, R. G. (1989). Culturally based reasoning. In A. Gellatly, D. Rogers, & J. A. Sloboda (Eds.), *Cognition and social worlds* (pp. 132–143). Oxford, UK: Clarendon Press.

del Galdo, E. (1990). Internationalization and translation. Some guidelines for the design of human–computer interfaces. In J. Nielsen (Ed.), *Designing user interfaces for international use* (pp. 1–10). Amsterdam: Elsevier.

del Galdo, E. (1996). Culture and design. In E. del Galdo & J. Nielsen (Eds.), *International user interfaces* (pp. 74–87). New York: Wiley.

Dillon, A., & Morris, M. (1999). Power, perception and performance: From usability engineering to technology acceptance with the P3 model of user response. *Proceedings of the 43rd Annual Conference of the Human Factors and Ergonomics Society*, 1017–1021.

Dillon, A., & Watson, C. (1996). User analysis in HCI: The historical lessons from individual differences research. *International Journal of Human–Computer Studies 45*, 619–637.

Dong, J. M., & Salvendy, G. (1999). Designing menus for the Chinese population: Horizontal or vertical? *Behaviour and Information Technology, 18*, 467–471.

Donskoy, M. V. (1997). The cultural adaptation versus localization. *Proceedings of the Seventh International Conference on Human–Computer Interaction*, 137–139.

Dormann, C., & Chisalita, C. (2002, September). *Cultural values in Web site design.* Paper presented at the Eleventh European Conference on Cognitive Ergonomics, Catania, Italy. Retrieved February10, 2003, from http://www.cs.vu.nl/~claire/Hofstede-dormann.pdf

Dray, S. (1996). Designing for the rest of the world: A consultant's observations. *Interactions, 3*, 15–18.

Dray, S., & Mrazek, D. (1996). A day in the life: Studying context across cultures. In E. del Galdo & J. Nielsen (Eds.), *International user interfaces* (pp. 242–256). New York: Wiley.

Duncker, E. (2002). Cross-cultural usability of the library metaphor. *Proceedings of the Second ACM/IEEE-CS Joint Conference on Digital Libraries*, 223–230.

Duncker, E., Theng, Y.-L., & Mohd-Nasir, N. (2000). Cultural usability in digital libraries. *ASIS Bulletin, 26*(4). Retrieved April 14, 2003, from http://www.asis.org/Bulletin/May-00/duncker__et_ al.html

Dunlap, W. (1999). *Reasons for success in international e-commerce: Speaking the customer's language.* Retrieved February 23, 2003, from http://glreach.com/eng/ed/art/reasonsfor global.html

Dzida, W. (1996). International usability standards. *ACM Computing Surveys, 28*, 173–175.

Eberts, R. E. (1995). *User interface design.* Englewood Cliffs, NJ: Prentice Hall.

Encyclopædia Britannica online. (n.d.) Culture. Retrieved September 18, 2000, from www.britanica.com/bYtricle/printable/6/0,5722,118246,00.html

Erickson, J. C. (1997). Option for presentation of multilingual text: Use of the Unicode standard. *Library Hi Tech, 59/60*, 172–188.

Ethnologue. (2000). *Geographic distribution of living languages.* Retrieved April 16, 2003, from http://www.ethnologue.com/ethno_docs/distribution.asp

Evers, V. (2001) *Cultural aspects of user interface understanding: An empirical evaluation of an e-learning Website by international User groups.* Unpublished doctoral dissertation, the Open University. Retrieved May 2, 2003, from http://www.swi.psy.uva.nl/usr/evers/publications.html

Evers, V. (2002) Cross-cultural applicability of user evaluation methods: A case study amongst Japanese, North American, English and Dutch Users. *Proceedings of the SIGCHI Conference on Human Factors in Computing Systems, CHI 2002*, 740–741.

Evers, V., & Day, D. (1997). The role of culture in interface acceptance. *Proceedings of INTERACT '97*, 260–267.

Fernandes, T. (1995). *Global interface design.* Chestnut Hill, MA: Academic Press.

Fussell, D., & Haaland, A. (1978). Communicating with pictures in Nepal: Results of practical study used in visual education. *Education Broadcasting International, 2*, 25–31.

Global Internet Statistics (by Language). (2003). Retrieved April 11, 2003, from http://www.glreach.com/globstats

Graddol, D. (1997). *The future of English.* London: The British Council.

Grefenstette, G., & Nioche, J. (2000). Estimation of English and non-English language use on the WWW. In *Proceedings of RIAO 2000.* Retrieved November 10, 2002, from http://133.23.229.11/~ysuzuki/Proceedingsall/RIAO2000/Wednesday/20plenary2.pdf

The GUI guide, international terminology for the Windows interface (European Ed.). (1993). Redmond, WA: Microsoft Press.

GVU's 8th WWW user survey. (1997). Retrieved April 14, 2002, from http://www.gvu.gatech.edu/user_surveys/survey-1997-10

Hall, E. T. (1990). *Understanding cultural differences.* Yarmouth, ME: Intercultural Press.

Helfrich, H. (1993). Methodologie kulturvergleicherender psychologischer Forschung [Methodology of psychological research into comparative cultures]. In A. Thomas (Ed.), *Kulturvergeicheder Psychologie: Eine Einfürung* (pp. 81–102). Göttingen, Germany: Hogrefe.

Herman, L. (1996). Toward effective usability evaluation in Asia: Cross-cultural differences. *Proceedings of the 6th Australian Conference on Computer–Human Interaction (OZCHI '96),* 135–136.

Hofstede, G. (1997). *Cultures and organizations: Software of the mind.* London: McGraw-Hill.

Hofstede, G. (2001). *Culture's consequences: International differences in work-related values.* Newbury Park, CA: Sage.

Hoft, N. (1996). Developing a cultural model. In E. del Galdo & J. Nielsen (Eds.), *International user interfaces* (pp. 41–73). New York: Wiley.

Holstius, K. (1995). Cultural adjustment in international technology transfer. *International Journal of Technology Management, 10,* 676–686.

Honold, P. (1999). Cross-cultural usability engineering: Development and state of the art. *Proceedings of HCI International '99 - the 8th Conference on Human Computer Interaction,* 1232–1236.

Ito, M., & Nakakoji, K. (1996). Impact of culture on user interface design. In E. del Galdo & J. Nielsen (Eds.), *International user interfaces* (pp. 105–126). New York: Wiley.

Johnson, S. (1997). *Interface culture.* New York: Basic Books.

Josephson, M. (2002). Globalization cost-out: How to reduce costs, improve quality, and launch multilingual Web sites faster. *Idiom Globalization Leadership Series.* Retrieved February 27, 2003, from http://www.idiominc.com/us/solutions/gls.asp

Kachru, B. B. (1982). *The other tongue: English across cultures.* Urbana: University of Illinois Press.

Karahanna, E., Evaristo, J. R., & Srite, M. (2002). Methodological issues in MIS cross-cultural research. *Journal of Global Information Management, 10,* 48–55.

Karat, J., & Karat, C. (1996). World-wide CHI: Perspectives on design and internationalization. *SIGCHI Bulletin, 28,* 39–40.

Keevil, B. (1998). Measuring the usability index of your Web site. *Proceedings on the Sixteenth Annual International Conference on Computer Documentation, SIGDOC '98,* 271–277.

Keniston, K. (1998). Language, power, and software. In C. Ess & F. Sudweeks (Eds.), *Culture, technology, communication* (pp. 283–306). Albany: State University of New York Press.

Kluckhohn, C., & Leighton, D. (1946). *The Navaho.* Cambridge, MA: Harvard University Press.

Kroeber, A. L., & Kluckhohn. C. (1952). *Culture: A critical review of concepts and definitions.* New York: Vintage Books.

Kukulska-Hulme, A. (2000). Communication with users: Insights from second language acquisition. *Interacting with Computers, 12*, 587–599.

Kurosu, M. (2001). Wind from east: When will it blow and how. In G. Salvendy, M. J. Smith, & R. J. Koubek (Eds.), *Design of computing systems: Cognitive considerations* (pp. 141–144). Amsterdam: Elsevier.

Kurosu, M., & Kashimura, K. (1995). Apparent ability vs. inherent usability: Experimental analysis on the determinants of the apparent usability. *CHI 1995 conference companion on human factors in computing systems* (pp. 292–293). New York: ACM Press.

Ladner, T. (1996, December 10). The French say *non* to English-language Web site. *Wired News*, Retrieved April 11, 2003, from http://www.wired.com/news/politics/0,1283, 911,00.html

Large, A., & Moukdad, H. (2000). Multilingual access to Web resources: An overview. *Program, 34*(1), 43–58.

Lepouras, G., & Weir, G. R. S. (1999). It's not Greek to me: Terminology and the second language problem. *SIGCHI Bulletin, 31*, 17–24.

Leung, Y. K., & Cox, K. (1997). Cross-cultural issues in user interface design. In G. Salvendy, M. J. Smith, & R. J. Koubek (Eds.), *Design of computer systems: Cognitive considerations* (pp. 181–184). Amsterdam: Elsevier.

Lin, R. (1999). Cultural differences in icon recognition. *Proceedings of HCI International '99 - the 8th Conference on Human Computer Interaction*, 726–729.

Madden, T. J., Hewett, K., & Roth, M. S. (2000). Managing images in different cultures: A cross-national study of color meanings and preferences. *Journal of International Marketing, 8*, 90–108.

Marcus, A. (1997). International user-interface standards for information superhighways: Some design issues. In G. Salvendy, M. J. Smith, & R. J. Koubek (Eds.), *Design of computing systems: Cognitive considerations* (pp. 189–192). Amsterdam: Elsevier.

Marcus, A. (2001). International and intercultural user interfaces. In C. Stephanidis (Ed.), *User interfaces for all: Concepts, methods, and tools* (pp. 47–63). Hillsdale, NJ: L. Erlbaum.

Marcus, A., Armitage, J., Volker, F., & Guttman, E. (1999, June 3). Globalization of user-interface design for the Web. *Proceedings of the 5th Conference on Human Factors and the Web.* Retrieved March 15, 2000, from http://www.amanda.com/resources/HFWEB99/ HFWEB99.Marcus.html_

Marcus, A. & Gould, E. W. (2000). Cultural dimensions and global Web user-interface design: What? So what? Now what? *Proceedings of the 6th Conference on Human Factors and the Web.* Retrieved June 15, 2001, from http://www.amanda.com/resources/hfweb 2000hfweb00.marcus.html

Masuda, T., & Nisbett, R. E. (2001). Attending holistically versus analytically: Comparing the context sensitivity of Japanese and Americans. *Journal of Personality and Social Psychology, 81*, 922–934.

May, J. (2000). Perceptual principles and computer graphics. *Computer Graphics, 19*, 271–279.

Mayhew, D. (1999a). *Deborah J. Mayhew & Associates, online.* Retrieved October 10, 2001, from http://drdeb.vineyard.net

Mayhew, D. (1999b). *The usability engineering lifecycle: A practitioner's guide to user interface design.* San Diego, CA: Morgan Kaufmann.

Mills, J. A., & Caldwell, B. S. (1997). Simplified English for computer displays. In G. Salvendy, M. J. Smith, & R. J. Koubek (Eds.), *Design of computing systems: Cognitive considerations.* (pp. 133–136). Amsterdam: Elsevier.

Mrazek, D., & Baldacchini, C. (1997). Avoiding cultural false positives. *Interactions 4*, 19–24.

Nielsen, J. (1990). Usability testing for international interfaces. In J. Nielsen (Ed.), *Designing user interfaces for international use* (pp. 41–44). Amsterdam: Elsevier.

Nielsen, J. (1993). *Usability engineering*. San Diego, CA: Morgan Kaufmann.

Nielsen, J. (1996). International usability engineering. In E. del Galdo & J. Nielsen (Eds.), *International user interfaces* (pp. 1–19). New York: Wiley.

Nielsen, J. (2002). *Coordinating user interfaces for consistency*. San Francisco: Morgan Kaufmann.

Nishida, H., Hammer, M. R., & Wiseman, R. L. (1998). Cognitive differences between Japanese and Americans in their perception of difficult social situations. *Journal of Cross-Cultural Psychology, 29*, 499–525.

Noiwan, J., & Norcio, A. F. (2000). A comparative analysis of Web heuristic usability between Thai academic Websites and US academic Websites. *Proceedings of the 4th Multiconference on Systemics, Cybernetics, and Informatics*, 536–541.

Norman, D. A. (1990). *The design of everyday things*. New York: Doubleday Dell.

Nunberg, G. (2000). Will the Internet always speak English? *The American Prospect Online, 11*. Retrieved April 18, 2003, from http://www.prospect.org/print/V11/10/nunberg-g.html

O'Connell, T. A. (2000). A simplistic approach to internationalization: Design considerations for an autonomous intelligent agent. *Proceedings of 6th ERCIM Workshop "User Interfaces for All."* Retrieved May 17, 2001, from http://ui4all.ics.forth.gr/UI4ALL-2000/files/Long_papers/O_Connell.pdf

Ossner, J. (1990). Transnational symbols. The rule of pictograms and models in the learning process. In J. Nielsen (Ed.), *Designing user interfaces for international use* (pp. 11–38). Amsterdam: Elsevier.

Packaging design in Japan. (1986). Tokyo: Kodansha International.

Pemberton, S. (1998). Views and feelings: Flags are not languages. *SIGCHI Bulletin, 30*(1). Retrieved February 9, 2004, from: http://www.acm.org/sigchi/bulletin/1998.1/views.html

Peterson, M. C. (1990). ARRIS: Redesigning a user interface for international use. In J. Nielsen (Ed.), *Designing user interfaces for international use* (pp. 103–109). Amsterdam: Elsevier.

Prabhu, R., & Harel, D. (1999). GUI design preference validation for Japan and China: A case for KANSEI engineering? *Proceedings of HCI International '99 - the 8th Conference on Human Computer Interaction*, 1218–1222.

Reeves, B., & Nass, C. (1996). *The media equation*. Cambridge, UK: Cambridge University Press.

Rey, L. (1998). Multiculturality and communication technologies in Switzerland. *The Electronic Journal of Communication*. Retrieved September 19, 2000, from http://www.cios.org/getfile/Rey_V8N398

Robbins, S. S., & Stylianou, A. C. (2002). A study of cultural differences in global corporate Websites. *Journal of Computer Information Systems, 42*, 3–9.

Rogers, E. M., & Steinfatt, T. M. (1999). *Intercultural communication*. Prospect Heights, IL: Waveland Press.

Russo, P., & Boor, S. (1993). How fluent is your interface? Designing for international users. *Proceedings of the SIGCHI Conference on Human Factors in Computing Systems, INTERCHI'93*, 342–347.

Sacharow, S. (1983). *Packaging design: The best of American packaging and international award winning designs*. New York: PBC International.

Satzinger, J., & Olfman, L. (1998). User interface consistency across end-user application: The effects on mental models. *Journal of Management Information Systems, 14*, 167–194.

Schank, R., & Abelson, R. (1977). *Scripts, plans, goals and understanding*. Hillsdale, NJ: L. Erlbaum.

Schmid-Isler, S. (2000). The language of digital genres: A semiotic investigation of style and iconology on the World Wide Web. *Proceedings of the 33rd Hawai'i International Conference on System Sciences (HICSS 2000)*, 1–9.

Schrier, J. R., Williams, E. L., MacDonell, K. S., Peterson, L. A., Strijland, P. F., Wichansky, A. M., & Williams, J. R. (1992). HCI standards on trial: You be the jury. *Proceedings of the SIGCHI Conference on Human Factors in Computing Systems, CHI '92*, 635–638.

Searle, J. R. (1995). *The construction of social reality*. London: Penguin.

Sheppard, C., & Scholtz, J. (1999). The effects of cultural markers on Web site use. *Proceedings of the 5th Conference on Human Factors and the Web*. Retrieved September 15, 2001, from http://zing.ncsl.nist.gov/hfweb/proceedings/sheppard

So now it gets interesting. (n.d.). Retrieved March 8, 2003, from http://www.accenture.com/xd/xd.asp?it=enweb&xd=aboutus%5Chistory%5Crebrand%headlines5.xml

Sprung, R. C. (1990). Two faces of America: Polyglot and tongue-tied. In J. Nielsen (Ed.), *Designing user interfaces for international use* (pp. 71–102). Amsterdam: Elsevier.

Sroka, M. (2000). Web search engines for Polish information retrieval: Questions of search capabilities and retrieval performance. *International Information and Library Review 32*, 87–98.

Statistics Canada. (1998). *Household facilities and equipment* (Catalogue no. 64-2002). Ottawa, Canada.

Steffensen, M., Joag-Dev, C., & Anderson R. C. (1979). A cross-cultural perspective on reading comprehension. *Reading Research Quarterly, 15*, 10–29.

Stephanidis, C., Savidis, A., & Paramythis, A. (1997). Addressing cultural diversity through unified interface development. In G. Salvendy, M. J. Smith, & R. J. Koubek (Eds.), *Design of computer systems: Cognitive considerations* (pp. 165–168). Amsterdam: Elsevier.

Stewart, E. C., & Bennett, M. J. (1991). *American cultural patterns: A cross-cultural perspective* (Rev. ed.). Yarmouth, ME: Intercultural Press.

Straub, D., Loch, K., Evaristo, R., Karahanna, E., & Srite, M. (2002). Toward a theory-based measurement of culture. *Journal of Global Information Management 10*, 13–23.

Sukaviriya, P., & Moran, L. (1990). User interface for Asia. In J. Nielsen (Ed.), *Designing user interfaces for international use* (pp. 189–217). Amsterdam: Elsevier.

Sun, H. (2001). Building a culturally-competent corporate Web site: An exploratory study of cultural markers in multilingual Web design. *Proceedings of the 19th Annual International Conference on Computer Documentation*, 95–102.

Taylor, S. E., & Crocker, J. (1981). Schematic bases of social information processing. In E. T. Higgins, C. P. Herman, & M. P. Zann (Eds.), *Social cognition: The Ontario Symposium* (Vol. 1, pp. 89–134). Hillsdale, NJ: L. Erlbaum.

Tractinsky, N. (1997). Aesthetics and apparent usability: Empirically assessing cultural and methodological issues. *Proceedings of the SIGCHI Conference on Human Factors in Computing Systems, CHI '97*, 115–122.

Travica, B., & Cronin, B. (1996). Business Web in Russia: Usability for the Western user. *Proceedings of the Annual Meeting of the American Society for Information Science*, 16–23.

Trompenaars, F. (1993) *Riding the waves of culture: Understanding diversity in global business*. London: Nicholas Brealey.

Tseligka, T. (2002, October). *Some cultural and linguistic implications of computer-mediated Greeklish*. Paper presented at Association of Internet Researchers conference Internet Research 3.0: NET / WORK / THEORY, Maastricht, The Netherlands.

U.S. Department of State. (2003). *Independent states in the world*. Retrieved April 16, 2003, from http://www.state.gov/s/inr/rls/4250pf.htm

van de Vijver, F. J. R., & Leung, K. (1997a). Methods and data analysis for comparative research. In J. W. Berry, Y. H. Poortinga, & J. Pandey (Eds.), *Handbook of cross-cultural psychology* (Vol. 1, pp. 257–300). Needham Heights, MA: Allyn & Bacon.

van de Vijver, F. J. R., & Leung, K. (1997b). *Methods and data analysis for cross-cultural research*. Thousand Oak, CA; Sage.

Walton, M., Vukovic', V., & Marsden, G. (2002). 'Visual literacy' as challenge to the internationalisation of interfaces: A study of South African student Web users. *Proceedings of the SIGCHI Conference on Human Factors in Computing Systems, CHI 2002,* 530–531.

Wang, J., Fu, D., Jin, Z., & Li, S. (1999). Design methodology of Chinese user interface. Lessons from Windows 95/98. *Proceedings of HCI International '99 - the 8th Conference on Human Computer Interaction,* 813–817.

Webster HyperText Interface. (1995). http://c.gp.cs.cmu.edu:5103/prog/webster

Wierzbicka, A. (1991). *Cross-cultural pragmatics: The semantics of human interaction.* Berlin, Germany: Mouton de Gruyter.

Whitford, M. (2000, July 3). It's a small World Wide Web: Industry expects multilingual Internet sites to grow. *Hotel & Motel Management, 215,* 58.

Woestenburg, J. C. (2003). **TAL?'s language technology: Hyphenators, spell checkers, dictionaries.* Bussum, The Netherlands: TAL?. Retrieved March 26, 2004, from http://www.techno-design.com/download/hypxt/language_book.pdf

Yeo, A. (1996). World-Wide CHI: Cultural user interfaces, a silver lining in cultural diversity. *SIGCHI Bulletin, 28,* 4–7.

Yeo, A. (2000). Are usability assessment techniques reliable in non-Western cultures. *The Electronic Journal on Information Systems in Developing Countries, 3*(1), 1–21.

Yeo, A. W. (2001). Global-software development lifecycle: An exploratory study. *Proceedings of the SIGCHI Conference on Human Factors in Computing Systems, CHI 2001,* 104–111.

CHAPTER 8

The Social Worlds of the Web

Caroline Haythornthwaite and Christine Hagar
University of Illinois

The World of the Web

We *know* this Web world. We *live in* it, particularly those of us in developed countries. Even if we do not go online daily, we *live with* it—our culture is imprinted with online activity and vocabulary: e-mailing colleagues, surfing the Web, posting Web pages, blogging,[1] gender-bending in cyberspace, texting[2] and instant messaging friends, engaging in e-commerce, entering an online chat room, or morphing[3] in an online world. We *use* it—to conduct business, find information, talk with friends and colleagues. We know it is something separate, yet we incorporate it into our daily lives. We *identify* with it, bringing to it behaviors and expectations we hold for the world in general. We approach it as explorers and entrepreneurs, ready to move into unknown opportunities and territory; creators and engineers, eager to build new structures; utopians for whom "the world of the Web" represents the chance to start again and "get it right" this time; utilitarians, ready to get what we can out of the new structures; and dystopians, for whom this is just more evidence that there is no way to "get it right."

The word "world" has many connotations. The *Oxford English Dictionary* (http://dictionary.oed.com) gives 27 definitions for the noun "world" including:

- The sphere within which one's interests are bound up or one's activities find scope; (one's) sphere of action or thought; the "realm" within which one moves or lives.

- A group or system of things or beings associated by common characteristics (denoted by a qualifying word or phrase), or considered as constituting a unity.

- Human society considered in relation to its activities, difficulties, temptations, and the like; hence, contextually, the ways, practices, or customs of the people among whom one lives; the occupations and interests of society at large.

- A vast quantity, an "infinity."

- Everybody in existence; in narrower sense, everybody in the community, the public.

To understand the many possible interpretations of the world of the Web, it is useful to stay grounded in its social construction; that is, to look at how we create this world. After all, it is a world created and sustained through the labors of people: computer hardware and software specialists; Web designers; content providers; search engine designers; policy makers; and the many who trawl the Web for information, conversation, and adventure. Its shape, form, and substance emanate from the actions and interactions of online participants. Its usefulness and appeal come from the way we enjoy this world, as well as the way it reflects, enhances, and supplements offline life.

In our years of using the Web, we have recognized, endorsed, and encouraged its world-like aspects. Early engagements with the "world of the Web" reveled in its *wholeness*, in its manifestation as a complete realm that could contain one's whole existence, one's whole life; and its *separateness* as a "world apart," different, special, detached from the "real" world. This approach supports views of the Web as a world of new possibilities, taking off from and transcending everyday life. Another approach has found the Web less "worldly," encouraging its supporting role in our own local worlds, in the spheres of interest, action, and community with which we engage every day. To understand contemporary engagement with the Web, we need to accommodate both views. We can no longer approach the world—or worlds—of the Web as something wholly separate from the rest of our lives. Although at many times we may approach the Web as a self-contained world, it also has a significant impact on how we organize and conduct our lives. Indeed, we find that although we are enjoying the separateness of online Web worlds, we are simultaneously integrating the Web into our offline worlds.

In making sense of the dual nature of the world of the Web, we find it useful to consider not the technological dividing line between online and offline worlds, but instead how the Web supports and constitutes *social worlds*, connecting individuals around common activities and interests and supporting the worlds of work, home, school, and play. These social worlds may be constituted completely online, but are more likely to be found crossing online and offline realms. (See Kazmer & Haythornthwaite [2001] for a longer discussion of the application of the social worlds concept to the Internet, and specifically to the social worlds of online learners; for other studies of technology from a social worlds perspective, see Covi [1996] on digital library use; Fitzpatrick, Kaplan, & Mansfield [1996] on systems administrators' perceptions and use of their work space; and Star, Bowker, & Neumann [2003] on the use of information systems.)

Social Worlds

Social worlds, as defined by Strauss (1978), consist of people who share activities, space, and technology, and who communicate with one another. A social world is identifiable by its coordination around a primary activity such as work, learning, or family support and around one site, for example, the workplace, school, or home (Clarke, 1991; Strauss, 1978). As individuals, we split our time between various social worlds, acting as a parent in one, a co-worker in another, an expert in a third, and even an anonymous game-player in a fourth, taking on roles, voices, and personae appropriate to each world (Bakhtin, 1986; Baym, 2000; Goffman, 1959; Merton, 1957; Turkle, 1995).

An important aspect of social worlds is that they intersect, for example, in home and school associations and in balancing work and family obligations. As we approach each world, we modify our behaviors and expectations about interactions, so that "a major analytic task is to discover such intersecting and to trace the associated processes, strategies, and consequences" (Strauss, 1978, p. 123).

The intersection of online and offline worlds is a particular area of interest, but it is insufficient to label them without understanding what it is about each arena that makes it a social world. Focusing on social worlds frees us from a media-bound definition of worlds. Instead of viewing the Web from a technological standpoint, defining it by protocols and network connections, setting it in opposition to the "real" world of offline activity, we can study how it constitutes worlds of activity and sustains computer-supported social networks of shared interest (Wellman, Salaff, Dimitrova, Garton, Gulia, & Haythornthwaite, 1996).[4] As we wander between our social worlds, we do not, in most cases, partition them; we act as a friend, colleague, and parent (or child) both online and offline. We interact with members of our communities of practice through listservs, bulletin boards, and chat room environments, publishing papers both online and offline, as well as gathering face-to-face in business meetings and conferences; and we act as local community members when we create and use community networks. Although we can identify both offline activities that appear to have no connection to the Web and totally Web-contained online social worlds that appear to be separated from the constraints of offline norms and conventions (e.g., some game-playing environments), more commonly we find activity spanning online and offline realms. Moreover, as we go online, we take with us much of our offline baggage in terms of language, behaviors, and expectations about interaction.

The Web has often been portrayed as a world separate from day-to-day life, but there is much that grounds it in our daily activities (Katz & Rice, 2002a; Wellman & Haythornthwaite, 2002). The Web has changed how we communicate, maintain contacts, search for information, stay in touch with people, gossip, and access the daily news. Yet, these are not new activities. Although we may go to e-mail instead of the phone, post a document on the Web rather than print and mail it, or search for information at our desktops rather than in a library, these are the same activities that we pursued before. Our aims are still to communicate, to interact with others, to store and retrieve information, and to participate in wider communities of practice.

In the rush to grasp the meaning of "the Web," many studies forgot to consider its use within individual and social contexts. An exclusive focus on "the Web" or single forms of computer-mediated communication (CMC) (e-mail, discussion groups, virtual reality, etc.) fails to see how multiple means are used to maintain our social worlds (Haythornthwaite, 2001). Similarly, an exclusive focus on tasks—getting work done, maintaining friendships, engaging in play—fails to address how aspects of multiple worlds affect individuals' lives, particularly when Web media bring work into the home (Kazmer & Haythornthwaite, 2001; Salaff, 2002), and new mobile technologies bring in anyone, anywhere, at any time (Rheingold, 2003). As we discuss in the next section, recent perspectives spend considerably more effort in trying to understand the place of the Web within other worlds, and vice versa.

It is sometimes confusing to talk about "the Web" when so much of the activity of the Web is based on "the Internet." Yet, as Miller and Slater (2000, online) note, "There is no such thing as the Internet. There are a number of different media and contents which people assemble into 'their' Internet. There is however a clear debate about the Internet." Here, we use the term "Web" to signify phenomena that go beyond computer network protocols; we include consideration of the "media and contents" of the Internet, notably different applications, posted information, and communication activity. In general, we are referring to the Internet when we refer to being online. Yet we do not mean to exclude consideration of computer-mediated communication that occurs through non-Internet networks, such as intranets that use internet protocols but are unconnected to the Internet. Intranets, and other proprietary networks, may themselves constitute and support social worlds, offering online capabilities and content to complement offline activity. We also note that the ability of devices to connect us to the Internet makes it difficult to separate the use of a device (laptop, PDA, cell phone; phone lines, cable connections, wireless connections) from our use of the Internet; using a desktop computer should not be considered synonymous with the use of either the Internet or the Web.

Situating Web Use

Although much early rhetoric about the Web stressed its world-transforming aspects, more reflective approaches analyzed its interplay with existing social and technical structures (Katz & Rice, 2002a; Wellman & Haythornthwaite, 2002). Thus, Giddens's structuration theory, which recognizes how structures emerge from interaction and are sustained and maintained by such interaction, has formed the basis for the work of a number of researchers and theorists in information and communication technology (Contractor & Eisenberg, 1990; DeSanctis & Poole, 1994;

Orlikowski, 2002; Orlikowski & Yates, 1994). These authors have recognized that the way social technologies, such as e-mail, listservs, and chat rooms, are used arises from group practices, including norms about what is discussed and the language and genres considered appropriate for communication. Moreover, these practices are not static, but emerge and evolve with group use, as social practices "constitute and reconstitute" (Orlikowski, 2002, p. 270) norms, conventions, and knowledge. In a similar vein, Warschauer (2003) addresses social inclusion and the Web by building on new institutionalism from the organization theory literature (Powell & DiMaggio, 1991), which identifies the way "routinized interaction" shapes "human activity in defined realms" (Warschauer, 2003, p. 208). Baym also follows a practice approach in her analysis of the r.a.t.s., the Usenet soap opera discussion group. She approaches it from a communities of practice perspective (Lave & Wenger, 1991), resting on the "assumption that a community's structures are instantiated and recreated in habitual and recurrent ways of acting and *practices*" (Baym, 2000, p. 22, italics Baym's). (For a more in-depth look at what constitutes communities of practice, and the relation with virtual communities, see E. Davenport and Hall [2002] and Ellis, Oldridge, and Vasconcelos [2004].)

The common theme in these approaches is that the use of technology derives from current practices. It is here we get the first clue that online life cannot be totally divorced from offline life. We have many years of offline practice to bring online, and we do. For many, online activity is not important unless it carries a connection to their offline lives.

Other perspectives from social studies of science, social studies of technology, and social informatics also identify potential influences in the availability and use of the Web, and thus the way worlds of the Web will look. The presence, absence, and configurations of physical infrastructures (e.g., electricity, telecommunications, and computer networks) affect who has access to the Web, from where, and when (Schement, 1998; Star, 1999; Warschauer, 2003). Whether or not work moves easily or rapidly onto the Web may depend on how separable it is from the artifacts, equipment, and means of production supporting the work—the actor networks (Latour, 1987) and activity systems (Engeström, Miettinen, & Punamäki, 1999)—that constitute the whole work effort. Such a view also shows us that the Web itself is changing the face of work, for example, making it not only possible to be connected at all hours, but also an expectation that we will be online. Finally, organizational and societal structures can affect where we allocate our attention, and then where and for what we choose to adopt and use the Web (T. Davenport & Beck, 2001; Goldhaber, 1997; Shapiro & Varian, 1999). Research in social informatics also shows how technologies and the people who use them are deeply embedded in socio-technical networks of action and constraint, with technologies both facilitating and limiting what can be done in various settings (Bishop & Star, 1996; Kling, 2000; Kling, Rosenbaum, & Hert, 1998; Lamb & Davidson, 2002; Sawyer & Eschenfelder, 2002).

Although this chapter does not review technology studies, it is important to bear in mind these kinds of influences as we approach the Web. They show that it is impossible to talk of "the Web" without understanding and acknowledging that the face of technology, its presentation in everyday life, will vary with the kind of "everyday life" into which it is introduced. Thus, it is a mistake to speak of a singular world of the Web, or even a singular presentation or embedding. Although there are certainly Web worlds that allow fantasies and creations that go beyond what is possible, offline and online groups that bridge geography and physicality in ways not possible face-to-face (e.g., by maintaining strict anonymity, or bringing in people from remote areas), there is not one world of the Web.

The Web is creating a global presence that may be interpreted as a world, but its ever-extending reach makes it embeddable in everyday life, pushing both its local and global characteristics and contributing to what Hampton and Wellman (1999, p. 12) have called "glocalization." The Web allows us to stay connected to our local worlds even when traveling globally, by using mobile devices (laptops, PDAs, cell phones), remote connectivity, Internet cafés, and public access points. Thus, we carry with us access to our personal contacts and archives, as well as instant access to the encyclopedic knowledge available on the Web. That knowledge also brings the global local, as, for example, organizations in developing countries obtain information about their own governments through postings to the Web made in other countries. Moreover, because no single country controls the exchange of information, even the concept of national boundaries—national "worlds"—is challenged by the Web (Frederick, 1993).

We explore further the emerging worlds of the Web, and Web influences in our worlds. We begin with a look at the growth of Web use and current views of what is important for access to the Web. We then examine cases of online and offline use of the Web, examining further the separateness, integration, overlaps, and synergies between Web worlds and social worlds. We conclude with a look at research issues for studying multi-modal, multiple world activity and practice.

Part I: Access to the World(s) of the Web

Statistics about the Web tell an undeniable story of growth: in the number of computers, access points, users, countries with users, Web sites, Internet service providers, and so on (see Table 8.1 for recent numbers of Internet users worldwide from Nua.com, and Figure 8.1 for numbers from 1995 to 2002 for the U.S.).[5] Thus, to speak of growth, or rapid deployment and development, with respect to "the Web" is to state a truism, but one that highlights important considerations.

The Web is in a growth stage: It has not leveled off in number of users or number of applications. It is still being adopted by newcomers and many applications are being tried and adopted by both new and

Table 8.1 How many online?

Region	Number of users in millions
Africa	6.31
Asia/Pacific (including Australia and New Zealand)	187.24
Europe	190.91
Middle East	5.12
Canada & USA	182.67
Latin America	33.35
World Total	**605.60**

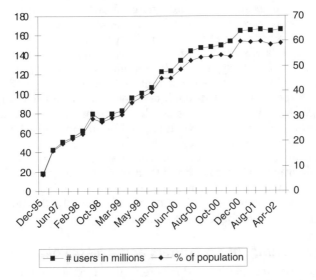

Figure 8.1   Number of U.S. Internet users and percent of population by year. Source: Nua.com (http://web.nua.com/surveys/how_many_online/n_ america.html). Note: Numbers to April 1999 are for adult users only (age 16 and over); statistics after April 1999 are from Nielsen/NetRatings Internet with users defined as all members (two years of age or older) of U.S. households that currently have access to the Internet.

experienced users (Haythornthwaite & Wellman, 2002, pp. 3–44). Even those involved in studying the Web admit defeat in keeping up with the rapid changes in number of users and sites, types of applications, and emergent uses. Researchers frequently note that in the time between writing, publication, and reading, many changes are likely to have occurred in modes of computer-mediated communication (Herring, 2002), the types and manifestations of online communities (Baym, 2000), and statistics on Internet access (Dodge & Kitchin, 2001; Haythornthwaite & Wellman, 2002, pp. 3–44; Katz & Rice, 2002a).

The Web is still unfolding and evolving: We have the *opportunity* (if not the burden) of watching a large-scale infrastructure as it becomes established. Although previous infrastructures have spread more slowly, growing across research "lifetimes," significant Internet growth and change can occur within a single research funding cycle of three to five years. Compare, for example, the rates of diffusion of the telephone, which took 38 years to reach 30 percent of U.S. households, and television, which took 17 years, to the very rapid diffusion of the Internet to the same level in just seven years (Katz & Rice, 2002a, p. 17).

Extreme dystopian and utopian views hold sway because we see the ability of this phenomenon to "tip" either way: And yet, when we look at what is really happening, we see that although it sometimes creates surprising outcomes, more often the Web supports daily routines (Katz & Rice, 2002a). This will dismay the utopians, particularly those who hoped to shed the binding constraints they perceive in contemporary distributions of status, power, and equity, associations between mind and body, and between identity and societal image. However, to ignore that these constraints exist even online, and to act as if the online and offline worlds operate independently of each other, with no baggage carried from one to the other, is to be disarmed and blinkered. As DiMaggio, Hargittai, Neuman, and Robinson (2001) have noted, there is a real need to include consideration of power and the full range of sociological impacts in studies of Internet use and behavior. These authors join others who have pointed out that gender, status, hierarchy, and similar considerations remain active and/or reemerge online, even in supposedly anonymous environments (Herring, 1993, 2002; Kolko, Nakamura, & Rodman, 2000; Shade, 2002).

More than Counts of Computers and Access Points

Access to the Web and its worlds is regularly assessed in terms of individual or household ownership of computers and Internet access. However, being online in a meaningful way depends on more that just connecting one's computer. Recent work shows how it is contingent on finding a connection to our everyday social lives and social worlds (Katz & Rice, 2002a). We join other researchers in recognizing that having a computer does not equate with being part of the world of the Web, nor does it automatically confer equal access. Thus, recent studies find that even when computer ownership begins to look the same across socio-economic categories, young, affluent, white males still dominate as Internet users because they spend more time online (Howard, Rainie, & Jones, 2002).

Access is about being aware that the Web is there *and* that online resources are of potential use, as well as about having the skills to make effective use of those resources. Being online in a meaningful way depends on individuals attaining a certain level of skill and experience (Howard, Rainie, & Jones, 2002; Katz & Rice, 2002a, 2002b; LaRose, Eastin, & Gregg, 2001; Nie, 2001), computer literacy (Warschauer,

2003), and familiarity and ease with the conventions and texts of the online world (Bregman & Haythornthwaite, 2003; Haythornthwaite, Kazmer, Robins, & Shoemaker, 2000; McLaughlin, Osborne, & Smith, 1995). Meaningful use also depends on finding content relevant to the user (Katz & Rice, 2002a; Warschauer, 2003).

Awareness

Katz and Rice (2002a) provide an in-depth look at statistics of Internet use, paying particular attention to where, in the stages from awareness of the Internet to expertise in Web use, issues of access play out (see also Katz & Aspden, 1997; Katz & Rice, 2002b; Katz, Rice, & Aspden, 2001). They identify awareness of the usefulness of the world of the Web and the information and contacts it supports as the key dividing line between users and non-users.

> Access barriers to the Internet are not primarily technical or financial, though those are real and difficult. Rather, the barriers seem to lie heavily in the realm of cultural perceptions about what is possible with the Internet and the nature of Internet activities. (Katz & Rice, 2002a, p. 99)

Individuals come to the worlds of the Web by "becoming aware of needs, interests, and desires ... confronting the limits of a local physical setting for addressing these needs ... learning that the Internet exists and identifying it as a potential resource," gaining tools and skills, searching and finding information or help, making contacts, then disseminating information to similar others, and further gaining skills and information (Katz & Rice, 2002a, p. 98).

Meaningful Use

Incorporating use into personal lives in meaningful ways is critical for belonging to the world of the Web. Providing hardware without also situating its use within local needs can be a wasted effort for all involved. Uses should emerge from people's needs and evolve by interacting with current practices (as noted in the social studies of technology approaches described here).

Warschauer (2003), for example, describes how a 1997 initiative to create a model "wired" town failed. Each household in an Irish town of 15,000 was given a computer and the town was set up to be Internet-ready, with ISDN (Integrated Services Digital Network) lines to every business and smart-card readers for businesses and smart-cards for families. Yet, three years later, it was reported that there was little to show for the money poured into equipment:

> Advanced technology had been thrust into people's hands with little preparation. Training programs had been run, but

they were not sufficiently accompanied by awareness programs as to why people should use the technology in the first place. And, in some instances, well-functioning social systems were disrupted in order to make way for the showcase technology. (Warschauer, 2003, p. 3)

By contrast, in three towns given only start-up funds, there was much more to show for the investment. Working with limited resources, "The towns built on existing networks among workers, educators, and business people to support grassroots uses of technology for social and economic development" (Warschauer, 2003, p. 4).

Bishop and colleagues (Bishop, Bazzell, Mehra, & Smith, 2001; Bishop, Mehra, Bazzell, & Smith, 2000, 2003; Merkel, 2002) also advocate a hands-on approach to providing community networking. Their work makes particular efforts to promote computer access and use among low income participants by providing computers and training and engaging in joint projects with local initiatives. Bishop and her colleagues, like Warschauer, stress the need for social inclusion and promote a participatory action approach that includes the community in creating their information infrastructure and Web worlds, building upon current practices and local needs.

Relevant Content

The reach and usefulness of the Web can also be limited by lack of culturally and linguistically relevant content. Warschauer (2000, 2003) describes how educators set up sites in Hawaiian as part of the revitalization of the language and because there was no content in the language for this community. These sites made it possible to have "computer education through Hawaiian," thus increasing the relevance of both language and content to the students. Johnson (2003) reports that users in Mongolia must work around the scarcity of sites in their language when looking for information. Users report looking at sites in other languages they know (e.g., Russian or other Asian languages), as well as using English sites to learn the language.

The Web's usefulness for information, activities, and contacts depends on there being a critical mass of resources online (Markus, 1990). As with other forms of social computing, success often depends on the altruistic behaviors of others (Connolly & Thorn, 1990; Markus, 1990) who seed the Web with relevant information (by creating interesting and useful community Web sites, starting interest groups on listservs, etc.), as well as model, behavioral norms.

Infrastructures

Along with linguistic and information infrastructures, technical infrastructures also affect access. Schement (1998) points out how disruption of continuous telephone connections (due to insufficient funds to cover telephone bills) constitutes an impediment to access to Web resources for some U.S. citizens. In other countries, the lack of personal or reliable telephone lines also limits use of Web services (Davidson, Sooryamoorthy, & Shrum, 2002). We must also question whether our Western world metric of computers in the household, predicated as it is on private ownership and access, accompanied by reliable telephone or cable connections, is appropriate for all countries. Where telephone access is limited, proximity to Internet cafés or other public access points may be more useful. Mobile telephone service is also playing an increasingly important role worldwide, but particularly where land lines are not available or reliable (see Rheingold, 2003, regarding cell phones in the Philippines). Moreover, as access to the Web via cell phones and PDA devices increases, and practices using these devices evolve, we encounter yet more variants of what it means to have "access to the Web" (Rheingold, 2003).

Individual Attributes

Finally, how the Web is adopted and used varies with a number of individual attributes. Age, skill, race, gender, socioeconomic status, marital status, and whether there are children in the household all affect adoption and use of the Web (for a summary, see Haythornthwaite and Wellman [2002, pp. 3–44] and the collected papers in Wellman and Haythornthwaite, [2002]; for more on the adoption of innovations in general, see Rogers [1995]). Cutting across these factors, many studies show that being online for over two years changes how the Web is used: experienced users engage in more kinds of activities online,[6] are less anxious about using computers and the Web, and feel higher levels of self-efficacy about using the Web (Howard, Rainie, & Jones, 2002; Katz & Aspden, 1997; Katz & Rice, 2002b; LaRose, Eastin, & Gregg, 2001; Nie, 2001; Nie & Erbring, 2000).

Howard, Rainie, and Jones (2002) have categorized users into four types based on their surveys of online use. These categories show differences linked to when individuals started using the Web and how often they go online. *Newcomers* (15 percent of the U.S. adult population as of September 2000) have been online one year or less. They are learning their way around and enjoying fun aspects of the Web, such as playing games, browsing, participating in chat rooms, finding information about hobbies, and listening to and downloading music. Newcomers are likely to have access in only one place—either at work or at home. *Experimenters* (13 percent) are of two types: either they started going online two to three years ago or they started about a year ago and go online from home every day. They have ventured beyond the fun activities that Internet novices enjoy and are interested in using the Web for information retrieval. *Utilitarians* (14 percent) started online more than three years ago, or gained access two or three years ago but log on from home every day. They exhibit a functional approach to the Web, using it as a tool for many tasks in their lives. As a group, they tend to see it as less useful and entertaining than more frequent users. *Netizens* (8 percent) started going online more than three years ago and go online from home every day. They have incorporated Web use into their work and home lives and are the most avid participants in most Web activities on an average day.

Part II: Separation, Integration, Overlaps, and Synergies

We turn now to exploring cases of Web use that demonstrate how the apparent separateness of worlds actually contains a tangle of online and offline threads. Although worlds may appear to be rooted in online or offline venues, for example, as family life is rooted in face-to-face contact in the home and an academic interest group may be rooted online, maintaining social contact often pushes a mix of media. Thus, we find families communicating via e-mail and cell phones and academic groups meet face-to-face at conferences. The social worlds perspective helps make sense of the inability to separate online from offline life clearly. It leads us to identify a world by its coordination around a primary activity: family life or academic interest. In the new uses of mixed media, we can also make sense of worlds by identifying a comparable coordination around a primary medium: the home for family life, the online environment for the interest group. Then we find that, although we mix our use of media, the identity of the world comes from what people do together as well as where they primarily situate the activity that gives them and their world its identity. To explain this further, consider the worlds of an online community and a community network.

Online communities provide means for members who are not co-located to congregate and create new, non-place-bound worlds, worlds that may be utopian, entertaining, educational, work-related, or present a combination of these attributes. The collective defines norms, at first emerging from interactions, later set by moderators, wizards, and others, and described in frequently asked questions (FAQ) lists and site mission statements. The community gains its identity from its online existence and definition. Without the online presence it would not exist. Offline interaction, if it occurs, is the supplement to the online world where the real definition of the community is forged. Examples include online communities such as the Well (Rheingold, 2000) or Electropolis (Reid, 1995; http://web.aluluei.com/work.htm); online educational communities (Haythornthwaite & Kazmer, in press; Haythornthwaite, Kazmer, Robins, & Shoemaker, 2000); and graphical worlds such as The Palace (http://www.thepalace.com) or ActiveWorlds (http://www.active-worlds.com; for a description, see Dodge and Kitchin [2001]).

By contrast, community network initiatives (CNIs) exist to supplement offline, geographic communities—"geo-communities" as they have recently been termed. Identity comes from the offline existence and definition (Gurstein, 2000, 2001; Keeble & Loader, 2001; Miller, 1999; Schuler, 1996). Their *raison d'être* is to supplement the offline community by strengthening its presence on the Web, thereby making it and its character and services visible in the online world, and by advertising the offline world, making its locality evident and promoting local service providers. From there new online activities can develop in support of the geo-community, providing, for example, health information sites, social support groups, means to extend offline activities to online (e.g., organizing face-to-face meeting times online), activist organization, and Internet accessibility to disadvantaged and/or marginalized groups. In all, the thrust is local support, local attention, and local focus: the world of interest is geographically bound, but technologically connected and empowered. Online is the supplement, augmenting the offline world. Examples include the Blacksburg Electronic Village (Cohill & Kavanaugh, 2000), and Prairienet (Bishop, Bazzell, Mehra, & Smith, 2001). (See also the Association for Community Networking, www.afcn. org, and a list of community networks worldwide compiled by Douglas Schuler at http://www.scn.org/commnet/appC.htm.)

Thus, although both configurations support a community, they are distinguished by a primarily online or offline world focus. We present next four cases of online and offline separation and integration, showing how the two modes mix in various settings to support, and sometimes create, social worlds. We then return to the discussion of overlaps and synergies to explore more subtle influences that integrate online and offline worlds.

Separation and Integration

The four examples discussed here each show, in different ways, how the Internet is crossing traditional boundaries to connect people to each other and to resources in support of common goals.

Nongovernmental Organizations and Netwar

Nongovernmental organizations (NGOs) comprise a diverse set of groups and institutions with a primary focus on humanitarian (rather than commercial) objectives, operating independently of government. Since the 1970s, NGOs have come to articulate, mobilize, and represent people's interests and concerns locally, nationally, and internationally. NGOs include:

> private agencies in industrial countries that support international development; indigenous groups organized regionally or nationally; and member groups in villages. NGOs include charitable and religious associations that mobilize private funds for development, distribute food and family planning services and promote community organization. They also include independent cooperatives, community associations, water-user societies, women's groups and pastoral associations. Citizen groups that raise awareness and influence policy are also NGOs. (World Bank, 2001)

Since the 1980s, as costs of accessing the Web have dropped, NGOs have begun to use the Web to pool information and coordinate activities (Ronfeldt, Arquilla, & Fuller, 1998). They use the Web to engage locally and explore globally, linking to each other through the Internet to share information, services, and support and to draw attention to issues of concern.

This increased use of the Internet, and widening access to the Web more generally, is blurring the boundaries between social movements and NGOs. Social movements can now engage in electronic activism: using the Web to conduct campaigns, prepare and plan protests, and carry out advocacy work, both locally and transnationally. The Web aids the establishment of effective support groups, comprised of people and organizations faced with similar problems, who can then mobilize more broadly and cost-effectively.

The Web has influenced more than isolated cases and has led to what has become known as "netwar:"

> [an] emerging mode of conflict (and crime) at societal levels, short of traditional military warfare, in which the protagonists use network forms of organization and related doctrines, strategies, and technologies attuned to the information age. (Ronfeldt & Arquilla, 2001, p. 4)

Ronfeldt & Arquilla (2001) note that netwar is principally an organizational dynamic, although it requires appropriate social and technological dynamics to work well. In netwars, small geographically dispersed individuals, groups, and organizations communicate, coordinate, and conduct their campaigns. The Web offers an important technological means of supporting such campaigns, although not the only means. Often there is no central command and the identities and loyalties of activists may cross regional boundaries, shifting from the nation state to the transnational level of "global civil society" (Ronfeldt & Arquilla, 2001, p. 4; see also Rheingold, 2003).

One of the earliest and most publicized cases of "netwar" involved the development of international solidarity in support of the Zapatista movement in Mexico. The "most striking thing about the sequence of events set in motion on January 1, 1994 has been the speed with which news of the struggle circulated and the rapidity of the mobilization of support which resulted" (Cleaver, 1994, online). Despite governmental spin control and censorship using computer communications, the Zapatistas were able to convey their message through the Internet and the Association for Progressive Communications networks (Cleaver, 1999). Discussion and debate via e-mail lists and online conferences facilitated the organization of protest and support activities in over 40 countries around the world. The Zapatista rebellion wove a global "electronic fabric of struggle" (Cleaver, 1994, online): "What began as a violent insurgency in an isolated region mutated into a nonviolent though no less disruptive 'social netwar' that engaged the attention of activists from far and wide and had nationwide and foreign repercussions" (Ronfeldt & Arquilla, 2001, p. 4).

Web activism also supported civil disobedience in connection with the Seattle protests during the third World Trade Organization (WTO) meeting in November and December 1999. The Web was used in preparing for the protest, spreading news of it globally, and creating an international network of solidarity. Many Web sites were created to prepare and coordinate the event: "A Global Day of Action" site posted a call for action in ten different languages and provided a directory of local contacts all over the world. The November 30th global day of action was to be

> organized in a non-hierarchical way, as a decentralized and informal network of autonomous groups ... Each event or action would be organized autonomously by each group, while coalitions of various movements and groups could be formed at the local, regional, and national levels. (N30 Global Day of Action November 30 1999, online)

Recognizing the importance of content in the language of potential contacts, People's Global Action (PGA) urged that the proposal be translated into as many languages as possible, "since the availability of translations will very much affect our chances of spreading it at a truly global level" (N30 Global Day of Action November 30 1999, online).

These examples show the power of the Web in creating and coordinating activities among many otherwise unconnected individuals, creating what Rheingold (2003) has termed "smartmobs." What world do such "smartmob" members belong to? They are online, but when they gather for protests they are offline. If we return to the social worlds perspective, we find they belong to the world of social activism because that is their common orientation. They may also, at times, belong to the world of the WTO protest movement, as they coordinate activity around that organization. We might then ask, what of the new "flashmobs?" These spontaneous gatherings involve "a large group of people who gather in a usually predetermined [physical] location, perform some brief action, and then quickly disperse" (http://www.wordspy.com/words/flashmob. asp). Although flashmobs can be enacted for political reasons, so far most of their communal gatherings have been for fun, for example as hundreds converge on a card store all claiming to be searching for a card for a friend called Bill, or on a Macy's store to gather around a $10,000 carpet claiming to be members of a commune needing a "love rug" (Gillin, 2003). The coordination around a primary event, the rapid flash event, occurs at a brief intersection between online and offline realms, but in support of a common social world. Perhaps it is the very lack of an established or persistent center, as well as coordination around the primary activity of play, that is the attraction of such activities.

With both global coordination and local action common to both activist groups and flashmobs, it is hard to label such activities as either online or offline, but it is less difficult to identify their reason for congregation. The connectivity provided by the Web has created a separate world of activity, yet one that is situated in local conditions.

Crisis Response: Support for Farmers during the U.K. 2001 Foot and Mouth Disease Outbreak and Beyond

The next example focuses on how a crisis precipitated the entry of the Web into farming households in the U.K., and how it is now being incorporated into farming worlds. The Web suddenly entered the lives of many farmers during the foot and mouth crisis which devastated rural communities in the U.K. during 2001. In order to contain the disease, movement in the countryside during the outbreak was restricted. Farmers were isolated, confined to their farms, often unable to leave the household for weeks. The Internet became a safety net for these British farmers: Solitary individuals came together online, creating a sense of community, with the Web as means to obtain news, sympathy, and support (Wall, 2002).

The Pentalk Network was one such response to the crisis. This community network was established during the outbreak as a rapid response scheme to assist farmers and their families in Cumbria. Pentalk, which was designed by farmers, focused on the needs of isolated farmers for information and community. Their needs included access to the most recent, accurate information from government and other sources on the spread of the disease, recent regulations, and attempts to halt the outbreak. They also needed a community with whom to discuss this information and their own situations. Pentalk supplied computers to farmers free of charge for six months during the crisis; since then the network has continued to grow. Along with Pentalk's original aim to reduce isolation across rural communities, its efforts are now also directed to ensuring that farmers have the computing skills necessary to improve their farm businesses and to cope with the national government's demands for online reporting.

The Pentalk information network was one of the few positive initiatives that emerged from what was one of the worst years in history for the rural community; it has helped bring about a change in the world of farming. As individuals and households forged new links because of the network, they gained a new sense of community within their locality and also beyond:

> to curb feelings of isolation and alienation and to take back control, people are developing novel communities. They are part of multiple communities where rural and urban begin to merge, boundaries break down, access to information widens

and ideas flourish. (Bennett, Carroll, Lowe, & Phillipson, 2002, online)

Farmers are now using the Web as an integral part of their business. The Internet is proving to be essential to the future success of farming: helping farmers communicate with each other, reduce paper-based bureaucracy, procure goods and services more cheaply, and market themselves more efficiently by selling produce directly on the Web. Also, as government moves to online reporting, so too must the farmers. However, integrating the Web into the farming world also requires sensitivity to context, which means more than merely disseminating information. As Oates (2003) reports, the information provided on local government sites during the crisis failed to address the psychological problems farmers were facing. Nor did the local government sites enhance participation and democracy in the communities, encourage active citizenship, or develop ways to empower the community. Such criticisms highlight the importance of multiple responses and multiple forms of support during crises, as well as the value of understanding the full context of needs during a crisis.

Similarly, it is important to understand the household context in a farming community. In the farm household, a key difficulty is the disconnect between hands-on farm work and the use of the Internet (Warren, 2000). Established patterns of farm work do not fit well with desktop, computer-based Internet use, particularly in the growing season. Warren (2000, p. 5) suggests "getting the computer out of the farm office and into the living room," mobilizing other members of the farming family to learn and make use of information and communication technologies on the family farm. "Farmers' spouses already make a significant contribution to the management of farm businesses, and aiming Internet services at them may be a powerful catalyst for change" (Warren, 2000, p. 53).

From this example, we can see that the social worlds of farmers shape uses of the Web and also how important it is to understand the context— crises, household-based businesses, farm work—into which the Web is introduced. Networks such as Pentalk are able to meet needs beyond information dissemination and retrieval. Such community networks can be designed to address the totality of farming experience, including discussion with other farmers, and the interpretive and supportive aspects of their lives that were hit so hard by enforced quarantine. Moreover, unlike city workers with their focus on the office, farming involves a household unit. Thus, it is farming families who are at work, not just farmers. Again, a view of single (male) workers, sitting at single terminals in offices separated from household activity, does not provide the right perspective for understanding this Web world.

Netville: Wired Neighborhoods

Netville, a "wired suburb" of Toronto, studied by Hampton and Wellman (1999, 2002; Wellman & Hampton, 1999), provides another view of overlaps between online and offline worlds, this time in a local geo-community. Netville was an experiment in providing broadband access installed at the time the houses were built and sold. "For 2 years, the local network provided residents with [free] high-speed Internet access (including electronic mail and Web surfing), computer desktop videophone, an online jukebox, entertainment applications, online health services, and local discussion forums." (Hampton & Wellman, 1999, p. 481). The 190 homes in the development were medium-priced, detached, and similar to others in the suburban area. Although it turned out that only 60 percent of the homes were actually connected to the network, the connected and nonconnected became a convenient sample for Hampton and Wellman's study of this neighborhood.

Contrary to the expectations of those who believe online connection takes us away from offline life, Hampton and Wellman (1999, 2002) found that being connected increased in-person connectivity among local residents. Access encouraged greater community involvement and expanded and strengthened local relationships with neighbors and family. It also helped maintain ties with friends and relatives living farther away. "Wired" residents recognized three times as many neighbors (25 versus eight) and talked to twice as many as did nonwired residents (six vs. three). Wired residents were also more active in contacting others locally, averaging five times as many local phone calls as nonwired residents and sending an average of four e-mail messages to other local residents each month. The greater access to neighbors facilitated by the local area network also meant that wired residents were much more likely to know neighbors living elsewhere in the suburb, rather than just those living near them. By contrast, nonwired residents' neighborhood contacts included only those in nearby households. The neighborhood e-mail list increased the amount of in-person socializing, as residents organized parties, barbecues, and other local events online. The same e-mail list also aided collective action and political involvement, as residents organized to protest housing concerns, collectively purchase goods, share information about burglaries, discuss a local teachers strike, and deal with their Internet service provider.

Thus, the construct of a neighborhood, something traditionally defined exclusively by its geographic boundaries, can be seen to exist in an augmented form through online communications. The social world of Netville, and of Netville residents, combines offline geography and online interaction to create and sustain a more active and participatory community.

rec.arts.tv.soaps (r.a.t.s.): Online Discussion and Community

Another variant in social worlds and their online and offline components is demonstrated in the world of "logged in" television soap-opera fans explored by Baym (2000). What is different here is that the overlap involves two media: television and the Web.

As Baym (2000) explains, for soap fans, social interaction around the viewing and discussion of the soap is integral to the overall experience and appeal of the genre. Examining the local context of soap viewing, including patterns of discussion at work, with friends, and with family, provides a way to understand the role of these soap operas in people's lives: "because of the value of this [pre and post viewing] talk ... the soap opera audience is interconnected" (Baym, 2000, p. 14). As a largely women's activity, others have argued that "soaps create a social space, enhancing women's social bonds to one another," engaging in "women's community building" (Baym, 2000, p. 16). Moreover, "By elaborating the complicated networks and video- and audiotape traders, fan conventions, fan publications, and so on, and by looking closely at the texts that fans produce around the media ... fan experience [can be situated] within social networks rather than an isolated relation to a television show" (Baym, 2000, p. 16). The discussion of television "texts" creates a connection between television watching and social interaction in local communities. Add to this the new dimension of online discussion about the shows, their plots, characters, and actors, and the worlds of television and the Web meet.

But, even as they meet, the phenomenon of the Web is also creating something new. The local community of soap fans is now constituted by those discussing the shows online. The Usenet group also shows evidence of being a self-contained world, replete with its own characters (e.g., frequent, humorous, or respected posters), texts (32,000 messages in 10 months), and structures (e.g., how message headers are defined and used, pre- and postshow discussion, temporal entrainment with television broadcasting schedules, and group practices in the use of names and signature files [Baym, 1995]).

It appears that for both online and offline soap fans, the television text plays only a supporting role to the community. We also note that although r.a.t.s. can be considered as a separate social world, it also fits within a larger social world of soap viewing, one that is also supported through multiple media, including television, online and offline discussion, magazines, and fan sites. However, what is key for both the more narrowly defined r.a.t.s. community and the larger soap-opera–watching community is discussion and the social connections that discussion fosters.

Summary

These cases have illustrated how the online realm sometimes supports, sometimes creates a social world. We could carry on with more examples, such as the Internet in Trinidad and Trinidadian culture (both local and in the diaspora [Miller & Slater, 2000]), the social worlds of online learners (Haythornthwaite et al., 2000; Haythornthwaite & Kazmer, in press), the business of newspapers (Boczkowksi, 1999, 2004), academic practices in the age of networked connectivity (Harrison & Stephen, 1996; Sudweeks, McLaughlin, & Rafaeli, 1998), and online publishing (Kling & Callahan, 2003). However, we turn now to more subtle interactions between worlds, ones that cannot be captured by the name of a particular space, work organization, or grouping.

Overlaps and Synergies: Switching Worlds and Living in Many

Our discussion in the previous sections highlighted activities and places that one can name and identify: activist sites, Pentalk, Netville, r.a.t.s. Here we discuss how Web-based activities and practices have seeped into every aspect of day-to-day life, and vice versa.

Lifestyle, career, and work changes in the second half of the 20th century, such as increased representation of women in the workforce, dual income families, and globalization, each contribute to changing our divisions and definitions of work, office, and home. Add to this a progression toward seamless communication and ubiquitous contact—from synchronous phone to asynchronous e-mail, e-mail at work to e-mail at home, phone to pager to cell phone—and our ability to keep worlds separate breaks down. Moreover, active, mobile life and work styles create and engender the need for ad hoc, person-to-person rather than place-to-place contact (Wellman, 2001). This in turn bootstraps yet more mobility and distance. Why should we not work with specialists and commune with like minds across distance? What is stopping us? Being on a different continent becomes much the same as on a different floor; at home, in the coffee shop, or in the car becomes no less connected and accessible than in the office. And so we do work and commune at a distance, nearly anywhere, anytime.

Just as our worlds stay with us when we roam, they also overlap when we are stationary. We do not need to change locations, or devices, to switch from work to home. We may find home "intruding" on work and vice versa: e-mail from friends and family arrives in the middle of work sessions; we read e-mail and surf the Web for work while at home. Who can resist opening work mail at home and vice versa? How many can actually keep these worlds apart? Salaff (2002), and Kazmer and Haythornthwaite (2001; Haythornthwaite & Kazmer, 2002; Kazmer, 2002), describe how Internet-based work intrudes not just virtually, but also physically into the home. Teleworkers (Salaff, 2002) and online learners (Kazmer & Haythornthwaite, 2001) create boundaries, some physical and some temporal, that define work "space" in the home. Rooms are set aside that signify the individual is "at work;" children are taught to understand that Mom or Dad is still "at work" even if visibly in the house, sometimes by time (e.g., until 5 P.M.), sometimes by a closed door. Home devices must be considered work devices at certain times of the day and treated accordingly by all family members, for example, deciding who answers the phone or who has access to the computer. A much less obvious manifestation of overlap between worlds exists in the way offline communication habits blend and merge into online contexts. As we take on roles in one world, we do not leave our baggage from other worlds behind. We carry behaviors, attitudes, and habits online and from one online world to another. Despite the anonymity of cyberspace, researchers find lingering effects of gender, race, age, and region in behavior and language online (see the work by Crystal [2001], Herring [1993, 1999, 2002], and Kolko, Nakamura, and Rodman [2000]). Baym (2000) also discusses how, despite (or perhaps because of) the anonymity of the online environment, participants in online conversations slip in self-disclosures about their personal, offline identity, thereby revealing information about such aspects as gender, marital status, and family.

Neither is the Web a cultural blank slate. The Web carries the legacy of its early, youth-oriented, play culture, as well as an orientation to democratic, public use. The computer as a tool was seen by many as an opportunity to play—play at programming, play with others. What has captured the attention of many regarding early computer use is the way the play culture, as well as commitment to open, "democratic" use, melded with programming to create a work-at-play/play-at-work composite (for a review of early Internet [U.S. Defense Advanced Research Project Agency Network (ARPANET)] culture, see King, Grinter, and Pickering [1997]). For example, Unsworth (1996), found the distinction between work, education, and play blurred in the PMC-MOO environment (created for discussing issues related to the journal *Postmodern Culture*); individuals invested time "playing" as they contributed to the MOO. Thus, not only do we cross worlds of work and home, but work may also cross with the "world" of play.

As well as dropping subtle clues online about our identity, we also actively reestablish ways of interacting we have learned offline. Thus, we reintroduce status to anonymous environments by using signature files or by determining and reinforcing community norms that separate "newbies" from netizens. Power relations are reintroduced in the way knowledge of online environments grants power to expert users and lack of power to newbies, and in the naming of roles within CMC environments with their hierarchical, or magical and superstitious connotations (e.g. list "owners" and MUD "wizards"). We reintroduce regulation of behaviors by granting control to moderators, adhering to community norms, and chastising or shunning those who violate the norms (Baym, 1995, 1997, 2000; McLaughlin, Osborne, & Smith, 1995; for reviews see Haythornthwaite, Wellman, & Garton, 1998; Herring, 2002; Wellman et al., 1996). This introduction of offline norms into online environments has come to many as a surprise. As noted earlier, both utopians and dystopians expected online activity to be something separate from offline life, whether allowing liberation from offline constraints (seen as a positive outcome) or aiding escape from offline responsibilities (seen as negative outcome). It is a surprise, and thus to some also a disappointment, to find that we are still the same social creatures, reinventing online our offline patterns of behavior.

Neither Good, nor Bad, nor Neutral

Although for some, the mix of work and home represents an ideal, an answer to the long commutes associated with work in city offices and transnational companies, for many the mix is less benign. Continuous electronic and mobile connectivity, combined with continuous operation of computer systems and the work operations they support, can take away the option of being disconnected or off work. On the other hand, the opportunity to be connected at home, at off hours, opens up opportunities for those who might otherwise be unable to participate (Sproull & Kiesler, 1991). It is clear that work and home now overlap, and both reach outside their single physical settings; what is not clear is whether this is good or bad or just different. Like many changes wrought by technology, increased connectivity and the ensuing overlap of worlds are turning out to be "neither good, nor bad, nor ... neutral" (Krantzberg, 1986, p. 545), and we might also add "nor singular" in impact.

As individuals spend time online chatting and forming relationships with (geographically) distant others, one burning question has been whether time online takes away from the quality of relationships with local family, friends, colleagues, and community. In other words, how do the different social worlds compete for time and attention? Who gains or loses in this competition? As we will see in the brief review that follows, current studies suggest a change in the pattern of interactions, but no decisive winner or loser (for a more complete review, see Haythornthwaite and Wellman [2002] and see Wellman and Haythornthwaite [2002] for a collection of papers that address this topic).

Nie (2001) eloquently points out a common feature, that those who gain online are those who are already richly engaged with others (see also Nie & Erbring, 2000; Nie, Hillygus, & Erbring, 2002). Kraut, Kiesler, Boneva, Cummings, Helgeson, and Crawford (2002) add further to this argument by finding that extroverts gain more than introverts in online connectivity. At the interpersonal level, Haythornthwaite (2002b) finds that those who are more closely tied tend to use more media to communicate (Haythornthwaite, 2000, 2001; Haythornthwaite & Wellman, 1998; Koku, Nazer, & Wellman, 2001). Where online means are added on to other means, again the advantage of online connection goes to those already well connected. This can be seen as another variant of the Matthew Effect, essentially that the rich get richer and the poor get poorer. This effect has been noted in how already famous scientists get more credit and recognition for their contributions (Merton, 1968), and for how children who are good readers early do better in all aspects of learning (Walberg & Tsai, 1983).

There are concerns that time spent online displaces attention to local social relations and has adverse effects on individual well-being (Kraut, Patterson, Kiesler, Mukhopadhyay, & Scherilis, 1998; Nie, 2001), but those who find themselves suddenly at a distance from others appear to benefit from online interactions. Such was the case for the isolated farmers discussed earlier. LaRose, Eastin, and Gregg (2001) find that an online connection helps reduce stress when college students are separated from family; and Hampton and Wellman (2002) find that an online connection allows people who have moved to a new home to maintain connections to those they have left behind.

Also in keeping with Nie's view about connectivity, recent research shows that offline integration with the community enhances engagement via the online world and vice versa. Ball-Rokeach and colleagues (Matei & Ball-Rokeach, 2002) find that the chance of making a friend online increases significantly the more an individual is engaged with the neighborhood, and increases substantially for those who know a neighbor well enough to talk about a personal matter. As described above, Hampton and Wellman (2002) find that online connectivity among neighbors increases the number who are recognized and talked to in a local neighborhood. Similarly, Kavanaugh and Patterson (2002) find that high levels of geo-community involvement in Blacksburg are associated with greater use of the Internet for interpersonal and group communication activities. They also suggest that local community involvement and social capital-building activities that resulted from engagement in the online Blacksburg Electronic Village may have come about because of the strong pre-existing offline social capital within the Blacksburg geo-community. And Boneva and Kraut (2002) find that women, the traditional connectors for family ties, continue their role online; they are more likely than men to maintain e-mail connections with distant friends and relatives and to maintain larger networks of distant contacts.

Experience online also enhances this online and offline connectivity. Those who have been online longer make more connections to others (Chen, Boase, & Wellman, 2002; Haythornthwaite & Kazmer, 2002; Kavanaugh & Patterson, 2002; Nie, 2001; Quan-Haase & Wellman, 2002). Their expertise and familiarity with the ways of the Web help them set up connections for others, including close and distant family. For example, Kiesler, Lundmark, Zdaniuk, and Kraut (2000) found teens played a major role in the household in seeking and giving help on Internet technical features. Kazmer and Haythornthwaite (2001; Haythornthwaite & Kazmer, 2002) found that as expertise and familiarity with the technology increased in the seemingly closed social world of an online distance education program, participants often found themselves helping others gain access to the Web. For this set of participants, who were largely women, connections were often set up for previously non-Internet-ready parents so the distance learners could communicate with their parents in the same way as with others in their online community.

Overall, we find changes in how people connect to others and where those others are. However, we do not find that life on the Web is leaving individuals isolated or communities abandoned. Instead, it appears to be taking its place in supporting individuals and communities in their own social worlds. We turn now to our final section, addressing some of the research challenges for studying the worlds of the Web.

Part III: Studying Overlapping Worlds

These new worlds, and new arrangements of traditional worlds, present special challenges for researchers (Haythornthwaite, Bowker, Bruce, Lunsford, Kazmer, Brown, et al., 2003; Lyman & Wakeford, 1999a, 1999b). They create new dilemmas (e.g., about the ethics of lurking [Heath, Koch, Ley, & Montoya, 1999]), challenge commonly expected associations (e.g., between community and geography, body and persona), provide multiple views of a single individual (Heath et al., 1999), and create previously unexpected associations, for example, connecting researchers around the globe (Sudweeks, McLaughlin, & Rafaeli, 1998; Sudweeks & Rafaeli, 1996), or a researcher with study participants seven time zones apart (Livia, 1999). Approaching the Web as a single world leads to the adoption of single approaches, a stance that limits the inclusion of sources, information, and methods. Let us take a moment to consider the range of inclusivity possible, and sometimes necessary, when examining and researching the worlds of the Web. We have loosely categorized the many aspects to be considered under the headings of surface characteristics, history and context, individual perspectives, technology.

Research Considerations

Surface Characteristics

Online activity is marked by multiple access methods, or modes of CMC (Herring, 2002), including Web pages, e-mail, listservs, bulletin boards, and virtual realities. Each mode can include one or more means of expression: text, audio, and graphics. Online environments may be relatively lean, text-based listservs, more comprehensive Web sites, or multimodal virtual environments such as collaboratories (Finholt, 2002) and gaming environments. Participants may go to these sites for work, play, school, social support, companionship, or information, although physically located at work or home, in a car, hotel room, airport lounge, library, and so on. The device used to access the Web may be a large-screen desktop computer, small laptop, handheld device, telephone, or cell phone; the device may run under one of a number of operating systems. Fellow online participants and correspondents may be co-workers, family, friends, adults, teens, children, strangers, or a combination. The audience may be a single individual or multiple others, with discussion carried on publicly or privately, with known or unknown others (Bakhtin, 1986; Bregman & Haythornthwaite, 2003; Clark, 1996). Individuals may present themselves as reflections of their true selves; as physically changed or exaggerated; as named, pseudonymous, or anonymous; through text, pictures, or avatars. The identity of information providers and correspondents may be easily verifiable and authenticated according to external standards (e.g., an .edu or .gov address for U.S. educational and government institutions), by standards of experts and novices internal to the group (e.g., in online communities of practice), or only by the judgment of the participant. Access to the information provided, or to the conversations and games, may be available free to the general public, only to those with paid subscriptions, or only to those with granted membership.

History and Context

Below the "surface characteristics" just discussed are hidden aspects of the Web that are not evident from lists of application capabilities. These are the forces and structures that shape the Web's creation and deployment, determine for whom the technologies are designed, and can limit or constrain the choices about who gets connected from where. As noted at the beginning of this chapter, how we use technology now, and how we communicate, are rooted in how we are already accustomed to doing things and talking with others (Kotamraju, 1999; Rice & Gattiker, 2001). We depend on "layers of technological artifacts" (Lyman & Wakeford, 1999a, p. 366) from infrastructure to present configurations (Star, 1999; Suchman, Blomberg, Orr, & Trigg, 1999), which are often dependent on corporate or government sponsorship for continued existence (Hampton & Wellman, 1999; Livia, 1999). Design emanates from developers' world views, so we must ask, who was involved in the design (e.g., corporate or open source development)? Whose metaphors (e.g., the "desktop" metaphor which continues assumptions about the organization of work activity rather than home or play), modes of behavior, subjects of interest, notions of community, and demonstrations of discourse influenced the development of the current presentation of the technology and the "world" it describes (Bowker & Star, 1999)? (See Markus and Bjorn-Andersen [1987] for a discussion of developers' power over users; see King, Grinter, and Pickering [1997] on the culture of early Internet developers and users.) Historical decisions about design can set trends that are unbreakable without further radical innovation. For example, as Warschauer (2003) points out, the ASCII character set does not support nonroman characters, precluding conversation in many languages; new standards had to be developed to expand the universe of languages and users.

Although some view development as being controlled by key players—developers, owners, dominant Western societies—others see Web use bootstrapped by the community using the technology, with use reinforcing practices and vice versa (Contractor & Eisenberg, 1990; DeSanctis & Poole, 1994; Fulk & Boyd, 1991; Markus, 1990; Orlikowski & Yates, 1994). One consequence of this view is that, because practices emerge from interactions among users, final presentations of interaction patterns cannot be predicted a priori from a developer's view of the system. Instead practices of interaction emerge from the uses that begin and sustain the community.

Individual Perspectives

We have noted here how demographic attributes such as experience online, age, and gender can affect adoption and use of the Web. Beyond demographics, behaviors can differ as individuals make choices about how to present themselves online, a decision that may change across environments. They may role-play in one context and maintain a straightforward public presence in another according to the venue. Thus, they may be newbies in one listserv, experts in another; mothers or medical patients when in support sites, mentors or students in another. The many roles we play in life we also play online.

We should also ask whose perspective we are using when examining the worlds of the Web (Lyman & Wakeford, 1999a). Are we trying to view the world from the perspective of the participants, by becoming part of their environment, as an anthropologist in cyberspace? Do we want to understand the designer's view, and/or the difference between designer and user views? Are we embedded as participant observers, or lurking on the periphery (Heath et al., 1999)? Are we interactive, both observing and asking questions, exploring design and history, or do we act externally and omnipotently, taking a view from above (Lyman & Wakeford, 1999a)?

Technology

Lastly, when examining worlds of the Web, it may be relevant to consider whether the technology in use (computer, network connection, software application) is highly experimental, available only to the technological elite; leading edge, available to those most adventurous in the market who are capable of buying and using the most up-to-date technology; mainstream, widely used, current technology available for every office or household; or older technology that runs at lower processing speeds, supports fewer features, available as cast-offs from those in the mainstream by those who cannot afford mainstream systems. Technologies may be visible—the computer on the desk—but also invisible as with infrastructures relating to electrical power grids, land or satellite telephone connections, or Internet service providers. Although important in Western worlds for subtle patterns of access to resources (Star, 1999), infrastructural considerations are far more basic in non-Western worlds where the unreliability of telephone connections means that access is variable in terms of speed, quality, and dependability rather than simply present or absent (Davidson, Sooryamoorthy, & Shrum, 2002; Warschauer, 2003).

Multiple Perspectives, Multiple Methods, Multiple Views

This emerging field of Web and Internet research demands and deserves innovative techniques and perspectives. It demands reopening definitions thought to be closed, embracing a definition of worlds that is more complex than just the division between online and offline, and acknowledging the emergent nature of Web phenomena. The Web may play a role in the background, connecting people who maintain local relationships, but it also may act in the foreground as the main means of connecting otherwise unconnected individuals, for example as individuals and organizations around the globe organize for netwar, or operate as smartmobs in pursuit of information, coordinated action, or fun (Rheingold, 2003).

Many see these new issues, and in particular the way worlds cross and are crossed by media, as grounds for combining and creating new methods of inquiry, as well as giving attention to new multimodal, and multilevel impacts (Haythornthwaite, 2001, 2002b; Heath et al., 1999; Howard, 2002; Lyman & Wakeford, 1999a; Rice & Gattiker, 2001). As researchers, we have to recognize and juggle multiple representations and multiple worlds (Heath et al., 1999; Kazmer & Haythornthwaite, 2001). Questions such as the following need to be asked:

- **What is the definition of local?** Is it the researchers' local, or the participants' local that matters? Is there one "local" when participants may be in multiple geographic locations? Is there a cyber-locality that is the "location" of such worlds? **What is the definition of global?** As the significance of national boundaries declines with the sharing of information, is everything local also global? If that one "local" where people meet brings in representatives from places around the globe, what does global mean?

- **Who is a person in the world?** How is membership defined: by an account on the system, a name listed in the chat room, active participation, or geographic residence (Haythornthwaite et al., 2003)? Is membership defined as an active poster to an online world? Does a lurker belong? Is the member the avatar or the avatar's operator? Given the potential for multiple personae in role playing, and also our multiple roles in life, we may even ask: **When is a person?** Is one a person in the online community only when online in the community? Or, does that membership extend its influence to other parts of life, making participants part of that community full time? Is one a gender-bending, body-redefining, nonearthly creature when sitting at the office desktop or only when connected to the off-world existence at home on one's laptop? **Where is a person in the world?** What time is it? Whose time is it—the local time or some measure of time associated with online activity? What is the impact on online worlds of functioning in different time zones (Haythornthwaite et al, 2003; Livia, 1999)?

- **Where is a community?** Do they have a common place, and is it real, virtual, or a combination? How do we find members of a community when they are no longer "centered [physically or singly] in the way auto workers are associated with Detroit, traders are associated with Wall Street, or IT workers are associated with Silicon Valley" (Howard, 2002, p. 571)? How do we do ethnography of online communities (Howard, 2002; Ruhleder, 2000)? Where is the locus of activity when we examine UNIX developers (Moon & Sproull, 2002), fan communities (Baym, 2000), or distributed groups and teams (Teigland, 2000)?

- **What does the world of the Web look like?** What structures are emerging on Usenet (Smith, 1999, 2002), via hyperlinks (Park, 2003), and from online social interactions (Haythornthwaite, 2001, 2002b)? How does online activity structure both online and offline worlds? What do online conversations look like (Donath, 2002; Donath, Karahalios, & Viégas, 1999; Erickson, Halverson, Kellogg, Laff, & Wolf, 2002; Erickson, Herring, & Sack, 2002; Nardi, Whittaker, Isaacs, Creech, Johnson, & Hainsworth, 2002)? How and what should we map to explore the "geographies of the information society" (Dodge & Kitchin, 2001, p. 32; see also http://web.cybergeography.org/ atlas/atlas.htm)?

Conclusion

We have discussed how approaching the Web from a social world's perspective helps ground our analyses in the purposes and identities of individuals, groups, and communities, freeing us from the constraint of a media-bound definition of worlds. Instead of trying to understand online life separately from offline, we follow how tasks and relationships lead use across modes of communication to define the social world. We have also stressed that Web integration into online and offline social worlds is an evolving phenomenon, but with its evolution and presentation rooted in what people do with the Web, and do together with the Web. Although there is a growing body of research on Web worlds and Web behavior, we are still at an early stage in understanding how the Web is affecting local, national, and global patterns of society (Haythornthwaite & Wellman, 2002; Rheingold, 2003; Rice & Gattiker, 2001). We have many more questions than answers, but that is the appeal of the Web.

Endnotes

1. Blog: "On a Web site, a blog, a short form for weblog, is a personal journal that is frequently updated and intended for general public consumption." (Whatis.com, http://whatis.techtarget.com/definition/0,289893,sid9_gci214616,00.html).
2. Texting: Communicating using text messaging feature on cell phones.
3. Morphing: "Short for *metamorphosing*, morphing refers to an animation technique in which one image is gradually turned into another." (webopedia, http://web.webopedia.com/TERM/M/morphing.html).
4. The social worlds perspective is complementary to a view of computer networks as social networks (Wellman et al., 1996). We will use the term social worlds here, but it is important to bear in mind that underlying any world are the social relations and interpersonal ties that characterize that world and hold it together. Both the social worlds and social network perspectives focus on the relations that tie people together in a social structure. Social network analysis can be used to define more specifically what it is people do together or have in common that constitutes their social world, as well as to reveal structures that are not readily apparent from a priori assumptions about group or community composition (Garton, Haythornthwaite, & Wellman, 1997; Haythornthwaite, 2002a). Both views allow a definition of a group, community, or world to emerge from what people do together, and thus are particularly apt for approaching the emergent phenomena of the Web.
5. See also Katz and Rice (2002a) for a comprehensive collection of statistics on Web use; the Pew Internet and American Life Project (www.pewinternet.org) for continuing surveys of Internet use; for other surveys, see UCLA Center for Communication Policy (www.ccp.ucla.edu); and Nie and Erbring (2000) regarding the survey by the Stanford Institute for the Quantitative Study of Society (SIQSS) (http://www.stanford.edu/group/siqss).
6. In 2002 the ten most popular Internet activities in the U.S. (UCLA Center for Communication Policy, 2003) were: e-mail and instant messaging (87.9 percent of Internet users); Web surfing or browsing (76.0); reading news (51.9); accessing entertainment information (46.4); shopping and buying online (44.5); hobbies (43.7); travel information (36.2); medical information (35.5); playing games (26.5); tracking credit cards (24.2).

References

Bakhtin, M. M. (1986). The problem of speech genres. In C. Emerson and M. Holquist (Eds.), V. W. McGhee (Trans.), *Speech genres and other late essays* (pp. 60–102). Austin: University of Texas Press. (Original work published in 1953)

Baym, N. K. (1995). From practice to culture on Usenet. In S. L. Star (Ed.), *The cultures of computing* (pp. 29–52). Oxford, UK: Blackwell.

Baym, N. K. (1997). Interpreting soap operas and creating community: Inside an electronic fan culture. In S. Kiesler (Ed.), *Culture of the Internet* (pp. 103–120). Mahwah, NJ: Lawrence Erlbaum.

Baym, N. K. (2000). *Tune in, log on: Soaps, fandom and online community*. Thousand Oaks, CA: Sage.

Bennett, K., Carroll, T., Lowe, P., & Phillipson, J. (2002). *Coping with crisis in Cumbria: Consequences of foot and mouth disease*. Centre for Rural Economy Research Report. University of Newcastle upon Tyne, UK. Retrieved November 5, 2003, from http://www.ncl.ac.uk/cre/publications/research_reports/rr02_01Cumbria.htm

Bishop, A. P., & Star, S. L. (1996). Social informatics of digital library use and infrastructure. *Annual Review of Information Science and Technology, 31*, 301–401.

Bishop, A. P., Bazzell, I., Mehra, B., & Smith, C. (2001). Afya: Social and digital technologies that reach across the digital divide. *First Monday, 6*(4). Retrieved November 5, 2003, from http://www.firstmonday.org/issues/issue6_4/bishop/index.html

Bishop, A. P., Mehra, B., Bazzell, I., & Smith, C. (2000). Socially grounded user studies in digital library development. *First Monday, 5*(6). Retrieved November 5, 2003, from http://www.firstmonday.org/issues/issue5_6/bishop/index.html

Bishop, A. P., Mehra, B., Bazzell, I., & Smith, C. (2003). Participatory action research and digital libraries: Reframing evaluation. In A. P. Bishop, N. Van House, & B. Buttenfield (Eds.), *Digital library use: Social practice in design and evaluation.* Cambridge, MA: MIT Press.

Boczkowski, P. (1999). Understanding the development of online newspapers. *New Media and Society, 1*(1), 101–126.

Boczkowski, P. (2004). *Digitizing the news: Innovation in online newspapers.* Cambridge, MA: MIT Press.

Boneva, B., & Kraut, R. (2002). Email, gender and personal relationships. In B. Wellman & C. Haythornthwaite (Eds.), *The Internet in everyday life* (pp. 372–403). Oxford, UK: Blackwell.

Bowker, G., & Star, S. L. (1999). *Sorting things out: Classification and its consequences.* Cambridge, MA: MIT Press.

Bregman, A., & Haythornthwaite, C. (2003). Radicals of presentation: Visibility, relation, and co-presence in persistent conversation. *New Media and Society, 5*(1), 117–140.

Chen, W., Boase, J., & Wellman, B. (2002). The global villagers: Comparing Internet users and uses around the world. In B. Wellman & C. Haythornthwaite (Eds.), *The Internet in everyday life* (pp. 74–113). Oxford, UK: Blackwell.

Clark, H. H. (1996). *Using language.* Cambridge, UK: Cambridge University Press.

Clarke, A. (1991). Social worlds/arenas theory as organizational theory. In D. Maines (Ed.), *Social organization and social process: Essays in honor of Anselm Strauss* (pp.119–158). New York: Aldine de Gruyter.

Cleaver, H. (1994, Marzo). The Chiapas uprising and the future of class struggle in the new world order. *Riff-Raff,* 133–145. Retrieved November 5, 2003, from http://www.eco.utexas.edu/facstaff/Cleaver/chiapasuprising.html

Cleaver, H. (1999). *Computer-linked social movements and the global threat to capitalism.* Retrieved November 5, 2003, from http://www.eco.utexas.edu/faculty/Cleaver/polnet.html

Cohill, A. M., & Kavanaugh, A. L. (2000). *Community networks: Lessons from Blacksburg, Virginia.* (2nd edition). Boston, MA: Artech House.

Connolly, T., & Thorn, B. K. (1990). Discretionary data bases: Theory, data and implications. In J. Fulk & C. W. Steinfield. (Eds.), *Organizations and communication technology* (pp. 219–234). Newbury Park, CA: Sage.

Contractor, N. S., & Eisenberg, E. M. (1990). Communication networks and new media in organizations. In J. Fulk & C. W. Steinfield. (Eds.), *Organizations and communication technology* (pp. 143–172). Newbury Park, CA: Sage.

Covi, L. (1996). Social worlds of knowledge work: Why researchers fail to effectively use digital libraries. *Proceedings of ASIS Mid-Year Meeting,* 84–100.

Crystal, D. (2001). *Language and the Internet.* Cambridge, UK: Cambridge University Press.

Davenport, E., & Hall, H. (2002). Organizational knowledge and communities of practice. *Annual Review of Information Science and Technology, 36,* 171–227.

Davenport, T. H., & Beck, J. C. (2001). *The attention economy: Understanding the new currency of business.* Cambridge, MA: Harvard Business School Press.

Davidson, T., Sooryamoorthy, R., & Shrum, W. (2002). Kerala connections: Will the Internet affect science in developing areas?. In B. Wellman & C. Haythornthwaite (Eds.), *The Internet in everyday life* (pp. 496–519). Oxford, UK: Blackwell.

DeSanctis, G., & Poole, M. S. (1994). Capturing the complexity in advanced technology use: Adaptive structuration theory. *Organization science, 5*(2), 121–47.

DiMaggio, P., Hargittai, E., Neuman, W. R., & Robinson, J. P. (2001). Social implications of the Internet. *Annual Review of Sociology, 27*, 307–336.

Dodge, M., & Kitchin, R. (2001). *Mapping cyberspace.* New York: Routledge.

Donath, J. (2002). A semantic approach to visualizing online conversations. *Communications of the ACM, 45*(4), 45–49.

Donath, J., Karahalios, K., & Viégas, F. (1999). Visualizing conversation. *Journal of Computer-Mediated Communication, 4*(4). Retrieved November 5, 2003, from http://www.ascusc.org/jcmc/vol4/issue4/donath.html

Ellis, D., Oldridge, R., & Vasconcelos, A. (2004). Community and virtual community. *Annual Review of Information Science and Technology, 38*, 145–186.

Engeström, Y., Miettinen, R., & Punamäki, R. (Eds.). (1999). *Perspectives on activity theory.* Cambridge, UK: Cambridge University Press.

Erickson, T., Halverson, C., Kellogg, W. A., Laff, M., & Wolf, T. (2002). Social translucence: Designing social infrastructures that make collective activity visible. *Communications of the ACM, 45*(4), 40–44.

Erickson, T., Herring, S., & Sack, W. (2002). *Discourse architectures: Designing and visualizing computer-mediated communication.* Workshop presented at the ACM SIGCHI Conference, Minneapolis, MN. Position papers available at http://www.pliant.org/personal/Tom_Erickson/DA_CHI02_WrkShp_Sum.html

Finholt, T. (2002). Collaboratories. *Annual Review of Information Science and Technology, 36*, 73–107.

Fitzpatrick, G., Kaplan, S., & Mansfield, T. (1996). Physical spaces, virtual places and social worlds: A study of work in the virtual. *Proceedings of the ACM Conference on Computer Supported Cooperative Work*, 334–343.

Frederick, H. H. (1993). *Global communication and international relations.* Belmont, CA: Wadsworth.

Fulk, J., & Boyd, B. (1991). Emerging theories of communication in organizations. *Journal of Management, 17*(2), 407–46.

Garton, L., Haythornthwaite, C., & Wellman, B. (1997). Studying online social networks. *Journal of Computer-Mediated Communication, 3*(1). Retrieved November 5, 2003, from http://www.ascusc.org/jcmc/vol3/issue1/garton.html

Gillin, B. (2003, August 7). Flash mob. *Daily Magazine, The Philadelphia Inquirer.* Retrieved November 21, 2003, from http://www.philly.com/mld/inquirer/news/magazine/daily/6474903.htm

Goffman, E. (1959). *The presentation of self in everyday life.* Garden City, NY: Doubleday.

Goldhaber, M. H. (1997). The attention economy and the Net. *First Monday, 2*(4). Retrieved November 5, 2003, from http://www.firstmonday.dk/issues/issue2_4/goldhaber

Gurstein, M. H. (2000). *Community informatics: Enabling communities with information and communications technologies.* Hershey, PA: Idea Group Publishing.

Gurstein, M. H. (2001). Community informatics for flexible networking. In L. Keeble & B. D. Loader (Eds.), *Community informatics: Shaping computer-mediated social relations* (p. 263–283). New York: Routledge.

Hampton, K. N., & Wellman, B. (1999). Netville online and offline: Observing and surveying a wired suburb. *American Behavioral Scientist, 43*(3), 475–492.

Hampton, K. N., & Wellman, B. (2002). The not so global village of Netville. In B. Wellman & C. Haythornthwaite (Eds.), *The Internet in everyday life* (pp. 345–371). Oxford, UK: Blackwell.

Harrison, T. M., & Stephen, T. (Eds.). (1996). *Computer networking and scholarly communication in the twenty-first-century university.* Albany: State University of New York Press.

Haythornthwaite, C. (2000). Online personal networks: Size, composition and media use among distance learners. *New Media and Society, 2*(2), 195–226.

Haythornthwaite, C. (2001). Exploring multiplexity: Social network structures in a computer-supported distance learning class. *The Information Society, 17*(3), 211–226.

Haythornthwaite, C. (2002a). Building social networks via computer networks: Creating and sustaining distributed learning communities. In K. A. Renninger & W. Shumar (Eds.), *Building virtual communities: Learning and change in cyberspace* (pp. 159–190). Cambridge, UK: Cambridge University Press.

Haythornthwaite, C. (2002b). Strong, weak and latent ties and the impact of new media. *The Information Society, 18*(5), 385–401.

Haythornthwaite, C., Bowker, G. C., Bruce, B., Lunsford, K. J., Kazmer, M. M., Brown, J., et al. (2003). *Research challenges in the study and practice of distributed collaboration.* Unpublished manuscript.

Haythornthwaite, C., & Kazmer, M. M. (2002). Bringing the Internet home: Adult distance learners and their Internet, Home and Work worlds. In B. Wellman & C. Haythornthwaite (Eds.), *The Internet in everyday life* (pp. 431–463). Oxford, UK: Blackwell.

Haythornthwaite, C., & Kazmer, M. M. (in press). *Learning, culture and community: Multiple perspectives and practices in online education.* New York: Peter Lang.

Haythornthwaite, C., Kazmer, M. M., Robins, J., & Shoemaker, S. (2000). Community development among distance learners: Temporal and technological dimensions. *Journal of Computer-Mediated Communication, 6*(1). Retrieved November 5, 2003, from http://www.ascusc.org/jcmc/vol6/issue1/haythornthwaite.html

Haythornthwaite, C., & Wellman, B. (1998). Work, friendship and media use for information exchange in a networked organization. *Journal of the American Society for Information Science, 49*(12), 1101–1114.

Haythornthwaite, C., & Wellman, B. (Eds.). (2002). *The Internet in everyday life.* Oxford, UK: Blackwells.

Haythornthwaite, C., Wellman, B., & Garton, L. (1998). Work and community via computer-mediated communication. In J. Gackenbach (Ed.), *Psychology and the Internet* (pp.199–226). San Diego, CA: Academic Press.

Heath, D., Koch, E., Ley, B., & Montoya, M. (1999). Nodes and queries: Linking locations in networked fields of inquiry. *American Behavioral Scientist, 43*(3), 450–463.

Herring, S. C. (1993). Gender and democracy in computer-mediated communication. *Electronic Journal of Communication, 3*(2). Retrieved November 5, 2003, from http://www.cios.org/getfile\Herring_v3n293

Herring, S. C. (1999). The rhetorical dynamics of gender harassment on-line. *The Information Society, 15*(3), 151–167.

Herring, S. C. (2002). Computer-mediated communication on the Internet. *Annual Review of Information Science and Technology, 36*, 109–168.

Howard, P. N. (2002). Network ethnography and the hypermedia organization: New media, new organizations, new methods. *New Media and Society, 4*(4), 550–574.

Howard, P., Rainie, L., & Jones, S. (2002). Days and nights on the Internet. In B. Wellman & C. Haythornthwaite (Eds.), *The Internet in everyday life* (pp. 45–73). Oxford, UK: Blackwell.

Johnson, C. A. (2003). *Information networks: Investigating the information behavior of Mongolia's urban residents.* Unpublished doctoral dissertation. University of Toronto, Toronto, Canada.

Katz, J. E., & Aspden, P. (1997). A nation of strangers? *Communications of the ACM, 40*(12), 81–86.

Katz, J. E., & Rice, R. E. (2002a). *Social consequences of Internet use: Access, involvement and expression.* Cambridge, MA: MIT Press.

Katz, J. E., & Rice, R. E. (2002b). Syntopia: Access, civic involvement and social interaction on the Net. In B. Wellman & C. Haythornthwaite (Eds.), *The Internet in everyday life* (pp. 114–138). Oxford, UK: Blackwell.

Katz, J. E., Rice, R. E., & Aspden, P. (2001). The Internet, 1995–2000: Access, civic involvement, and social interaction. *American Behavioral Scientist, 45*(3), 405–419.

Kavanaugh, A. L., & Patterson, S. (2002). The impact of computer networks on social capital and community involvement in Blacksburg. In B. Wellman & C. Haythornthwaite (Eds.), *The Internet in everyday life* (pp. 325–344). Oxford, UK: Blackwell.

Kazmer, M. M., & Haythornthwaite, C. (2001). Juggling multiple social worlds: Distance students online and offline. *American Behavioral Scientist, 45*(3), 510–529.

Kazmer, M. M. (2002). *Disengagement from intrinsically transient social worlds: The case of a distance learning community.* Unpublished doctoral dissertation, University of Illinois at Urbana-Champaign.

Keeble, L., & Loader B. D. (Eds.). (2001). *Community informatics: Shaping computer-mediated social relations.* New York: Routledge.

Kiesler, S., Lundmark, V., Zdaniuk, B., & Kraut, R. E. (2000). Troubles with the Internet: The dynamics of help at home. *Human Computer Interaction, 15*, 323–351.

King, J. L., Grinter, R. E., & Pickering, J. M. (1997). The rise and fall of Netville: The saga of a cyberspace construction boomtown in the great divide. In S. Kiesler (Ed.), *Culture of the Internet* (pp. 3–33). Mahwah, NJ: Lawrence Erlbaum.

Kling, R. (2000). Learning about information technologies and social change: The contribution of social informatics. *The Information Society, 16*, 217–232.

Kling, R., & Callahan, E. (2003). Electronic journals, the Internet, and scholarly communication. *Annual Review of Information Science and Technology, 37*, 127–177.

Kling, R., Rosenbaum, H., & Hert, C. (1998). Social informatics in information science: An introduction. *Journal of the American Society for Information Science, 49*(12), 1047–1052.

Koku, E., Nazer, N., & Wellman, B. (2001). Netting scholars: Online and offline. *American Behavioral Scientist, 44*(10), 1752–1774.

Kolko, B. E., Nakamura, L., & Rodman, G. B. (Eds.). (2000). *Race in cyberspace.* New York: Routledge.

Kotamraju, N. P. (1999). The birth of Web site design skills: Making the present history. *American Behavioral Scientist, 43*(3), 464–474.

Krantzberg, M. (1986). Technology and history: Krantzberg's laws. *Technology and Culture, 27*(3), 544–560.

Kraut, R., Kiesler, S., Boneva, B., Cummings, J., Helgeson, V., & Crawford, A. (2002). Internet paradox revisited. *Journal of Social Issues, 58*(1), 49–74.

Kraut, R., Patterson, V. L., Kiesler, S., Mukhopadhyay, T., & Scherilis, W. (1998). Internet paradox: A social technology that reduces social involvement and psychological well-being? *American Psychologist, 53*(9), 1017–1031.

Lamb, R., & Davidson, E. (2002). *Social scientists: Managing identity in socio-technical networks.* Proceedings of the Hawai'i International Conference on System Sciences, Big Island, Hawaii. Retrieved November 23, 2003, from http://lamb.cba.hawaii.edu/pubs/socialscientists.pdf

LaRose, R., Eastin, M. S., & Gregg, J. (2001). Reformulating the Internet paradox: Social cognitive explanations of Internet use and depression. *Journal of Online Behavior, 1*(2). Retrieved November 5, 2003, from http://www.behavior.net/JOB/v1n2/paradox.html

Latour, B. (1987). *Science in action: How to follow scientists and engineers through society*. Philadelphia: Open University Press.

Lave, J., & Wenger, E. (1991). *Situated learning: Legitimate peripheral participation*. Cambridge, UK: Cambridge University Press.

Livia, A. (1999). Doing sociolinguistic research on the French Minitel. *American Behavioral Scientist, 43*(3), 422–435.

Lyman, P., & Wakeford, N. (1999a). Going into the (virtual) field. *American Behavioral Scientist, 43*(3), 359–376.

Lyman, P., & Wakeford, N. (Eds.). (1999b). Analyzing virtual societies: New directions in methodology [Special issue]. *American Behavioral Scientist, 43*(3).

Markus, M. L. (1990). Toward a "critical mass" theory of interactive media. In J. Fulk & C. W. Steinfield (Eds.), *Organizations and communication technology* (pp. 194–218). Newbury Park, CA: Sage.

Markus, M. L., & Bjorn-Andersen, N. (1987). Power over users: Its exercise by system professionals. *Communications of the ACM, 30*(6), 498–504.

Matei, S., & Ball-Rokeach, S. (2002). Belonging in geographic, ethnic and Internet spaces. In B. Wellman & C. Haythornthwaite (Eds.), *The Internet in everyday life* (pp. 404–427). Oxford, UK: Blackwell.

McLaughlin, M. L., Osborne, K. K., & Smith, C. B. (1995). Standards of conduct on Usenet. In S. G. Jones (Ed.), *CyberSociety: Computer-mediated communication and community* (pp. 90–111). Thousand Oaks, CA: Sage.

Merkel, C. (2002). *Uncovering the hidden literacies of "have-nots": A study of computer and Internet use in a low-income community*. Unpublished doctoral dissertation, University of Illinois at Urbana-Champaign.

Merton, R. K. (1957). *Social theory and social structure*. New York: Free Press.

Merton, R. K. (1968). The Matthew effect in science. *Science, 159*(3810), 56–63.

Miller, D. (1999). Community information networks: Definitions and a review of the developments during the 1990s. In S. Pantry (Ed.), *Building community information networks: Strategies and experiences* (pp. 1–10). London: Library Association.

Miller, D., & Slater, D. (2000). *The Internet: An ethnographic approach*. Oxford, UK: Berg. Retrieved November 24, 2003, from http://ethnonet.gold.ac.uk

Moon, J. Y., & Sproull, L. (2002). Essence of distributed work: The case of the LINUX kernel. In P. J. Hinds & S. Kiesler (Eds.), *Distributed work* (p. 381–404). Cambridge, MA: MIT Press.

N30 Global Day of Action, November 30 1999. Retrieved August 5, 2003, from http://infoshop.org/news4/lob_fly1.pdf

Nardi, B., Whittaker, S., Isaacs, E., Creech, M., Johnson, J., & Hainsworth, J. (2002). Integrating communication and contact via ContactMap. *Communications of the ACM, 45*(4), 89–95.

Nie, N. H. (2001). Sociability, interpersonal relations, and the Internet: Reconciling conflicting findings. *American Behavioral Scientist, 45*(3), 420–435.

Nie, N. H., & Erbring, L. (2000). *Internet and society: A preliminary report*. Stanford, CA: Stanford Institute for the Quantitative Study of Society (SIQSS), Stanford University, and InterSurvey Inc. Retrieved November 5, 2003, from http://www.stanford.edu/group/siqss/Press_Release/Preliminary_Report.pdf

Nie, N. H., Hillygus, D. S., & Erbring, L. (2002). Internet use, interpersonal relations and sociability: A time diary study. In B. Wellman & C. Haythornthwaite (Eds.), *The Internet in everyday life* (p. 215–243). Oxford, UK: Blackwell.

Oates, B. J. (2003). Foot and mouth disease: Informing the community? *Informing Science Journal, 6*, 103–114.

Orlikowski, W. J. (2002). Knowing in practice: Enacting a collective capability in distributed organizing. *Organization Science, 13*(3), 249–273.

Orlikowski, W. J. & Yates, J. (1994). Genre repertoire: The structuring of communicative practices in organizations. *Administrative Science Quarterly, 39*, 541–574.

Park, H. W. (2003). What is hyperlink network analysis?: New method for the study of social structure on the Web. *Connections, 25*(1), 49–61.

Powell, W. W., & DiMaggio, P. J. (Eds.). (1991). *The new institutionalism in organizational analysis*. Chicago: University of Chicago Press.

Quan-Haase, A., & Wellman, B. (2002). Capitalizing on the Net: Social contact, civic engagement and sense of community. In B. Wellman & C. Haythornthwaite (Eds.), *The Internet in everyday life* (pp. 291–324). Oxford, UK: Blackwell.

Reid, E. (1995). Virtual worlds: Culture and imagination. In S. G. Jones (Ed.), *CyberSociety: Computer-mediated communication and community* (pp. 164–183). Thousand Oaks, CA: Sage.

Rheingold, H. (2000). *The virtual community: Homesteading on the electronic frontier* (2nd ed.). Cambridge, MA: MIT Press.

Rheingold, H. (2003). *Smart mobs: The next social revolution*. Cambridge, MA: Perseus.

Rice, R. E., & Gattiker, U. E. (2001). New media and structuring. In F. M. Jablin & L. L. Putnam (Eds.), *The new handbook of organizational communication* (pp. 544–581). Thousand Oaks, CA: Sage.

Rogers, E. M. (1995). *Diffusion of innovations* (4th ed.). New York: The Free Press.

Ronfeldt, D., & Arquilla, J. (2001). Networks, netwars, and the fight for the future. *First Monday, 6*(10). Retrieved November 5, 2003, from http://www.firstmonday.org/issues/issue6_10/ronfeldt/index.html

Ronfeldt, D., Arquilla, J., & Fuller, G. E. (1998). *The Zapatista social netwar in Mexico*. Santa Monica, CA: Rand. Retrieved November 23, 2003, from http://www.rand.org/publications/MR/MR994

Ruhleder, K. (2000). The virtual ethnographer: Fieldwork in distributed electronic environments. *Field Methods, 12*(1), 3–17.

Salaff, J. (2002).Where home is the office: The new form of flexible work. In B. Wellman & C. Haythornthwaite (Eds.), *The Internet in everyday life* (pp. 464–495). Oxford, UK: Blackwell.

Sawyer, S. & Eschenfelder, K. R. (2002). Social informatics: Perspectives, examples, and trends. *Annual Review of Information Science and Technology, 36*, 427–465.

Schement, J. R. (1998). Thorough Americans: Minorities and the new media. In A. Gramer (Ed.), *Investing in diversity: Advancing opportunities for minorities and the media* (pp. 87–124), Washington, DC: Aspen Institute.

Schuler, D. (1996). *New community networks: Wired for change*. Reading, MA: Addison-Wesley.

Shade, L. R. (2002). *Gender & community in the social construction of the Internet*. New York: Peter Lang.

Shapiro, C., & Varian, H. R. (1999). *Information rules*. Boston, MA: Harvard Business School Press.

Smith, M. (2002). Tools for navigating large social cyberspaces. *Communications of the ACM, 45*(4), 51–55.

Smith, M. A. (1999). Invisible crowds in cyberspace: Mapping the social structure of the Usenet. In M. A. Smith & P. Kollock, (Eds.), *Communities in cyberspace* (pp. 195–219). London: Routledge.

Sproull, L., & Kiesler, S. (1991). *Connections: New ways of working in the networked organization*. Cambridge, MA: MIT Press.

Star, S. L. (1999). The ethnography of infrastructure. *American Behavioral Scientist, 43*(3), 377–391.

Star, S. L., Bowker, G. C., & Neumann, L. J. (2003). Transparency beyond the individual level of scale: Convergence between information artifacts and communities of practice.

In A. Bishop, N. Van House, & B. Buttenfield (Eds.), *Digital library use: Social practice in design and evaluation* (pp. 241–270). Cambridge, MA: MIT Press.

Strauss, A. L. (1978). A social world perspective. *Studies in Symbolic Interactions, 1,* 119–128.

Suchman, L., Blomberg , J., Orr, J., & Trigg, R. (1999). Reconstructing technologies as social practice. *American Behavioral Scientist, 43*(3), 392–408.

Sudweeks, F., McLaughlin, M. L., & Rafaeli, S. (Eds.). (1998). *Network and netplay.* Cambridge, MA: MIT Press.

Sudweeks, F., & Rafaeli, S. (1996). How do you get a hundred strangers to agree?: Computer mediated communication and collaboration. In T. M. Harrison & T. Stephen (Eds.), *Computer networking and scholarly communication in the twenty-first-century university* (pp. 115–136). Albany: State University of New York Press.

Teigland, R. (2000). Communities of practice at an Internet firm: Netovation vs. on-time performance. In E. L. Lesser, M. A. Fontaine, & J. A. Slusher (Eds.). *Knowledge and communities* (pp. 151–178). Boston, MA: Butterworth Heinemann.

Turkle, S. (1995). *Life on the screen: Identity in the age of the Internet.* New York: Simon & Schuster.

UCLA Center for Communication Policy. (2003). *The UCLA Internet report: Surveying the digital future, year three.* Retrieved November 23, 2003, from http://www.ccp.ucla.edu/pdf/UCLA-Internet- Report-Year-Three.pdf

Unsworth, J. (1996). Living inside the (operating) system: Community in virtual reality. In T. M. Harrison & T. Stephen (Eds.), *Computer networking and scholarly communication in the twenty-first-century university* (pp. 137–150). Albany: State University of New York Press.

Walberg, H. J., & Tsai, S. L. (1983). Matthew effects in education. *American Educational Research Journal, 20,* 359–373.

Wall, M. (2002, November 10). Online lifeline for farmers. *The Sunday Times,* pp. 51–52.

Warren, M. F. (2000). *E -farming or e-folly? Adoption of Internet technology by farmers in England.* Retrieved November 5, 2003, from http://www.sh.plym.ac.uk/ResearchPapers/Efarming.pdf

Warschauer, M. (2000). Language, identity, and the Internet. In B. E. Kolko, L. Nakamura, & G. B. Rodman (Eds.), *Race in cyberspace* (pp. 151–170). New York: Routledge.

Warschauer, M. (2003). *Technology and social inclusion.* Cambridge, MA: MIT Press.

Wellman, B. (2001). Physical place and cyber place: The rise of personal networking. *International Journal of Urban and Regional Planning, 25*(2), 227–252.

Wellman, B., & Hampton, K. (1999). Living networked on and off line. *Contemporary Sociology, 28*(6), 648–654.

Wellman, B., & Haythornthwaite, C. (Eds.). (2002). *The Internet in everyday life.* Oxford, UK: Blackwell.

Wellman, B., Salaff, J., Dimitrova, D., Garton, L., Gulia, M., & Haythornthwaite, C. (1996). Computer networks as social networks: Collaborative work, telework, and virtual community. *Annual Review of Sociology, 22,* 213–238.

World Bank. (2001). *Nongovernmental organizations.* Retrieved April 30, 2001, from www.worldbank.org/ngo

CHAPTER 9

Children, Teenagers, and the Web

Andrew Large
McGill University

Introduction

The rapid penetration of the Web into schools, libraries, and homes since the mid-1990s has opened many opportunities for children and teenagers to enrich their educational, leisure, and social activities. It has also raised issues concerning young people's access to the Web; its educational efficacy; and its potential to foster social isolation, undermine moral well-being, and threaten personal safety. These developments have encouraged research in a number of disciplines to answer such questions as who is using the Web, for which purposes is it being used, what educational role can it play, how can it most effectively be exploited as an information resource, and how can its potentially negative aspects be countered. To date, however, little has been done to synthesize such research, although a good start has been made by Abbas (2003) in her article published in the *Encyclopedia of Library and Information Science*. Such a synthesis is complicated by the very broad scope of the topic, the multidisciplinary nature of the research literature, and the uneven international distribution of research data.

This chapter is organized into seven parts. The first part establishes the scope of the chapter, indicating exactly what aspects from this very broad research area are included and excluded. Part two considers large-scale national surveys of access to and use of the Web by children and teenagers. The third part deals with research on the information-seeking behavior of this user community. Part four looks at design criteria for Web sites intended to be used by children and teenagers. Part five focuses on the three main Web applications: education, leisure, and social interaction. The sixth part looks at issues relating to Web content, especially pornographic material, and the personal safety of young Web users in a networked and uncontrolled environment. The final part considers future research agendas in this rapidly developing area.

Scope

In order to make a review of the literature relating to young people and the Web manageable within the limitations of a single *ARIST* chapter, certain constraints have been imposed. First, we must consider the Web itself. In this review, attention is focused on that part of the Web that is freely accessible via general-purpose search engines and subject directories; other areas of undoubted importance are omitted. Research studies of digital libraries accessible through the Web are not considered here, even though these libraries may be intended specifically for young people, as in the case, for example, of the International Children's Digital Library (Druin, Bederson, Weeks, Farber, Grosjean, Guha, et al., 2003) or the ARTEMIS Digital Library project (Abbas, Norris, & Soloway, 2002). Web-based reference services have also been excluded (a good overview can be found in Lankes, 2003). A considerable research literature can also be found on educational software products that are delivered via the Web (see, for example, Luchini, Quintana, & Soloway, 2002), but it is not considered at all in this chapter. Studies targeted specifically at communication tools such as e-mail and instant messaging generally have not been included unless it was difficult to isolate them within a more general presentation. An example is provided by the Pew Internet and American Life Project (Lenhart, Rainie, & Lewis, 2001); its survey results do not always distinguish between different Internet-based activities, but it would be wrong to omit it from any discussion of Web usage by American teenagers, as it is a large-scale national study undertaken by a respected agency.

Any consideration of children in relation to the Web may well include other actors: parents, teachers, librarians, and so on. Literature that focuses squarely on adults, and where children occupy a secondary role, has not been included. An example would be literature that discusses teachers' proficiencies and needs in relation to Web skills for the classroom (such as a survey of "Teacher Use of the Computer and the Internet in Public Schools" undertaken by the National Center for Education Statistics [2000], or a study of primary school teachers' use of computers in the classroom as reported by Drenoyianni and Selwood [1998]).

Children's and teens' use of the Web is of interest to researchers and practitioners in a growing number of disciplines. The literature covered here is drawn mainly from two areas: library and information science, and education. To a smaller extent and more selectively, however, literature from other disciplines—such as sociology, psychology, and law— has also been included where considered especially relevant to readers of *ARIST*. In terms of national origin, much of the published research on children, teens, and the Web emanates from the U.S. and the review reflects this; material from Canada and the U.K., and to a lesser extent a number of other countries, however, is included when identified. All the research discussed here has been published in English.

The literature reviewed in this chapter is largely, though not entirely, research based, although this orientation represents a minority of publications dealing with use of the Web by the young. The majority comprises short articles in the professional journals of librarianship and education that describe how the Web is used in individual libraries or schools, review current legal and political developments relating to content and access, evaluate individual Web portals and filtering software, and so on. These articles contain much of value and their omission from this review should not be interpreted as dismissive. But their sheer quantity precludes their listing, let alone discussion. Furthermore, many are intended to play a current awareness rather than an archival role; only the reader interested in an exhaustive study of the professional literature would need retrospective recourse to them.

The research listed in this chapter covers children in the early grades of elementary school through the later grades of high school, approximately from the ages of five to 18 years. Age cannot be directly equated with school grade (some students are advanced to grades normally reserved for their elders and some are held back in lower grades) and the authors cited here use both categorizations (a majority of the studies report by grade level, as they have been carried out in schools and have drawn their students from particular classes in particular grades). When does a child become a teen? In practice most of the small-scale research studies reported in the chapter deal with students from one school grade (grades five through seven have proven most popular). Surveys that have dealt with large samples of young people do not exhibit any consistent way of categorizing the young. For example, the Pew surveys that have investigated teenagers (Lenhart, Rainie, & Lewis, 2001; Levin & Arafeh, 2002) defined them as being between 12 and 17 years of age; the UCLA Center for Communication Policy (2003) survey defined teens as between the ages of 12 and 18; and a Canadian survey for the Media Awareness Network (Environics Research Group, 2001) chose ages nine to 17. School administrations in different regions and countries compound the problem by choosing different age divisions between elementary and high school as well as in some cases opting for middle schools between these two. In the cited studies, the ages of children are expressed in the terms employed by their authors, whether this is by grade or age. The term "elementary" rather than "primary" has been employed, although the latter commonly is used, for example, in the U.K.

The educational sector in particular tends to use "Internet" not only as a generic term to include e-mail and chat facilities as well as the Web, but also as a synonym for the latter term. This is reflected not only in the titles of many of the works cited here but also in their texts; to that extent, both terms will be encountered in this chapter.

National User Surveys

Survey Objectives and Methodologies

Large-scale national surveys attempt to provide an overview of Web use (and nonuse) by children and teens, and as such are a valuable resource. The U.S. is relatively well served by such surveys, but for most other countries national surveys of Web use by children and teenagers are sketchy or nonexistent. Relying mainly upon questionnaire surveys, telephone interviews, and discussion groups, such surveys sample relatively large numbers of children (and often parents) to determine which kinds of young people have access to the Web, how they make use of it, and what they think about it. Unfortunately, these surveys pose problems of analysis and interpretation, especially when attempting to compare the findings of one with another: The survey objectives, user populations, sampling techniques, data collection methods, definitions of Internet use, and time periods covered are likely to differ.

For example, Becker (2000) analyzes U.S. national survey data describing computer access from school and home by students of all school ages and the varying conditions that affect such access. Most of his data for schools are taken from a national survey of more than 4,000 teachers, whereas his data for home use relies upon the U.S. Census Bureau's Current Population Survey of U.S. Households in 1997 and 1998. In contrast, the Pew Internet and American Life Project, in its national surveys in 2000 and again in 2002, focused upon teens aged 12 to 17, although the methods adopted were somewhat different in each case. In the earlier study Lenhart, Rainie, and Lewis (2001) undertook a telephone survey of 754 teenagers—all of whom were established Internet users—and their parents to explore "teenage life online." These data were supplemented by a bulletin board discussion with 21 teens and preliminary e-mail interviews with 16 more. The second Pew Project (Levin & Arafeh, 2002), which was concerned with the "digital disconnect" between Web-savvy teens and their schools, was conducted between November 2001 and July 2002 and used 14 focus groups involving 136 middle school and high school students from 36 schools, supplemented by essays about the educational use of the Internet submitted by nearly 200 additional students. The National School Boards Foundation (2002) involved yet another age group of young people. It conducted a national survey of children and parents to learn about children's Internet use. It surveyed 1,735 U.S. households nationwide, conducting telephone interviews with parents of children aged two to 17 and also with their children aged nine to 17 as a "reality check" to compare with parents' comments. The U.S. National Center for Education's statistics (Infoplease.com, 2000) unsurprisingly focus upon Internet use from schools.

The UCLA Center for Communication Policy (2003) conducted Internet use surveys in the U.S. in 2000, 2001, and 2002, contacting a sample of 2,000 households across the U.S. Although these surveys provide a basis for assessing changes in usage over time, it is difficult to compare their results with those from the Pew studies. The UCLA Center polls Internet nonusers as well as users and relies upon information from parents about their offspring's use of the Internet rather than direct reporting from the young people themselves. Some skepticism concerning this approach is warranted, because parents may not always be a reliable source of information about their children's Internet habits.

A report from the Kaiser Family Foundation (Roberts, Foehr, Rideout, & Brodie, 1999) takes a broader sweep than these studies. It attempts to assess the impact on U.S. children of media in general, not just the Internet. It points out that few nationwide studies of children and media use "are truly representative of U.S. children" (Roberts et al., 1999, online). This study relied upon data from 3,155 children aged two to 18, together with week-long media use diaries kept by 621 children, questionnaires for the older children, and interviews with the parents of the younger children (aged two to seven). In addition to a wealth of survey data, the report provides a useful history of media in relation to children.

Moving outside the U.S., a few user surveys involving young people have been undertaken in Canada. In March 2000 a nationwide sample of 5,682 Canadian students between the ages of nine and 17 was surveyed for the Media Awareness Network in February and March 2001. This survey investigated patterns of Internet use in Canadian families and parental attitudes and perceptions about the nature, safety, and value of children's online activities (Environics Research Group, 2001; Taylor, 2001). Clark (2001), in contrast, relied upon nationwide interviews with parents of children aged five to 18, carried out as part of the Canadian 2000 General Social Survey, to determine Internet use of children and teens aged 5 to 18 living with their parents; here again, reliance upon parents rather than their children for information on the behavior of the latter should be noted.

Turning to England, a review by the U.K. Office for Standards in Education (1998) of English primary schools between 1994 and 1998 included the use of information technology in teaching in general. Holloway and Valentine (2003) conducted a two-year study of English children aged 11 to 16 in just three high schools. Although not a national survey, their work is interesting for its depth of analysis. Questionnaires were completed by 753 students, observations were undertaken by the researchers in a number of classes, the students were asked to keep diaries, and teachers were interviewed. In addition, ten children from each school and their families, plus another ten children in families where the former were deemed to be "high-end" users were interviewed in depth. The researchers, from a university-based department of geography, were interested in inequalities of access, gender effects, and dangers to children from Internet access. This book-length study focuses upon what happens as children have increasing access to information technology.

Under the aegis of the World Internet Project (http://www.worldinternetproject.net), data on children are included in its surveys conducted in Sweden (Findahl, 2001) and Hungary (Istri-Tarki, 2002). Subrahmanyam, Kraut, Greenfield, and Gross (2000) are critical of much of the research they have reviewed. As they comment, most time-use data have been gathered through self-reports or, in the case of children, self-reports and reports by parents, usually in telephone surveys. Despite their overall usefulness for sampling large numbers of people, self-report survey data encounter problems of accuracy and reliability stemming from memory limitations and inaccurate estimations by respondents, especially children. Subrahmanyam et al. recommend more reliable methods of data collection, such as computer log analysis; they say that such methods have not been widely used because they are more expensive and time-consuming to carry out and also raise concerns regarding privacy.

For the U.S., at any rate, despite the pitfalls in making comparisons within and between the studies, a picture is emerging of access to and use of the Internet. Internet use is "the norm" for U.S. youth according to the Pew Project (Levin & Arafeh, 2002), involving 78 percent of those aged between 12 and 17. They estimate 30 to 40 percent of teens are "Internet savvy," technologically literate and relying heavily upon the Internet for both their school and social lives. The UCLA Center for Communication Policy (2003) puts the figure even higher. It reports for 2002 that 97 percent of teens aged 12 to 18 are Internet users. Access to the Internet by Canadian teens is comparable, according to the Media Awareness Network (Environics Research Group, 2001).

Web Access from Schools, Libraries, and Homes

Where do young people gain access to the Internet? This question has been addressed by several of the national surveys. The Pew survey (Lenhart, Rainie, & Lewis, 2001) found in 2000 that most U.S. teens (87 percent) went online from multiple locations, including home, school, and library. UCLA's Center for Communication Policy (2003) survey reported that, of the children under 18 who went online, 84 percent did so from home, 73 percent from school, and 60 percent from elsewhere (unspecified but presumably including libraries). In Canada, a nationwide survey conducted in March 2000 on behalf of the federal government found that children (aged six to 16) accessed the Internet primarily from home, followed by school, a friend's home, and the public library (Environics Research Group, 2000a). Becker (2000) extracted from the U.S. Census Bureau's Current Population Survey for 1997 and 1998 that some children were much more likely than others to have home access to the Internet. The key predicators were parental income, education, ethnicity, and work-based computer use.

It is not surprising, given the investment in school-based information technologies as well as the potential of the Web to support teaching and learning, that national surveys have focused much attention on the school environment. The U.S. National Center for Education reported that 97 percent of elementary and 100 percent of secondary schools had Internet access by 2000 (Infoplease.com, 2000). These figures may not be as encouraging as they first appear, however, because they do not reveal the number of individual Internet access points available within a school. A large difference exists between one access point per school and multiple access points per classroom. Nor can it be assumed that the information and communication technologies needed to deliver the Web into schools and classrooms always are functioning. Soloway, Norris, Blumenfeld, Fishman, Drajcik, and Marx (2000) noted that in the Detroit public schools, for example, on any given day the probability was less than 50 percent that students would be able to access the Internet from their computers. The Center for Children and Technology at the Benton Foundation and Education Development Center (Dickard, 2003) reported that 99 percent of U.S. schools were Internet-connected by 2001, but nevertheless concluded that many schools were not using the Internet to maximum advantage. Schools attended by students from low-income families were less likely to have Internet access than those attended by students from average income or, more especially, from high income families. Becker (2000) believes that the divide between the socio-economically advantaged and disadvantaged children in the U.S. is quite large and may be getting larger. A report from the Corporation for Public Broadcasting (2003), based on a nationwide study of American households undertaken in 2002, focused upon children from under-served populations, concluding that they still lagged significantly behind more advantaged children both in school and home access to the Internet, even though their situation had improved. Wilhelm, Carmen, and Reynolds (2002) commented upon the "formidable gap" in computer use and Internet access between the "haves" and the "have nots." Generally, they found, children who were already disadvantaged in other respects were least likely to have access to new technology— minority children, those in poor families, and those in high poverty neighborhoods—and schools closed only some of this gap.

Schofield and Davidson (2002) focused more narrowly on 29 elementary, middle, and high schools in one U.S. urban district. Although somewhat dated (the study terminated in 1998), it is valuable for its detail and also unusual in providing longitudinal data over five years (1993–1998). The project was designed to bring the Internet to schools and classrooms in the district, to offer technical and collegial support to teachers, and to investigate the results of these efforts. Data were gathered by observation and interviews. The authors could say at the time (and this remains largely true now) that "there has been little systematic research about the academic consequences of Internet use for precollegiate students" (Schofield & Davidson, 2002, p. 11).

Surprisingly little research has been published about young people's use of the Web in public libraries. Dresang, Gross, and Holt (2003) comment on the irony that so much importance is given to networked digital resources for children, and yet it was not until 2001 or 2002 that the first efforts were made to collect and analyze data to evaluate children's use of them. Dresang et al. have developed an outcomes-based research evaluation model to guide a study of children aged nine to 13 using resources and services in the Saint Louis Public Library. They raised four critical questions: To what extent are the resources invested in technology in public libraries reducing the digital divide for children? What level of knowledge, skills, behaviors, and attitudes do children have with networked technology, as evidenced by public library use? What effect does children's use of technology have on library policies and service responses? What is the impact of children's use of technology in public libraries over time? These questions await clear answers. Turner and Kendall (2000) conducted a survey to identify the use being made of the recently established Internet facilities in an English public library. Unfortunately, only three percent of the 178 questionnaire respondents were under the age of 16 (41 percent were between the ages of 16 and 25) and among the nine people interviewed only one was under 16 years old.

The Internet is almost universally available in U.S. schools, but its availability in individual classrooms is patchy. Access from homes also has increased significantly, although falling short of universal availability. Despite such achievements, a divide remains in schools and homes in terms of Internet access, especially high-speed access, and this divide is defined by race and family wealth. The need for repeated surveys in which standardized methodologies are applied is important not only because accessibility is changing from year to year but also because we require longer-term studies in which data can be meaningfully compared over time. Outside the U.S. the situation is different; large-scale studies, whether conducted by government agencies or private foundations, appear to be rare in most countries, making international comparisons difficult.

Information-Seeking Behavior

Information-Seeking Studies

The Web did not emerge in a technological vacuum. Much of the research to explore the information-seeking behavior of young people in non-Web digital environments such as dial-up online services, CD-ROMs, non-Web-based hypertext systems, and stand-alone online catalogs certainly are relevant to understanding Web use. Authors whose work is highly pertinent include Borgman, Hirsh, and Walter (1995); Hirsh (1997); Large, Beheshti, Breuleux, and Renaud (1994); Large, Beheshti, and Breuleux (1998); Lawless, Mills, and Brown (2002); Marchionini (1989); Neuman (1995); Perzylo and Oliver (1992); Revelle, Druin, Platner, Bederson, Hourcade, and Sherman (2002); and Solomon (1993). Nevertheless, space precludes their discussion here. Almost no work has been undertaken to compare young people's information seeking on the Web with any of these other technologies, although the sixth-grade students in Large and Beheshti's (2000) study did comment briefly on CD-ROMs as an alternative to the Web as a source of information for school projects. In contrast, Hirsh (1999) included print and CD-ROM as well as the Web in her investigation of information evaluation by elementary school students, and Shenton and Dixon (2003) compared students' use of CD-ROM and the Web as information resources. Shenton and Dixon found that the search strategies adopted were similar for both technologies.

The research by Luckin, Rimmer, and Lloyd (2001) is worth mentioning here, although its focus was on how well children understood networked computer technology, and specifically the Internet, rather than on information-seeking behavior. They undertook two studies with a class of nine- and ten-year-olds to learn how they conceptualized the Internet and how these images changed when they used it. Participants drew pictures (a technique used in different circumstances by Druin, Large, and Bilal), completed a questionnaire, and talked in semi-structured interviews about the Internet. In general, their conceptions remained remarkably stable across the two studies and they did not perceive the Internet in terms of connections even when they talked about or drew features that meant a connection must exist. The researchers concluded that if children did not understand the basic features of networked computing they were likely to be less effective users of the Internet. More research is needed to explore the mental models children of different ages have formed of the Internet in general and the Web in particular and the extent to which the accuracy of these models influences their success in using the Web to find information.

A relatively active research agenda has been pursued, especially in the last few years, in relation to the information-seeking behavior of young Web users. An observer could report with justice in 1999 that "few research studies have examined youth and their information seeking on the Internet" (Dresang, 1999, p. 1123). This would no longer be a fair comment; Hsieh-Yee (2001), for example, has provided a useful overview of early Web-based information-seeking studies, including research with children as well as adults. Abbas (2003) summarized several studies of children's information-seeking behavior on the Web as well as other digital environments such as online catalogs and CD-ROMs. Kuhlthau's (1991) elaboration of the information-seeking process, while neither confined to children nor targeted at the Web, nevertheless will prove useful preliminary reading for any consideration of this topic.

The studies themselves focus mainly on users in a particular age group, typically defined in terms of school grades (an exception is the study by Shenton and Dixon [2003], previously mentioned, that involved students from 14 grades distributed across elementary and high schools). We shall first consider elementary school students, followed by middle school and then high school students.

Little attention has been directed at information seeking by very young users. A notable exception is the research of Kafai and Bates (1997) with elementary school students in grades one through six. They found that all the children were able to use Web sites to advantage in their learning and could scroll and use hypertext links. Only the older children, however, could effectively use search engines and the rudiments of Boolean logic. Although the content of Web sites overall was not child-friendly (too much text and too difficult vocabulary) the children in general were enthusiastic about it. In the absence of other research studies, the article by Burgstahler (1999), although based solely on her experience with her son in kindergarten, is of interest and offers advice on Internet use with very young children. She recommends adults choose activities that encourage collaboration with their child or children, discuss different tasks that an activity involves and help to determine specific roles for each member of group, listen to kids working together and intervene with questions and suggestions where necessary, and after online activity ask the child to describe the experience aloud and follow this up with questions.

Hirsh (1999) explored the searching behavior of ten randomly selected fifth graders using print material, CD-ROMs, and the Web. She was particularly interested in their ability to make relevance judgments on retrieved information. She concluded that children encountered difficulties in formulating and revising search queries and did not use advanced search features. Students had problems in evaluating the authority and accuracy of information, and their opinions were mixed about the Web as an information resource. Enochsson (2001) also studied how students made relevance judgments, although her students (aged nine to 11) were Swedish rather than American. She categorized them as credulous, unreflective, or reflective.

Schacter, Chung, and Dorr (1998) worked with 32 students from grades five and six. They were reactive information seekers rather than planners and overwhelmingly browsed rather than searched. Like Hirsh's students, they did not exploit the advanced search features available. They performed better on ill-defined than well-defined tasks. Large, Beheshti, and Moukdad (1999) collected data from a series of searches conducted over several weeks by 53 sixth graders working in groups of two or three on a class project. As in the previous studies, the students preferred browsing to searching and made little use of Boolean operators or other advanced search features. When interviewed after the searches were completed, the students expressed frustration at the difficulty in finding a few highly relevant pages and at determining relevance from the information displayed in their results sets (Large & Beheshti, 2000). Bowler, Large, and Rejskind (2001) looked at the captured search data and interview transcripts from just one of these student groups. They reported in some detail on the problems encountered by one girl and two boys in searching, interpreting, and using the retrieved information. They concluded that the search engines used were unsuitable for children and the Web pages for the most part were unsuitable in content and vocabulary. Wallace, Kupperman, and Krajcik (2000), in their study of sixth graders, found that information seeking seemed to be an unfamiliar activity, and the students were unable to improve their strategies in the light of experience.

Bilal has published several studies based on her work with students in middle school (grade seven). She gave the students a series of experimental tasks to complete in order to explore their behavior when looking for information using a Web portal designed specifically for children (in contrast to the studies previously cited, where the children used adult portals). She argues that portals must build on children's cognitive and physical behavior (Bilal, 2000), recommends various ways in which one such portal—Yahooligans!—might be improved (Bilal, 2001), and believes that children need training in order to identify an information need (Bilal, 2002b).

Students in the upper grades of high school, like those in the lower grades of elementary school, have not yet received much attention from researchers. Fidel, Davies, Douglass, Holder, Hopkins, Kushner, et al. (1999) used observation and interviews to investigate the Web searching behavior of eight students in grades 11 and 12. They concluded that, despite the students appearing to be satisfied with their results, they needed formal training in Web searching as well as support from the portal in order to make more effective use of the Web in their class assignments. Agosto (2002a) interviewed 32 students from grades nine and ten about a Web surfing session. The students found the Web overwhelming and considered many sites to be of low quality. Agosto thinks that a better understanding of young people's decision-making processes will help intermediaries assist them more effectively and help designers build better sites with increased youth appeal.

These studies, by focusing on users in narrowly defined age bands, have tacitly considered age to be a likely variable in differentiating the information-seeking behaviors of the young. Both cognitive and motor skills are age-related. Piaget's (Piaget & Inhelder, 1969) structuralist views hold that cognition develops in a series of discrete stages (approximately identified as birth to two years, two to seven years, seven to 11 years, and 11 to 16 years). Although Piaget is cited in some studies, his proposed stages are not considered adequate to capture what actually occurs in child development (Bjorklund, 2000). In fact, many similarities are observed in the information-seeking behavior of students at least from the higher grades in primary school through to high school.

Gender has also been identified by a number of researchers as a factor influencing use of information technology (Mumtaz, 2001; Orleans & Laney, 2000; Siann, Durdell, Macleod, & Glissov, 1988) or media in general (Roberts et al., 1999). Several studies suggest that gender can affect user reactions. A study of high school students in Taiwan found that males had more positive Internet attitudes than females, but both shared similar perceptions about its usefulness (Tsai, Lin, & Tsai, 2001). An Israeli study of students in a Tel Aviv high school reported that boys (56 percent) used the Internet more than girls (38 percent) and spent more time online (Nachmias, Mioduser, & Shemla, 2000). Large, Beheshti, and Rahman (2002b) were interested specifically in gender effects on searching behavior: To what extent do sixth grade boys and girls differ when working collaboratively in same-sex groups to find information on the Web? The boys' groups used significantly fewer words in a query, spent less time viewing retrieved pages, clicked on more hypertext links, and tended to perform more page jumps per minute than the girls. In other respects, however, significant differences were not found. Schacter, Chung, and Dorr (1998) found that boys browsed significantly more than girls. Other researchers, however, either have not chosen to examine gender effects or have found none worth reporting. For example, Miller, Schweingruber, and Brandenburg (2001) were interested in gender effects among middle school student users of the Internet, but were unable to identify any significant gender differences.

An interesting research question is the extent to which the information-seeking behavior of young users differs from that of adults. Surprisingly, Bilal and Kirby (2002) are the only authors explicitly to examine the similarities and differences between children (seventh graders in their case) and adults (graduate students in information science) as they looked for information on the Web using a children's portal (all were novices). The latter were more successful than the former in finding answers to a factual question, although the researchers identified many similarities in the behavior of these two very different groups. It would be instructive to see more information-seeking studies that compare children with adults, particularly if the adults were not selected from university undergraduates (typically the case in studies of adult searching), and especially not from those in library and information science programs, but rather were drawn from the wider adult population. Slone (2002) presents a study involving 31 participants of varying ages: 11 were children (as young as seven) or teens, and the remainder were adults. The article unfortunately does not sharply identify the responses from the children, teens, and adults respectively, although it reports that motivation and experience influenced specific information-seeking behavior. It therefore does not shed light on the differences between adult and child searchers. Furthermore, Slone observed and interviewed the users of a Web-based online catalog in a U.S. public library rather than users of the Web at large.

In discussing prior studies on the information-seeking behavior of children and teenagers, Fidel et al. (1999, p. 24) comment that summarizing the findings "is not an easy task because each study examined users of a certain age, ability, and socioeconomic level. These factors are likely to affect searching behavior and therefore prevent a comparison among the studies." Several others could be added to this list. Some studies have been conducted in operational settings related to actual class assignments (for example, Large, Beheshti, & Moukdad, 1999; Wallace, Kupperman, & Krajcik, 2000), whereas others opted for an experimental environment where the children were presented with a task by the researchers (these include Bilal's studies [2000, 2001]). Large and Beheshti (1999) discuss various methodological approaches and present the relative advantages and disadvantages of experimental and operational studies. In both cases, the children have normally been assigned the search task by an adult. However, Bilal (2002b) allowed the children themselves to generate their own search tasks. The children were more motivated, challenged, and engaged in completing their tasks when they selected topics that interested them than they were when topics were assigned to them. Bilal emphasized, however, that they were still operating in an experimental mode where a researcher asked them to produce a search topic out of context; as she says, the children may have chosen topics on which they did not really need information. She further (Bilal, 2002c) discussed the differences between children's approaches to assigned and self-generated tasks; and although the children preferred the latter, they did not find them easier.

What can we conclude from these studies about young people's information-seeking behavior on the Web? Young users encounter problems in selecting appropriate search terms and orienting themselves when browsing. They have a tendency to move from page to page, spending little time reading or digesting information, and have difficulty making relevance judgments about retrieved pages. Information seeking does not appear to be intuitive, and practice alone does not make perfect. All too often children fail to satisfy their information needs from the Web, even though they—and in many cases also their teachers—are optimistic about their success. They find it hard to express their information needs in the kind of query formulations required by Web-based search engines and encounter problems in revising unsuccessful strategies.

The research evidence suggests that in certain respects, at least, gender influences searching and browsing behavior on the Web; in particular, boys scan pages more rapidly and click on more hyperlinks than girls. In terms of search success, however, it remains unclear as to whether gender plays a significant role. The relationship between age and information-seeking behavior deserves more research. Although differences between young children and high school students can be deduced from various studies (for example, in the accuracy with which keywords are spelled), these differences do not appear to be especially strong. In comparisons between young people and adults, a major problem lies with the adult research literature; it has focused very heavily on university students, who may not be the most interesting group to compare with children and teens. It would be revealing to see how children compare with less well-educated adult populations whose language manipulation skills, for example, may be different from those in the university community. Haycock, Dobor, and Edwards (2003, p. 4), then, may be premature in stating that "by virtue of their developmental level, children demonstrate information needs and search behaviors quite different from those of adults."

Portal Design

Several of the information-seeking studies reported here commented specifically on the problems typically encountered by children and teenagers when using Web portals, whether these portals were specifically designed for that age group (as in the case, for example, of

Ask Jeeves for Kids, KidsClick, or Yahooligans!) or not. They suggest that searching and browsing could be improved by designing more appropriate portals. This provides empirical evidence to support pleas made by observers like Jacobson (1995) and Druin (1996) that children be treated as a separate user community when it comes to interface design. Such pleas were not immediately answered by the research community. Marchionini and Komlodi (1998) reviewed in *ARIST* the literature on Web-based interfaces and on interfaces for information seeking by children, but at that time there was little overlap between the two topics. Indeed, Najjar (1998), discussing educational multimedia user interface design in general, commented that guidelines were almost entirely based upon the opinions of experts rather than on the results of empirical research.

The first steps in improving portal design for young users were taken when researchers began to evaluate systematically Web portals that had been designed with children in mind. Broch (2000), for example, examined Yahooligans! and Ask Jeeves for Kids in terms of children's cognitive and mechanical skills. Stevenson (2001) discussed several educational portals and assessed them in the light of 11 categories that she considered critical for children. McDermott (2002) reviewed a variety of specialized subject portals relevant to students with homework assignments. Haycock, Dobor, and Edwards (2003, p. 17) provided detailed evaluations of the 20 "most highly recommended and popular" portals designed explicitly for children's use on the Web, as well as short annotations on 11 others. Kuntz (2000), the manager of a children's portal, identified five broad criteria that can be used to evaluate children's search tools: database size, accountability, categorization, search access methods, and other features (such as help, spell checking, and layout). Large, Beheshti, and Cole (2002) described a matrix containing 51 design characteristics that would link a Web portal's purpose and design objectives with its design architecture in such a way as to support personalization. In discussing how the matrix would work, they used the example of a children's Web portal to provide access to museum information.

A number of researchers have argued, however, that it is insufficient to design interfaces with children in mind: children themselves should be involved in the process (see, for example, Druin, 1999, 2002). According to Hanna, Risden, and Alexander (1997, p. 10), "as the body of literature on children's use of computer products grows, a necessary step is to flesh out the details of exactly how to include children in computer product design." An overview of the literature on the design theories underpinning this movement, as well as examples of their application in a Web context, can be found in Nesset and Large (2004).

Druin and her colleagues in human–computer interaction are pursuing intergenerational, interdisciplinary team design of interfaces, for example, in their work to develop a Web-based children's digital library (Druin, Bederson, Hourcade, Sherman, Revelle, Platner, et al., 2001; Druin et al., 2003). Druin is a strong advocate of intergenerational design as well as low-tech prototyping that includes encouraging children to express their ideas through drawings. Two library and information science research teams have followed Druin's procedures in the context of children's Web portal design. Large, Beheshti, and Rahman (2002a) initially organized several focus groups whose members were between 10 and 13 years of age. The focus groups evaluated four children's Web portals (Ask Jeeves for Kids, KidsClick, LycosZone, and Yahooligans!), saying what they liked and disliked, as well as any changes they would make to the portals in order to improve their effectiveness and attractiveness. Their comments were grouped under four main categories: goal diversification, visual design, information architecture, and personalization. The focus group members clearly indicated their interest in Web portals designed specifically for younger users, being attracted by their use of color, graphics, and animation. When children were asked to rate the four portals, a gender effect emerged: the girls preferred LycosZone and the boys Yahooligans!. Nielsen (2002) observed bigger differences between girls and boys than between men and women in matters such as Web site design (boys were put off by large amounts of text) and Web access behavior (girls spent more time using computers with a parent; boys spent more time alone). As a consequence, he recommended that Web usability studies should include equal numbers of boys and girls.

After working with focus groups, Large and Beheshti (2001) went on to discuss the wider involvement of children in the portal design process. They organized two intergenerational teams—the researchers with sixth- or third-grade students. Each team worked over a number of sessions to design a Web portal suitable for elementary school students, starting with simple drawings and working toward a low-tech prototype (Large, Beheshti, Nesset, & Bowler, 2003a, 2003b). The team plans to build a high-level working prototype of each portal to be evaluated by elementary school students. Bilal (2002a; 2003) is pursuing somewhat similar lines. She asked individual seventh-grade students to draw an initial Web portal interface and then to modify their designs, if they so chose, after looking at two existing children's Web portals. She, too, concluded that children can be effective design partners.

It is still too early to say where this line of research will lead, but the initial steps appear to have been productive and are likely to be followed by others. Nevertheless, the idea of involving children as designers (rather than simply as testers) of information technology applications remains controversial (Nesset & Large, 2004).

Web Site Design

There is no shortage of advice on how best to design software interfaces (for example, Shneiderman, 1998) nor on the more specific aspect of interface design considerations when constructing Web sites (see, for example, Nielsen, 2000). The general design principles elaborated in such works may well apply to all sites, including those intended for young users. But what do researchers have to say about the specific needs of children and teenagers?

One approach is to recommend particular Web sites that can be used by children, targeting a specific age range of users or a specific subject content or both. Only a few examples will be cited here because they tend to be based on the personal opinions of their authors rather than on research findings; moreover, such recommendations become quickly outdated as sites change or disappear and new ones replace them. Auxier (2001) provided brief reviews of Web sites she considered helpful for students doing homework. Minkel (2000), based on his work as a children's librarian in a public library, suggested sites suitable for young children. In general he considered CD-ROMs more effective than the Web for children under seven (because CD-ROMs rely heavily on point and click graphics, reducing the need to read text, while maintaining the fast response time that children demand). Harbeck and Sherman (1999) shared Minkel's doubts about the efficacy of Web sites for children aged seven and under. They did not subscribe to the "strong and pervasive faith" that younger and younger children benefit at home and school from "the marvels of electronic communication technologies" and were critical of the vast majority of Web sites they reviewed that had been designed for young children (Harbeck & Sherman, 1999, p. 39). They proposed seven principles that should be followed when designing Web sites for young children.

Nielsen (2002) noted that very little was known about how children actually used Web sites or how to design sites that would be easy for them to use. He observed that most Web site designs were "based on pure folklore about how kids supposedly behave, or at best are insights gleaned when designers observe their own children" (Nielsen, 2002, online). Nielsen only conducted usability studies in the U.S. and Israel (he gives no reason for this particular choice) with 55 children between the ages of six and 12. He found that children often had the greatest success when using Web sites intended for adults rather than children, as long as such sites had been simply designed and in compliance with Web design conventions. Many of the children's sites had not followed such guidelines, being so complex and convoluted as to confound the test users. Nielsen identified several general problem areas for children, and offered best practice suggestions.

Several research studies involving children and teenagers have considered questions of Web site design. Kafai and Bates (1997) found that Web sites for children in grades one through six were generally not child friendly. They incorporated too much text with difficult vocabulary. The author reported that children preferred Web sites with high visual content and short, simple texts, liked animation and interactivity, and had a low tolerance for download delays. Large and Beheshti (2000) agree on intolerance for slow downloads but conclude that children will exploit Web-based video and sound sequences (in contrast to text and still images) only if there is sufficient motivation behind their information seeking. Enochsson (2001) found that fourth graders' judgments of Web pages were based on criteria such as content, currency, layout, usability, and interactivity. Furthermore, fewer girls than boys reflected on the reliability of Web content or thought it illegal "to put lies on the Net" (Enochsson, 2001, p. 159). Reinforcing the importance of design, Montgomery (2000, p. 161), based on preliminary findings from a major study on Internet content and services for children, argued that "the goal of creating a quality digital media culture for children must be placed at the forefront of public debate."

In the European Union, a project with partners from six member countries explored Web-based library services for children aged nine to 12 and created a multimedia virtual simulation of a children's library on the Web. This Children's Library—Information—Animation—Skills (CHILIAS) project involved children in designing the virtual library and in generating content. Children liked graphics, images, and sound as well as text, combined in an interactive environment (Bolger, Bussmann, & Fieguth, 1997; Bussmann, 1999). Ormes (1997) discussed a Web site based on R. L. Stevenson's classic, *Treasure Island* (chosen because it is well known, has strong themes, and is out of copyright). The site was evaluated by children aged 9 and 10 from two elementary schools. The children liked using the site, but were critical of its confusing navigation and lack of sound and interactivity. The quiz was the most visited area and the text itself was not read by the children (both a summary of the plot and the entire book were available from the site). Another U.K.-based project looked at the attractions and faults of Web sites for children aged 10 and 11 (Williams, 1999; see also Benn, 1996).

In the U.S., the Center for Media Education, a national, nonprofit organization whose goal is to create quality electronic media culture for children and youth, has issued a number of reports. One provided an overview of popular teenage Web sites, identifying key features that characterize content and activities for teens (Center for Media Education, 2001). It drew upon publications, interviews, and participation in industry meetings and conferences as well as its own analysis of Web sites.

Other research into Web content for young users has focused on more specific themes. Hoffman, Kupperman, and Wallace (1997) emphasize a Web site's potential to support inquiry-based learning, where students over several weeks ask questions, design and implement investigations, participate in learning communities, use scientific tools, and create artifacts. In such a learning environment teachers can encounter problems in making adequate resources available. Hoffman et al. worked with 500 students in four middle schools and three high schools, developing materials through cycles of design, modification, testing, and analysis. Their report elaborates a number of general Web site design criteria that they believe will support learning. Agosto (2001), worked with female high school students to discover what aspects of Web sites are likely to attract or repel the greatest number of young women. She has also developed a model of the criteria used by young people to evaluate individual Web sites (Agosto, 2002b). Although she again worked only with girls in grades nine and ten, she believes that her findings are likely to apply to boys and to other age groups.

Arnone and Small (1999) developed WebMAC (Motivational Analysis Checklist) Junior to capture the perspective of very young children—in grades one through four. The children were first given an overview/demonstration of the specific site they were to evaluate, followed by 20 minutes in which to explore and interact with it. Their subsequent evaluations were intended to provide information for designers to improve their Web sites. Loh and Williams (2002) also were interested in the effect of motivation in Web site design. They developed a Motivation Analysis Rating Kit, based on WebMAC, to assess children's perceptions of Web design elements and features they considered "cool." Loh and Williams studied 72 children from two sixth grade classes in a Singapore elementary school. The researchers concluded that content was more important for children than presentation; the novelty color, sound, and animation may initially draw children to a Web site, but after the novelty effect faded, it was interesting content that motivated children to return to the site.

The study by Lazarus and Mora (2000) had a different focus: to identify groups of Americans underserved by Internet content, to describe what these groups wanted in the online world, to identify the barriers they faced, to analyze online content available for low income and underserved Americans, and to provide a road map for action. They organized discussion groups with more than 100 low-income Internet adults, children, and teenagers, and nearly 100 technology experts, as well as analyzing 1,000 Web sites. The children and teenagers wanted participation and self-expression, high-impact interactive games (mention of school or learning caused interest to drop), multimedia, and user-friendly tutorials on how to create, for example, animation or how to do programming.

As with the research on information seeking and on portal design, these studies are beginning to shed light on the kind of content and design that children and teens really want and will use, rather than on what adults think they want. A consensus is emerging on certain issues, such as the importance of age-suitable vocabulary, rapid response times, and the appropriate use of graphics and sound. Yet the studies remain too fragmented in approach as well as participating age groups to draw firm conclusions on Web site design. More work remains to be done.

Education, Leisure, and Social Interaction

Children and teens can use the Web for many purposes. These purposes can usefully be grouped into three broad categories: education, where the Web is a source of information to support school projects and assignments; leisure, where the Web is a place to find music, pictures, and text to support hobbies and provide entertainment, as well as every-day information to help young people live their lives more easily; and social interaction with friends and relatives (as well as, perhaps, strangers) via e-mail and chat rooms. Of course, in any one online ses-sion all three of these objectives may be pursued. Much research has focused upon educational uses, but it is clear from user studies that the Web, for young people, is much more than an educational tool.

Several of the large-scale national surveys of Internet use that were reviewed earlier (under "National User Surveys") concerned the kinds of Internet activities engaged in by children and teenagers. In Canada, the Media Awareness Network survey found that in 2001 the most popular Internet activity among students aged nine to 17 was playing and down-loading music, followed by e-mail, surfing for fun, playing and down-loading games, instant messaging, chat rooms, and, lastly, homework (Environics Research Group, 2001). When asked to name the major ben-efits, however, respondents placed easier access to information second, below communicating with people but above entertainment or enjoy-ment; nevertheless, educational benefits were placed at the end of the list. Parents surveyed (Environics Research Group, 2000a), however, placed much more weight on the educational advantages than enter-tainment benefits (61 percent compared with just 11 percent).

In the U.S., the Pew Internet and American Life Project surveyed teenagers who were already online and their parents (Lenhart, Rainie, & Lewis, 2001). They found strong agreement among both groups that the Internet helped at school: 91 percent of the online teens used it to research school assignments and 71 percent said they had used it for their last big school project (compared with 24 percent for library resources). Another U.S. survey also found that parents, more than their offspring, believed the Internet to be a powerful tool for learning and communication within the family (National School Boards Foundation, 2002). The analysis of U.S. Census data by Becker (2000) led him to con-clude that the most common reason for home computer access to the Internet was schoolwork, but if all recreational uses—e-mail, games, music, etc.—were added together they would account for more online time than homework. Infoplease.com (2000) provided a breakdown of Internet activities by teens aged 13 to 17 that placed e-mail first (86 per-cent) followed by research for school projects/homework (78 percent). But totaling the various leisure activities—visiting music sites; downloading music files; playing games; visiting sites on apparel and fashion; down-loading free software; and accessing information on sports, movies, con-certs, television programs, and so on—would far exceed homework. In 1999, Sandvig (2001) investigated Internet use by children under the age of 14 at a public library in an underprivileged area of San Francisco. An analysis of sites visited by the children revealed that game playing was most popular, followed by chat and e-mail services. Explicitly educational content often was avoided, and in general the Internet was more a medium for play and leisure.

Williams (1999), based on observations of Web use and also informal conversations with teachers and students in an elementary school in England, could find no pattern of use for leisure vs. education, although he does say that the latter seemed to be more dominant (this particular conclusion appears to be based on discussions with teachers rather than children). The leisure use included extensive game playing. Holloway and Valentine (2003) concluded from their British study that children viewed information and communication technologies as social and leisure tools rather than educational tools (here they are talking about computer use in general, however, and not simply the Web).

The Center for Media Education (2001) found that a majority of sites targeted at teens presents popular culture rather than "information." It also noted that teens were a key target for online marketing and that collection of personal information was pervasive on teen sites (in the U.S. only children under the age of 13 are protected by the Children's Online Privacy Protection Act [COPPA]). Watson (2001) says that the high school students she interviewed expressed greater self-confidence and competence in using the Internet for personal than school purposes. Miller, Schweingruber, and Brandenburg (2001) reported their middle school respondents used the Web from home mostly to play games, although the researchers said that Web sites appeared to be an effective delivery mechanism for educational content. Of the tenth graders who completed questionnaires for Vansickle (2002), 73 percent said that they used the Web for personal purposes from several times per week to several times per day, whereas only 56 percent reported similar use for academic purposes, and 44 percent said they rarely or never used it for school-related assignments.

The Web and Education

The Web is now available in almost every school throughout North America and also widely available in many other countries. For example, a Canadian newspaper reported that in the U.K. by 1999–2000, at least 93 percent of secondary schools and 62 percent of primary schools had Internet connections (Wood, 2000). In Canada, SchoolNet, an Industry Canada project, claimed in 1999 to have made the country the first in the world to connect all its public schools to the Internet, and the goal for 2001 was 250,000 connected computers, the equivalent of one connected computer per classroom (Foss, 2000). This phenomenon is not confined to the English-speaking world. By 2000 Laanpere (2000) could report that all Estonian secondary schools and most primary schools had Internet access. In France, by the academic year 2000–2001, all secondary schools were connected to the Internet (Embassy of France in the United States, 2001). Large investments have been made in providing the necessary facilities to support such access in the face of competing educational demands. Many educators are enthusiastic about the Web's potential to promote learning by exposing students to more resources than could any school or public library (see, for example, Soloway et al., 2000). What do researchers have to say about the payoff from such an investment? Students may use the Web in support of class assignments and homework, but how effective is it as an educational and learning tool?

Wallace (1999, online) noted that "learning is a process of constructing understanding by grappling with problems, seeking solutions, and organizing knowledge for oneself;" technology alone cannot perform miracles. Mistler-Jackson and Songer (2000) observed that the prevalence of technology in classrooms by no means implied that it was either well used or well understood. Windschitl (1998) argued for a stronger research focus on the aspects of learning and teaching in K–12 classrooms that are being influenced by the Web. As he says, the Web is a conduit to other people's information and its advantages are matters of efficiency and scope rather than offering ways fundamentally to alter relationships among learners, teachers, and the curriculum. For him, the critical question is "in what contexts is the Web useful as an inquiry tool, and how are students learning in these contexts?" (Windschitl, 1998, p. 29). He concluded that researchers had a responsibility to investigate and document the changes occurring as a result of Internet-based teaching and learning: "We will not find answers unless we start asking more important questions and pursuing them in a systematic fashion" (Windschitl, 1998, p. 32).

Researchers have only partially responded to such a call. A body of research findings is accumulating on the role of the Web in facilitating learning, but there is still a long road to travel. How, if at all, can the Web become a powerful educational force rather than merely a source of information? Persin (2002) discussed Web-assisted instruction in secondary schools—a Web page can be constructed to supplement conventional instruction. Web-assisted instruction offers a way for students to remain connected to their instructor even when absent from school. Persin also reviews research studies related to Web-assisted instruction in high schools.

Lento, O'Neill, and Gomez (1998) go further in their advocacy of learning via the Web through collaborative visualization. Behind this suggestion lies the idea that learning is facilitated by participation in communities of practice—groups of people who share a common purpose through language, work practices, tools, and intellectual values. Their project was based on using communication technology to create distributed learning opportunities where teachers and students could participate in and learn about multiple communities of practice. They worked with students in 40 middle and high schools involved in science learning, two universities, and a science museum. They admitted that, as impressive as ease of information access may be via the Web, the challenge is to leverage Web resources in support of curricular goals. The authors identify a set of conditions for success with educational networking of this kind.

Churach and Fisher (2001) looked at high school science students. They took a constructivist approach, where knowledge is constructed (or generated) within learners' minds as they draw upon their existing knowledge to make sense of perplexing new experiences (in contrast to objectivism, where knowledge is an independent commodity of unquestionable truth that can be conveyed through language from mind to mind). They surveyed 431 students in five Hawaiian high schools, conducted some follow-up interviews, used classroom observation, and interviewed teachers. They concluded that student Web usage has a positive effect on science classes and that science classes with higher student Web usage took a more constructivist approach.

Mistler-Jackson and Songer (2000) were also interested in the Web as a tool to help science students, though in this case the students were in grade six of a middle school. They adopted a case study approach to explore learning and motivational outcomes in one class, where the students were actively involved in an Internet-rich science program called Kids as Global Scientists. This allowed the students over eight weeks to collaborate with other students and scientists across North America on an atmospheric science program. Mistler-Jackson and Songer concluded that the students reaped both learning and motivational benefits—an increase in empowerment, and greater confidence and interest in learning.

Wallace, Kupperman, and Krajcik (2000) looked at sixth graders carrying out an assignment on the Web. This research was part of the University of Michigan Digital Library project and the Middle Years Digital Library project. Although the authors found the students to be engaged and involved in their work, they remain skeptical about the Web as a learning resource. They say that Web tools were not yet designed to support learning, gave almost no support for finding content based on meaning, and provided no means for using information once it was found. They raised a number of crucial issues, including how tasks and tools should be defined to take advantage of the unique features of the Web and what were the real information needs of students that can be satisfied by the Web.

Many articles have been written by individual teachers who described how they used the Internet to teach a particular subject; just two will be cited here. A detailed and good example is offered by Zukas (2000) who described how he incorporated the Web into his teaching of history and set out clearly the benefits as well as the problems he encountered. With thought and care, he said, teachers can minimize most of the problems. Ristau, Crank, and Rogers (2000) reported a study to identify how well the Internet could be used by business teachers in Wisconsin high schools. The teachers believed that student motivation was enhanced. The main obstacles were economic, not educational: lack of computers, slow speed connections, and access costs. In contrast, Feldman, Konold, and Coulter (2000) concluded from their extensive study of the Internet's impact on learning science that it was less effective in improving learning than expected and that face-to-face communication played a central role in learning.

A few researchers have reported work on how students with various kinds of learning problems can use the Web. Hasselbring and Glaser (2000) reported that approximately one in six students in U.S. schools cannot benefit fully from a traditional educational program because they have a disability that impairs their participation in classroom activities; and over 5 million students aged between 6 and 17 were receiving special education services in 1997–1998. Nevertheless, they were placed in the regular classroom (a full inclusion policy). A majority had problems that were primarily academic, emotional, social, or behavioral, rather than physical. Teachers found that technological innovations could help special needs students by expanding the learning environment beyond the classroom. Hasselbring and Glaser contended, however, that the Internet would be a powerful tool for learning only if it offered students opportunities to exchange their thoughts and ideas with others in collaborative learning environments—this interaction was especially beneficial for students with learning disabilities, as it could actively engage them in learning. Hyperlinks were especially helpful for students with mild learning disabilities, although they could be overwhelming, and multimedia helped by providing alternative ways to present information. The authors also discussed how students with hearing and visual disorders could use the Internet with special equipment. They concluded, "Technology has the potential to act as an equalizer by freeing many students from their disability in a way that allows them to achieve their true potential" (Hasselbring & Glaser, 2000, pp. 118–119).

Castellani (2000) also believed that information on the Web had the potential to enhance the development of materials to support students with emotional and learning disabilities, but found few examples. His article was based on a synthesis of teachers' interview responses and offered practical strategies for using the Internet with students with emotional and learning disabilities.

Gardner and Wissick (2002) extolled the virtues of thematic units as one way that the Web could support the curriculum and enhance instruction for students with mild learning disabilities. They presented the key principles to consider when using Web-based activities and recommended strategies and resources to develop thematic units (sets of related learning activities and experiences that can effectively support the teaching of multiple content areas and skills organized around a central topic, idea, or theme). They emphasized that using the Web to promote meaningful learning is much more than simply having students browse sites that have common content. When planning Web activities for students with mild disabilities, one should consider the structure of the learning experience, the amount of guidance the learner will receive, the kinds of activities that emphasize higher order thinking skills, and the interactivity of the sites selected (a key principle behind meaningful learning on the Web).

The Web has a potentially important role to play in distance and distributed learning; and researchers are turning attention to this aspect. Most reports on the characteristics of online students have relied primarily on anecdotal evidence (Wang & Newlin, 2000). Roblyer and Marshall (2002–2003) discuss the thriving virtual high school (VHS) movement in the U.S., where credit courses have been taught primarily on the Internet, some within existing physical schools and others completely online. These schools aim to provide better educational access to students regardless of location; however, they suffer from comparatively high dropout and failure rates. The authors developed an instrument to predict educational success based on studies of successful and unsuccessful online learners and on characteristics posited through direct observations of VHS teachers. This instrument proved very successful in predicting whether students would pass or fail a course.

Several researchers, unsurprisingly, have commented on the critical role played by teachers in determining whether the Web can promote learning by their students. Churach and Fisher (2001), for example, believed that the teacher played a large part in determining how valuable the Web was to the students in the science classes they studied. Large and Beheshti (2000) argued that the potential of the Web to provide multimedia content—especially sound and video clips—would be fully realized only when teachers gave their students assignments that allowed them to exploit these media. Traditional assignments that asked students to collect factual information and present it in writing did little to encourage multimedia use. Goldsby and Fazal (2001) offered teachers one solution to this problem—digital portfolios incorporating Web-based materials—and provided guidance on how to evaluate such portfolios.

Williams (1999) found in his visits to an elementary school that teachers expressed concerns about their own abilities to incorporate the Internet into their work in a way that would maximize its potential; he concluded that formal, structured training for teachers was needed. Sutherland-Smith (2002) noted that teachers as well as students must be comfortable and competent with the Web if they were to use it successfully. She argued that special teaching techniques were required to teach students how to read Web-based text, including developing mechanisms to overcome frustration with technology and search guidelines to avoid random text scanning. Wallace, Kupperman, and Krajcik (2000) also believed that teachers needed strategies to help students learn from the Web. A Pew study conducted in the U.S. during 2001 and 2002 found that many teachers had not yet recognized the new ways in which students access information, identified wide variation in teacher policies on Internet use by students, and concluded that professional development and technical assistance for teachers was crucial for the effective integration of the Web into curricula (Levin & Arafeh, 2002).

How can the librarian or media specialist help to ensure that the Web plays a positive learning role? Walter (1997) observed that librarians serving youth in public and school libraries have been early and active adopters of information technologies of all kinds. She recommended that libraries respond to the challenge of the Web by offering user education programs. Kafai and Bates (1997) believed there were abundant opportunities for school library media specialists to be involved in Internet instruction in schools. Schofield and Davidson (2002) found that school librarians took a leading role in devising and managing Internet activities in the school district they investigated. Librarians introduced the Internet to students and teachers who did not have classroom access, shared resources from the Internet with teachers, and helped with curricular activities. Shantz-Keresztes (2000) summarized a conversation between a group of Canadian teacher-librarians in the Calgary Board of Education. They discussed, among other topics, how teacher-librarians promoted their role as information literacy specialists to teachers, and what strategies could be used by teacher-librarians in their cooperative planning with teachers to ensure that students had the necessary background to evaluate Web sites. Hart (2000) reports on a study using questionnaires and interviews of 65 children's librarians in metropolitan Cape Town (South Africa). It found that children's librarians were already performing a strongly educational role, but that librarians themselves lacked computer literacy (at that time only four were familiar with the Internet), only seven of the 63 libraries involved had Internet access, and only two allowed user access.

Meyer, Middlemiss, Thodorou, Brezinski, and McDougall (2000) have explored the possibilities for intergenerational tutoring programs to help students develop reading skills and especially to recall what they have read. Their program linked retired adult professionals (aged 60 to 80) with fifth grade students through an instructional Web site. The students followed lessons on the Web and sent their work to their older tutors. The researchers found this procedure increased recall of the lesson material among both the children and the tutors, suggesting that it could both help older adults maintain improved memory skills and help children learn to read and to recall expository text.

Caskey (2002) also considered intergenerational instruction, this time involving parents. She evaluated the effects of such instruction on the attitudes of young adolescents and their parents toward school-based use of the Internet. Internet instruction was given separately to parents and students assigned to one group, and in the other group parents and their children worked together in family pairs. Both students and parents increased their positive attitudes regardless of group; the parents preferred family pairing and the students had no preference.

Kupperman and Fishman (2001–2002) studied how computers were adopted and used by Latino families as part of a larger project developing technology-enhanced curricula for middle school science classes in urban public schools. The families were provided with access to the

Internet from home. The Web was used both as a source of information for school assignments and as a place to look at pictures, for example, of cars or youth idols. The researchers found that the children normally took the lead in using the computer and were much more skilled than their parents, but they admit that it is impossible to make neat generalizations from their limited study about urban Latino families as Internet users.

These various research studies have tried to answer critical questions: under what conditions, if at all, does the Web enhance learning; how can teachers effectively integrate the Web's resources into their teaching; is the Web an important tool for students with physical and learning disabilities, and what role can school librarians and media specialists play? Some preliminary answers have emerged, but to date the results are too fragmented and in some cases contradictory for school administrators, teachers, and librarians to formulate policies with confidence. It is hard to dispute Schofield and Davidson's (2002, p. 28) conclusion that "without teacher interest in using the Internet in their classrooms, efforts to link schools to it are likely to be pointless." They say that teachers face twin obstacles: widespread lack of knowledge about computers and the Internet as well as limited time in the classroom to use the Internet. Moreover, the researchers assert, little solid evidence exists about the impact of Internet use on either teachers or students.

The Web and Social Interaction

Is the Web having detrimental effects on the social and psychological well-being of the young? A few research studies have considered this question. Sanders, Field, Diego, and Kaplan (2000) asked whether high levels of Internet use were associated with depression and social isolation among adolescents. Based on their findings from 89 seniors in a Florida high school, no significant differences related to the students' relationships with their fathers or with levels of depression; low Internet users, however, did have significantly better relationships with mothers and friends. Orleans and Laney (2000) concluded that boys were more likely to socialize than girls, probably because networked computer games promoted socialization and such games were more popular with boys than girls. Kraut, Patterson, Lundmark, Kiesler, Mukopadhyay, and Scherlis (1998) concluded from a field trial of the Internet involving 169 adults and teens over the age of 10 that on balance the participants had less social engagement and poorer psychological well-being during their first year or two online. Furthermore, teenagers were more likely to report loneliness than adults. On the other hand, they commented that the Internet had turned out to be far more social than television. In contrast to Kraut et al.'s findings, a British study of youngsters aged 11 to 16 concluded that "they appear to use technology in balanced and sophisticated ways to develop online and offline social relationships which can open their minds to a wider, if Americanized, world" (Holloway & Valentine, 2003). It should be emphasized that these studies involved relatively small numbers of young people and, as Kraut et al. commented, generalization from them is difficult. Here, clearly, is an area requiring more systematic research. A larger scale survey (1,735 U.S. households) by the National School Boards Foundation (2002) concluded that the Internet did not disrupt children's "healthy activities"; Internet users apparently watched less television and read more newspapers, magazines, and books as well as played outdoors more.

These findings raise a number of related questions. Do parents and teachers really know what their children are doing online, and does this matter? If the children are to be believed, adults may be exaggerating the educational uses to which their offspring put the Web, at least in the home environment (although in the UCLA Center for Communication Policy [2003] survey, parents were skeptical about its effect on school grades, with nearly 75 percent saying that their children's grades have stayed the same despite the household acquiring Internet access). Of course, the children may consider that by emphasizing the educational benefits they will more easily persuade their parents to provide the necessary resources to enjoy all the advantages of Internet access.

The Student Perspective on the Web as an Information and Learning Resource

What do the students themselves have to say about the Web as an information and learning resource? Surprisingly little research has focused on students' own perceptions of the Web despite the emphasis on a user-centered approach to the design and deployment of information systems. As a consequence, the findings are piecemeal as well as scattered across various age groups.

Large and Beheshti (2000) interviewed 53 students from two sixth grade classes in a Canadian elementary school to ascertain their reaction to using the Web to find information for a school project. On the basis of the responses, the students were categorized by the researchers as technophiles who strongly favored the Web as an information resource, traditionalists who favored print materials, or pragmatists who used whichever medium best suited the kind of information they needed to find. Many of the students appreciated the amount of relevant information they could find on the Web compared with books in their school or public library and found it a good place for both current and obscure information. They said they could usually understand the vocabulary and syntax used on Web pages (in contrast to the conclusions of many adult researchers, as discussed previously). Nevertheless, the students reported frustrations with the difficulty in finding a few highly relevant articles and were intolerant of slow response times.

Watson (1998) adopted a similar approach with a smaller sample of nine students in the eighth grade, asking them to reflect on their personal experiences—levels of confidence, pleasure, and frustration. They were positive and self-confident about their Internet use, although Watson cautioned that this may not indicate success in information retrieval or content evaluation. Two and a half years later she interviewed four of the same students, now in grade 11. She warned readers to be careful in generalizing from such small numbers, but found that the students expressed greater self-confidence and competence in using the Internet, especially for personal rather than school purposes (Watson, 2001). Vansickle (2002) had 136 volunteers from tenth grade complete a questionnaire about their knowledge and use of the Web. The responses dealt with how they learned to use the Web, its advantages, and the purposes for which it was used (personal or academic).

Hirsh (1999) was especially interested in fifth grade students' criteria for judging the relevance of information retrieved from the Web. The students rarely mentioned authority as an evaluation criterion and generally did not question the accuracy or validity of retrieved information. These findings were supported by Lorenzen (2001), who interviewed older high school students about the ways they used the Web to find information for school assignments. Most said they used both the library and the Web, although a few relied exclusively on the Web. None relied exclusively on the library. The Web was preferred for current events. It became clear that the students had given little thought to how information from the Web might be evaluated, and many struggled with the question of how they knew whether information on a site was good; the most common answer was that they did not know. The high school students in a study by Bos (2000) had difficulties identifying potential biases in scientific resources they found on the Web. He concluded that if the Web was to be used by students to practice evidence-based reasoning, then modifications were needed in how students were taught to recognize evidence.

Information overload (the moment when the amount of available information exceeds a user's ability to process it) was the focus of a study in a Texan elementary school (Akin, 1998). The majority of students said they had experienced information overload, and it was more marked in the younger than the older students. Akin described the strategies employed by the students to deal with this problem and discussed ways in which a school library media specialist could assist.

Mitra and Rana (2001) installed an outdoor Internet kiosk in the exterior wall of an office bordering a slum in New Delhi, India. The kiosk was used immediately by local children aged five to 16 (adults made no attempt to use it), even though most of them did not attend school, had only a limited understanding of the Roman alphabet, and could not speak English. Within a few days they had learned how to browse using Internet Explorer, teaching each other the necessary skills. Kiosks were installed by the authors in two other locations with similar results. Simply by using the Web in this way, these Indian children had expressed their views about it in a more dramatic fashion than any number of interviews or questionnaires could do. The authors conceded, however, that generalizations from these experiences were inadvisable without more rigorous research. Unfortunately, this is one of the very few published studies of children's Web use in a developing country.

Any evaluation of research on young people's attitudes concerning the Web must take into account when that research was undertaken. As each year goes by the Web becomes more commonplace in schools and homes. Children now in the early grades of elementary school may have grown up with the Web as an alternative to other educational, informational, and entertainment resources. Research conducted just a few years ago, however, may have involved children and teenagers for whom the Web was a new phenomenon alongside other technologies with which they had greater familiarity. This makes it difficult to know whether findings from studies conducted, for example, in the late 1990s describe current conditions. Shenton and Dixon (2003) alluded to this problem in discussing the findings from their study conducted with British children and teenagers during 1999 and 2000; at that time the younger students in particular compared CD-ROMs very favorably with the Web as information resources, but with increased familiarity, their counterparts in 2003 may think differently. Unless and until the Web reaches a plateau, both in terms of its availability and its technical development, it will be necessary, at the very least, to take account of when individual studies were conducted and to rerun them with new generations.

Content and Personal Safety Issues

Children might use the Web in ever growing numbers, and express positive reactions overall to it, but not everyone, child or adult, is thereby convinced that its influence is benign. The increasing reach of the Web highlights controversy in the published literature surrounding content that is deemed inappropriate for children and related personal safety issues. Surprisingly little research (in contrast to anecdote or opinion) could be found in the literature on other issues such as copyright and plagiarism as they relate to young Web users. Shantz-Keresztes (2000), in her account of a conversation between Canadian teacher-librarians, did touch on these topics.

Web Content

As any Web surfer is aware, just about any kind of material can be located, ranging from the erudite and rarified to the popular and vulgar. The easy availability of pornographic text and images has provoked widespread discussion regarding the pros and cons of content control (Cronin & Davenport, 2001). To a lesser extent, hate literature and extremist political missives also have attracted attention. For example, Oravec (2000) discussed Web-based violent and hate-related materials.

The Internet can create a separate enclave for troubled youth in which violent fantasies can be acted out easily and possibly become reality. Hate group sites have proved fertile ground for recruitment. Oravec provided ideas for a course addressing these issues, to be offered in teacher education classes. The Web has become an important advertising venue for many companies, and the intrusive, ubiquitous online advertisements (especially pop-ups) have provoked anger (for an exploratory investigation of children's perceptions of advertising on the Web, see Henke, 1999). When children, and to a lesser extent teens, are the intended or actual audience for such content, the volume of protest has proven difficult for parents, teachers, librarians, and politicians to ignore. Interest groups have emerged to fight for and against censorship on the Web; it is hardly surprising that a body of literature has appeared reflecting these concerns.

The debate over Web content and security issues has generated numerous opinion pieces on both sides of the censorship divide. While these outspoken comments can make for lively reading, they have not been included in this literature review. The American Library Association Office for Intellectual Freedom (2003) provides a list of Web sites that contain rules for and advice on safety, help organizations, guidelines for parents and children, and a list of sites dealing with privacy issues.

At the outset, it is prudent to heed the reminder from Wartella and Jennings (2000) that each new mass media innovation has prompted debate on the effects of new technology, especially concerning young people. They locate the current controversy on children and computers in an historical context, examining motion pictures, radio, and television as well as computers. For Wartella and Jennings, proponents of media innovation generally have argued that new technology benefits children by opening new worlds, while opponents have warned of the danger that new media might be used to substitute for real life, reduce respect of ethical principles, undermine children's morality, and cause them to engage in illicit sexual and criminal behavior. They believe that many lessons for today can be learned from the history of media research, while conceding a greater degree of urgency about the need to monitor and improve the quality of Web content.

Much of the discussion on Internet content comes from North America. Oswell (1999) offered a European perspective. He looked at ways in which the European Commission has sought to protect children from the Internet through content regulation, drawing upon what he called the "advanced liberal" form of governance that hold both parents and industry responsible for regulating content (Oswell, 1999, p. 43). He commented that the children's own views, however, have been largely ignored, and argued that they should be able to participate in the decision-making process.

Two relevant large-scale surveys have been conducted in Canada for the Media Awareness Network—*Canada's Children in a Wired World:*

The Parents' View (Environics Research Group, 2000a) and *Young Canadians in a Wired World: The Students' View* (Environics Research Group, 2001). Both were summarized by Taylor (2001). Overall, parents were optimistic about their children's use of the Internet. Their greatest concern was inappropriate content on the Web (47 percent), by which they meant pornography, violence, and hate sites. They wanted online safety issues to be addressed in schools and libraries. An overwhelming majority thought online advertising aimed at children should be regulated. In general they thought their child's Internet use was under control (Environics Research Group, 2000a). The students told a rather different story (Environics Research Group, 2001). Only 16 percent said they thought their parents knew a great deal about the sites they visited and 38 percent said their parents knew very little or nothing about this. A follow-up qualitative study used eight focus groups with children aged nine to 16 and four focus groups of parents with children between these ages; participants recommended that more be learned about how young people used the Internet, and that media Web sites young people visit frequently could play a role in communicating information about Internet safety (Environics Research Group, 2000b).

In the U.S., the study conducted by Penn, Schoen, and Berland Associates (2000) included many findings of relevance to any discussion of Web content and security issues. Interviews were conducted with teenagers as well as parents. As in the Canadian study, parents were concerned that their children would visit inappropriate sites, but did not necessarily know what their teens were doing on the Web; nor did they know what was available to help them guide and regulate children's Internet access.

The U.S. Congress has held several inquiries into children's vulnerability to harm on the Internet. The objectives were to consider the legal, educational, technological, and social contexts and to provide information useful to various decision-making communities, including schools and libraries, about possible courses of action to help children be safer in their use of the Internet (Commission on Child Online Protection, 2000; Iannotta, 2001). Thornburgh and Lin (2002) concluded that although there has been a concentration on both technological solutions (especially filters) and legal solutions, neither can offer a complete answer. Appropriate social and educational strategies also are required.

Web Filtering

Much has been written on filters, including reviews and assessments of individual products. This is unsurprising as filters are quite widely installed on computers likely to be used by young people. Curry and Haycock (2001) surveyed a sample of subscribers to *School Library Journal* to find out how widespread Web filtering was in North American schools and libraries. From the responses (only 24 percent replied), 53 percent of school libraries and 21 percent of public libraries were using filters. The researchers were surprised that so many respondents in both types of libraries understood little about how filtering software actually worked.

Pownell and Bailey (1999) worked to present a balanced discussion of pornography on the Web, in the context of Web filtering software. They emphasized how frustrating it was for educators to appreciate the great potential of the Web as a teaching and learning environment while at the same time accepting the problems of students inadvertently accessing inappropriate information. Pownell and Bailey put forward the views of filter proponents and opponents alike, though it is fair to say that overall the article is opposed to filtering. They emphasized that schools need to have control over their curricula and to decide for themselves what is appropriate. It is difficult, they argue, for any generic software to take into consideration all the variables necessary to decide what content should and should not be accessed.

A British study of 274 students and 64 teachers in four high schools found that 9 percent of the students' parents had installed filtering software at home and an additional 31 percent did not know whether this had been done (Allbon & Williams, 2002). Their study offers insight into teenagers' views rather than those of adults. Although 60 percent of males and 28 percent of females reported seeing unpleasant or offensive material, 62 percent in total felt there was no need for such concern. Only 14 percent said their parents had talked to them about possible Internet threats. Turning to the teachers, 68 percent thought their schools operated a system to block controversial material and 82 percent strongly believed filtering was necessary in schools. An earlier pilot study by Williams (1999) found little evidence in a primary school of material having an adverse effect on the children (aged 10 to 11) or likely to concern their parents.

Diaz (1998), from a librarian's standpoint, took an overtly anti-filtering position. She believed that "while many hoped for a technical solution ... such solutions have proven to be less than effective and [were] often chosen by libraries to placate boards and communities or provide some stopgap measure to limit offensive materials on their public machines" (Diaz, 1998, p. 147). She emphasized, perhaps somewhat disingenuously, the difference between filtering, which seeks to keep a user from finding or viewing certain types of material, and selection, which seeks to aid discovery of useful information by including only that which is determined to be worthwhile. As an alternative to filters, Diaz suggested software such as the Library Channel, with guided access to the Web through a subject directory using three access levels: preselected sites accessed via menus; any URL; or sites not blocked by domain name or word in the domain name, which are added individually by the library. She conceded that this was labor intensive for librarians to build and maintain. Oravec (2000) agrees that filters did not provide a magic solution in this case. Shuman (2001) provided a balanced summary of the issues surrounding filtering and censorship on the Web in relation to children. He offered as an alternative for libraries mandatory parental monitoring and supervision—one of the child's parents must sign an Internet policy form, must be present and supervising the child's searching, or both. Arrighetti (2001) worried that, as a way of preventing children from accessing inappropriate Web sites, public libraries might exclude unattended children as an alternative to installing filtering software.

Reilly (2001) looked at filtering as a political and legal issue. He discussed U.S. congressional attempts to regulate the Web by mandating that filters be employed to block certain material: the Communications Decency Act, the Child Online Protection Act, and the Children's Internet Protection Act. He identified several problems with all three. For example, they restricted adult as well as minor access, unconstitutionally limited free speech, were unconstitutionally broad in mandate, and violated constitutionally protected anonymity and privacy. In any case, he said, technology protection measures simply did not work and such acts would prove ineffective. He justified this by listing the shortcomings of filtering software and pointing to the high percentage of erroneous blocking that typically resulted from their application.

Hunter (2000) reported that many studies of filtering software had been undertaken by journalists or anti-censorship groups that applied largely unscientific methods to conclude that filters were deeply flawed. He set out to change this in his evaluation study. The first problem was to determine exactly what constituted objectionable content that any filter should exclude (and exclude only). He relied on the Recreational Software Advisory Council (RSAC) Internet rating system, which he said was the most popular. This was used to rate 200 sample Web sites. Four filters were then applied to these sites. On the basis of his results, he concluded that, indeed, filters were highly problematic both in terms of correctly blocking objectionable sites and incorrectly blocking unobjectionable sites. He conceded, however, that his sample was small and recommended a large, truly randomized sample. He also wondered whether RSAC would be the best rating system to use, as it excluded, for example, alcohol and gambling sites that some filters explicitly seek to eliminate.

Burt (1997) offered advice, going beyond the installation of filters, for public libraries with publicly accessible Internet stations. The advice is based upon his examination of policies formulated by public libraries across the U.S. McKechnie (2001) asked a random sample of Canadian public libraries to provide copies of all policies that applied to services for children; the most frequently provided policies dealt with Internet access and with circulation. Filtering software was prescribed by 21 percent. McKechnie (2001, p. 52) specifically referred to Burt's earlier U.S. survey, commenting that her findings were "somewhat comparable" but that substantially more Canadian libraries required written parental permission or parental presence for Internet use by their children. Campbell (1998) listed preliminary questions to be answered before a library wrote a policy and a checklist to determine how the Internet policy would fit into established policies and procedures. She also suggested actions the library could take to strengthen its position on Internet access for children.

Legal Solutions

Legal solutions have been perceived as a solution to Web content problems (especially in the case of young people) more in the U.S. than elsewhere. In particular, the Children's Internet Protection Act (CIPA) has been the focus of attention from the librarianship community, as is evidenced, for example, by a visit to the American Library Association's Web site (http://www.ala.org). The crux of this issue was the requirement under CIPA that filtering software be installed on every publicly accessible computer in public libraries as a condition for receiving government discounts for Internet access. CIPA was challenged by the American Library Association and the American Civil Liberties Union (schools also were required to filter under CIPA but were not included in the lawsuit) on several grounds: filters do not work, CIPA is unconstitutional, libraries should not be forced to choose between funding and censorship, and CIPA abolishes local decision making. In 2002 CIPA was invalidated under appeal (Oder, 2002), but in June 2003 the U.S. Supreme Court (2003) narrowly upheld CIPA. More generally, Lessig and Resnick (1999) pointed out the jurisdictional problems in seeking national legal solutions to problems in international cyberspace. Sandvig (2001) provided a good description of government attempts to regulate Internet access by children and concluded that it has been largely ineffective.

Personal Safety Issues

Personal safety, especially in the case of young people, is another area of concern. However, the issues relate primarily to e-mail, especially chat room and instant messaging use, rather than Web usage per se and, therefore, will be discussed only briefly. The National Center for Missing and Exploited Children (1994), in its brochure "Child Safety on the Information Highway" emphasized that children can benefit considerably from being online but they can also be targets of crime and exploitation. It explains to parents how to reduce the risks. The National Center's "Teen Safety on the Information Highway" (National Center for Missing and Exploited Children, 1998) is another brochure warning of dangers lurking on the Internet—contacting strangers, advertising, harassment, and pornography—and offers basic rules of online safety for teens.

The Children's Online Privacy Protection Act (COPPA) in the U.S. is designed to protect the online privacy of children under 13 years of age. Commercial Web sites must obtain parental consent to collect, use, or disclose personal information from children, and privacy policies must be posted on Web sites whenever personal information is requested. Bushong (2002) pointed out problems for commercial Web sites in adhering to these requirements, especially the extra work, which caused some sites to eliminate Web features or close down permanently. Based on a review of online safety sites and print publications, Bushong suggested several safety measures to parents.

The University of New Hampshire's Crimes against Children Research Center conducted a Youth Internet Safety survey that focused on sexual solicitations and approaches, unwanted exposure to sexual material, and harassment. It concluded that many young people were subjected to dangerous and inappropriate experiences on the Internet and made several recommendations to counter this (Finkelhor, Mitchell, & Wolak, 2001).

Turow and Nir (2000) found that young people aged between 10 and 17 were much more likely to give sensitive personal information to Web sites than their parents, especially when a gift was offered. They based this on a national survey conducted for the Annenberg Public Policy Center in early 2000. The study explored parents' concerns about their teenagers' release of information to the Web and how parents dealt with this challenge. The Annenberg study argues for a social policy that helps families establish clear norms for information privacy.

It is hardly surprising that adults have striven to ensure that the virtues of the Web, as they see them, are promoted for young people while these same young people are safeguarded from the lurking dangers of pornography, pedophiles, hate mongers, and purveyors of myriad commercial products advertised on the Web. It is also not surprising, of course, that there is little agreement either on the identification of problematic Web content or on the means to combat it. Most evaluations of filtering software, for example, have commented upon its shortcomings, and confidence in legal solutions is far from universal. Librarians in particular are uneasy about restrictions on information access, even for the young, but cannot afford to ignore safety issues for their patrons. Perhaps on this topic systematic research will always have to do battle with personal opinions; it is by no means clear that the former will have the advantage in this encounter.

Research Agendas for the Future

Given the relatively short time that children and teenagers have been exposed in any great numbers to the Web, it is surprising that so much research has been completed on so many facets of Web use. We now have a good idea, at least in North America, of the extent to which young people access the Web and a real sense of where, why, and how they access it. We are beginning to understand the problems they encounter when searching for information on the Web and are able to formulate plausible responses to these problems. Children are beginning to collaborate with researchers in a meaningful way rather than being merely the objects of study, in the design and development of tools such as portals. We understand that learning is more than accumulating information and are at the early stages of determining how best to exploit the Web as an educational rather than a purely informational asset. We are conscious of the negative as well as the positive side of the Web and the challenges it poses in terms of inappropriate content and threats to personal safety. Researchers are beginning to assemble evidence that can counter the wilder assertions on either side of the censorship debate concerning the need for control mechanisms and the most suitable types. There is much to be proud of here, and the rapid progress made in several academic and professional domains should not be minimized.

Notwithstanding the growth in research interest, it would be unwise to exaggerate the extent to which young people are deemed worthy of serious study. Compared with adults of university and working age, children and teenagers (as well as senior citizens, at the opposite end of the age spectrum) remain on the research sidelines. It is noteworthy, for example, that the author of a recent book devoted to surveying research findings on information seeking concedes that children's use of electronic sources has been the subject of a number of recent reports and commentaries, but goes on to say that he will not review them (Case, 2002).

Furthermore, despite the expanding number of studies on information seeking and use by young people, the research surface has been scratched only lightly. A major reason for this situation lies in the dramatic differences among the developmental stages humans traverse from birth to adulthood. These stages have a major impact on how information is sought, selected, evaluated, and used. As a consequence, a study of, say, how young children in elementary school seek information will not necessarily shed insight into the information-seeking behavior of high school students. The aggregate research output dealing with young people and their use of the Web would constitute a more significant platform for informational, educational, political, and economic decision making were it not for the fact that it is scattered over very different developmental stages. The body of established research theories on children's and teens' behavior on the Web at any one stage remains disappointingly small.

Subrahmanyam et al. (2000) highlighted another research shortcoming. They noted that most data were gathered by self-reports that were beset by problems of accuracy and reliability—memory limitations, inaccurate estimations—especially when done by children. They argued for more computer tracking of what children do. This is a fair comment on many of the surveys conducted to explore children's and teenagers' use of the Web (in fact, as mentioned earlier in the chapter, in too many cases it was not the children but their parents who were asked to comment on the former's behavior). It is less true, however, in other research areas, such as investigations of information-seeking behavior.

Where information technology is concerned, researchers in any case struggle to keep pace with developments. It seems likely that, before too long, the Web will be accessed not just from institutional settings such as schools, homes, and libraries, but also via handheld devices from just about anywhere (Norris & Soloway, 2003). To what purposes will these devices be put, how will adults monitor their use, and what kinds of new content will appear on them? Will more sophisticated retrieval algorithms from portal providers simplify information seeking by young and old alike? Will higher bandwidths change the kinds of information delivered by the Web and the ways in which it is found (for example, through 3-D portals)? The Web is likely to remain a fruitful environment for researchers interested in the interface between information technology and the young.

Where, then, should investigators devote their energies in the future? They might start with the following list of research questions:

- To what extent are young people without access to the Web from home and/or school disadvantaged educationally or socially?

- Does access to the Web improve academic performance and, if so, how might this be facilitated?

- For what purposes do young people use the Web, especially outside the school, and are these purposes correctly perceived by parents and teachers?

- How do young people look for information on the Web? And does this differ according to age?

- Do young people seek and use information on the Web differently from adults?

- How might Web portals and Web pages be better designed to meet the cognitive and affective needs of young people?

- How do young people evaluate information retrieved from the Web?

- What kinds of mental models do young people themselves construct about the Web and how do such models affect their information-seeking behavior?

- How can young people be given access to the wealth of resources available via the Web while ensuring their emotional and personal safety?

Research has been undertaken that edges us close to answering all of these questions, but none of them can be answered with any degree of certainty. Of course, it is not enough that researchers find answers; they should then be put into practice. Norris, Smolka, and Soloway (2000) remind us that only limited use has been made of research by teachers and decision makers, and the same comments might have been made of librarians and even software designers. As these authors say, "all this research effort is taking place with little benefit to the actual practice of education" (Norris, Smolka, & Soloway, 2000, p. 46). They offer practitioners tips on how to sift through the research to find the practical information. They also offer researchers some advice: Include in your research team practitioners who know what they need, even if they do not always know how to get it.

References

Abbas, J. (2003). Children and information technology. In M. A. Drake (Ed.), *Encyclopedia of Library and Information Science* (2nd ed.) Vol. 1, pp. 512–521. New York: Marcel Dekker.

Abbas, J., Norris, C., & Soloway, E. (2002). Middle school children's use of the ARTEMIS Digital Library. *Proceedings of the 2nd ACM/IEEE-CS Joint Conference on Digital Libraries,* 98–105.

Agosto, D. E. (2001). Propelling young women into the cyber age: Gender considerations in the evaluation of Web-based information. *School Library Media Research* 4. Retrieved November 19, 2003, from http://www.ala.org/Content/NavigationMenu/AASL/Publications_and_Journals/School_Library_Media_Research/Contents1/Volume_4_(2001)/Agosto.htm

Agosto, D. (2002a). Bounded rationality and satisficing in young people's Web-based decision making: Study of female high school students in New Jersey. *Journal of the American Society for Information Science and Technology, 53*(1), 16–27.

Agosto, D. E. (2002b). A model of young people's decision-making in using the Web. *Library & Information Science Research 24*(4), 311–341.

Akin, L. (1998). Information overload and children: A survey of Texas elementary school students. *School Library Media Quarterly Online, 1*. Retrieved January 29, 2004, from http://www.ala.org/ala/aasl/aaslpubsandjournals/slmrb/slmrcontents/volume11998slmqo/akin.htm

Allbon, E., & Williams, P. (2002). Nasties in the Net: Children and censorship on the Web. *New Library World, 103*(1/2), 30–38.

American Library Association Office for Intellectual Freedom. (2003). *Especially for children and their parents: Online safety rules and suggestions.* Retrieved November 19, 2003, from http://www.ala.org/Content/NavigationMenu/Our_Association/Offices/Intellectual_Freedom3/ForYoung_People/For_Young_People.htm

Arnone, M. P., & Small, R. V. (1999). Evaluating the motivational effectiveness of children's Websites. *Educational Technology, 39*(2), 51–55.

Arrighetti, J. (2001). The challenge of unattended children in the public library. *Reference Services Review, 29*(1), 65–71.

Auxier, T. (2001). Helpful homework Web sites. *Illinois Libraries, 83*(2), 29–31.

Becker, H. J. (2000). Who's wired and who's not: Children's access to and use of computer technology. *The Future of Children, 10*(2), 44–75.

Benn, S. (1996). An island of treasure in the Net: UKOLN's Treasure Island Web site. *Electronic Library, 14*(5), 414–416.

Bilal, D. (2000). Children's use of the Yahooligans! Web search engine: I. Cognitive, physical, and affective behaviors on fact-based search tasks. *Journal of the American Society for Information Science, 51*(7), 646–665.

Bilal, D. (2001). Children's use of the Yahooligans! Web search engine: II. Cognitive and physical behaviors on research tasks. *Journal of the American Society for Information Science and Technology, 52*(2), 118–136.

Bilal, D. (2002a). Children design their interfaces for Web search engines: A participatory approach. *Proceedings of the 30th Annual Conference of the Canadian Association for Information Science, 204–214.*

Bilal, D. (2002b). Children's use of the Yahooligans! Web search engine. III. Cognitive and physical behaviors on fully self-generated search tasks. *Journal of the American Society for Information Science and Technology, 53*(13), 1170–1183.

Bilal, D. (2002c). Perspectives on children's navigation of the World Wide Web: Does the type of search task make a difference? *Online Information Review, 26*(2), 108–117.

Bilal, D. (2003). Draw and tell: Children as designers of Web interfaces. *Proceedings of the 66th ASIST Annual Meeting, 135–141.*

Bilal, D., & Kirby, J. (2002). Differences and similarities in information seeking: Children and adults as Web users. *Information Processing & Management, 38*(5), 649–670.

Bjorklund, D. F. (2000). *Children's thinking: Developmental function and individual differences* (3rd ed.). Belmont, CA: Wadsworth.

Bolger, P., Bussmann, I., & Fieguth, G. (1997). The CHILIAS Project: A new concept for children's libraries. *Program, 31*(4), 365–371.

Borgman, C., Hirsh, S., & Walter, V. (1995). Children's searching behavior on browsing and keyword online catalogs: The Science Library Catalog Project. *Journal of the American Society for Information Science, 46*(9), 663–684.

Bos, N. (2000). High school students' critical evaluation of scientific resources on the World Wide Web. *Journal of Science Education and Technology, 9*(2), 161–173.

Bowler, L., Large, A., & Rejskind, G. (2001). Primary school students, information literacy and the Web. *Education for Information, 19*(3), 201–223.

Broch, E. (2000). Children's search engines from an information search process perspective. *School Library Media Research, 3.* Retrieved November 19, 2003, from http://www.ala.org/Content/NavigationMenu/AASL/Publications_and_Journals/School_Library_Media_Research/Contents1/Volume_3_(2000)/childrens.htm

Burgstahler, S. (1999). Surfing the Internet with the younger set. *Learning and Leading with Technology, 26*(5), 25–29.

Burt, D. (1997). Policies for the use of public Internet workstations in public libraries. *Public Libraries, 36*, 156–159.

Bushong, S. (2002). Parenting the Internet. *Teacher Librarian, 29*(5), 12–16.

Bussmann, I. (1999). The European Virtual Children's Library on the Internet: A new service to foster children's computer literacy. *Exploit Interactive, 1*, 1–6.

Campbell, S. (1998). Guidelines for writing children's Internet policies. *American Libraries, 29*(1), 91–92.

Case, D. O. (2002). *Looking for information: A survey of research on information seeking, needs, and behavior.* Amsterdam: Academic Press.

Caskey, M. M. (2002). Using parent–student pairs for Internet instruction. *Journal of Research on Technology in Education, 34*(3), 304–317.

Castellani, J. D. (2000). Strategies for integrating the Internet into classrooms for high school students with emotional and learning disabilities. *Intervention in School and Clinic, 35*(5), 297–305.

Center for Media Education. (2001). *TeenSites.com: A field guide to the new digital landscape.* Washington DC. Author. Retrieved November 19, 2003, from http://www.cme.org

Churach, D., & Fisher, D. (2001). Science students surf the Web: Effects on constructivist classroom environments. *The Journal of Computers in Mathematics and Science Teaching, 20*(2), 221–247.

Clark, W. (2001, Autumn). Kids and teens on the Net. *Canadian Social Trends*, 6–10.

Commission on Child Online Protection (COPA). (2000, October 20). Report to Congress. Retrieved November 19, 2003, from http://www.copacommission.org/report

Corporation for Public Broadcasting (2003). *Connected to the future: A report on children's Internet use from the Corporation for Public Broadcasting.* Retrieved November 19, 2003, from http://www.cpb.org/ed/resources/connected

Cronin, B. & Davenport, E. (2001). E-rogenous zones: Positioning pornography in the digital economy. *The Information Society, 17*(1), 33–48.

Curry, A., & Haycock, K. (2001). Filtered or unfiltered? *School Library Journal, 47*(1), 42–47.

Diaz, K. (1998). Filtering, selection, and guided access. *Reference & User Services Quarterly, 38*(2), 147–150.

Dickard, N. (Ed.). (2003). *The sustainability challenge: Taking edtech to the next level.* Washington, DC: Benton Foundation and Education Development Center. Center for Children and Technology. Retrieved November 19, 2003, from http://www.benton.org/publibrary/sustainability/sus_challenge.html

Drenoyianni, H. & Selwood, D. I. (1998). Conceptions or misconceptions? Primary teachers' perceptions and use of computers in the classroom. *Education and Information Technologies, 3*(2), 87–99.

Dresang, E. T. (1999). More research needed: Informal information-seeking behavior of youth on the Internet. *Journal of the American Society for Information Science, 50*(12), 1123–1124.

Dresang, E. T., Gross, M., & Holt, L. E. (2003). Project CATE. Using outcome measures to assess school-age children's use of technology in urban public libraries: A collaborative research process. *Library & Information Science Research, 25*(1), 19–42.

Druin, A. (1996). A place called childhood. *Interactions, 3*(1), 17–22.

Druin, A. (1999). Cooperative inquiry: Developing new technologies for children with children. Proceedings of the SIGCHI Conference on Human Factors in Computing Systems *CHI '99*, 592–599.

Druin, A. (2002). The role of children in the design of new technology. *Behaviour and Information Technology, 21*(1), 1–25.

Druin, A., Bederson, B., Hourcade, J. P., Sherman, L., Revelle, G., Platner, M., et al. (2001). Designing a digital library for young children: An intergenerational partnership. *Proceedings of the first ACM/IEEE-CS joint conference on Digital libraries,* 398–405.

Druin, A., Bederson, B. B., Weeks, A., Farber, A., Grosjean, J., Guha, M. L., et al. (2003). The International Children's Digital Library: Description and analysis of first use. *First Monday 8*(5). Retrieved November 19, 2003, from http://www.firstmonday.dk/issues/issue8_5/druin/index.html

Embassy of France in the United States. (2001). *Internet in schools.* Retrieved January 7, 2004, from http://www.info-france-usa.org/atoz/internet_schools.asp

Enochsson, A. (2001). Children choosing Web pages. *The New Review of Information Behaviour Research, 2*, 151–166.

Environics Research Group. (2000a). *Canada's children in a wired world: The parents' view.* Ottawa, Canada: Media Awareness Network. Retrieved November 19, 2003, from http://www.education-medias.ca/english/special_initiatives/surveys/parents_survey.cfm

Environics Research Group. (2000b). *Young Canadians in a wired world: Parents and youth focus groups in Toronto and Montreal.* Ottawa, Canada: Media Awareness Network. Retrieved November 19, 2003, from http://www.education-medias.ca/english/special_initiatives/surveys/focus_groups.cfm

Environics Research Group. (2001). *Young Canadians in a wired world: The students' view.* Ottawa, Canada: Media Awareness Network. Retrieved November 19, 2003, from http://www.education-medias.ca/english/special_initiatives/surveys/students_survey.cfm

Feldman, A., Konold, C., & Coulter, B. (2000). *Network science a decade later: The Internet and classroom learning.* Hillsdale, NJ: L. Erlbaum.

Fidel, R., Davies, R., Douglass, M., Holder, J., Hopkins, C., Kushner, E., et al. (1999). A visit to the information mall: Web searching behavior of high school students. *Journal of the American Society for Information Science, 50*(1), 24–37.

Findahl, O. (2001). *Swedes and the Internet. Year 2000.* Gävle, Sweden: World Internet Institute. Retrieved November 19, 2003, from http://www.worldinternetinstitute.org

Finkelhor, D., Mitchell, K., & Wolak, J. (2001). Highlights of the youth Internet safety survey. *OJJDP Fact Sheet 4.* Retrieved November 19, 2003, from http://www.ncjrs.org/txtfiles1/ojjdp/fs200104.txt

Foss, K. (2000, September 11). Canadians at forefront of Web use in schools. *Globe and Mail,* A1, A10.

Gardner, J. E., & Wissick, C. (2002). Enhancing thematic units using the World Wide Web: Tools and strategies for students with mild disabilities. *Journal of Special Education Technology, 17*(1), 27–38.

Goldsby, D., & Fazal, M. (2001). Now that your students have created Web-based digital portfolios, how do you evaluate them? *Journal of Technology and Teacher Education, 9*(4), 607–616.

Hanna, L., Risden, K., & Alexander, K. (1997). Guidelines for usability testing with children. *Interactions, 4*(5), 9–14.

Harbeck, J. D., & Sherman, T. M. (1999). Seven principles for designing developmentally appropriate Web sites for young children. *Educational Technology, 39*(4), 39–44.

Hart, G. (2000). A study of the capacity of Cape Town's children's librarians for information literacy education. *Mousaion, 18*(2), 67–84.

Hasselbring, T. S., & Glaser, C. H. W. (2000). Use of computer technology to help students with special needs. *The Future of Children, 10*(2), 102–122.

Haycock, K., Dobor, M., & Edwards, B. (2003). *The Neal-Schuman authoritative guide to kids' search engines, subject directories, and portals.* New York: Neal-Schuman.

Henke, L. (1999). Children, advertising and the Internet: An exploratory study. In D. W. Schumann (Ed.), *Advertising and the World Wide Web* (pp. 73–80). Mahwah, NJ: L. Erlbaum.

Hirsh, S. G. (1997). How do children find information on different types of tasks? Children's use of the Science Library Catalog. *Library Trends, 45*(4), 725–745.

Hirsh, S. G. (1999). Children's relevance criteria and information seeking on electronic resources. *Journal of the American Society for Information Science, 50*(14), 1265–1283.

Hoffman, J. L., Kupperman, J., & Wallace, R. (1997). *On-line learning materials for the science classroom: Design methodology and implementation.* EDRS Report.

Holloway, S. L., & Valentine, G. (2003). *Cyberkids: Children in the information age.* London: Routledge Falmer.

Hsieh-Yee, I. (2001). Research on Web search behavior. *Library & Information Science Research, 23*(2), 167–185.

Hunter, C. D. (2000). Social impacts: Internet filter effectiveness: Testing over- and under-inclusive blocking decisions of four popular Web filters. *Social Science Computer Review, 18*(2), 214–222.

Iannotta, J. G. (Ed.). (2001). *Nontechnical strategies to reduce children's exposure to inappropriate material on the Internet: Summary of a Workshop. Committee to Study Tools and Strategies for Protecting Kids from Pornography and Their Applicability to Other Inappropriate Internet Content.* Washington, DC: National Academy Press.

Infoplease.com. (2000). *Public schools with access to the Internet, 1994–2000.* Retrieved November 19, 2003, from http://www.infoplease.com/ipa/a0764484.html

Istri-Tarki. (2002). *Mapping the digital future: Hungarian society and the Internet.* Budapest: Hungary. Retrieved November 19, 2003, from http://www.worldinternetproject.net

Jacobson, F. F. (1995). From Dewey to Mosaic: Considerations in interface design for children. *Internet Research, 5*(2), 67–73.

Kafai, Y., & Bates, M. (1997). Internet Web-searching instruction in the elementary classroom: Building a foundation for information literacy. *School Library Media Quarterly, 25*(2), 103–111.

Kraut, R., Patterson, M., Lundmark, V., Kiesler, S., Mukopadhyay, T., & Scherlis, W. (1998). Internet paradox: A social technology that reduces social involvement and psychological well being? *American Psychologist 53*(9), 1017–1031.

Kuhlthau, C. C. (1991). Inside the search process: Information seeking from the user's perspective. *Journal of the American Society for Information Science, 42*(5), 361–371.

Kuntz, J. (2000). Criteria for comparing children's Web search tools. *Library Computing, 18*(3), 203–207.

Kupperman, J., & Fishman, B. (2001–2002). Academic, social, and personal uses of the Internet: Cases of students from an urban Latino classroom. *Journal of Research on Technology in Education, 34*(2), 189–215.

Laanpere, M. (2000). Tools and methods for Internet usage in Estonia's system of general education. *Baltic IT Review, 2.* Retrieved January 7, 2004, from http://www.dtmedia.lv/raksti/EN/BIT/200007/00071112.stm

Lankes, R. D. (2003). Current state of digital reference in primary and secondary education. *D-Lib Magazine 9*(2). Retrieved November 19, 2003, from http://www.dlib.org/dlib/february03/lankes/02lankes.html

Large, A., & Beheshti, J. (1999). Children's information seeking behavior: A laboratory versus an operational research environment. *Information Science: Proceedings of the 27th Annual Conference of the Canadian Association for Information Science,* 134–143.

Large, A., & Beheshti, J. (2000). The Web as a classroom resource: Reactions from the users. *Journal of the American Society for Information Science, 51*(12), 1069–1080.

Large, A., & Beheshti, J. (2001). Focus groups with children: Do they work? *The Canadian Journal of Information and Library Science, 26*(2/3), 77–89.

Large, A., Beheshti, J., & Breuleux, A. (1998). Information seeking in a multimedia environment by primary school students. *Library & Information Science Research, 20*(4), 343–376.

Large, A., Beheshti, J., Breuleux A., & Renaud, A. (1994). A comparison of information retrieval from print and CD-ROM versions of an encyclopedia by elementary school students. *Information Processing & Management, 30*(4), 499–513.

Large, A., Beheshti, J., & Cole, C. (2002). Architecture for the Web: The IA matrix approach to designing children's portals. *Journal of the American Society for Information Science and Technology, 53*(10), 831–838.

Large, A., Beheshti, J., & Moukdad, H. (1999). Information seeking on the Web: Navigational skills of grade-six primary school students. *Proceedings of the 62nd ASIS Annual Meeting,* 84–97.

Large, A., Beheshti, J., Nesset, V., & Bowler, L. (2003a). Children as designers of Web portals. *Proceedings of the 66th ASIST Annual Meeting,* 142–149.

Large, A., Beheshti, J., Nesset, V., & Bowler, L. (2003b). Children as Web portal designers: Where do we start? *Proceedings of the 31st Annual Conference of the Canadian Association for Information Science,* 139–152.

Large, A., Beheshti, J., & Rahman, T. (2002a). Design criteria for children's Web portals: The users speak out. *Journal of the American Society for Information Science and Technology, 53*(2), 79–94.

Large, A., Beheshti, J., & Rahman, T. (2002b). Gender differences in collaborative Web searching behavior: An elementary school study. *Information Processing & Management, 38*(3), 427–443.

Lawless, K. A., Mills, R., & Brown, S. W. (2002). Children's hypertext navigation strategies. *Journal of Research on Technology in Education 34*(3), 274–284.

Lazarus, W., & Mora, F. (2000). *Online content for low-income and underserved Americans: The digital divide's new frontier.* Santa Monica, CA: The Children's Partnership.

Lenhart, A., Rainie, L., & Lewis, O. (2001). *Teenage life online: The rise of the instant-message generation and the Internet's impact on friendships and family relationships.* Washington, DC: Pew Research Center.

Lento, E., O'Neill, K., & Gomez, L. (1998). Integrating Internet services into school communities. *Yearbook of the Association for Supervision and Curriculum Development,* 141–168.

Lessig, L., & Resnick, P. (1999). Zoning speech on the Internet: A technical and legal model. *Michigan Law Review, 98*(2), 395–431.

Levin, D., & Arafeh, S. (2002). *The digital disconnect: The widening gap between Internet savvy students and their schools.* Washington, DC: Pew Research Center.

Loh, C. S., & Williams, M. D. (2002). What's in a Web site? Student perceptions. *Journal of Research on Technology in Education, 34*(3), 351–363.

Lorenzen, M. (2001). *The land of confusion? High school students and their use of the World Wide Web for research.* Retrieved November 19, 2003, from http://www.michaellorenzen.net

Luchini, K., Quintana, C., & Soloway, E. (2002). ArtemisExpress: A case study in designing handheld interfaces for an online digital library. In F. Paterno, (Ed.), *Mobile Human–Computer Interaction: Proceedings of the 4th International Symposium, Mobile HCI, Pisa, Italy, 18–20 September, 2002.* Heidelberg, Germany: Springer-Verlag, 306–310.

Luckin, R., Rimmer, J., & Lloyd, A. (2001). What is the Internet and how can it help us learn? Exploring children's conceptions of what the Internet is and does. *ICCE 2001: International Conference on Computers in Education, Seoul, Korea, November 2001.* Retrieved November 19, 2003, from http://www.icce2001.org/cd/pdf/p13/UK008.pdf

Marchionini, G. (1989). Information-seeking strategies of novices using a full-text electronic encyclopedia. *Journal of the American Society for Information Science, 40*(1), 54–66.

Marchionini, G., & Komlodi, A. (1998). Design of interfaces for information seeking. *Annual Review of Information Science and Technology, 33,* 89–130.

McDermott, I. E. (2002). Homework help on the Web. *Searcher, 10*(4), 10–18.

McKechnie, L. (2001). Children's access to services in Canadian public libraries. *Canadian Journal of Information and Library Science, 26*(4), 37–55.

Meyer, B., Middlemiss, W., Thodorou, E., Brezinski, K., & McDougall, J. (2000). Effects of structure strategy instruction delivered to fifth-grade children using the Internet with and without the aid of older adult tutors. *Journal of Educational Psychology, 94*(3), 486–519.

Miller, L., Schweingruber, H., & Brandenburg, C. (2001). Middle school students' technology practices and preferences: Re-examining gender differences. *Journal of Educational Multimedia and Hypermedia, 10*(2), 125–140.

Minkel, W. (2000, January). Young children AND the Web: A Boolean match, or NOT? *Library Journal, 125*(1, supplement), 10–11.

Mistler-Jackson, M., & Songer, N. B. (2000). Student motivation and Internet technology: Are students empowered to learn science? *Journal of Research in Science Teaching, 37*(5), 459–479.

Mitra, S., & Rana, V. (2001). Children and the Internet: Experiments with minimally invasive education in India. *British Journal of Educational Technology, 32*(2), 221–232.

Montgomery, K. C. (2000). Children's media culture in the new millennium: Mapping the digital landscape. *The Future of Children, 10*(2), 145–167.

Mumtaz, S. (2001). Children's enjoyment and perception of computer use in the home and the school. *Computers and Education, 36*(4), 347–362.

Nachmias, R., Mioduser, D., & Shemla, A. (2000). Internet usage by school students in an Israeli school. *Journal of Educational Computing Research, 22*(1), 55–73.

Najjar, L. (1998). Principles of educational multimedia user interface design. *Human Factors, 40*(2), 311–323.

National Center for Education Statistics. (2000). *Teacher use of the computer and the Internet in public schools.* Retrieved November 19, 2003, from http://nces.ed.gov/surveys/frss/publications/2000090

National Center for Missing and Exploited Children. (1994). *Child safety on the information highway.* Retrieved November 19, 2003, from http://www.missingkids.com

National Center for Missing and Exploited Children. (1998). *Teen safety on the information highway.* Retrieved November 19, 2003, from http://www.missingkids.com

National School Boards Foundation. (2002). *Safe and smart: Research and guidelines on children's use of the Internet.* Retrieved November 19, 2003, from http://www.nsbf.org/safe-smart/full-report.htm

Nesset, V., & Large, A. (2004). Children in the information technology design process: A review of theories and their applications. *Library & Information Science Research, 26*(2), 140–161.

Neuman, D. (1995). High school students' use of databases: Results of a national Delphi study. *Journal of the American Society for Information Science, 46*(4), 284–298.

Nielsen, J. (2000). *Designing Web usability.* Indianapolis, IN: New Riders.

Nielsen, J. (2002). *Kids' corner: Website usability for children.* Retrieved November 19, 2003, from http://www.useit.com/alertbox/20020414.html

Norris, C. A., Smolka, J., & Soloway, E. (2000). Extracting value from research: A guide for the perplexed. *Technology & Learning, 20*(11), 45–48.

Norris, C. A., & Soloway, E. (2003). How handhelds can change the classroom. *Harvard Education Letter 19*(5), 8–17.

Oder, N. (2002, June 15). CIPA overturned due to overblocking protected speech. *Library Journal, 127*(11), 14.

Oravec, J. A. (2000). Countering violent and hate-related materials on the Internet: Strategies for classrooms and communities. *The Teacher Educator, 35*(3), 34–45.

Orleans, M., & Laney, M. C. (2000). Children's computer use in the home: Isolation or sociation? *Social Science Computer Review, 18*(1), 56–72.

Ormes, S. (1997). Treasure Island on the World Wide Web. *Information Services & Use, 17*(2/3), 125–131.

Oswell, D. (1999). The dark side of cyberspace: Internet content regulation and child protection. *Convergence, 5*(4), 42–62.

Penn, Schoen, & Berland Associates. (2000). *Web savvy and safety: How kids and parents differ in what they know, whom they trust.* Retrieved November 19, 2003, from http://www.microsoft.com/presspass/features/2000/nov00/11-27staysafe-study.doc

Persin, R. (2002). Web-assisted instruction in physics: An enhancement to block scheduling. *American Secondary Education, 30*(3), 61–69.

Perzylo, L., & Oliver, R. (1992). An investigation of children's use of a multimedia CD-ROM product for information retrieval. *Microcomputers for Information Management, 9*(4), 225–239.

Piaget, J., & Inhelder, B. (1969). *The psychology of the child.* New York: Basic Books.

Pownell, D., & Bailey, G. (1999). Electronic fences or free-range students? Should schools use Internet filtering software? *Learning and Leading with Technology, 27*(1), 50–57.

Reilly, R. (2001). Filtering, protecting children, and shark repellant. *Multimedia Schools, 8*(4), 68–72.

Revelle, G., Druin, A., Platner, M., Bederson, B., Hourcade, J. P., & Sherman, L. (2002). A visual search tool for early elementary science students. *Journal of Science Education and Technology, 11*(1), 49–57.

Ristau, R., Crank, F., & Rogers, H. (2000). Wisconsin public high school business teachers use of the Internet as a teaching and learning tool. *Delta Pi Epsilon Journal, 42*(3), 171–188.

Roberts, D. F., Foehr, U. G., Rideout, V. J., & Brodie, M. (1999). *Kids & media @ the new millennium: A comprehensive national analysis of children's media use.* Menlo Park, CA: Kaiser Family Foundation. Retrieved November 19, 2003, from http://www.kff.org/content/1999/1535

Roblyer, M. D., & Marshall, J. C. (2002–2003). Predicting success of virtual high school students: Preliminary results from an educational success prediction instrument. *Journal of Research on Technology in Education, 35*(2), 241–255.

Sanders, C. E., Field, T. M., Diego, M., & Kaplan, M. (2000). The relationship of Internet use to depression and social isolation among adolescents. *Adolescence, 35*(138), 237–242.

Sandvig, C. (2001). Unexpected outcomes in digital divide policy: What children really do in the public library. In B. Compaine & S. Greenstein (Eds.), *Communication policy in transition: The Internet and beyond* (pp. 265–293). Cambridge, MA: MIT Press.

Schacter, J., Chung, G., & Dorr, A. (1998). Children's Internet searching on complex problems: Performance and process analysis. *Journal of the American Society for Information Science, 49*(9), 840–849.

Schofield, J. W., & Davidson, A. L. (2002). *Bringing the Internet to school: Lessons from an urban district.* San Francisco: Jossey-Bass.

Shantz-Keresztes, L. (2000). "In conversation:" Students as Internet users in school libraries. *School Libraries in Canada, 20*(2), 13–16.

Shenton, A. K., & Dixon, P. (2003). A comparison of youngsters' use of CD-ROM and the Internet as information resources. *Journal of the American Society for Information Science and Technology 54*(11), 1029–1049.

Shneiderman, B. (1998). *Designing the user interface: Strategies for effective human–computer interaction* (3rd ed.). Reading, MA: Addison-Wesley Longman.

Shuman, B. A. (2001). *Issues for libraries and information science in the Internet age.* Englewood, CO: Libraries Unlimited.

Siann, G., Durdell, A., Macleod, A., & Glissov, P. (1988). Stereotyping in relation to the gender gap in computing. *Education Research, 30*, 98–103.

Slone, D. (2002). The influence of mental models and goals on search patterns during Web interaction. *Journal of the American Society for Information Science and Technology, 53*(13), 1152–1169.

Solomon, P. (1993). Children's information retrieval behavior: A case analysis of an OPAC. *Journal of the American Society for Information Science, 44*(5), 245–264.

Soloway, E., Norris, C., Blumenfeld, P., Fishman, B., Krajcik, J., & Marx, R. (2000). Log on education: K–12 and the Internet. *Communications of the ACM, 43*(1), 19–23.

Stevenson, S. (2001). K–12 education portals on the Internet. *Multimedia Schools, 8*(5), 40–44.

Subrahmanyam, K., Kraut, R. E., Greenfield, P. M., & Gross, E. F. (2000). The impact of home computer use on children's activities and development. *The Future of Children, 10*(2), 123–144.

Sutherland-Smith, W. (2002). Weaving the literacy Web: Changes in reading from page to screen. *The Reading Teacher, 55*(7), 662–669.

Taylor, A. (2001). Young Canadians in a wired world: A new survey on how Canadian kids are using the Internet. *Education Canada, 41*(3), 32–35.

Thornburgh, D., & Lin, H. S. (2002). *Youth, pornography and the Internet*. Washington, DC: National Academy Press.

Tsai, C.-C., Lin, S., & Tsai, M.-J. (2001). Developing an Internet attitude scale for high school students. *Computers & Education, 37*(1), 41–51.

Turner, K., & Kendall, M. (2000). Public use of the Internet at Chester Library, UK. *Information Research, 5*(3). Retrieved November 19, 2003, from http://informationr. net/ir/5-3/paper75.html

Turow, J., & Nir, L. (2000). *The view from parents; The view from kids*. Philadelphia: Annenberg Public Policy Center of the University of Pennsylvania.

UCLA Center for Communication Policy. (2003). *Internet report: Survey the digital future, year three*. Retrieved November 19, 2003, from http://www.ccp.ucla.edu

U.K. Office of Standards in Education. (1998). *A review of primary schools in England, 1994–1998*. Retrieved November 19, 2003, from http://www.official-documents.co.uk/doc-uments/ofsted/ped.htm

U.S. Supreme Court. (2003). *United States et al. v. American Library Association, Inc., et al.* Retrieved November 19, 2003, from http://www.supremecourtus.gov/opinions/02pdf/02-361.pdf

Vansickle, S. (2002). Tenth graders' search knowledge and use of the Web. *Knowledge Quest, 30*(4), 33–37.

Wallace, R. (1999) *Learners as users, users as learners: What's the difference?* Retrieved November 19, 2003, from http://www.msu.edu/~ravenmw/pubs/HCIC.LCDBoaster.htm

Wallace, R. M., Kupperman, J., & Krajcik, J. (2000). Science on the Web: Students online in a sixth-grade classroom. *Journal of the Learning Sciences, 9*(1), 75–104.

Walter, V. A. (1997). Becoming digital: Policy implications for library youth services. *Library Trends, 45*(4), 585–601.

Wang, A. Y., & Newlin, M. H. (2000). Characteristics of students who enroll and succeed in psychology Web-based classes. *Journal of Educational Psychology, 92*(1), 137–143.

Wartella, E., & Jennings, N. (2000). Children and computers: New technology–old concerns. *The Future of Children, 10*(2), 31–43.

Watson, J. S. (1998). "If you don't have it, you can't find it." A close look at students' per-ceptions of using technology. *Journal of the American Society for Information Science, 49*(11), 1024–1036.

Watson, J. S. (2001). Students and the World Wide Web. *Teacher Librarian, 29*(1), 15–19.

Wilhelm, T., Carmen, D., & Reynolds, M. (2002, June 1–8). Connecting kids to technology: Challenges and opportunities. *Kids Count Snapshot*. Retrieved November 19, 2003, from http://www.aecf.org/publications/pdfs/snapshot_june2002.pdf

Williams, P. (1999). The net generation: The experiences, attitudes and behaviour of chil-dren using the Internet for their own purposes. *Aslib Proceedings, 51*(9), 315–322.

Windschitl, M. (1998). The WWW and classroom research: What path should we take? *Educational Researcher, 27*(1), 28–33.

Wood, L. (2000, November 3). E-Xchange. *National Post*, E02.

Zukas, A. (2000). Active learning, world history, and the Internet: Creating knowledge in the classroom. *International Journal of Social Education, 15*(1), 62–79.

National Intelligence

CHAPTER 10

Intelligence, Terrorism, and National Security

Blaise Cronin
Indiana University

Introduction

This chapter deals with the nature of extreme terrorism and the threat it poses to national security. It analyzes the challenges faced by intelligence and counterintelligence services in the U.S., focusing in particular on the comparative strengths and weaknesses associated with the dominant organizational forms favored by, on one hand, radical terrorist groups (network) and, on the other, national intelligence agencies (bureaucracy).

"Intelligence, Terrorism and National Security" is not the first *ARIST* chapter to treat the subject of national intelligence. That honor goes to Philip Davies's "Intelligence, Information Technology, and Information Warfare," which appeared in volume 36; an earlier chapter by Cronin and Davenport (1993) on social intelligence addressed only tangentially classical notions of governmental intelligence and statecraft. P. H. J. Davies (2002, p. 313) rightly noted in his introduction that "it is indeed difficult to find any topic in information science and technology not relevant to intelligence, information warfare, and national security, or conversely." Recent events have merely underscored the validity of his claim, a claim that is buttressed in the present volume by Lee Strickland in his compendious chapter on domestic security surveillance and civil liberties.

Theoretical insights and professional developments in information science have had, as one might expect, an effect on the practice of intelligence work; and there have been close ties since World War II between the intelligence and information science professions (Farkas-Conn, 1990, chapter 6; B. Wilson, 2001). Moreover, many of the concepts, tools, and techniques developed by information scientists are used routinely in the gathering, organizing, retrieval, and dissemination of information and data by intelligence specialists in a wide variety of institutional and organizational contexts (Cronin, 2000; Cronin & Crawford, 1999b; Sigurdson & Tågerud, 1992). But this chapter is less about information and communication technologies (ICTs) than the ways, subtle and profound, in which organizational structure and organizational culture can influence both individual and institutional information behaviors—including the adoption and use of information systems—and, ultimately, the effectiveness of national intelligence and security services.

Connecting the Dots

The stanza that follows has been quoted more than once since the events of September 11th, 2001, but it bears repetition:

Upon this gifted age, in its dark hour
Rains from the sky a meteoric shower
Of facts...they lie unquestioned, uncombined.
Wisdom enough to leech us of our ill
Is daily spun, but there exists no loom
To weave it into fabric. (Edna St. Vincent Millay, *Collected Sonnets*, CXL)

The poet's words found plangent echo in many quarters following the spectacular terrorist attacks on the cities of New York and Washington. At first glance September 11th was a textbook illustration of strategic surprise, an attack that was not, and probably could not have been, foreseen in its specifics (see Dearth & Goodden [1995] on the nature of strategic surprise in warfare and the challenges it poses for intelligence services, Harmon [in press] on the predictive modeling of terrorist incidents, and Berkowitz & Goodman [2000, p. 12] on "instantaneous threat"). The scale and audacity of the attacks on U.S. soil—a traumatizing penetration of the "zone of interior" and loss of sanctuary (Molander, Riddile, & Wilson, 1996, pp. 12–13)—caused not just physical but also economic and emotional shockwaves (see Silke [2003b] for discussion of the psychological dimensions of terrorism). More particularly, it underscored Gerald Holton's (2002, online) observations on both the goals and consequences of new forms of terrorism:

> Cataclysmic events, the perpetration of enormities, and other precipitous changes in the human condition, stretch personal and historic memory beyond the limits of the accommodations of the ordinary plastic range; they make a plastic deformation, leaving the psyche different, distorted...

Criticism of the nation's intelligence community came from all sides, most authoritatively in the shape of an 850-page report written after an extensive and unprecedented inquiry conducted by the U.S. Congress (U.S. Senate. Select Committee on Intelligence & U.S. House of Representatives. Permanent Select Committee on Intelligence, 2002). The committee (2002, p. 608) excoriated the "deep background of analytical and organizational dysfunction and mismanagement in the national security arena." At the time of writing, a federal commission is completing another investigation into the attacks with a target date of Summer 2004 for the publication of its eagerly awaited report (Isikoff, 2004). In addition, independent commissions have been established (early 2004) in both the U.S. and U.K. to investigate the production, quality, and use of the intelligence that was invoked to support the still unproven claims that Iraq possessed weapons of mass destruction (WMD) and thus constituted a clear and present danger to national and international security (e.g., Barry & Hosenball, 2004a, 2004b).

Senator Richard Shelby, a long-serving member of the Senate Committee on Intelligence, was notably vocal on the subject of deficiencies within the intelligence establishment and wrote separately on the need for reform in the wake of the attacks (Shelby, 2002). Shelby pulled no punches, highlighting the calamitous effects of "the centrifugal tendencies of bureaucratic politics" on the performance of the major agencies, in general, and also identified particular shortcomings within the system:

> The CIA's [Central Intelligence Agency] chronic failure, before September 11th, to share with other agencies the names of known Al-Qa'ida terrorists who it knew to be in the country allowed at least two such terrorists the opportunity to live, move and prepare for the attacks without hindrance from the very federal officials whose job it is to find them. Sadly, the CIA seems to have concluded that the maintenance of its information monopoly was more important than stopping terrorists from entering or operating within the United States. Nor did the FBI [Federal Bureau of Investigation] fare much better, for even when notified in the so-called "Phoenix Memo" of the danger of Al-Qa'ida flight school training, its agents failed to understand or act upon this information in the broader context of information the FBI already possessed about terrorist efforts to target or use U.S. civil aviation. (Shelby, 2002, p. 5)

Shelby's indictment of the intelligence community is widely shared, the evidence hard to dismiss. The picture that is painted calls to mind Alvin and Heidi Toffler's (1993, p. 61) term, "diseconomies of complexity." Others sought to provide a larger historico-political context, relating criticisms of the intelligence establishment to the policies and actions, formal and informal, overt and covert, of the Bush administration, its neoconservative core in particular (e.g., Chossudovsky, 2002; Clarke, 2004; Vidal, 2002).

The U.S. intelligence community is a sprawling and enormous enterprise, in terms of its personnel and budgets ($40 billion is probably not an unreasonable estimate of the aggregate annual expenditure on national intelligence activities) (see Richelson [1999] for a comprehensive and highly detailed overview of both civilian and military agencies). The 14 member organizations (see http://www.intelligence.gov) are: Air Force Intelligence, Army Intelligence, Central Intelligence Agency, Coast Guard Intelligence, Defense Intelligence Agency, Department of Energy, Department of Homeland Security, Department of State, Department of Treasury, Federal Bureau of Investigation, Marine Corps Intelligence, National Geospatial-Intelligence Agency, National Reconnaissance Office, National Security Agency, and Navy Intelligence. The community's evolution post-World War II and subsequent maturation during the Cold War era have given it a particular set of institutional and behavioral characteristics, which does not equip it for the challenges of the twenty-first century (Berkowitz & Goodman, 2000; Treverton, 2001).

Two of the behemoths of the intelligence community, the CIA and FBI, were seen as "the bumbling rivals that dropped the ball many times in the years and months before September 11, 2001" (H. Thomas, 2002, online). In their book, *The Age of Sacred Terror*, Daniel Benjamin and Steven Simon, former director and senior director, respectively, for counterterrorism at the National Security Council, note that the FBI—unlike, for instance, the U.K.'s MI5—is a law enforcement agency, not a domestic intelligence or national security body (Benjamin & Simon, 2002, p. 395; see also Strickland in this volume). Historically, the FBI has been both constitutionally and culturally ill-suited to the counterterrorism role it is currently expected to perform, not just within the U.S., but, of late, extra-territorially (Frankling, 2002). That said, the Bureau does have a counterintelligence arm that investigates espionage and other national security crimes, as well as crimes perpetrated against Americans overseas (e.g., terrorist incidents that result in the death of U.S. servicemen). Although not referring specifically to the FBI's situation, Michael Herman (2001, p. 230) has made the point that "terrorism should be recognized as intelligence's special and distinctive target, distinguished from its more generalized back-up to law enforcement, and not confused by it." Role confusion is not, however, the whole story; the FBI has "evolved into an anarchic patchwork of fiefdoms ... a disorganized jumble of competing and unruly power centers" (Benjamin & Simon, 2002, pp. 299–300) with "antique information technology" (Hitz & Weiss, 2004, p. 22), which has seriously impaired its operational effectiveness. The CIA, for its part, has been roundly criticized for its lack of relevant foreign language capability (far too few Arab linguists) and also for its inability to infiltrate Al Qaeda and acquire the kind of human-source, clandestine intelligence needed to deal with the adversary. Of course, penetrating a close-knit terrorist group or cell calls for a very special mix of talents and resources and inevitably requires the services of unsavory types, individuals whose motives, morality, and actions may leave much to be desired (see Berkowitz & Goodman, 2000, Chapter 5, on the nature of and need for covert action in intelligence work). This kind of covert operation differs in certain respects from Cold War espionage and counterespionage and also, in other fundamental respects, from the sophisticated functions performed by the NRO (National Reconnaissance Office) and comparable technology-intensive agencies. It involves long-term immersion in a hostile environment well beyond the *cordon sanitaire* of diplomatic exchange and immunity, unusually high levels of personal risk, and almost invariably a degree of moral funambulism.

With hindsight, the dots can often be connected with ease: History is littered with so-called intelligence failures and post hoc rationalizations (see Finley, 1994, and Lanning, 1996). Indeed, "failure to connect the dots" has become the media's and others' metaphor of choice to explain blunders within the nation's sprawling intelligence and counterintelligence system: "too many missed opportunities, too many missed clues, and too much systemic blindness" (Benjamin & Simon, 2002, p. 385). A year after the World Trade Center attack, there was widespread conviction that "little fundamental restructuring" (Risen & Johnston, 2002, p. 1) had taken place within the nation's intelligence system, despite efforts to enhance interagency cooperation and bolster counterterrorism analysis capability (e.g., belated consolidation of the twelve different terrorist watch lists maintained by nine different federal agencies). Yet, as the second anniversary of September 11th approached, Robert Mueller, Director of the FBI, declared that the Bureau was a changed organization and that fighting terrorism had been made its top priority (Mueller, 2003). But connecting the dots is not the end of the matter: "The fact that the dots are connected must be persuasively explained and communicated to decision- and policymakers" (Popp, Armour, Senator, & Numrych, 2004, p. 41). Intelligence consumers often hear what they want to hear and ignore evidence that does not fit or support their view of the world (see Dearth & Gooden, 1995, Part II, "The politics of intelligence"), or, more charitably, they have a richer sense of the context to which the intelligence relates.

Criticism of the two high-profile agencies has featured continually in both the general media and the scholarly/professional press over the past couple of years. One of the most stinging indictments of the status quo came from John Perry Barlow, cofounder of the Electronic Frontier Foundation:

> Bureaucracies naturally use secrecy to immunize themselves against hostile investigation, from without or within. This tendency becomes an autoimmune disorder when the bureaucracy is actually designed to be secretive and is wholly focused on other, similar institutions. The counterproductive information hoarding, the technological backwardness, the unaccountability, the moral laxity, the suspicion of public information, the arrogance, the xenophobia (and resulting lack of cultural and linguistic sophistication), the risk aversion, the recruiting homogeneity, the inward-directedness, the preference for data acquisition over information dissemination, and the uselessness of what is disseminated—all are

the natural, and now fully mature, whelps of bureaucracy and secrecy. (Barlow, 2002, pp. 45–46)

However justified the Jeremiahs are by the evidence (for instance, the President's Daily Briefs [PDB] prior to September 11th contained warnings about possible attacks by Al Qaeda, including the hijacking of planes [Isikoff, 2003, 2004]), their criticisms sometimes constitute an oversimplification of the facts. In fairness, the highlighting of failures should be viewed in relation to the number of successful interdictions achieved by intelligence services in the face of a continuous and escalating global terrorist threat, for, as is frequently observed, the terrorist has to be successful only once, but the security services have to be successful every time. Malcolm Gladwell (2003, online) has challenged knee-jerk September 11th postmortems, arguing that "there is no such thing as a perfect intelligence system" and suggesting that the rivalry between the FBI and CIA should be viewed not as an instance of dysfunctionality, but rather as "a version of the marketplace rivalry that leads to companies working harder and making better products"—an issue explored at length by Berkowitz and Goodman (2000, chapters 2 and 3). Gladwell's (2003, online) most telling point, though, has to do with connecting the dots, and specifically his invocation of Baruch Fischoff's notion of "creeping determinism," defined thus: "the occurrence of an event increases its reconstructed probability and makes it less surprising than it would have been had the original probability been remembered." Beware 20/20 (re)vision(ism), in other words.

Connecting the dots may seem fairly straightforward to armchair intelligence analysts, but, as Gladwell reminds us, at the time of Senator Shelby's remarks the FBI's counterterrorism division reportedly had 68,000 outstanding and unassigned leads dating back to 1995. That translates into a potentially gargantuan population of connectable dots or "an information management problem of a new order of magnitude" (Hitz & Weiss, 2004, p. 25), one exacerbated by a host of information technology (IT) and information systems-related shortcomings (e.g., inadequate IT infrastructure, outmoded equipment, noninteroperable databases, lack of analytic tools) facing the FBI, CIA, NSA (National Security Agency), and others, to say nothing of the persistent political rivalries and turf clashes that resulted in suboptimal data exchange and information sharing (e.g., Berkowitz, 2003; Verton, 2003). Turf clashes are nothing new, of course; John Keegan (2003, p. 336) talks of the "disdain evinced by the 'hard' agencies—NSA, GCHQ [Government Communications Headquarters]—for the 'soft'—CIA, SIS [Secret Intelligence Service]" in both the U.S. and U.K., but the negative externalities caused by this friction can have grave consequences for national security. Of course, which dots are actually seen, heeded, and subsequently connected is, as organizational theorists would point out, partly a function of organizational structure and human sense-making processes. In the words of Karl Weick (1995, p. 133), sense making "is about the enlargement of small cues. It is a search for contexts within which small details fit together and make sense." Believing is seeing, in other words. Thomas Hammond (1993, p. 131), in reconstructing CIA intelligence gathering prior to the Cuban missile crisis, makes the point that structure materially affects behavior and outcomes:

> In sum, we see that even when an organization's field agents are seeing exactly the same raw data about the organization's environment and so are sending identical reports upward, different kinds of comparisons among different sets of information can be expected to take place in different structures; the different things that are thereby learned can have important political considerations.

The race is on to learn from past mistakes: The title of Peter Bergen's (2002) review essay in *Foreign Affairs*, "Picking up the pieces: What we can learn from—and about—9/11," is indicative of the prevailing sentiment. As far as IT is concerned, the goal is not only to enhance existing systems but also to create a new generation of data fusion centers and powerful databases ("extremely large, omni-media, virtually-centralized, semantically-rich information repositories," in the language of DARPA [Defense Advanced Research Projects Agency: http://www.darpa.mil/baa/baa02-08.htm]) along with innovative software tools to better track, analyze, and predict terrorist actions and threats. Less ambitious and more concrete initiatives are also underway across the nation to improve federal law enforcement authorities' intelligence collection and processing capabilities (e.g., Chen, Zeng, Atabakhsh, Wyzga, & Schroeder, 2003). At one level the failure to prevent some or all of the September 11th hijackings of commercial airliners can be accounted for in terms of a series of more or less discrete events, "an infinity of petty circumstances," to appropriate Carl von Clausewitz's (1982, p. 164) phrase: human error, cognitive overload, selective bias, organizational misunderstandings, and lapses of judgment exacerbated by inadequate or incompatible information systems and historical rivalries between (and within) key agencies. But pettiness is by no means the whole story; the root problem is systemic in nature, and something more than particularistic analysis and ad hominem recrimination is called for, if meaningful reform is to be instituted and progress made in addressing the omnipresent threat of "postmodern terror" (Juergensmeyer, 2003, p. 228). Double-loop, not single-loop learning from recent mistakes and oversights is required.

Organizational Forms and Cultures

Although the historical missions and organizational cultures of the CIA and FBI differ in many important respects (Riebling, 1994), the two leviathans have much in common. They are hierarchically structured (with, for instance, predictable career paths, clear lines of reporting, and unity of command) and generally favor functional specialization: see S. T. Thomas (1999, p. 407) on the CIA's "long history of bureaucratic entrenchment." And according to Berkowitz and Goodman (2000, p. 67), who provide concrete illustrations of organizational failings and operational failures:

> The intelligence community is a classical bureaucracy, characterized by centralized planning, routinized operations, and a hierarchical chain of command. All of these features leave the intelligence organization ill suited for the Information Age.

Few would nominate the CIA and FBI as examples of risk-taking cultures. Both organizations—high profile instances of normative drift and moral delinquency notwithstanding (e.g., Chomsky, 1988)—are publicly accountable, and as such are expected to behave in certain ways. To resort to cliché, changing the culture of either the CIA or FBI is akin to making a U-turn in a supertanker, only much more difficult, as many an ex-DCI (Director of Central Intelligence) has learned to his cost: DCIs come and go, but the core culture remains seemingly resolute (Baer, 2002; Berkowitz & Goodman, 2000; Johnson, 2000; Jones, 2001). The CIA and its Soviet era equivalent, the KGB (Committee for State Security), were children of their time. But the enemy (henceforth for convenience Al Qaeda, which means "the base" in Arabic, will be used metonymically for the extreme terrorism threat spectrum) has changed: it is no longer a nation state, territorially defined, physically embodied, bureaucratically organized, and possessed of a conventional military. Rather it is a network, albeit one that is at various times and in various ways state-sponsored (e.g., Bodansky, 2001; Daskal, 2003)—more accurately, a network or coalition of networks, structurally and dynamically different in many respects from colossi such as the CIA or the U.S. military. These differences are not merely academic; they, in fact, create potentially powerful advantages for the networked organization and raise plausible concerns about fitness for purpose within the intelligence establishment. As Berkowitz and Goodman (2000, p. 63) wrote *prior* to September 11th, the traditional (late twentieth century) model for producing intelligence "is outdated ... hierarchical, linear, and isolated."

Among the first to grasp the competitive advantages accruing to networked organizations in an informational society was Manuel Castells (but see also, for example, Ouchi [1980] and Thompson, Frances, Levacic, & Mitchell [1991]). In volume one of his monumental trilogy, *The Information Age: Economy, Society and Culture* (Castells, 2000, p. 187), he provides a generic definition of an organization as "a system of means structured around the purpose of achieving particular goals." Castells next distinguishes between bureaucracies, enterprises, and network enterprises. For bureaucracies, "reproduction of their system of means becomes the main organizational goal," whereas in enterprises, "goals, the change of goals, shape and endlessly reshape the structure of means" (Castells, 2000, p. 87): fixity vs. flexibility and fluidity, in other words. The third type of organization is the network enterprise, "that specific form of enterprise whose system of means is constituted by the intersection of segments of autonomous systems of goals" (Castells, 2000, p. 187)." Castells (2000, p. 187) goes on to state that "the components of the network are both autonomous and dependent vis-à-vis the network, and may be part of other networks, and therefore of other systems of means aimed at other goals." That is to say, organizations may at times share goals, have dovetailing objectives, and be mutually supportive at the operational level, while at the same time maintaining independence of mission and action.

In short, bureaucracies are in the business of self-reproduction/self-replication; it does not matter that their particular form of organizational design may not best satisfy prevailing needs and strategic aims (see Beetham [1991] on models of bureaucracy and Hammond [1986] on agenda control and bureaucratic politics). By way of contrast, contemporary networked (terrorist) organizations are loosely coupled assemblies of operationally independent enterprises, with more or less convergent goals—ideological commitment to universal jihad in the case of pan-Islamic fundamentalists; pragmatism in the case of Al Qaeda's links with organized crime in various parts of the world (see Bodansky, 2001, p. 322; J. Gray, 2003, pp. 90–91; Gunaratna, 2002, p. 115)—that are able to reconfigure themselves rapidly in response to the demands and challenges of the moment. Their inherent flexibility allows them, in boxing parlance, to punch above their weight, while classical bureaucracies, to stick with the pugilistic metaphor, have the fleetness of foot of a super-heavyweight. The question, in short, is whether large, bureaucratically organized intelligence agencies are structurally disadvantaged when confronted by postmodern terrorist coalitions, such as Al Qaeda—described by Magnus Ranstorp (2002, p. 1) as an organization that "mutates and transforms itself according to operational requirements," by Rohan Gunaratna (2002, p. 58) as "a fluid and dynamic, goal-oriented rather than rule-oriented organization," by Benjamin and Simon (2002, p. 27) as akin to "a team from McKinsey or Boston Consulting" when it comes to planning and managing its distributed operations, and by Bruce Berkowitz and Allan Goodman (2000, p. 10) as "a network organization in the purest sense."

To what extent, in the so-called "War against Terror," does the defining organizational form of the protagonists have a bearing on events? To what extent can, or should, the intelligence community attempt to reorganize itself to adapt to the new realities of asymmetric, substate conflict? What are the potential downsides of initiating large-scale organizational reform for publicly accountable bodies? There is a broadly based conviction that new institutions and initiatives are required to deal with the nature and scale of the threat posed by Al Qaeda and its international affiliates, as the following sampling of conventional wisdom demonstrates. Gunaratna (2003a, online), of the Centre for the Study of Terrorism and Political Violence at the University of St. Andrews, maintains that "there is no standard textbook for fighting al-Qaeda ... new structures and institutions will have to be built and shaped to fight a rapidly evolving but cunning and ruthless foe, willing to kill and die." Vice Admiral Lowell Jacoby (2003, online), Director of the (U.S.) Defense Intelligence Agency, is on record as saying that the Al Qaeda "network is adaptive, flexible, and, arguably, more agile than we are." Similar sentiments can also be found in the popular press. Howard Rheingold (2002) has described the emergence of "smart mobs," and Thomas Stewart (2001a, p. 58) claims that we will "have to think like a street gang, swarm like a soccer team, and communicate like Wal-Mart" if we are to deal with the terrorist threat. Such notions are not altogether new, however. As we shall see, the theoretical antecedents of current thinking on networked terrorism are to be found in a number of domains.

The Concept of Netwar

Researchers, consultants, and policy analysts at the RAND Corporation (see http://www.rand.org) have long been interested in new organizational forms (NOFs) and the implications of NOFs for the conduct of terrorism, insurgency, and organized crime. RAND's theoretical insights on the nature and likely significance of netwar—sometimes referred to as net-centric war (Scott & Hughes, 2003)—have had a profound effect on military and national security thinking both within the U.S. and farther afield. In their seminal study, *The Advent of Netwar*, John Arquilla and David Ronfeldt (1996) provide a detailed analysis of the organizational and doctrinal dimensions of netwar—not to be confused with information or cyber warfare (e.g., Cronin & Crawford, 1999a)—which they define thus:

> An archetypal netwar actor consists of a web (or network) of dispersed, interconnected "nodes" (or activity centers)—this is the key defining characteristic ... these nodes may be individuals, groups, formal or informal organizations, or parts of groups or organizations. The nodes may be large or small in size, tightly or loosely coupled, and inclusive or exclusive in membership. ... The organizational structure is quite flat. There is no single leader or commander. ... There may be multiple leaders. ... Decisionmaking and operations are decentralized ... the structure may be cellular for purposes of secrecy or substitutability. (Arquilla & Ronfeldt, 1996, p. 9)

Although Al Qaeda is not mentioned by name, the authors note that this form of warfare will be most attractive to nonstate actors, such as terrorists, revolutionaries, and transnational criminals. As John Gray (2003, p. 76) has put it: "Al Qaeda resembles less the centralized command structures of twentieth-century revolutionary parties than the cellular structures of drug cartels and the flattened networks of virtual business networks." The organization's structural characteristics are certainly very different from those of either the FBI or CIA. In another RAND report written three years later, the authors (Lesser, Hoffman, Arquilla, Ronfeldt, & Zanini, 1999, p. 67) note that an "increasing number of terrorist groups are adopting networked forms of organization and relying on information technology to support such structures." In a subsequent monograph, Arquilla and Ronfeldt update and summarize their thinking on the nature and relevance of netwar:

> Hierarchies have a difficult time fighting networks. ... It takes networks to fight networks. ... Whoever masters the network form first and best will gain major advances. ... Counternetwar may thus require very effective interagency approaches. (Arquilla & Ronfeldt, 2001a, pp. 15–16)

The emphasis, here and in other RAND studies, is much more on the organizational than the technical aspects of networking—which is not to say that terrorists, revolutionaries, social activists, and other groups do not make extensive and often imaginative use of information and communication technologies to support their recruitment, propaganda, planning, communication, and operational activities (see, for example, Zanini & Edwards, 2001).

Of course, networked and cellular structures are not without precedent in the history of revolutionary struggle (one has only to think of the FLN [National Liberation Front] in Algeria or the IRA [Irish Republican Army] in Ulster), but Al Qaeda differs from these and other revolutionary/guerrilla groups in a number of key respects. First, the struggle is not defined territorially (as is the case with, say, the IRA in Northern Ireland, ETA [Basque Fatherland and Liberty] in Spain, or the PLO [Palestine Liberation Organization] in the Middle East), but globally. Second, the objective is not to become part of the political status quo and lay down arms when key sociopolitical demands have been met. Third, the scale and scope of the organization (and its resource base) are manifestly much greater than those of any traditional revolutionary or terrorist group: Al Qaeda is a globally distributed network of networks, the first transnational terrorist enterprise. Fourth, Al Qaeda's implacable opposition to the West, its mores, and value system permits "no compromise or coexistence" (Bodansky, 2001, p. 388): It is an all-or-nothing struggle. The millennial or anomic form of radical Islamic fundamentalism promoted by Al Qaeda is not amenable to deterrence; proportionality of action and reaction simply does not come into play, so the kinds of strategic gaming and risk calculation that characterized the Cold War standoff between the U.S. and the Soviet Union have no relevance today.

As Paul Davis and Brian Jenkins (2002, p. xviii) concluded in their study of deterrence and influence in the war on Al Qaeda, the "concept of deterrence is both too limiting and too naïve to be applicable to the war on terrorism." Reason, force, and concessions will all come to naught. While the prospect of mutually assured destruction (MAD) may have helped keep the two superpowers' ICBMs (inter-continental ballistic missiles) in their silos, the concept of MAD does not enter the calculus of radical Islamic fundamentalist groups, as the popularity of suicide bombing among young, rational, and well-educated recruits in the Middle East and elsewhere (e.g., Sri Lanka, Chechnya) demonstrates compellingly (Krueger & Malecková, 2003). Significantly, it appears that poverty is not "a necessary, let alone sufficient, condition for suicidal terrorism" and, further, that "the bomber has no clear profile" (Martyrdom and murder, 2004, p. 21; see also Ganor, 2003). Recruits—male and female—are as easy for terrorists to attract as they are difficult for intelligence analysts to identify; and their motivations are as likely to be secular as religious (Merari, 2002; see Hudson, 2002 for a comprehensive review of the research literature on terrorist profiling). It is probably only a matter of time before suicide bombing—"self-chosen martyrdom," in the language of Hamas (Juergensmeyer, 2003, p.74)—is exported to the U.S. homeland and elsewhere by sleeper cells acting under instructions from the leadership of Al Qaeda or local sympathizers, be they resident aliens or U.S. citizens, acting autonomously or in concert.

Al Qaeda (both the core elements and the constellation of geographically dispersed groups that subscribe, self-interestedly or otherwise, to its ideology or draw inspiration from its leadership) thus constitutes an unprecedented threat, both ideologically and organizationally, for traditionally minded and conventionally structured intelligence services. The enemy (the target) is not simply scattered across the globe and organizationally nimble; it has its own sophisticated intelligence and counterintelligence capability. Yossef Bodansky has described how Osama bin Laden's right hand man, Ayman al-Zawahiri, reviewed with his multinational commanders the organizations' entire chain of command "to make sure that the new redundant, resilient modus operandi would withstand future onslaughts by hostile intelligence service, including the arrest and interrogation of senior leaders" (Bodansky, 2001, p. 393). Al Qaeda is a good if idiosyncratic example of a learning organization (Senge, 1990), though Silke (2003a, online) would argue that it is as much "a state of mind" as a formal organization:

> To begin with, Al-Qaeda is not a traditional terrorist organization. It does not have a clear hierarchy, military mindset and centralized command. At best it is a network of affiliated groups sharing religious and ideological backgrounds, but which often interact sparingly. Al-Qaeda is a state of mind, as

much as an organization; it encompasses a wide range of members and followers who can differ dramatically from each other.

This is a revealing statement. There is a growing tendency, reinforced by certain genres of political rhetoric, to reify Al Qaeda. Kimberly McCloud and Adam Dolnik (2002, online), of the Monterey Institute of International Studies, refer to this as the "phenomenon of 'exaggerated enemy'" and note how the public packaging of Al Qaeda presents it as "the top brand name of international terrorism." They go on to say:

> The United States and its allies in the war on terrorism must defuse the widespread image of Al Qaeda as a ubiqui-tous, super-organized terror network and call it as it is: a loose collection of groups and individuals that doesn't even refer to itself as "Al Qaeda." Most of the affiliated groups have distinct goals within their own countries or regions, and pose little threat to the United States. Washington must also be careful not to imply that any attack anywhere is by defin-ition, or likely, the work of Al Qaeda. (McCloud & Dolnik, 2002, online)

Their cautionary views have been endorsed by others (e.g., Burke, 2004; Laughland, 2004; O'Neill, 2003). It is important, without in any way diminishing the nature and scale of the threat presented by postmodern terrorist groups, to consider the extent to which careless or lazy use of the label, Al Qaeda, helps construct in the popular imagination an entity that does not in fact exist. Al Qaeda is an admittedly useful portmanteau term, but this kind of invocation is double-edged; it powerfully encapsu-lates the threat while at the same time simplifying and mythologizing it.

At the risk of resorting to caricature, hierarchically structured bureaucracies are characterized by rigidities, risk-aversion, status bar-riers, accountability, and rule deference (one of the recurrent criticisms of the status quo has been the lack of integrated counterterrorism capa-bility, presumably exacerbated by the fact that the U.S. Congress dis-perses antiterrorism funding to almost fifty federal bureaus and offices in some twenty agencies [Hitz & Weiss, 2004, p. 24]). There is a connec-tion between the bureaucratic structures adopted by the intelligence community and the kinds of systemic failings that have been portrayed in the press and vividly attested to by insiders and whistleblowers (e.g., Rowley, 2002). The FBI is deluged with paperwork; the CIA is top-heavy with desk-tied analysts, while the NSA, with roughly 40,000 employees, is seriously stressed: "As tens of millions of communications continue to be vacuumed up by NSA every hour, the system has become over-whelmed as a result of too few analysts" (Bamford, 2002a, p. 647).

Generally speaking, there is inadequate sharing and coordination of information across the many different agencies and departments (including state and local law enforcement) with a mandated interest in intelligence and counterintelligence activities. Lack of system integration and interoperability are persistent problems. Interagency turf clashes, deep-seated, institutional loyalties, and differences in organizational culture have seriously impeded effective cooperation. Such factors suggest that bureaucracies are likely to have a difficult time fighting networks, which, to stay in stereotyping mode, are fluid, redundant, self-organizing, distributed, and flexible. Thus, as much current thinking would have it (e.g., Lesser et al., 1999), networks will be needed to fight networks (the Law of Requisite Variety), just as "white hat" (benign) hackers help us understand the mindset and methods of "black hat" (malign) hackers (see Schwartau, 1996).

According to conventional wisdom, today's intelligence community bureaucracies will have to metamorphose into flatter, networked enterprises if they are to contain the threat of hyper-terrorism. This is, of course, a somewhat crass characterization of the issue, one that tends to ignore the many constitutional, political, juridical, and cultural challenges that such wholesale organizational transformation would entail. In practice, it may be that something less than mirroring—controlled organizational innovation, for the sake of argument—will be the most that can reasonably be expected, a view held, for example, by Frederick Hitz and Brian Weiss (2004, p. 24), who argue that "reorganizing the entire federal government to fight international terrorism—a threat of unknown size and unknown duration—is imprudent." In the rush to adopt and adapt networked forms of organization it is important, as Duncan Watts points out (2003, p. 274) in an admittedly different context, to remember that "as much as we need to get away from an exclusively hierarchical view of firms, the hierarchy is not just an endemic feature of modern business, but an important one."

Means and Ends

History has shown that combating terrorism means abandoning any pretense of playing by the Queensbury rules. It is also the case that the penumbral world of covert operations remains effectively off limits to most of us for most of the time. That said, there is clear evidence that the U.S. and other nations are adapting their military and counterterrorism tactics in the Middle East and elsewhere to the new realities. In Afghanistan, Iraq, and elsewhere (e.g., Yemen, Somalia) the U.S. "is now taking on terror cells with its own furtive counter cells made up of spooks, paramilitaries, and G-men" (Hirsch, 2002, p. 19). Of late, both the CIA's Special Operations Group and Special Activities Division and the FBI's Special Operations Command have considerably expanded their manpower. In Iraq, the U.S. has been recruiting former officers of the Mukhabarat (Saddam Hussein's secret police) to support its intelligence gathering and counterintelligence efforts. Within the Pentagon, there is a growing recognition of the need for better human intelligence, deeper penetration on the ground, and more extensive use of covert operations or "special activities" (a.k.a. "black ops") to gather both tactical and operational intelligence. However, the U.S. is disadvantaged on a number of counts, for instance, lack of trust among the Muslim diaspora, which makes the recruiting of agents difficult, and, more concretely and immediately, a critical shortage of relevant language skills within all sections of the intelligence community. In other parts of the world, the situation may be different, as in the case of the Israelis, "whose intelligence agencies enjoy the advantage of being able to recruit agents among refugees from ancient Jewish communities in Arab lands, colloquial in the speech of the countries from which they have fled but completely loyal to the state in which they have found a new home" (Keegan, 2003, p. 316). The U.S. has a long way to go on the human intelligence front if it is to contain or neutralize the threat posed by postmodern terrorists.

In 2002, the Defense Science Board recommended the creation of a super-Intelligence Support Activity—an organization known as the Proactive, Preemptive Operations Group (P2OG), or less formally by the sobriquet, Gray Fox—to "bring together C.I.A. and military covert action, information warfare, intelligence and cover and deception" (Isenberg, 2002, online). One of the most contentious proposals was that P2OG would stimulate reactions among terrorist groups, i.e., goad terrorist cells into action, thereby exposing them to rapid counterattack. History has long shown that there are risks attendant to the use of source operations, rogue actions, and agents provocateurs, namely, "blowback" and loss of control (see Todd & Bloch, 2003, Chapter 3). One has only to consider the case of U.S. involvement in Central and Latin America in the latter half of the twentieth century (e.g., Chomsky, 1988), or the alleged collusion of the FRU (Force Research Unit)—a covert British military intelligence unit—in Northern Ireland with Loyalist death squads in the 1980s (e.g., N. Davies, 1999).

William Daugherty (2004) has recently analyzed review procedures for, and congressional oversight of, covert action programs in the U.S. since the Reagan presidency. There is a need, demonstrated authoritatively by Strickland in the present volume, for explicit controls to ensure that we do not end up betraying certain basic principles—what Michael Ignatieff (2003, online) refers to as "constitutional precommitments ... the cords we tie ourselves with to prevent us from becoming our own worst enemies" when our societies are faced with potentially destabilizing terrorist threats (cases in point being the internment of IRA suspects and sympathizers without trial in Northern Ireland during the 1970s and, currently, the disputed legality of the U.S. detention policy at Guantánamo [e.g., Lobel, 2004; Toobin, 2004]). In the struggle against terrorism, it is important "to keep the post-September 11 security imperatives in perspective" (Carothers, 2003, p. 97) and not permit the erosion of basic civil rights (Ignatieff, 2003).

Related concerns are being expressed at the expanding of the U.S. defense establishment's use of PMCs (private military corporations)

(Berrigan, 2001; see Mayer [2004] for a discussion of the role of private contractors in the war on, and reconstruction of, Iraq; and Berkowitz & Goodman [2000, chapters 2 and 3] on exploiting the private sector). One of the largest of a growing band of PMCs is DynCorp, which boasts annual revenues in excess of $2 billion (Baum, 2003). The rapid corporatization and globalization of the age-old mercenary business allows a government to control fixed costs by outsourcing certain military and quasi military functions and at the same time to create arms-length separation between itself and the contractor, which can be most convenient when things go awry, such as when U.S. civilians in Columbia working for DynCorp were involved in a firefight between FARC (Revolutionary Armed Forces of Colombia) guerrillas and Colombian police. Under such an arrangement, the federal government can, in theory at least, maintain plausible deniability as far as the direct involvement of U.S. forces is concerned (Berrigan, 2001).

Network Advantage

To understand why bureaucracies are ill-fitted to the task of combating contemporary terrorist threats, it is necessary to examine in more detail prevailing conceptions of netwar and also consider the ways in which networked terrorist groups go about their business. There is a burgeoning literature on the nature and dynamics of networked organizations in general and on terrorist and organized crime networks in particular (e.g., Arquilla & Ronfeldt, 1996, 2001c; Barabási, 2003; Castells, 2000; Zabludoff, 1997). Arquilla and Ronfeldt (2001a, p. 7) take pains to distinguish among, and provide illustrations of, the three generic forms of network: chain, star, and all-channel. By way of illustration, in a chain network message flow is uni- or bi-directional, whereas in an all-channel network communication (increasingly facilitated by pervasive cellular phone technology, global internetworking, and the World Wide Web) will be many-to-many and typically dense (hence the frequent media allusions to high levels of terrorist "chatter," although chatter can just as easily be misinformation). More specifically, Arquilla and Ronfeldt (2001a, p. 10) acknowledge the influence on their thinking of the early research on social movements carried out by Luther Gerlach and Virginia Hine, in particular the notion of segmented, polycentric, ideologically integrated networks (SPINSs), defined by Gerlach as follows:

> By *segmentary* I mean that it is cellular, composed of many different groups of varying size, scope, mission and capability. ... By *polycentric* I mean that it has many different leaders or centers of direction. ... Leaders are often no more than "first among equals" and some act chiefly as traveling evangelists, criss-crossing the movement network. ... By *networked* I mean that the segments and the leaders are reticulated systems of networks through various structural,

personal, and ideological ties ... this acronym [SPIN] helps us picture this organization as a fluid, dynamic expanding one, spinning out into mainstream society. (Gerlach, 1987, p. 115)

What becomes apparent from reading descriptions of Al Qaeda's activities and governance, however, is that the group successfully exploits elements of different organizational forms (for example, a vertical leadership structure at the core coupled with webs of semi-autonomous cells that undertake local entrepreneurial action) and continuously evolves and adapts to the needs of the hour. Al Qaeda also exemplifies an important characteristic identified by Castells (2000), namely, local flexibility conjoined with international complexity. According to Peter Bergen (2001, p. 196), Al Qaeda "is as much a creation of globalization as a response to it." The group's members are often cosmopolitan and highly mobile, its seemingly atavistic ideology notwithstanding. In somewhat similar vein, J. Gray (2003, p. 1) has characterized the organization as "a byproduct of globalization ... a privatized form of organized violence worldwide," one that on closer inspection has more in common with some "modern European revolutionaries than it does with anything in medieval times" (J. Gray 2003, p. 21).

One of the key tenets of netwar is that there should be no single center of gravity (i.e., no single leader; no single center of command and control) so that when a node is destroyed or degraded the overall performance of the network is not thereby compromised. Building in redundancy and resilience makes a network difficult to disrupt: This, of course, is the axial design principle underpinning the Internet. And yet, it could be argued plausibly that Al Qaeda constitutes the antithesis of this condition, in that its iconic leader, Osama bin Laden, has become synonymous in the popular mind with the organization. In reality, much of what is enacted under the name of Al Qaeda (and its various known or presumed affiliates) takes place without direct input or approval from the top. Bin Laden's real and enduring value to the Islamic fundamentalist movement is (and will continue to be) as a unifying and universally recognized symbol, the recruiting value of which is probably inestimable. He is more likely to inspire others than personally conceive, oversee, or conduct terrorist actions. In the words of a recent *Economist* editorial (Still out there, 2004, p. 9):

> Mr. Bin Laden operated more like a venture capitalist than a CEO, sponsoring operations with varying degrees of control. ... Some of the bombings committed under the al-Qaeda banner may take little more than inspiration from Mr. Bin Laden himself. Nobody really knows how large that loose network is, nor whether it is growing or shrinking.

If Osama bin Laden were to be captured or killed by Western or other forces, it is unlikely that Al Qaeda's operational capability would be impaired in the way that those of Sendero Luminoso (the Shining Path movement) in Peru were following the capture by government troops of its charismatic founder, Abimel Guzman. In Al Qaeda, "authority [is] determined by knowledge or function rather than position" (Todd & Bloch, 2003, p. 13). The organization is at once leaderless and leader-driven (leader-inspired, is perhaps a better way of putting it), centralized and distributed or, in the terminology of Arquilla and Ronfeldt (1996, p. 9), "acephalous (headless) and polycephalous (Hydra-headed)." This structural flexibility—schizophrenia, one is tempted to say—finds its echo in the organization's modus operandi, notably the use of surge/unsurge and pulse/withdraw tactics to put a designated target off balance (see Arquilla & Ronfeldt, 2001c; Stewart, 2001a). Al Qaeda, broadly understood, has shown repeatedly that it has the ability to mount simultaneous attacks on carefully selected targets (e.g., the 1998 East African embassy bombings and, in 2003, the tightly coordinated bombing attacks in both Saudi Arabia and Turkey) and then swiftly disperse and regroup. To complicate matters, Al Qaeda does not limit itself to targeting American assets, and/or those of American allies; it will strike at any target or asset it deems to be contaminated by infidel values. From a counterterrorism perspective, the breadth of the threat spectrum, and the political and ideological mix of nations thus affected, can mightily complicate intelligence gathering and sharing activities and also counterintelligence efforts.

The tactics associated with small-unit swarming (an approach used by gangs of muggers on London's Oxford Street and the New York subway at various times in the recent past) are very different from the mass maneuvers and waves of attacks associated with conventional military thinking. Al Qaeda's preferred mode of operation (in addition to "postcard" target selection, i.e., targets with high symbolic value or "high human intensity," such as the World Trade Center [Miller & Stone, 2002, p. 264]) is the use of sustainable, pulsing force across a wide theater of operations. Mark Juergensmeyer (2003, p. 121; p. 124) talks revealingly of a "theater of terror" and "performance violence." Additionally, Al Qaeda's willingness to devolve operational autonomy to the local level means that intelligence and security forces struggle to come to grips with a sprawling, self-synchronizing, loosely coupled terrorist collectivity apparently capable of launching attacks in almost any place at any time, as demonstrated by the 2003 bombings in Bali and Casablanca and the March 2004 bombings in Madrid. As Gerlach (1987, p. 16) observed generally of social movements, the "autonomy and self-sufficiency of local groups make it difficult for its opposition to gather intelligence information about it and to know what all of its groups will do." For the reasons just adumbrated, bureaucratic organizations—both traditional intelligence agencies and conventional armed forces—are reassessing their effectiveness and the appropriateness of their structures in the face of a new and escalating threat to not only national but also international security.

Determinants of Netwar Effectiveness

Developing and exploiting a network structure is a *conditio sine qua non* of engaging in netwar, but organizational design (the nature, breadth, flexibility, and resilience of the network itself) is only one of several dimensions identified by Ronfeldt and Arquilla (2001) in their theoretical analysis of the phenomenon. This section of the chapter draws liberally on their thinking and summarizes the key dimensions of the construct with Al Qaeda specifically in mind (the authors illustrate current thinking and developments with reference to a variety of revolutionary, social, and other movements). Designing for optimal netwar effectiveness extends well beyond network formation and includes structural, cultural, technological, and behavioral dimensions. Al Qaeda has considerably refined the cellular or cluster model: many cells that are functionally autonomous and whose members are unknown to one another (Gunaratna, 2002, p. 76). In addition to organizational design, Ronfeldt and Arquilla note the importance of narrative assurance, doctrinal coherence, technological infrastructure, and social/cultural solidarity.

By narrative assurance they mean the story the organization tells and retells about itself, its goals, its achievements, and, in the particular case of Al Qaeda, the enemy (the Great Satan, a.k.a. the United States). Such behavior—the weaving of organizational lore, the systematic reinforcing of collective memory, and the basing of "organizational realities" [Weick, 1995, p. 127] on narration—is by no means peculiar to networked enterprises; corporations and governments, including various U.S. administrations, have shown themselves to be adept at symbolic representation (see Stone, 2002, Chapter 6): crafting and spinning stories of their own and deploying attenuated discourse to reinforce the simplified message. To quote Jerome Bruner (2002, p. 103): "We come to conceive of a 'real world' in a manner that fits the stories we tell about it." In the case of a multinational, multiethnic, virtual terrorist organization, the narrative glue becomes an absolutely critical element in forging groupthink and ideological solidarity. "Human beings united in commitment," is how Ranstorp (2002, p. 1), in fact, describes Al Qaeda.

Doctrinal coherence refers to the organization's preferred way of operating, the portfolio of strategies and tactics it has honed and deploys in pursuit of its short- and long-term goals. As far as Al Qaeda is concerned, that includes highly selective recruiting, formidable rites of passage, decentralized target selection, swarming and pulsing, meticulous operational planning, and functional compartmentalization.

Much more is implied by technological infrastructure than the computing and communications capabilities at the organization's disposal. The term encompasses an array of technical domains: bomb-making skills, the orchestration of suicide attacks (sometimes referred to as "martyrdom operations"), actual or presumed expertise in weapons of mass destruction and their delivery, the wherewithal to mount effective guerrilla attacks, and, no less important, the organization's money laundering mechanisms, exceptionally complex and well developed in the case of Al Qaeda (de Borchgrave, 1998; The money trail, 2002).

The fifth and final dimension that Ronfeldt and Arquilla (2001) identify is social/cultural solidarity, an especially important factor as far as Al Qaeda is concerned. This refers to the networked enterprise's ability to create or capitalize on (through, for instance, intra- or inter-group marriage) kinship and other solidifying sociocultural ties (ethnicity, language, religion), an example, according to J. Gray (2003, p. 80), of the organization's "pre-modern values." A similar phenomenon has been noted recently by U.S. military personnel serving in Iraq: "It's the Arabic rule of five. … The insurgency is self-replicating, like a virus through the vengeance of brothers, sons, cousins and nephews" (Hirsch, 2004, p. 39). In the Middle East, Hamas recruits suicide bombers for its struggle against Israel "through friendship networks in school, sports, and extended families" (Juergensmeyer, 2003, p. 79). In similar vein, Gunaratna (2002, p. 96) describes Al Qaeda as being organized along the lines of "a broad-based family clan with its constituent multi-national members designated as 'brothers'." The denser these kinship and cultural ties, the more robust the overall network; and the more resistant it is to infiltration by spies.

Sociologists (e.g., Granovetter, 1973, 1983) have long recognized the communicative value of weak ties:

> Weak ties are asserted to be important because their likelihood of being bridges is greater than (and that of strong ties less than) would be expected from their numbers alone. This does not preclude the possibility that most weak ties have no such function. (Granovetter, 1983, p. 229)

However, sociocultural cohesiveness may well be hard to maintain as the network (of networks) expands over time and space and the number of weak ties or links increases. The disadvantage of having a large numbers of weak ties is that the network becomes potentially vulnerable to penetration—after all, it is only as strong as the weakest link.

A sixth dimension might be added to the RAND list: mass support. Revolutionary groups such as the IRA, Tamil Tigers, or PLO depend for their existence on both the active and passive support of their heartland constituencies. If actions are too extreme or target selection grossly insensitive (e.g., the Real IRA's 1998 bombing in Omagh, Northern Ireland, which claimed 28 lives and injured 200) the group runs the risk of alienating public opinion and losing critical support within its own ranks. With Al Qaeda things are quite different. Its brand of radical Islamic fundamentalism seeks the overthrow of the West, its value system, and all those who sympathize, collaborate, or ally themselves with the West, including the Saudi royal family and the ruling elites in other Arab nations, most notably Egypt. Al Qaeda does not wish to become part of the established order or participate in the overall political process; it has nothing to lose; and there is nothing it fears losing. In the words of Fareed Zakaria (2002, p. 44), "The new Islamic terrorists want a lot of people dead and even more watching." Many of the organization's supporters are economically and politically disenfranchised; and they, in turn, have little if anything to lose by encouraging the kind of universal jihad enacted by Al Qaeda. This globally distributed mass of support—a base that is being continuously refreshed by the output of *madrassas*, Islamic religious schools that in some cases promote virulent anti-U.S. sentiments (Armanios, 2003; Lamb, 2003)—cushions and nurtures the organization in unique fashion. The nature of the foe and its unconventional modus operandi pose a stern challenge for counterterrorism forces in Western and other nations.

Counterterrorism Challenges

In February 15, 2001, the United States Commission on National Security/21st Century produced a report, *Road Map for National Security: Imperative for Change*, which made a number of predictions and recommendations covering the next quarter century. The Hart-Rudman Report, as it has come to be known, correctly identified the imminent threat of attack against the U.S. homeland (it actually occurred seven months later), but at the same time concluded that the "basic structure of the U.S. intelligence community does not require change. ... We recommend no major structural change" (U.S. Commission on National Security/21st Century, 2001, pp. 81–82). One wonders whether the authors would have produced a different recommendation had their report appeared *after* the September 11th attacks. In any event, it seems likely that meaningful reforms of the status quo can be carried out that do not necessitate wholesale organizational restructuring, but that could nonetheless alleviate some of the dysfunctionalities arising from cultural conflicts and institutional rivalries within the intelligence establishment.

During the Cold War, the CIA was tasked with gathering intelligence against a monolithic government and a large military–industrial infrastructure, the principal elements and instrumentalities of which were knowable and, in the main, fixed. The stabilities of the bi-polar world (United States *vs.* Soviet Union) are no more, and the U.S. has become "the lonely superpower" (Huntington, 1999, p. 35). Today's threat spectrum includes numerous nonstate/substate actors, groups that are operationally scattered and often very loosely coupled, examples, one might say, of distributed cognition (see Hutchins, 1995) at the organizational level. These challenges cannot be met by "a conventional force-on-force military campaign" (Daskal, 2003, online). The nature (indeed, the very existence) of some terrorist groups is unknown, or at best little known, and the complex of relationships among the various affiliates and factions is only partially apprehended. For example, Al Qaeda's use of sleeper cells and of sympathizers in the wider Muslim population makes it very difficult for security and intelligence forces to fully grasp the contours of the enemy. More specifically, it has necessitated greater degrees of cooperation and intelligence sharing, not just with Western allies but also with other, sometimes unfriendly, nations (e.g., Syria, Iran, Pakistan, Indonesia, Malaysia), a trend that will likely grow (Herman, 2002). Al Qaeda's elaborate cellular structure, culture of secrecy, and use of deception strategies all pose major challenges for the intelligence community, recent captures and killings of senior Al Qaeda figures notwithstanding. It is extremely difficult to develop a clear picture of the organism as a whole, as the critical elements are sometimes virtual, sometimes only dimly perceived, sometimes mutating. Tellingly, in the words of Magnus Ranstorp (2002, p. 1): "The way we think of Al Qaeda today is the day-before-yesterday's version." Caveat lector!

This realization has led some scholars and analysts to call for major organizational change within the intelligence community. Gregory Treverton, in his book, *Reshaping National Intelligence in an Age of Information*, argues that the traditional design of intelligence services ("stovepipes") around the different forms of intelligence—human intelligence (HUMINT), signals intelligence (SIGINT), image intelligence (IMINT), etc.—has resulted in the different organizations operating like "baronies" (Treverton, 2001, p. 8). He goes on to say that the intelligence community, were it a commercial enterprise, would logically restructure itself, establishing "a distributed network or a loose confederation in which the different parts of intelligence would endeavor to build very close links to the customers each served" (Treverton, 2001, p. 8). Robert Steele has also argued the case for structural reform following September 11th and made a number of specific recommendations: the breaking off of "a new Clandestine Services Agency out of the overly bureaucratized Central Intelligence Agency" and the conversion of the National Security Agency into the "National Processing Agency, a single point where all imagery, signals, human intelligence, and open source intelligence can be sliced, diced and visualized, time-stamped, and molded into usable grist for the all-source analysts and their consumers" (Steele, 2002, p. 172). These views are strikingly different from the conclusions and recommendations put forward by the Hart-Rudman commission.

Specific remedial steps that fall far short of wholesale structural reorganization might include concrete reforms aimed at reducing both intra- and interagency rivalry and segmentalism (to use Rosabeth Moss Kanter's [1983] term) such as the mixing and rotation of staff members from one agency to another to another—a practice not without precedence in the U.S. intelligence community. There is a felt (and real) need for the creation of fusion centers for intelligence pooling that would be staffed by analysts from different agencies. Greater use can be made of collaboration tools in ways that challenge the tendency to centralization:

Collaboration tools permit the formation of high-performance agile teams from a wide spectrum of organizations. These tools must support both top-down, hierarchically organized and directed, "center-based" teams as well as bottom-up, self-organized and directed adhocracies—"edge-based" collaboration. These two modes of operation must also be able to interoperate: "center-edge" coexistence. (Popp et al., 2004, p. 40)

Generally speaking, there is a need to develop more agile multisource intelligence capability, which would include making greater use of open source information (e.g., from the World Wide Web) and also more use of "unclassified lookers," business travelers, NGO (nongovernmental organizations) employees, and others, whose eyes and ears could be used in the harvesting of information and insights on specific countries, regions, or subjects. These and other ideas have been explored in detail by Treverton (2001), although the case for open source intelligence (making better use of unclassified information in the public domain) and also citizen intelligence has been made persistently and cogently by Steele (2000), and also more recently by Barlow (2002).

Comparative Organizational Strengths and Weaknesses

Al Qaeda is the prototypical global terrorist network. In certain respects the organization instantiates what Gerald Holton (2002, online) has termed Type III terrorism, a form of terrorism that "enables the two previously distinct types of terrorist agencies—states with potentially biblical scales of terror, and relatively independent small groups with limited powers of devastation—to collaborate, merge, or act, in secret or in more or less open collusion." Much has been written about Al Qaeda's origins, ethos, structure, ideology, and operations (e.g., Bergen, 2001; Bodansky, 2001; Gunaratna, 2002; Stern, 2003), but despite the not inconsiderable amount of information available (see the Council on Foreign Relations' Web site on Al Qaeda and other terrorism-related trends and developments [http://www.cfrterrorism.org/groups/alqaeda.html#Q]), it remains a stealth organization, loosely articulated, and, as such, inherently difficult to comprehend, infiltrate, and dismantle. As Gunaratna (2002, p. 10) has observed, the organization has "a proven capacity to regenerate new cells: its networks are intertwined in the socioeconomic, political and religious fabric of Muslims living in at least eighty countries." Al Qaeda's structural resilience, its international span, webs of alliances, dynamically evolving character, and growing diffuseness are in marked contrast to the defining organizational features of the intelligence community—bureaucratic structures, a fondness for massification, and risk aversion (e.g., Benjamin & Simon, 2002; S. T. Thomas, 1999; Treverton, 2001).

Moreover, the threat itself is an order of magnitude greater than prior threats posed by revolutionary, insurgency, or terrorist groups, which typically were mindful of the public's limited appetite for loss of human, particularly civilian, lives. Al Qaeda has propelled terrorism into an age of post-proportionality, in the process dramatically raising the intelligence and counterintelligence stakes. Its followers, particularly members of the Egyptian Islamist movement (such as Mohammed Atta, leader of the September 11th group of hijackers), have "become weapons in the fight to the death against technology" (Scruton, 2002, p. 13). How, then, do national security agencies deal with an enemy that is perceived to be "culturally and temporally perverse" (Arquilla & Ronfeldt, 2001b, p. 366), one that is committed to creating a pervasive climate of terror and cognitive dissonance?

Clearly, improvements can be made in both the operating efficiency and effectiveness of intelligence and counterintelligence activities without wholesale organizational restructuring. Greater emphasis can be placed on gathering human intelligence, developing automatic text translation capability, painstakingly penetrating terrorist cells and their support networks—both indigenously and extraterritorially—and on trying to reduce the massive backlog of information (e.g., cell phone traffic records; captured, but untranslated planning documents) and potential leads produced by both high-technology surveillance systems and by more traditional forms of intelligence tradecraft (Bamford, 2002b). More time and effort will have to be invested in grappling with the motivational drivers of Al Qaeda leaders and adherents; "red teaming," in military parlance, or "getting inside the enemy's head," to use the vernacular. Much more can be done to improve the ICT infrastructure across the intelligence community. The FBI's $458 million Trilogy network project now allows agents at almost 600 sites to receive multimedia case files at high speed (Mark, 2003), ensuring that criminal records, immigration databases (Strickland & Willard, 2002a, 2002b), terrorist watch lists, and other pertinent datasets (Allen-Mills, Fielding, & Leppard, 2004) are accessible on appropriate, need-to-know bases across agency and other lines. New software tools will be developed to help predict terrorist behavior, facilitate foreign language machine translation, and model network behavior (see the DARPA Information Awareness Program: http://www.darpa.mil/body/tia/tia_report_page.htm; and see also Shachtman, 2004). The creation of fusion centers, cross-organizational teams of specialists, and joint intelligence initiatives will not only help reduce interagency suspicions, but also enable analysts to identify and connect the relevant dots more swiftly (Popp et al., 2004). And, as already stated, more can be done to exploit the mass of information that is in the public domain. But, ultimately, what matters most is how the government's political agenda and policy thrusts dictate or constrain the professional activities of the intelligence community, and how government, as consumer, chooses to source, view, frame, and use both the raw and finished intelligence available to it, as the continuing controversy surrounding the intelligence used to legitimate the invasion of Iraq in 2003 has demonstrated (see Tenet, 2004).

Nonetheless, there are those who believe that patchwork solutions, system upgrades, and incremental reforms are simply insufficient to deal with the scale and nature of the threat posed by the current generation of transborder terrorists (e.g., Lesser et al., 1999; Treverton, 2001). Al Qaeda is a highly distributed enterprise, capable of directly or indirectly orchestrating or promoting terrorist attacks across an almost unbounded theater of operations. By way of contrast, the principal components of the intelligence community are often highly centralized in nature. Organizations such as the CIA, NSA, and FBI were not conceived or designed with Al Qaeda in mind. They are hierarchically organized, tightly managed, and expected to conform to the norms and laws of a democratic nation. And it is not just in the realm of counterterrorism that these tensions and contradictions are being experienced. Law enforcement organizations dealing with drug smuggling have found that they are disadvantaged by being slower than the narco-traffickers when it comes to changing their operations and procedures and in communicating information across the organization (Kenney, 2003, pp. 232–233). On the other hand, networks—especially leaderless ones—are not without an Achilles' heel of their own. All-channel networks in particular need constant interaction to hold the distributed membership together, if they are to function effectively, and that, in turn, may result in their communications circuits or critical nodes being compromised by intelligence services, a particularly pertinent observation given what is known of Al Qaeda's use of the Internet and other digital media (T. L. Thomas, 2003) and the rise, more generally, of "techno-terrorism" (see Zanini, 1999, p. 251).

Reformists would argue that the principal intelligence institutions should be demassified and made more flexible, distributed, mobile, proactive, and responsive to the emerging threat spectrum. Unquestionably, the intelligence community could exhibit greater structural flexibility, promote greater autonomy of action within the ranks, and encourage greater decentralization of both power and assets (human and technical) without abandoning the legal and moral standards to which they are required to conform, but there is a fine line between progressive devolution of authority (to, for instance, permit covert or black operations of one kind or another) and loss of institutional control such as when mavericks within the organization take the law into their own hands. But one should be careful not to reify "structure" and "network." The relationship between structure and action in a bureaucracy is as complex as the nostrums proposed by arms-length reformists are sometimes jejune. Structure is dynamic, not static. Social and political forces can shape the form and functioning of even the most sclerotic bureaucracy. One has only to imagine how leaks must affect social relations and reporting requirements inside the FBI or consider the impact on organizational climate within the CIA of a government-initiated inquiry into the efficacy of its intelligence gathering and analysis capability.

Although networks may offer certain inbuilt advantages over other forms of organizational design, as Castells (2000) maintains, they are certainly not indestructible. Fareed Zakaria (2002, p. 4) has tried to inject some balance into the current debate, noting as follows: "Things are not as bleak as they look. Terrorists today have few substantive advantages. Their resources are pitiful when compared with the combined power of governments all working together." This point has been made more recently and generally by the doyen of military historians, John Keegan (2003, p. 349):

> Foreknowledge is no protection against disaster. Even real-time intelligence is never real enough. Only force finally counts. As the civilized states begin to chart their way through the wasteland of a universal war on terrorism without foreseeable end, may their warriors shorten their swords. Intelligence can sharpen their gaze. The ability to strike sure will remain the best protection against the cloud of unknowing, prejudice and ignorance that threatens the laws of enlightenment.

But there is more to it than that. Zakaria goes on to quote (2002, p. 44) Edward Luttwak, who makes the perhaps obvious but all too easily overlooked point that "the terrorists are at an inherent disadvantage. Every time they attack they emerge. So you hunt them down, find them and kill them." In other words, every terrorist action is a possible window into the network, or at least into the workings of a particular cell or node, which, of course, is precisely the reason why both military and covert intelligence forces are keen to stimulate reactions; to force terrorists to show their faces, both literally and figuratively (e.g., Vistica, 2004). The challenge for intelligence analysts is to construct from the meticulous collection and examination of fragmentary evidence a viable picture of the network's overarching organizational structure and the important nodes. One way of doing this is to use and adapt techniques developed by social network analysts (e.g., Krebs, 2002; Stewart, 2001b). Formal network theory is now beginning to inform our understanding of how terrorist networks organize themselves. According to Albert-László Barabási (2003, p. 223), even if Mohammad Atta had been taken out, the hijacker cell would not have been crippled:

> Many suspect that the structure of the cell involved in the September 11 attack characterizes the whole terrorist organization. Because of its distributed self-organized topology, Al Qaeda is so scattered and self-sustaining that even the elimination of Osama bin Laden and his closest deputies would

not eradicate the threat they created. It is a web without a true spider.

The more one can visualize the structure, the better one is placed to penetrate Al Qaeda with undercover agents or "turn" a disaffected member of the organization. In fact, techniques and tools are in place to do just that. The capture of Saddam Hussein was reportedly facilitated by software called Analyst's Notebook (see http://www.i2inc.com and also Coffman, Greenblatt, & Marcus, 2004) which helped intelligence operatives piece together tribal and kinship connections and also identify weaknesses within extended social networks: The software is being used at present in the quest for Osama bin Laden (Yousafzai & Hirsch, 2004, p. 60)—even though it is now the fashionable view in some quarters that his arrest or elimination is not a critical issue (see Mayer, 2003). Of course, data mining, data tracking, and information visualization software are only as good as the information fed into the computer programs and the ability of analysts to make the necessary contextualizations and inferences.

But there is another dimension to be factored into the intelligence calculus. Human nature being what it is, a sprawling conglomerate—in this instance a transnational enterprise comprising a leadership cadre, battle-hardened fighters (the "Afghans"), core followers, associates, affiliates, agents, "unaligned mujahideen" (Taylor, 2004, online), and sympathizers—no matter how tightly bound its polyglot members are by their common belief system, will inevitably be faced with internal rivalries, schisms, and ideological splintering. J. Gray (2003, p. 79), in fact, describes Al Qaeda's ideology as "a highly syncretic construction." In addition, Al Qaeda and its networked partners could themselves, as do classical bureaucracies, become self-perpetuating organizations: "professional" terrorist groups (Stern, 2003, p. 1). Such developments could seriously compromise the organization's overall effectiveness and lead to debilitating, internecine warfare. It is hard to imagine that Al Qaeda and its many affiliates would be entirely immune to such weaknesses, although in making that casual statement one is mindful of a remark made to Rohan Gunaratna by his former mentor, Gerard Chaliand (Gunaratna, 2003b, p. 21): "You can only say you are a specialist on a terror group if you have cooked for them."

Finally, history may be able to provide some lessons for counteracting twenty-first century networks. Andrew Roach and Paul Omerod (2002) have suggested tantalizingly that grasping the secret of scale-free networks may help security forces combat terrorism, just as it helped the Inquisition stamp out the rapidly spreading Cathar heresy in late twelfth century Europe. It seems likely that combating the threat of extreme terrorism at the practical level, if not at the international, socioeconomic, policy-making level, will require a combination of critical self-reflexivity (grasping the limitations of prevailing organizational structures and judiciously adapting them to the new reality) and a deeper theoretical appreciation of the structural characteristics and evolutionary dynamics of social networks: understanding the enemy's strengths and vulnerabilities is, after all, a prerequisite of effective engagement, whether the battlefield is physical or virtual.

But, as we know, Al Qaeda is quick to learn. It has its own intelligence network and counterintelligence capability. Its operatives are deep inside enemy targets, security, and intelligence forces, as the assassination of Egypt's Anwar al-Sadat in 1981 and multiple attempts on Egypt's President Hosni Mubarak in 1995 and on Pakistan's President Pervez Musharraf in late 2003 convincingly demonstrate. Al Qaeda listens and watches as national security and military forces adapt themselves to deal with a new kind of enemy, and the terrorists, in turn, make organizational and procedural adaptations in light of what they learn from open sources, their operational failures, and their own agents. The battle is as much a battle of minds and ideas as of bombs and bullets: Think of it as a perpetual dialectic (or as a high stakes game of poker) as each side tries to pull and remain one step ahead of the other.

Summary and Conclusions

The so-called Revolution in Military Affairs (RMA) (Campen, Dearth, & Goodden, 1996; Herman, 1998) has helped foreground the axial role ICTs play in contemporary warfare (and also provided us with a new lexicon: netcentric warfare, digital warfare, postmodern war, meta-war, strategic information warfare, softwar, etc. [e.g., Adams, 1998; Cronin, 2001; de Landa, 1991; C. H. Gray, 1997; Molander et al., 1996; Rattray, 2001]). As the invasions of both Afghanistan and Iraq have demonstrated, brute force, when augmented with high-technology, precision-guided weapon systems, and intelligence-based battlespace planning, makes for a lethal combination. The RMA has challenged established theories about the conduct of war within the curricula of the nation's war colleges and also among the armed services' top brass (e.g., Libicki, 1995).

Antipathy in some quarters toward the new thinking, which places greater emphasis on intelligence; digital technologies; smaller, nimbler, and more flexible fighting units; aggressive deployment of special forces; and the like, has been both predictable and dogged. But recent U.S. military successes around the globe (whether or not credible cases of *jus ad bellum*, or, for that matter, evidence of sage foreign policy) and the growing terrorist threat suggest that the proponents of "force transformation" will eventually prevail. Military might and muscle will indubitably still matter, but the enemy—to use the terminology associated with Fourth Generation Warfare (4GW)—will resort to neither massed firepower nor massed manpower:

> It will be nonlinear, possibly to the point of having no definable battlefields or fronts. The distinction between

"civilian" and "military" may disappear. Actions will occur concurrently throughout all participants' depth, including their society as a cultural, not just a physical entity. (Lind, Nightengale, Schmitt, Sutton, & Wilson, 1989, p. 23)

Intriguingly, this definition of Fourth Generation Warfare was produced almost fifteen years ago, before the nature of the Al Qaeda threat had been evidenced and appreciated. Since then, the doctrine of 4GW, which has been described parsimoniously as "all forms of conflict where the other side refuses to stand up and fight" (Grossman, 2001, p. 1), has been taken on board enthusiastically by the Pentagon and has driven much of the current thinking on the need for transformation of the U.S. military. Donald Rumsfeld, U.S. Secretary of Defense and architect-in-chief of the force transformation program, argues that defense planning should begin with a frank assessment of a country's vulnerabilities and also address the critical question posed by Frederick the Great, namely: "What design would I be forming if I were the enemy?" (Rumsfeld, 2002, p. 25). This of course, begs the question whether the enemy does, in fact, have *a design*, in the sense of a consciously mapped out set of strategic objectives. Despite all the empirical evidence and informed speculation reviewed or alluded to in this chapter, it is still not easy to describe Al Qaeda with precision. Is it an exemplar of the networked enterprise described by Castells (2000), is it a more evolved example of the networked enterprise, or is it something altogether novel for which a new label is required? Although Al Qaeda resists pigeonholing, emergence is one of its defining properties.

It may well be that the intelligence community would benefit from a quiet revolution of its own, similar in some regards to the RMA, to ensure that the nation's national security services are optimally organized and strategically equipped to deal with the threats posed by what Jessica Stern (2003, online) has labeled the "protean enemy" and the associated spread of "totalitarian Islamist revivalism." But one should be careful not to over-focus on Al Qaeda, as J. Gray (2003, p. 98) cautions:

Al Qaeda is unlikely to be at the centre of resistance to US power for more than a decade or so. Radical Islam is likely to be only the first of a number of challenges to American hegemony. Asymmetric warfare will undoubtedly continue, with new protagonists we cannot foresee.

If an enormous bureaucracy such as the U.S. military can accommodate large-scale conceptual and organizational reengineering, albeit grudgingly, to deal with the threat of "post-Clausewitzian wars" (J. Gray, 2003, p. 73), there is no logical reason why the same should not hold for the intelligence community, the best efforts of the "gorillas in the stovepipes"—the various agency directors (Johnson, 2003, p. 643)—notwithstanding. What James Bamford (2002a, p. 650) said of the NSA, namely, to "succeed against the targets of the twenty-first century, the agency will have to undergo a metamorphosis, changing both its culture and technology," could equally be said of the intelligence community as a whole.

But changing the culture of the intelligence community will entail much more than transforming its information technology and systems. As Castells (2000, p. 186) has observed of the manufacturing industry, the major obstacles to achieving the kind of structural flexibility needed to operate effectively in a global economy are "rigidity of corporate cultures" and "bureaucratization." These, as we have seen, are also the charges most frequently leveled against the intelligence community, the CIA and FBI, in particular. Merely combining staffs in joint intelligence centers and reorganizing the key agencies will not solve the fundamental problem; organizational restructuring needs to be accompanied by "retooling" of analysts and the ways they are taught to think about issues, threats, and the mind-set of the enemy (see Treverton [2001] and Berkowitz & Goodman [2000] for critiques of the status quo and also for a raft of recommendations for both paradigmatic and procedural change). There is an irresistible case for organizational and procedural reform, but one should retain a sense of balance and perspective in the quest for what Ruth David (quoted by Berkowitz & Goodman, 2000, p. 81) has termed an "agile intelligence enterprise." As James Wilson (2000, p. 377) noted in his classic study of how and why government bureaucracies work the way they do:

> America has a paradoxical bureaucracy ... we have a system laden with rules ... we also have a system suffused with participation. ... Public bureaucracy in this country ... is rule-bound without being overpowering, participatory without being corrupt.

The intelligence community is neither perfect nor a law unto itself. Agencies such as the FBI and CIA exemplify the paradoxical, bureaucratic system described in such detail by J. Wilson (2000). They are subject to continuous oversight and, on occasion, intense public scrutiny. Their failures are trumpeted, their successes typically unrevealed or taken for granted. The immediate challenge will be to enhance the performance and accountability of the various national security agencies by inducing greater degrees of flexibility, customer-responsiveness, and decentralization without destroying the structural and cultural elements that have served them well over the years. Ultimately, it may be impossible to fashion an inherently bureaucratic organization that is free of internal contradictions and cultural tensions (see Richelson, 1999, p. 470). As Hammond and Miller (1985, p. 1) put it two decades ago: "It appears that the design of a bureaucracy requires making some unpalatable trade-offs among several desirable, but incompatible, organizational principles." For the intelligence community, the time has come to make some critical trade-offs and adaptations to the new reality without in the process abandoning proven and necessary elements of the dominant organizational form. The alternative, in the language of Berkowitz and Goodman (2000, p. 166), is likely to be "reform by catastrophe."

There might also be several lessons for information science in the foregoing summary or continuing narrative of intelligence, terrorism, and national security. A reading of recent events might raise questions about the focus and foundations of information science research, broadly construed, which has typically been oriented to systems user satisfaction approaches, with an emphasis on information storage and retrieval, information user profiling and usage patterns, bibliometrics, and the like. Have developments in information science and technology served largely to reinforce some of the shortcomings addressed here, such as bureaucratic rigidity, the creation of information fiefdoms and hoarding, "relevance" filtering, preferences for data collecting rather than dissemination, false feelings of security or outright arrogance, risk aversion, inward-directedness, and inappropriate secrecy? It seems that the nascent, and frankly still fuzzy, field of social informatics (e.g., Kling, 2000), augmented with cultural anthropology might at least partially provide a platform for reformed intelligence and information gathering and processing to deal with the widespread and still incipient global terrorist movement.

Acknowledgments

This chapter grew out of a public lecture, "Netcentric Terrorism," originally delivered in May 2003 at St. Anthony's College, University of Oxford, as part of the Trinity term seminar series, "Intelligence Services in the Modern World." An updated version was presented at Napier University, Edinburgh, in November 2003. Feedback received at both events is gratefully acknowledged, as are the many helpful comments of Elisabeth Davenport, Glynn Harmon, Kathryn La Barre, Major Craig Normand (U.S. Army), Alice Robbin, Yvonne Rogers, and Debora Shaw.

References

Adams, J. (1998). *The next world war: Computers are the weapons and the front line is everywhere*. New York: Simon & Schuster.

Allen-Mills, T., Fielding, N., & Leppard, D. (2004, January 11). Database duel lets terrorists through. *The Sunday Times*, 10.

Armanios, F. (2003). *Islamic religious schools, Madrasas: Background*. CRS Report for Congress. October 29, 2003. Retrieved January 24, 2004, from http://www.dec.org/pdf_docs/PCAAB157.pdf

Arquilla, J., & Ronfeldt, D. (1996). *The advent of netwar*. Santa Monica, CA: RAND.

Arquilla, J., & Ronfeldt, D. (2001a). The advent of netwar (revisited). In J. Arquilla & D. Ronfeldt (Eds.), *Networks and netwars: The future of terror, crime, and militancy* (pp. 1–25). Santa Monica, CA: RAND.

Arquilla, J., & Ronfeldt, D. (2001b). Afterword (September 2001): The sharpening fight for the future. In J. Arquilla & D. Ronfeldt (Eds.), *Networks and netwars: The future of terror, crime, and militancy* (pp. 363–371). Santa Monica, CA: RAND.

Arquilla, J., & Ronfeldt, D. (Eds.). (2001c). *Networks and netwars: The future of terror, crime, and militancy.* Santa Monica, CA: RAND.

Baer, R. (2002). *See no evil: The true story of a ground soldier in the CIA's war on terrorism.* New York: Crown.

Bamford, J. (2002a). *Body of secrets: Anatomy of the ultra-secret National Security Agency.* New York: Anchor Books.

Bamford, J. (2002b, September 8). Eyes in the sky, ears to the wall, and still wanting. *The New York Times*, 5.

Barabási, A.-L. (2003). *Linked: How everything is connected to everything else and what it means for business, science, and everyday life.* New York: Plume.

Barlow, J. P. (2002, October 7). Why spy? *Forbes ASAP*, 42–47.

Barry, J., & Hosenball, M. (2004a, February 16). The tale of the lying defector. *Newsweek*, 34.

Barry, J., & Hosenball, M. (2004b, February 9). What went wrong. *Newsweek*, 24–31.

Baum, D. (2003, February). This gun for hire. *Wired.* Retrieved January 23, 2004, from http://www.wired.com/wired/archive/11.02/gunhire.html

Beetham, D. (1991). Models of bureaucracy. In G. Thompson, J. Frances, R. Levacic, & J. Mitchell (Eds.), *Markets, hierarchies and networks: The coordination of social life* (pp. 128–140). London: Sage.

Benjamin, D., & Simon, S. (2002). *The age of sacred terror.* New York: Random House.

Bergen, P. (2001). *Holy war: Inside the secret world of Osama bin Laden.* New York: Free Press.

Bergen, P. L. (2002). Picking up the pieces. What we can learn from—and about—9/11. *Foreign Affairs, 81*(2), 169–175.

Berkowitz, B. (2003). Failing to keep up with the information revolution. *Studies in Intelligence, 47*(1). Retrieved January 26, 2004, from http://www.cia.gov/csi/studies/vol47no1/article07.html

Berkowitz, B. D., & Goodman, A. E. (2000). *Best truth: Intelligence in the information age.* New Haven, CT: Yale University Press.

Berrigan, F. (2001, February). The "ugly American problem" in Colombia. *Foreign Policy in Focus.* Retrieved January 26, 2004, from http://www.fpif.org/pdf/gac/0102colombia.pdf

Bodansky, Y. (2001). *Bin Laden: The man who declared war on America.* New York: Random House.

Bruner, J. (2002). *Making stories: Law, literature, life.* New York: Farrar, Straus & Giroux.

Burke, J. (2004). *Al-Qaeda: Casting a shadow of terror.* London: I. B. Tauris.

Campen, A. D., Dearth, D. H., & Goodden, R. T. (Eds.). (1996). *Cyberwar: Security, strategy and conflict in the information age.* Fairfax, VA: AFCEA International Press.

Carothers, T. (2003). Promoting democracy and fighting terror. *Foreign Affairs, 82*(1), 84–97.

Castells, M. (2000). *The rise of the network society* (2nd ed.). Oxford, UK: Blackwell.

Chen, H., Zeng, D., Atabakhsh, H., Wyzga, W., & Schroeder, J. (2003). Coplink: Managing law enforcement data and knowledge. *Communications of the ACM, 46*(1), 28–34.

Chomsky, N. (1988). *The culture of terrorism.* Boston: South End Press.

Chossudovsky, M. (2002). *War and globalisation: The truth behind September 11.* Shanty Bay, Ontario: Global Outlook.

Clarke, R. A. (2004). *Against all enemies: Inside America's war on terror.* New York: Free Press.

Coffman, T., Greenblatt, S., & Marcus, S. (2004). Graph-based technologies for intelligence analysis. *Communications of the ACM, 47*(3), 45–47.

Cronin, B. (2000). Strategic intelligence and networked business. *Journal of Information Science, 26*(3), 133–138.

Cronin, B. (2001). Information warfare: Peering inside Pandora's postmodern box. *Library Review, 50*(6), 279–294.

Cronin, B., & Crawford, H. (1999a). Information warfare: Its application in military and civilian contexts. *The Information Society, 15*(4), 257–263.

Cronin, B., & Crawford, H. (1999b). Raising the intelligence stakes: Corporate information warfare and strategic surprise. *Competitive Intelligence Review, 10*(3), 58–66.

Cronin, B., & Davenport, E. (1993). Social intelligence. *Annual Review of Information Science and Technology, 28*, 3–44.

Daskal, S. E. (2003, November 7). Changing the paradigm of war. Retrieved January 24, 2004, from http://www.d-n-i.net/fcs/daskal_changing_paradigms.htm

Daugherty, W. J. (2004). Approval and review of covert action programs. *International Journal of Intelligence and Counterintelligence, 17*(1), 62–80.

Davies, N. (1999). *Ten-thirty-three: The inside story of Britain's secret killing machine in Northern Ireland.* Edinburgh, UK: Mainstream Press.

Davies, P. H. J. (2002). Intelligence, information technology, and warfare. *Annual Review of Information Science and Technology, 36*, 313–353.

Davis, P. K., & Jenkins, B. M. (2002). *Deterrence and influence in counterterrorism: A component in the war on al Qaeda.* Santa Monica, CA: RAND.

de Borchgrave, A. (1998, August 25). Terrorism's cached assets. *The Washington Times,* A14.

de Landa, M. (1991). *War in the age of intelligent machines.* New York: Zone Books.

Dearth, D. H., & Goodden, R. T. (Eds.). (1995). *Strategic intelligence: Theory and application* (2nd ed.). Carlisle Barracks, PA: United States Army War College, Center for Strategic Leadership; Washington, DC: Defense Intelligence Agency, Joint Military Intelligence Training Center.

Farkas-Conn, I. S. (1990). *From documentation to information science: The beginnings and early development of the American Documentation Institute-American Society for Information Science.* Westport, CT: Greenwood Press.

Finley, J. (1994). Nobody likes to be surprised: Intelligence failures. *Military Intelligence, 20*(1), 15–21.

Frankling, D. (2002, October 14). Spooks vs. suits. Why the FBI and the CIA don't cooperate, and why they shouldn't. *Slate.* Retrieved January 18, 2004, from http://slate.msn.com/id/2072266

Ganor, B. (2003, April 2). The aim of suicide bombers—and how to beat them. *The Guardian.* Retrieved January 25, 2004, from http://www.ict.org.il/articles/mediaarticledet.cfm?articleid=12

Gerlach, L. E. (1987). Protest movements and the construction of risk. In B. B. Johnson & V. T. Covello. (Eds.), *The social and cultural construction of risk.* (pp. 103–145). Boston: Reidel.

Gladwell, M. (2003, March 10). Connecting the dots. The paradoxes of intelligence reform. *The New Yorker.* Retrieved January 24, 2004, from http://www.gladwell.com/2003/2003_03_10_a_dots.html

Granovetter, M. (1973). The strength of weak ties. *American Journal of Sociology, 78*, 1360–1380.

Granovetter, M. (1983). The strength of weak ties: A network theory revisited. *Sociological Theory, 1*, 201–233.

Gray, C. H. (1997). *Postmodern war: The new politics of conflict.* New York: Guilford Press.

Gray, J. (2003). *Al Qaeda and what it means to be modern.* New York: The New Press.

Grossman, E. M. (2001, October 4). Key review offers scant guidance on handling "4th generation" threats. *Inside the Pentagon,* 1. Retrieved January 26, 2004, from http://www.d-n-i.net/fcs/ITP_QDR.htm

Gunaratna, R. (2002). *Inside Al Qaeda: Global network of terror.* New York: Columbia University Press.

Gunaratna, R. (2003a). *Al-Qaeda's trajectory in 2003.* Retrieved January 23, 2004, from http://www.ntu.edu.sg/idss/Perspective/research_050303.htm

Gunaratna, R. (2003b, February 14). Cooking for terrorists. *The Times Higher Education Supplement,* 20–21.

Hammond, T. H. (1986). Agenda control, organizational structure, and bureaucratic politics. *American Journal of Political Science, 30*(2), 379–420.

Hammond, T. H. (1993). Toward a general theory of hierarchy: Books, bureaucrats, basketball tournaments, and the administrative structure of the nation-state. *Journal of Public Administration Research and Theory, 3*(1), 120–145.

Hammond, T. H., & Miller, G. (1985). A social choice perspective on expertise and authority in bureaucracy. *American Journal of Political Science, 29*(1), 1–28.

Harmon, G. (2002). Aircraft delivery platforms for biological attack: Systems for prevention, detection and countermeasures. *Proceedings of BTR 2002: Unified Science and Technology for Reducing Biological Threats and Countering Terrorism,* 215–223.

Harmon, G. (in press). Predicting major terrorist attacks: An exploratory analysis of predecessor event intervals in timelines. *Proceedings of BTR 2003: Unified Science and Technology for Reducing Biological Threats and Countering Terrorism.*

Herman, M. (1998). "Where hath our intelligence been?" The revolution in military affairs. *RUSI Journal, 143*(6), 62–68.

Herman, M. (2001). *Intelligence services in the information age: Theory and practice.* London: Cass.

Herman, M. (2002). 11 September: Legitimizing intelligence? *International Relations, 16*(2), 227–241.

Hirsch, M. (2002). Bush and the world. *Foreign Affairs, 81*(5), 18–43.

Hirsch, M. (2004, February 2). Blood & honor. *Newsweek,* 38–40.

Hitz, F. P., & Weiss, B. J. (2004). Helping the CIA and FBI connect the dots in the war on terror. *International Journal of Intelligence and Counterintelligence, 17*(1), 1–41.

Holton, G. (2002, February 4). Reflections on modern terrorism. *Edge, 97.* Retrieved January 27, 2004, from http://www.edge.org/3rd_culture/holton/holton_p6.html

Hudson, R. A. (2002). *Who becomes a terrorist and why: The 1999 government report on profiling terrorists.* Guilford, CT: The Lyons Press.

Huntington, S. P. (1999). The lonely superpower. *Foreign Affairs, 78*(2), 35–49.

Hutchins, E. (1995). *Cognition in the wild.* Cambridge, MA: MIT Press.

Ignatieff, M. (2003). The ethics of emergency. Lecture four in the Gifford Lectures 2003/04, *The lesser evil: Political ethics in an age of terror.* Edinburgh, UK: The University of Edinburgh. Retrieved January 19, 2004, from http://www.facultyoffice.arts.ed.ac.uk/Gifford/gifford_lectures.htm

Isenberg, D. (2002, November 5). 'P2OG' allows Pentagon to fight dirty. *Asia Times.* Retrieved January 21, 2004, from http://www.atimes.com/atimes/Middle_East/DK05Ak02.html

Isikoff, M. (2003, June 2). Censoring the report about 9-11? *Newsweek,* 8.

Isikoff, M. (2004, February 16). A new fight over secret 9/11 docs. *Newsweek,* 5.

Jacoby, L. E. (2003). *Global threat. Statement for the record. Senate Select Committee on Intelligence, 11 February.* Retrieved January 15, 2004, from http://www.nti.org/e_research/official_docs/dod/2003/dod021103.pdf

Johnson, L. K. (2000). The DCI vs the eight-hundred-pound gorilla. *International Journal of Intelligence and Counterintelligence, 13*(1), 35–48.

Johnson, L. K. (2003). Preface to a theory of strategic intelligence. *International Journal of Intelligence and Counterintelligence, 16*(4), 638–663.

Jones, C. M. (2001). The CIA under Clinton. *International Journal of Intelligence and Counterintelligence, 14*(4), 503–528.

Juergensmeyer, M. (2003). *Terror in the mind of god: The global rise of religious violence* (3rd ed.). Berkeley: University of California Press.

Kanter, R. M. (1983). *The change masters: Innovation for productivity in the American corporation.* New York: Simon & Schuster.

Keegan, J. (2003). *Intelligence in war: Knowledge of the enemy from Napoleon to Al-Qaeda.* New York: Knopf.

Kenney, M. C. (2003). Intelligence games: Comparing the intelligence capabilities of law enforcement agencies and drug trafficking enterprises. *International Journal of Intelligence and Counterintelligence, 16*(2), 212–243.

Kling, R. (2000). Learning about information technologies and social change: The contributions of social informatics. *The Information Society, 16*(3), 217–232.

Krebs, V. E. (2002). Uncloaking terrorist networks. *First Monday, 7*(4). Retrieved January 27, 2004, from http://www.foreignaffairs.org/20030701faessay15403/jessica-stern/the-protean-enemy.html

Krueger, A. B., & Malecková, J. (2003, June 6). Seeking the roots of terrorism. *The Chronicle of Higher Education,* B10–B11.

Lamb, C. (2003, March 20). 'Nurseries of terror' surge in Pakistan. *The Sunday Times,* 29.

Lanning, M. L. (1996). *Senseless secrets: The failures of U.S. military intelligence from George Washington to the present.* New York: Barnes & Noble.

Laughland, J. (2004, January 17). Do you believe in conspiracy theories? *The Spectator,* 14.

Lesser, I. O., Hoffman, B., Arquilla, J., Ronfeldt, D., & Zanini, M. (1999). *Countering the new terrorism.* Santa Monica, CA: RAND.

Libicki, M. C. (1995). *What is information warfare?* Washington, DC: National Defense University, Institute for National Strategic Studies.

Lind, W. S., Nightengale, K., Schmitt, J. F., Sutton, J. W., & Wilson, G. I. (1989, October). The changing face of war: Into the fourth generation. *Marine Corps Gazette,* 22–26. Retrieved January 26, 2004, from http://www.d-n-i.net/fcs/4th_gen_war_gazette.htm

Lobel, J. (2004, February 6). Winning lost causes. *The Chronicle of Higher Education,* B7–B9.

Mark, R. (2003, March 31). FBI says Trilogy program is complete. *Business.* Retrieved January 25, 2004, from http://www.internetnews.com/bus-news/article.php/2172131

Martyrdom and murder. (2004, January 10). *The Economist,* 20–22.

Mayer, J. (2003, August 4). The search for Osama. *The New Yorker,* 26–34.

Mayer, J. (2004, February 16). Contract sport. *The New Yorker,* 80–91.

McCloud, K. A., & Dolnik, A. (2002, May 23). Debunk the myth of Al Qaeda. *Christian Science Monitor.* Retrieved February 11, 2004, from http://www.csmonitor.com/2002/0523/p11s02-coop.html

Merari, A. (2002, April 19). Suicide bombers. *The Chronicle of Higher Education,* B4–B5.

Miller, J., Stone, M., with Mitchell, C. (2002). *The cell: Inside the 9/11 plot, and why the FBI and the CIA failed to stop it.* New York: Hyperion.

Molander, R. C., Riddile, A. S., & Wilson, P. A. (1996). *Strategic information warfare: A new face of war.* Santa Monica, CA: RAND.

The money trail: Investigating al-Qaeda. (2002). Retrieved January 13, 2004, from http://news.bbc.co.uk/1/shared/spl/hi/world/02/september_11/investigating_al_qaeda/money_trail/html/default.stm

Mueller, R. (2003, August 4). Quoted in: Isikoff, M., & Klaidman, D. Failure to communicate. *Newsweek,* 34–36.

O'Neill. B. (2003, November 28). Does al-Qaeda exist? *Spiked.* Retrieved February 11, 2004, from http://spiked-online.com/Printable/00000006DFED.htm

Ouchi, W. (1980). Markets, bureaucracies and clans. *Administrative Science Quarterly, 25*, 124–141.

Popp, R., Armour, T., Senator, T., & Numrych, K. (2004). Countering terrorism through information technology. *Communications of the ACM, 47*(3), 36–43.

Ranstorp, M. (2002, September 8). Quoted in: D. Frantz. Learning to spy with allies. *The New York Times*, Section 4: 1, 7.

Rattray, G. R. (2001). *Strategic warfare in cyberspace*. Cambridge, MA: MIT Press.

Rheingold, H. (2002). *Smart mobs: The next social revolution*. New York: Perseus Books.

Richelson, J. T. (1999). *The U.S. intelligence community* (4th ed.). Boulder, CO: Westfield.

Riebling, M. (1994). *Wedge: The secret war between the FBI and CIA*. New York: Knopf.

Risen, J., & Johnston, D. (2002, September 8). Not much has changed in a system that failed. *The New York Times*, Section 4: 1, 7.

Roach, A., & Omerod, P. (2002, June 13). Get medieval with al-Qaida. *The Times Higher Education Supplement*, 21.

Ronfeldt, D., & Arquilla, J. (2001). What next for networks and netwars? In J. Arquilla & D. Ronfeldt. (Eds.), *Networks and netwars: The future of terror, crime, and militancy* (pp. 311–361). Santa Monica, CA: RAND.

Rowley, C. (2002). Memo to FBI director Robert Mueller. Retrieved January 24, 2004, from http://www.time.com/time/covers/1101020603/memo.html

Rumsfeld, D. H. (2002). Transforming the military. *Foreign Affairs, 81*(3), 20–32.

Schwartau, W. (Ed.). (1996). *Information warfare: Protecting your personal security in the electronic age*. New York: Thunder's Mouth Press.

Scott, W. B., & Hughes, D. (2003, January 27). Nascent net-centric war gains Pentagon toehold. *Aviation Week & Space Technology*, 50.

Scruton, R. (2002, Fall). The political problem of Islam. *The Intercollegiate Review*, 3–15.

Senge, P. (1990). *The fifth discipline: The art and practice of the learning organization*. New York: Doubleday.

Shachtman, N. (2004, February). The bastard children of Total Information Awareness. *Wired*, 26.

Shelby, R. C. (2002). *September 11 and the imperative of reform in the U.S. intelligence community. Additional Views of Senator Richard C. Shelby Vice Chairman, Senate Select Committee on Intelligence*. Retrieved January 15, 2004, from http://intelligence.senate.gov/shelby.pdf

Sigurdson, J., & Tågerud, Y. (1992). (Eds.). *The intelligent corporation: The privatisation of intelligence*. London: Taylor Graham.

Silke. A. (2003a). Profiling terror. *Police Review*. Retrieved January 26, 2004, from http://www.janes.com/security/law_enforcement/news/pr/pr030807_1_n.shtml

Silke, A. (2003b). *Terrorists, victims and society: Psychological perspectives on terrorism and its consequences*. New York: Wiley.

Steele, R. D. (2000). Possible presidential intelligence initiatives. *International Journal of Intelligence and Counterintelligence, 13*(4), 409–423.

Steele, R. D. (2002). Crafting intelligence in the aftermath of disaster. *International Journal of Intelligence and Counterintelligence, 15*(2), 161–178.

Stern, J. (2003, July/August). The protean enemy. *Foreign Affairs*. Retrieved January 26, 2004, from http://www.foreignaffairs.org/20030701faessay15403/jessica-stern/the-protean-enemy.html

Stewart, T. A. (2001a). America's secret weapon. *Business 2.0. 2*(10), 54–63.

Stewart, T. A. (2001b). Six degrees of Mohamed Atta. *Business 2.0. 2*(10), 63.

Still out there. (2004, January 10). *The Economist*, 9–10.

Stone, D. (2002). *Policy paradox: The art of political decision making* (Rev. ed.). New York: Norton.

Strickland, L. S., & Willard, J. (2002a). Re-engineering the immigration system: A case for data mining and information assurance to enhance homeland security. Part I: Identifying the current problems. *Bulletin of the American Society for Information Science and Technology, 29*(1), 16–21.

Strickland, L. S., & Willard, J. (2002b). Re-engineering the immigration system: A case for data mining and information assurance to enhance homeland security. Part II: Where do we go from here? *Bulletin of the American Society for Information Science and Technology, 29*(1), 22–26.

Taylor, P. (2004, February 10). The secret war against al-Qaeda. *BBC News Magazine*. Retrieved February 10, 2004, from http://news.bbc.co.uk/1/hi/magazine/3476121.stm

Tenet, G. J. (2004, February 5). Iraq and weapons of mass destruction. Retrieved February 11, 2004, from http://www.cia.gov/cia/public_affairs/speeches/2004/tenet_georgetown-speech_02052004.html

Thomas, H. (2002, December 16). Assigning blame for 9/11 not so easy. Retrieved January 25, 2004, from http://www.wesh.com/helenthomas/1841170/detail.html

Thomas, S. T. (1999). The CIA's bureaucratic dimensions. *International Journal of Intelligence and Counterintelligence, 12*(4), 399–413.

Thomas, T. L. (2003, spring). Al Qaeda and the Internet: The danger of "cyberplanning." *Parameters*, 112–123.

Thompson, G., Frances, J., Levacic, R., & Mitchell, J. (Eds.). (1991). *Markets, hierarchies and networks: The co-ordination of social life*. London: Sage.

Todd, P., & Bloch, J. (2003). *Global intelligence: The world's secret services today*. London: Zed Books.

Toffler, A., & Toffler, H. (1993). *War and antiwar: Survival at the dawn of the 21st century*. Boston: Little, Brown.

Toobin, J. (2004, February 9). Inside the wire. *New Yorker*, 36–41.

Treverton, G. F. (2001). *Reshaping national intelligence for an age of information*. Cambridge, UK: Cambridge University Press.

U.S. Commission on National Security/21st Century. (2001). *Road Map for National Security: Imperative for Change*. Retrieved January 24, 2004, from http://www.nssg.gov/Phase IIIFR.pdf

U.S. Senate. Select Committee on Intelligence, & U.S. House of Representatives. Permanent Select Committee on Intelligence. (2002). *Joint inquiry into intelligence community activities before and after the terrorist attacks of September 11, 2001*. Washington, DC: Government Printing Office.

Verton, D. (2003, July 26). IT deficiencies blamed in part for pre-9/11 intelligence failure. *Computerworld*. Retrieved January 21, 2004, from http://www.computerworld.com/governmenttopics/government/story/0,10801,83469,00.html

Vidal, G. (2002, October 27). The enemy within. *The Observer*. Retrieved January 25, 2004, from http://www.mindfully.org/Reform/2002/The-Enemy-Within-Vidal27oct02.htm

Vistica, G. L. (2004, January 12–18). 'Kick down the doors' everywhere? *The Washington Post National Weekly Edition*, 6–7.

von Clausewitz, C. (1982). *On war* (A. Rapoport, Ed.). London: Penguin.

Watts, D. J. (2003). *Six degrees: The science of a connected age*. New York: Norton.

Weick, K. E. (1995). *Sensemaking in organizations*. Newbury Park, CA: Sage.

Wilson, B. (2001). Editorial: Challenging times. *D-Lib Magazine, 7*(11). Retrieved January 22, 2004, from http://www.dlib.org/dlib/november01/11editorial.html

Wilson, J. Q. (2000). *Bureaucracy: What government agencies do and why they do it*. New York: Basic Books.

Yousafzai, S., & Hirsch, M. (2004, January 5). The harder hunt for Bin Laden. *Newsweek*, 69–61.

Zabludoff, S. (1997). Colombian narcotics organizations as business enterprises. *Transnational Organized Crime*, *3*(2), 20–49.

Zakaria, F. (2002, December 9). How to fight the fanatics. *Newsweek*, 44.

Zanini, M. (1999). Middle Eastern terrorism and netwar. *Studies in Conflict & Terrorism*, *22*(3), 247–256.

Zanini, M., & Edwards, S. J. A. (2001). The networking of terror in the information age. In J. Arquilla & D. Ronfeldt, (Eds.), *Networks and netwars: The future of terror, crime, and militancy* (pp. 29–60). Santa Monica, CA: RAND.

CHAPTER 11

Domestic Security Surveillance and Civil Liberties

Lee S. Strickland
with David A. Baldwin and Marlene Justsen
University of Maryland

Introduction

Surveillance is a key intelligence tool that has the potential to contribute significantly to national security but also to infringe civil liberties. This potential is especially important because information science and technology have expanded dramatically the mechanisms by which data can be collected and knowledge extracted and thereafter disseminated. Moreover, in times of national or social threat, history has demonstrated that governments often expand surveillance and other powers at the expense of citizen rights; this expansion is accompanied by arguments that the innocent have nothing to fear, that mistakes can be corrected, and that the status quo will return when the danger is past. All too often, history also confirms that these powers tend to become a new and diminished baseline of legal rights.

This chapter examines the evolution of government surveillance in the U.S. from the emergence of organized policing, through the early efforts addressing sociopolitical threats, to the passage of the U.S.A. Patriot Act and additional proposals addressing the new threat of terrorism. It does so in the context of the very real threat presented to the nation today, the need for government to have the necessary intelligence to defend the public order, and the concomitant need for effective checks and balances to guarantee individual rights. It proposes that oversight and transparency can best protect national security and individual liberty.

The contributions to American national security made by information professionals are well documented. Indeed, during the era of the Cold War, information science concepts and tools contributed substantially to the defeat of an intractable enemy. Whether threats (e.g., the Khrushchev boast at a diplomatic reception in 1956 that "We will bury you") or obstructions (e.g., the famous shoe-banging episode at the United Nations General Assembly in 1960), there was little doubt then that the security of the nation and its people was at stake (Andrews, Biggs, & Seidel, 1996). And nowhere was the information science contribution more evident than in the context of intelligence. The word *intelligence* has various connotations, some negative; it can be properly defined as a process, a product, or an organization. It is characterized by the collection of information (ranging from raw, technical data to individual, expert knowledge) from all available sources (open as well as covert), the processing of that information (e.g., decryption or data reduction), the analysis of that information (i.e., validation, integration, and assessment of meaning), and lastly the creation of a product known as *finished intelligence* that is made available to national policymakers in order to inform them of relevant events, threats, or developments (*Factbook on intelligence*, 2003; Heuer, 1999). But the most salient fact about intelligence is that it is the necessary factual predicate to any successful national defense, as noted by President Eisenhower (1960, online): "During the period leading up to World War II we learned from bitter experience the imperative necessity of a continuous gathering of intelligence information ... [and] there is no time when vigilance can be relaxed." What, then, was the intersection between information science and intelligence? In summary terms, intelligence was a massive and largely secret developer of new hardware and software technologies key to its information storage, retrieval, and exploitation needs with examples including the development of the first large-scale, online retrieval system, keyword-in-context indexing, automatic indexing through statistical analysis, selective dissemination of information tools, and the conversion of digital data from technical platforms into relevant information for the analytical process (Bowden, Hahn, & Williams, 1999; Williams, n.d.).

Critical policy issues are at stake when considering information science and intelligence, given that an essential element of the intelligence mission is the collection of information concerning individuals through surveillance—officially defined by the U.S. Government (U.S. Joint Chiefs of Staff, 2001, p. 438) as "the systematic observation or monitoring of ... places, persons, or things by visual, aural, electronic, photographic or other means." Because surveillance encompasses a broad array of techniques, ranging functionally from covert to overt and legally from those requiring judicial warrants to being unrestricted by the law, it follows that information professionals must be as concerned with the civil liberties issues presented as with the technology and techniques utilized. Consider the spectrum of surveillance activities: from the ostensibly benign, such as the overt viewing of public events, to the most intrusive, such as the covert electronic acquisition of private conversations. Or consider the balance of equities: from the individual's right to privacy to the needs of the state to have the necessary information for the public defense. Indeed the rights of the people are complex and far reaching, including both traditional concepts of privacy vis-à-vis activities not exposed to the public and a right of anonymity as one travels from place to place, associates with others, or simply reads in a library.

Finally, consider the potential misuse of the powers of government surveillance. One example is from just forty years ago (King, 2001, pp. 342–343): "The ultimate measure of a man is not where he stands in moments of comfort, but where he stands at times of challenge and controversy." These and many other words by Martin Luther King, Jr. not only galvanized civil rights efforts but also presented an essential thesis of the American system of government—that the advocacy of even unpopular ideas must be rigorously protected under our Constitution. Yet, as documented by Congressional investigations (Church, 1976, Book III, pp. 81–82, 92), government law enforcement and intelligence, unrestricted by law or policy, engaged in a campaign of surveillance and harassment "to discredit and destroy Dr. King ... employing nearly every intelligence-gathering technique at the Bureau's disposal"—despite the absence of any evidence of criminal conduct or that he "was a communist, or that he was being influenced to act in a way inimical to American interests."

Nor was this an isolated exception, as evidenced thirty years ago when the Federal Bureau of Investigation's (FBI) library surveillance efforts were active and librarian Zoia Horn was jailed after an FBI informant working in her college library convinced Horn to hold meetings with associates of then-jailed priest Philip Berrigan. When FBI special agents demanded information from Horn regarding those meetings, she fearfully complied at first but later refused to testify at the trial of Berrigan and others (the infamous Harrisburg Seven) in 1972. As a result, she was convicted of contempt, jailed, and released only when the prosecution collapsed after the informant's criminal background, actions, and conflicting testimony became evident and the jury was unable to reach a verdict. In fact, even this intrusion into libraries was not unusual given the FBI's wide-ranging Library Awareness Program that involved visits, surveillance, and nonjudicial requests for documents at institutions ranging from the University of Maryland (where first discovered) to the New York Public Library (Egelko, 2002; Foerstel, 1991).

At one level, these cases express the clear need for a firm legal framework regulating the power of government surveillance and infiltration. The essential point, however, is regulation, not prohibition, because the government does require effective tools to protect national security from very real threats. As Strickland (2002a, 2002b, 2002c, 2002d, 2002e, 2002f, 2002g, 2003a, 2003c) has detailed in a series of articles in the *Bulletin of the American Society for Information Science and Technology*, the nation today faces a new enemy—ad hoc groups of international terrorists—where there are neither strategic infrastructure targets nor substantial government or military targets, and where the terrorist target hides in the shadows of foreign countries or even individual cities and states within the U.S. Quite clearly, this is not a contest between symmetrical forces where the nation can effectively use the methods and mechanisms of historical military contests but an asymmetric war against a multichannel network (see the chapter by Cronin in this volume). Here, terrorist nodes are highly interconnected, and the organization as a whole is highly collaborative in nature, quick acting, and effective. And because it is the model network of the information age, it can be, when organized by terrorists, very effective in that it can launch repeated attacks from different points, very difficult to identify key directors in that it is typically polycephalous, and very difficult to destroy in its entirety given that nodes are highly redundant. It follows that the U.S. requires new tools, foci, and strategies to address this new form of threat—where information will be the key to overall success. Such efforts are often termed *information warfare* or *Netwar*, with the latter term first utilized by the Rand Corporation and defined by Hoffman (1999, p. 19) as "an emerging mode of crime and conflict, short of traditional war, in which the protagonists use network forms of organization and related strategies and technologies attuned to the information age."

Accepting the propositions that information is the key to victory, that intelligence is the essential governmental activity necessary for ascertaining and disrupting these threat networks, and that surveillance is the primary intelligence tool, then an immediate question follows: *Exactly what legal framework is required to collect and act on the needed information but to do so in conformity with Constitutional principles and, hence, with the concomitant support of the public and the Congress?* Part of the answer concerning this instrument of state power comes from the fascinating historical role of domestic surveillance that will be detailed subsequently. Another part comes from the political and policy perspectives presented by the very statement of the issue. When we consider the collection of information in the U.S. for national security purposes, do we speak of the time-honored need for "intelligence" so that the system of government may endure? Or do we deride the equally time-honored abuses of "spying" on citizens? Indeed, what actions constitute these functional terms? Is it government agents "infiltrating" community groups or is it merely "observing" public events? Does it matter which community groups, citizens, or public events? And does it include collection of any information that may prove of interest or only information relating clearly to a criminal act or terrorism? The final and most substantive part of the answer has been presented by the passage of the U.S.A. Patriot Act, as well as the issuance of a number of policy guidelines by Attorney General Ashcroft that greatly enhance the domestic information collection authorities of the government—a subject that will be addressed later in this chapter.

Significant factual evidence demonstrates, however, that the legal framework in place today is being met with declining levels of public and Congressional support. Without that support American intelligence cannot succeed in its national security mission. In point of fact, the opposition that began in the library and civil liberties communities has spread to encompass a bipartisan spectrum of Americans who have not been satisfied by mere assurances that the government should simply be trusted with these authorities. That argument—specifically, that debate is unnecessary—began during the extraordinarily short debate over the U.S.A. Patriot Act when the Attorney General, covered by the media (e.g., CNN.com/US, 2001, online), nevertheless admonished the Senate Judiciary Committee that "those who scare peace-loving people with phantoms of lost liberty ... only aid terrorists for they erode our national unity and diminish our resolve ... give ammunition to America's enemies and pause to America's friends." Although Senator Leahy (D. Vt.) then offered some criticism by noting that government power must be balanced against civil liberties, subsequent months saw a growing expression of public sentiment against these authorities with the critical fuel being the "near-total information vacuum" as to their exercise (Goldstein, 2003, A1).

First, consider the data released to date detailing that exercise of new powers:

- The Attorney General's annual report on the use of the Foreign Intelligence Surveillance Act (FISA) authority (Ashcroft, 2003) stated only that it was used 1,228 times in the year 2002.

- Two reports by the Department of Justice (Brown, 2003; Bryant, 2002) to the House Judiciary Committee that have been criticized for lack of detail and broad claims of classification.

- A vitriolic statement by the Attorney General in September 2003 (Eggen, 2003) castigated members of the library community and asserted that no § 215 FISA court orders had been issued against libraries or bookstores.

These disclosures have done little to quell public concern; the Attorney General's statement, in particular, has been criticized for what it did not say. Specifically, concern continues to be expressed about the truthfulness of the statement (Krug, 2004) and the fact that it did not detail any use of § 215 against other targets or any use of other provisions authorized by the U.S.A. Patriot Act. Moreover, the harsh words and tenor of that statement—"The charges of the hysterics are revealed for what they are: castles in the air built on misrepresentation; supported by unfounded fear; held aloft by hysteria"—are unlikely to be a basis for public consensus (Eggen, 2003, p. A02).

Second, consider the data relevant to public and Congressional opinion:

- Literally hundreds of local jurisdictions have enacted ordinances in opposition to the U.S.A. Patriot Act including the states of Alaska, Vermont, and Hawaii as well as 142 local governments; some oppose the U.S.A. Patriot Act in principle and some make it an offense for local officials to cooperate with federal agents

(Holland, 2003; Nieves, 2003). As a matter of constitutional law, however, no such legislation could affect the performance of federal officers in these jurisdictions.

- Litigation under the Freedom of Information Act (FOIA), 5 U.S.C. § 552, includes a case brought by a number of public interest groups, the *American Civil Liberties Union, et al., v. Department of Justice*, seeking the number of subpoenas or other legal demands for bookstore and library records issued under the U.S.A. Patriot Act. The plaintiff's motion for summary judgment was denied in May 2003, after the release of a few highly redacted documents.

- In another FOIA case, the *Center for National Security Studies, et al. v. U.S. Department of Justice*, the plaintiffs sought the release of the names of the September 11th detainees. Although the District Court ordered release in August 2002, it stayed the decision pending appeal. In June 2003, the U.S. Court of Appeals for the D.C. Circuit reversed in a bitter two-to-one decision with the majority holding (*Center for National Security Studies, et al. v. U.S. Department of Justice*, 2002, p. 928) that the release "would give terrorists a composite picture of investigative efforts" and harm national security, but the minority found (*Center for National Security Studies, et al. v. U.S. Department of Justice*, p. 937) that the withholding "eviscerates" the FOIA and the well-established principles of openness.

- The most notable of other lawsuits to learn more about the use of U.S.A. Patriot Act authorities have concerned the blanket closure of immigration proceedings in 2002—with the 3rd Circuit finding closure appropriate in *North Jersey Media Group v. Ashcroft* and the 6th Circuit finding the opposite in *Detroit Free Press v. Ashcroft*. This divergence resulted even though the focus in both cases was the two-part "experience and logic" test established by the U.S. Supreme Court in *Richmond Newspapers Inc. v. Virginia* (1980), which examines whether specific types of proceedings have traditionally been open to the public and whether openness plays a significant, positive role in this process. When the U.S. Supreme Court denied review in the 3rd Circuit decision at the urging of the Department of Justice, the role of openness in judicial proceedings of this nature was left undecided, given that the vast majority of the immigration cases had been concluded.

- Senator Hatch and Representative Sensenbrenner, Chairmen, respectively, of the Senate and House Judiciary Committees, have refused to endorse additional powers set forth in the proposed

Domestic Security Enhancements Act, a follow-on to the U.S.A. Patriot Act (Goldstein, 2003).

- The Otter Amendment to the 2004 Appropriations Act for the Departments of Commerce, Justice, and State, passed in the House by a vote of 309 to 118. The amendment denies funding to execute any delayed notice (otherwise known as sneak-and-peak) search warrants. The use of such delayed notice warrant (the correct legal term) must be authorized by courts and was specifically authorized by § 213 of the U.S.A. Patriot Act that amended 18 U.S.C. § 3103a. To do so, a court must find that there is reasonable cause to believe that providing immediate notice would have an adverse result as defined in 18 U.S.C. § 2705, that there is a showing of reasonable necessity for the seizure, and that a specified time for ultimately giving notice is provided. In fact, there is a long history of such warrants that have been consistently recognized by the courts over the years (e.g., *United States v. Villegas*, 1990, and *United States v. Freitas*, 1986), which makes the Congressional opposition all the more emblematic of growing bipartisan opposition to certain provisions of the U.S.A. Patriot Act.

- A host of other legislation is pending in the House. For example, H.R. 1157, the Freedom to Read Protection Act of 2003 (Sanders, D. Vt.), would eliminate § 215 business record FISA orders directed toward libraries and bookstores but would still allow criminal warrants and subpoenas as well as FISA search and intercept warrants. The bill had 144 co-sponsors as of early 2004 but had been excluded from immediate legislative consideration by a procedural move. H.R. 3352, the Security and Ensured Freedom (SAFE) Act, introduced by Rep. Otter with 38 bipartisan co-sponsors would, inter alia, return the requirement that § 215 business record FISA orders pertain to a foreign agent (e.g., a spy or terrorist), restrict the granting of delayed notice search warrants in criminal cases, and define domestic terrorism to exclude political or other protestors.

- Proposed legislation in the Senate includes S. 1507, the Library, Bookseller and Personal Records Privacy Act (Feingold, D. Wis.), that is somewhat similar to H.R. 1157 and would also amend § 215 by requiring the government to show some individualized suspicion; specifically, the standard would become "specific and articulable facts" (Library, Bookseller and Personal Records Privacy Act, § 2, p. 2) that warrant an individual being suspected of being "an agent of a foreign power" (Library, Bookseller and Personal Records Privacy Act, § 3, p. 4). The Protecting the Rights of Individuals Act (S. 1552, Murkowski, R.

Alaska, and Wyden, D. Ore.) would authorize a substantial roll-back of U.S.A. Patriot Act authorities by redefining domestic terrorism to protect political protesters, by requiring a higher standard of proof for § 215 orders, by prohibiting agencies from engaging in data mining without explicit congressional authorization, and by reverting the standard for FISA orders primarily concerned with foreign intelligence. And there is S. 1709, the Senate companion to the House SAFE Act.

All of these points should not suggest a tsunami of opposition. Rather, there is survey evidence of a deeply conflicted public as documented by an ABC News public opinion poll on privacy and the war on terrorism (ABC News, 2003a, 2003b, 2003c). It shows, for example, that 58 percent of those surveyed believe that agencies such as the FBI are intruding on the privacy rights of Americans, but 78 percent (of the entire sample) nevertheless prefer proactive investigations of terrorism threats even at the expense of their individual privacy. This compromise of interests is more specifically demonstrated by certain surveillance-specific questions where 65 percent of the public tended to approve of the monitoring of public spaces such as libraries or of easy access to electronic communications.

Congressional sentiment is also mixed. Despite the legislation previously considered, an October 21, 2003, hearing of the Senate Judiciary Committee saw some members call for modification, but others, such as Senator Joseph Biden (D. Del.), asserted that opposition to the U.S.A. Patriot Act was "ill-informed and overblown" and Senator Orrin Hatch (R. Utah) said that the Act had been the victim of "extremists on both ends who seem to be dominating the debate in the media today." (U.S. Senate, Committee on the Judiciary, 2003, online; see also Schmidt, 2003). From the hearing, it appeared clear that the members were bedeviled by a swell of criticism largely devoid of factual evidence of abuse and unclear as to the legal specifics, but driven, of course, by the secrecy surrounding many of the powers.

This conflict as to the role and exercise of government surveillance powers in a time of national threat presents the rationale for this examination of the information-centric world of domestic surveillance. Given that more than half of the public believe their privacy is being invaded and more than a third do not support broad surveillance practices, this chapter will suggest the need for balance and openness, the need to manage the surveillance tool consistent with the historic principles of a representative democracy, as well as the need, in the words of Rep. Otter (R. Idaho), to revisit the newest government surveillance authorities and "brick by brick take the most egregious parts out of the Patriot Act" (Goldstein, 2003, p. A1).

The first section of the chapter briefly discusses the scientific discipline of intelligence including the key role of surveillance as a collection tool. The next considers the factual history as well as judicial recognition (e.g., *Handschu v. Special Services Division (NYPD)*, online) of surveillance as "a legitimate and proper practice of law enforcement" that is "justified in the public interest ... to prevent serious crimes of a cataclysmic nature" but that such actions may become so extreme or involve such direct injury as to transgress legitimate constitutional rights of Americans. The third part addresses the current threat of terrorism and the increase in domestic law enforcement and intelligence surveillance authorities flowing from the enactment of the U.S.A. Patriot Act, the issuance of new investigatory guidelines by the Attorney General, the recent judicial decisions of the Foreign Intelligence Surveillance courts, and recent proposals for additional powers. The next section considers the impact of new information collection technologies on citizen privacy and whether such technology presents a grave threat or the opportunity not only to identify terrorist threats but also to reduce inherent human bias and thus enhance civil liberties.

The final section of this chapter considers some critical questions that define domestic intelligence collection in a democratic society. Although various calls have been made for the creation of a new domestic investigative agency as well as the reformation of the FBI mission, including improvements in information focus and sharing, it will suggest that the most important requirement is the inclusion of an effective right of individual challenge as well as a comprehensive system of public oversight and reporting. In other words, the question presented is how the nation avoids the political abuses of the past yet acquires the needed information to protect its citizens.

A Brief Look at Government Intelligence

To begin to answer the critical governance issue presented—how best to manage government surveillance in a democracy—an understanding of the business of intelligence as well as of the appropriate performance expectations is required. Simply stated, intelligence bears a very strong resemblance to other information-based businesses, such as a major news organization or research center in the academic community, and all have three principal characteristics. First, all undertake the same primary activities of collecting information through overt and covert means, analyzing that information to decide what it means (turning information into intelligence or news through a validation process), and disseminating the resulting product (informing those who need to know of what "they need to know"). The ultimate mission of intelligence is different, however, in terms of the customer: Government intelligence exists to inform the President and policymakers about strategic and tactical issues and to provide warning as necessary. The second characteristic is that all rely increasingly on technology to enable the mission, from collection of information to analysis and dissemination, where one of the most notable impacts of modern information technology has been the ability to analyze vast quantities of data and put results—up-to-the-minute intelligence—in the hands of users. And third, the most important characteristic, all are constrained by the inherent limitations of information businesses—the inability to collect the totality of the required information as well as the inability to interpret the certain meaning from incomplete or deceptive data. Stated differently, the intelligence analysis process will lack some relevant factual data, cannot be assured that validation (e.g., asking questions with known answers, tasking others, or analyzing for logical inconsistencies) will remove all deceptive and inaccurate information, and must include certain tactical and strategic assumptions. As a result, intelligence analysis, although rigorous in the identification of evidence and the development and testing of hypotheses, does not move inexorably from ambiguity to certainty.

Moreover, the criticism of intelligence almost without exception has failed to distinguish between the types of problems that are presented and the role that information collected through surveillance or other methods can play. We term one type of intelligence problem a *puzzle*—a question it may be able to answer if the necessary factual data are collected. A classic example of a puzzle would be the identity of the members of a given terrorist cell that could be ascertained through effective surveillance of a range of individuals and institutions. The other type of intelligence problems is a *mystery*—a question that cannot be answered because it is beyond our ability to understand and predict, for it depends on imponderables such as undisclosed human intentions or the chance convergence of factors. At most, intelligence may collect information that informs us about the mystery but it will not and cannot provide an answer. A classic example of a mystery would be the success in recruitment over time by a terrorist cell; here, all of our intelligence collection methods, including surveillance, will allow us to understand the situation and perhaps provide a range of predictions but will not provide a definitive answer (Heuer, 1999; Treverton, 2001).

Several critical points follow for our examination on the proper role and conduct of government surveillance. First, surveillance is a primary method of intelligence collection; it is key to the intelligence process because of its inherent reliability. Second, surveillance must be managed to limit data collection to high-value information and avoid dissipation of resources. Third, surveillance must also be regulated in order to ensure a balance between the needs of the state and the rights of individual citizens. And fourth, although the intelligence process—collection, collation, validation, and analysis—is rigorous, the noted limitations in the tools and processes must be acknowledged.

A History of Government Surveillance

The History and Issues in General

The history of government surveillance, whether for political, law enforcement, or intelligence purposes, is long and complex, as are the arguments of necessity and in opposition. Some readers, perhaps from personal recollection, will date this issue to the 1960s and 1970s and recall it as a time of unparalleled political unrest and law enforcement response arising from demands for civil rights and withdrawal from Vietnam. Arson, bombings, rioting, and lesser forms of civil disobedience were widespread, engulfing major cities as well as college campuses. In response, police and National Guard units battled on the streets and intelligence units within these organizations attempted through various forms of surveillance (including infiltration) to identify the individuals and suspected foreign organizations behind this threat to domestic stability.

Although a visceral reaction against surveillance directed at political expression often arises, particularly in the U.S. given the First Amendment, the line between lawful, peaceful expression and unlawful, planned or actual conduct can, in fact, be fluid and difficult to ascertain. As a result, the courts, including for example *Handschu* (1972) and *Kenyatta* (1974), have rather consistently upheld the right of surveillance, recognizing the necessity for law enforcement not only to resolve effected crimes but also proactively to prevent crime and thus to collect information from direct or indirect observation—in other words, surveillance and the use of informers. Indeed, the courts have specifically noted that without the collection, analysis, and dissemination of information (i.e., intelligence), law enforcement would be impossible.

Crime solution and intelligence differ, however, in their critical information objectives: The former focuses on the collection of specific information in the context of an actual crime and individualized, suspected wrongdoing, but the latter focuses on the collection of generalized information that may prove relevant to future investigations often without any evidence of specific wrongdoing. Stated differently, the purpose of intelligence is to collect that totality of information relevant to a mission—in this case, to develop knowledge of actions, events, and/or threats that might affect domestic stability or national security. Thus, although surveillance activities for law enforcement and intelligence purposes appear not only logical but also necessary, equally powerful countervailing considerations drive political opposition, especially in the context of intelligence. As Chevigny (1984, p. 735) in his comprehensive consideration of this time and these issues observed, overt or covert surveillance of public events, the introduction of informants, and the maintenance of files in the nature of political dossiers quite simply creates an "atmosphere of fear and intimidation" and results in a chilling impact on the First Amendment rights of citizens, some or all of whom may be innocent of wrongdoing.

The Early History of Surveillance

The era of the 1960s was not the first threat to domestic tranquility nor would it be the last. Indeed, the historical record of political surveillance and infiltration by law enforcement authorities in the U.S.—much like the railroadman's overcoat of pioneering undercover operative Allan Pinkerton—is a long and checkered affair. These activities disclose a cyclical pattern of cause and effect—an actual threat to state security and order is first perceived (war, labor violence, civil unrest, terrorism), followed by increased law enforcement intelligence activities, eventually leading to perceived civil rights abuses and calls for or implementation of greater oversight.

One of the first documented incidents of law enforcement surveillance and infiltration of domestic dissident groups occurred on the eve of the Civil War and involved Allan Pinkerton and his pioneering National Detective Agency. Retained in 1861 by the Lincoln administration for executive security and protection, the agency began to investigate sabotage and other threats to national security by infiltrating pro-rebel groups in Baltimore. Their clandestine operations uncovered a conspiracy to kill President Lincoln in Baltimore as he traveled by rail from Philadelphia to Washington; the assassination was averted by an unannounced schedule change and a presidential disguise, both suggested by Pinkerton (Cohen, 2002; Milles, 1995; Moffett, 1894).

Pinkerton surveillance and infiltration also figured prominently in the defeat of the Molly Maguires and the suppression of miner unrest in post–Civil War Pennsylvania. By 1873, as documented in detail by Dives, Pomeroy, and Stewart (1911, p. 34), the workers in the secret organization known by that name had committed so many violent acts and created so much financial loss that the management of the Philadelphia & Reading Railway Company as well as the Philadelphia & Reading Coal Company retained Pinkerton to "bring evidence before the Courts of this and adjoining Counties whereby convictions could be successfully prosecuted against this organization, which has committed murders innumerable, and every time the cases are brought before the courts, convenient alibis are produced whereby the criminals escape justice." In short order, Pinkerton Detective James McParlan of New York arrived in Schuylkill County disguised as a tramp and gained the confidence of the Irish workers believed to be part of the organization. Until his cover was disclosed three years later, he used surveillance and infiltration to secure evidence of every murder and prevented a number of other attempts. At that juncture, with his true identity known, he then testified in open court against the leadership of the organization and secured a number of convictions resulting in substantial jail sentences and multiple hangings. Although labor troubles continued to surface, the Molly Maguires had been so damaged that the organization collapsed (Broehl, 1964; Dives, Pomeroy, & Stewart, 1911; Kenny, 1998; Moffett, 1894; Pinkerton, 1877/1973).

Pinkerton set the early standards for intelligence practice (e.g., surveillance, infiltration, and comprehensive information collection and reporting), but early American policing during the entire nineteenth century was highly individualistic. A county sheriff or city marshal enforced the law largely as the individual saw fit, with little in the way of processed intelligence, oversight, technology, or systematic planning. Even the advent of modern police departments (e.g., Boston in 1838, New York City in 1844, and Philadelphia in 1856) did not immediately improve the environment—local politics, corruption, and minimal training and planning remained characteristic (Berg, 1999; Vila & Morris, 1999). In sum, early law enforcement was at best reactive rather than proactive in assessing vulnerabilities, identifying threats, and minimizing community risks. Intelligence, thus, played little if any role.

Surveillance at the Beginning of the Twentieth Century

Paradoxically, the rise of police department professionalism at the start of the twentieth century changed the very nature of policing and presented the opportunity for the development of intelligence units employing the classic methods of surveillance, infiltration, and recruitment of agents within target organizations. This professionalism was first evidenced by the establishment of the International Association of Chiefs of Police (IACP) in 1893, which promoted organization, planning, and the importance of information collection and management. The IACP created the National Bureau of Criminal Identification in 1897, widely encouraged the use of fingerprint technology at the St. Louis World's Fair in 1904, began work on a uniform crime records reporting system in 1922, and transitioned those identification files and crime records systems to the FBI later in that decade (International Association of Chiefs of Police, 2003a, 2003b). Leaders in establishing the importance of information in law enforcement included Richard Sylvester in Washington (who introduced intelligence functions), August Vollmer in Berkeley (who introduced scientific and information methods, including the establishment of the Uniform Crime Reporting System), Theodore Roosevelt in New York City (who reinvented the detective bureau), and O. W. Wilson in Wichita and Chicago (who introduced key management models, but also presided over the 1968 debacle of police response at the Democratic Convention) (W. Andrews, n.d.; Berg, 1999; Vila & Morris, 1999)

Donner (1990, p. 30) dates organized police intelligence units engaging in systematic surveillance to the time immediately after the 1886 Chicago Haymarket bombing when the police recognized that "the revolutionary movement must be carefully observed and crushed if it showed signs of growth." Other cities followed, including New York, where former Police Commissioner Patrick Murphy (Chevigny, 1984, p. 735) had confirmed the existence of intelligence units with functional, organized surveillance as early as 1904. A so-called "Italian Squad" worked aggressively against a criminal gang of Italian immigrants known as the "Black Hand Society," which was present in Brooklyn as early as 1903 as a typical criminal enterprise (Court News, 1907).

The threats to public order only increased in the first two decades of the twentieth century: from more organized criminal enterprises, continued and even more violent advocates for communism, radical labor, international anarchism, and even nationalist movements. By the time of American involvement in World War I, scores of government officials, including two U.S. presidents, had been assassinated, and the Bolshevists' revolution in Russia had announced an international revolutionary agenda. The lines separating these groups and their intentions were often amorphous and overlapping. For example, Nedjelko Cabrinovic, a participant in the assassination of Austro-Hungarian Archduke Ferdinand at the beginning of World War I, was self-described as a socialist, anarchist, and nationalist (Simic, 1995).

The certain results of this turbulence were new federal laws restricting dissident speech, the establishment of special political or subversive squads within more municipal police departments, and the creation of J. Edgar Hoover's Alien Radical Division within the U.S. Department of Justice (DoJ) (Powers, 1987). Specifically, we see the Espionage Act of 1917 and the Sedition Act of 1918 that proscribed not only language "to incite curtailment of war production ... [or] obstruct the draft" but also language "disloyal, scurrilous ... about the form of Government of the U.S" (Sedition Act of 1918, amending the Espionage Act of 1917, 40 Stat. 553–554). More than 2,000 people were convicted, largely of speech violations, not espionage in the classic sense. What of the boundaries between the First Amendment and prohibited speech? The answer comes in part from the convictions for anti-draft circulars affirmed in *Schenck v. United States* (1919, p. 52) with adoption of a "clear and present danger" test and language by Justice Holmes:

> We admit that in many places and in ordinary times the defendants in saying all that was said in the circular would have been within their constitutional rights. But the character of every act depends upon the circumstances in which it is done. ... The question in every case is whether the words used are used in such circumstances and are of such a nature as to create a clear and present danger that they will bring about the substantive evils that Congress has a right to prevent. It is a question of proximity and degree. When a nation is at war many things that might be said in time of peace are such a hindrance to its effort that their utterance will not be endured so long as men fight and that no Court could regard them as protected by any constitutional right.

Less than a year later, and in different times, Holmes dissented in the *Abrams* case (1919) involving the convictions of Russian Communists who had criticized American intervention in the Russian Revolution.

A well-defined political dimension to intelligence gathering and subsequent prosecutions emerged after the formal end of the war because of the continuation of the radical threat. One outcome was the Palmer Raids of 1919 where, in responding to dozens of mail bombings and assassinations, the Attorney General ordered raids, the arrests of thousands, and the deportation of many aliens largely without warrants (Watson, 2002). Characterized as a war against crime by Palmer (1920, p. 174) himself: "The 'Reds' were criminal aliens and ... the American Government must prevent crime." The rule of law was not restored until the courageous actions of Assistant Labor Secretary Louis Post, who canceled a group of deportation orders and successfully defended his actions before the House of Representatives as they attempted to impeach him (Watson, 2002). The term "rule of law" is used frequently in the literature; it highlights a fundamental difference between the events of 1920 and today. The Palmer raids were uniformly based on generalized sweeps and arrests of *groups* often in the exercise of their First Amendment rights (e.g., a labor union meeting), but the arrests after the events of September 11, 2001, were all occasioned on *individualized determinations* including the existence of outstanding criminal warrants, significant immigration violations, or pursuant to the federal material witness statute. Thus, as of June 2002, the DoJ reported that 751 individuals were held on immigration charges, 129 on outstanding criminal charges, and a small but unspecified number pursuant to material witness warrants (Stern, 2002). This said, contrary views remain and equate the time of Palmer and Ashcroft (Cole, 2002).

Throughout the 1920s the continued fear of speech viewed as disloyal and harmful prompted aggressive police intelligence operations, arrests, and prosecutions under newly enacted state criminal anarchy statutes. For example, in 1919, the state of New York formed the Joint Legislative Committee to Investigate Seditious Activities that probed hundreds of people and organizations with raids and seizures of papers, produced a four-volume report: *Revolutionary Radicalism: Its History, Purpose and Tactics*, and proposed a criminal anarchy statute that became law in 1921. Convictions under these laws were generally affirmed on appeal, although the Supreme Court in *Gitlow v. New York* (1925) held for the first time that the First Amendment applied to the states through the 14th Amendment. In this case, the defendant's attorney, Clarence Darrow, had eloquently argued that the matter involved pure speech, not criminal action, merely abstract doctrine. Indeed, it should be noted that such state laws, as well as the previously discussed federal law, prohibiting seditious speech and activities (variously termed seditious libel, criminal libel, or seditious anarchy) have a long and ignoble history beginning with the English common law offense of seditious libel that criminally punished any "false, scandalous and malicious writing" that had "the intent to defame or to bring into contempt or disrepute" a private party or the government; truth was no defense (The Sedition Act of 1798, officially titled "An Act for the Punishment of Certain Crimes Against the United States." Approved 14 July 1798. 1 Stat. 596–597). Seditious libel remained part of the U.S. common law despite the First

Amendment (American Civil Liberties Union, 1997). Indeed, only seven years after approving the Bill of Rights, Congress enacted the Alien and Sedition Acts that incorporated the language quoted here. Although the federal law was subsequently repealed, the states maintained their laws and later enacted statutes to restrict speech contrary to general government interests or specific issues (e.g., restricting anti-slavery speech in Southern states prior to the Civil War). In historical terms, the emergence of the "Red Scare" provided for both new statutory enactments as well as the enforcement of existing statutes in other states.

Surveillance by Mid-Century

By 1940, the conflict between protected First Amendment activities and law enforcement surveillance (often resulting in arrests and prosecutions) remained unabated as the threats of international communism continued and fascism emerged. Although some recognized this as a compelling legal issue—as a congressional subcommittee detailed in hearings on the "violations of free speech and rights of labor" by the Intelligence Bureau of the Los Angeles Police Department (U.S. Senate. Committee on Education and Labor, 1940)—the Congress in general continued to address security threats. The year 1940 also saw the passage of the Smith Act, requiring that all aliens be registered and fingerprinted, and making it a crime to advocate the violent overthrow of the U.S. or to belong to an organization advocating such activity—thus establishing a new legal basis for the investigation of individuals deemed to present actual national security threats.

The Smith Act was applied extensively against domestic threats, primarily Communists, and was held constitutional by the Supreme Court (*Dennis v. United States*, 1951), although it was judicially modified six years later to require direct action rather than mere advocacy (*Yates v. United States*, 1957). In the interim between the passage of the Smith Act and these decisions, the Congress moved to establish even more control of dissidents. Specifically, in 1950 the Internal Security Act created the Subversive Activities Control Board, required loyalty tests for federal employees, and denied passports to and required registration of subversive organizations (including individual members). In sum, the nation had by this time established a comprehensive legal scheme for the surveillance of and action against those deemed an ideological or actual threat to national security.

Although the registration provision for organizations was upheld by the Supreme Court in 1961 (*Communist Party v. SACB*), there were subsequent moves toward greater protection for individual rights. In 1964 the passport provision was found to be unconstitutional (*Aptheker v. Secretary of State*), as were the individual registration provisions in 1965 (*Albertson v. SACB*). Indeed, this move toward individual rights continued in the seminal criminal incitement case of *Brandenburg v. Ohio* (1969, p. 447) when the Supreme Court fully protected political speech unless it was "directed to inciting or producing imminent lawless action and is likely to incite or produce such action."

This move toward individual protection continued in Congress when Title II of the Internal Security Act of 1950, also known as the Emergency Detention Act, was repealed in 1971. That section, although never used, contained draconian provisions that authorized the President in time of war to declare an "internal security emergency" and detain without due process anyone deemed to be a potential spy or saboteur. Interestingly, the DoJ supported the repeal because any potential advantages were outweighed by citizen concerns. Two years later, in 1973, Congress denied further funding for the enforcement of all remaining Title I provisions and the Internal Security Act passed into oblivion. The replacement for this legislation, codified at 18 U.S.C. § 4001(a), prohibits the detention of American citizens except pursuant to an act of Congress and today plays a central role in the litigations challenging the detention of aliens and U.S. citizens as enemy combatants pursuant to the President's foreign relations and national defense authority.

But, these enhancements of constitutional interpretation were paradoxical when compared to the political and social change then convulsing the nation and the governmental response. In essence, one could argue that the judiciary and Congress, with their oversight roles, and the executive department, through its law enforcement authorities, were pursuing conflicting agendas. Hence, we see an array of cases striking down arrests for the use of intemperate political language, from *Cohen v. California* (1971) protecting a jacket emblazoned with the message "Fuck the Draft" to *Texas v. Johnson* (1989) protecting the burning of the American flag; at the same time aggressive surveillance and use of arrest powers by law enforcement continued, even when advocacy was the offense.

But why this dichotomy between judicial and executive position? One answer is that there is not so much a divergence of views but rather a critical line between advocacy and action. The Executive Branch has clearly advised Congress (U.S. Department of Justice, 1997, p. 37), that the First Amendment and *Brandenburg's* "imminence" requirement generally pose little obstacle to the punishment of speech that constitutes criminal aiding and abetting because "culpability in such cases is premised, not on defendants' 'advocacy' of criminal conduct, but on defendants' successful efforts to assist others by detailing to them the means of accomplishing the crimes." The increasing public unrest occasioned by civil rights discrimination and anti-war activism—notwithstanding these judicial decisions—also provided a functional empowerment of law enforcement to use the tools of intelligence to meet the threat to public peace and safety: "The FBI's counterintelligence program came up because there was a point—if you have anything in the FBI, you have an action-oriented group of people who see something happening and want to do something to take its place" (Church, 1976,

Book III, testimony of George C. Moore, Chief, Racial Intelligence Section, FBI, p. 11).

If this latter analysis is correct, even in part, there is little doubt that the empowerment was facilitated by the loose legal standards regulating intelligence investigations, and surveillance in particular, at this time. For example, before March 1965, the FBI was authorized to carry out audio surveillance (e.g., use of covertly installed microphones) on its own volition and without other approval in any case it deemed related to the national security. Communications between the Attorney General and the Director of the FBI made clear that the use was not restricted to national security in the modern sense ("national defense" and "foreign relations" interests of the U.S.—foreign not wholly domestic threats [Executive Order 12958 § 1.1(a), 1995]). Rather the targets included "subversive persons" and efforts were to support both the intelligence function and the law enforcement function against "major criminal activities," even if trespass without court approval were necessary (Church, 1976, Book III, testimony of Attorney General Brownell, 1954 and FBI Director Hoover, 1961, pp. 112–113). Indeed, the surreptitious use of microphones in the homes of suspects did not end until the U.S. Supreme Court in *Irvine v. California* (1961, p. 133) described the practice as "obnoxious." As for wiretaps, the only additional requirement was an initial, but otherwise unlimited, approval from the Attorney General (Church, 1976, Book III, testimony of Attorney General Katzenbach, p. 112). However, notwithstanding the desire for action and the lack of controls, some in the Government did recognize the dangers to civil liberties: "The risk was that you would get people who would be susceptible to political considerations as opposed to national security considerations, or would construe political considerations to be national security considerations, to move from the kid with a bomb to the kid with a picket sign, and from the kid with the picket sign to the kid with the bumper sticker of the opposing candidate. And you just keep going down the line" (Church, 1976, Book III, p. 27 and fn 114). This testimony is especially insightful coming from Tom Charles Huston, Assistant to President Nixon, the architect of the so-called Huston Plan for broadly ranging and clearly illegal domestic intelligence collection.

Notwithstanding these voices of concern, albeit expressed long after the fact, the apparent and growing threat of civil unrest led to growing institutionalization of intelligence units at all levels of government. In the summer of 1967, President Johnson formed the National Advisory Committee on Civil Disorders to address the rioting and civil disobedience sweeping the nation. Chaired by Governor Otto Kerner (1968, p. 1), the committee's 1968 report is perhaps most remembered for the finding that the nation was becoming two societies, "one black, one white—separate and unequal." But on the specific topic of disorder and a national response, it noted that "the absence of accurate information both before and during a disorder has created special control problems for police" and advocated that police departments develop intelligence squads using undercover police personnel and informants in order "to gather ... and disseminate information on potential as well as actual civil disorders" (Kerner, 1968, p. 269). Within a short time, the Law Enforcement Assistance Administration, an element of the DoJ formed as part of the national war on crime, began to make grants for the development and operation of local police intelligence units where surveillance and infiltration would become primary tools (Chevigny, 1984).

The Critics Emerge: The Public Interest Litigation and Congressional Investigations

Concomitant with this institutionalization of and financial support for intelligence units came organized opposition from the political activists who were targeted and the civil libertarians who opposed on obvious principle any governmental action that infringed upon citizen First Amendment rights. Emerging first from evidence introduced in open court in various prosecutions of many prominent dissidents in the late 1960s and early 1970s (Berry, 1982) and then from the post-Watergate Presidential and Congressional investigations into governmental excesses (including the Rockefeller Commission, the Senate Select Committee to Study Intelligence Activities chaired by Frank Church of Idaho, and the corresponding committee in the House chaired by Otis Pike of Georgia) the extent of surveillance became clear. Documented in extensive detail were the facts that federal, state, and local law enforcement had mounted an extensive intelligence operation against the American people ranging from the collection of the names of people who supported given causes to the compilation of political dossiers utilizing physical, mail, and electronic surveillance; informants; and other tools. Indeed, the details in the Pike report were so specific that the House of Representatives suppressed the document because of national security considerations by a vote of 246 to 124, with the Democrats divided on the issue and the Republicans almost unanimous in their opposition (Rosenbaum, 1976). Nevertheless, the report was ultimately leaked to and published by the *Village Voice*.

Perhaps most informative of the scope of law enforcement surveillance and the constitutional issues was the public interest litigation pursued at the federal, state, and local levels. These challenges addressed such practices as overt collection, covert collection through infiltration, and harassment through various provocations. Each alleged that the focus of the law enforcement action was protected political expression, not criminal action. The most significant and productive actions took place in six cities and two states: New York City, Chicago, Los Angeles, Seattle, Memphis, Detroit, Michigan, and New Jersey. The reason for this local focus, according to Chevigny, was that favorable precedent challenging federal action never developed. This is so because the U.S. Supreme Court (e.g., *Hoffa v. United States*, 1966 and *Laird v. Tatum*, 1972) has held consistently that there is no constitutional objection to the presence of and collection of information by officers at public meetings, much less the surveillance of a citizen individually through the use of an informer under either the First or Fourth Amendment. Other federal courts following this analysis have found no constitutional protection for other aspects of law enforcement surveillance including the collection of personal information held by third parties (*Reporters Committee for Freedom of the Press v. AT&T*, 1978) or the dissemination of collected information at least to the extent the disclosure is made to other law enforcement agencies (*Philadelphia Yearly Meeting of the Religious Society of Friends v. Tate*, 1975). This holds even when the only basis for government collection was the fact that the subjects were engaged in political or other expressive activity.

The local cases that were successful, although they began earlier, were buoyed by a comment by the Supreme Court in 1972 in *Laird v. Tatum* (1972) that governmental conduct affecting First Amendment rights would be actionable if it were *proscriptive*—meaning it had a tangible, adverse effect on an individual beyond emotional distress *and* if there had been an intent to violate the individual's constitutional rights. Essentially, the courts thereafter created a narrow cause of action requiring actual injury and malicious intent that could be established directly by proof of deliberate disruption of political activity or indirectly by the extensive nature of the surveillance and the lack of any legitimate law enforcement rationale.

One of the earliest challenges was *Andersen v. Sills* (1969), where the activist plaintiffs charged the New Jersey State Police Central Security Unit with the retention of political intelligence information from police Security Incident Forms in violation of First Amendment protections of political speech. Although the case reached the New Jersey State Supreme Court, a significant precedent with judicially ordered relief was avoided by preemptive, self-imposed regulation in the form of newly adopted police guidelines in 1976. This case set political precedent for other groups to challenge perceived police overreaching by various means—including judicial intervention and legislative change—and to seek specific relief in the form of regulations and the introduction of civilian oversight.

The landmark case that put First Amendment–based freedom from political surveillance, infiltration, and police harassment on the map, *Handschu v. Special Services Division*, began in New York City in 1971 (discussed comprehensively by Chevigny, 1984, and Eisenberg, 2003) Here, multiple plaintiffs (including prominent radicals such as Youth International Party founder Abbie Hoffman) sought relief from infiltration, surveillance, and the retention of political intelligence dossiers by the political intelligence unit of the New York City Police Department (NYPD). Factually, the case was predicated on one of the most egregious actions of the NYPD when, in the course of the prosecution of members of the Black Panther Party for conspiring to destroy government buildings and transportation facilities, it was revealed that the Manhattan

District Attorney had ordered police infiltration and incitation to such a degree that the jury could not "distinguish between the felonious impulses of the Panthers and the undercover cops" (Powell, 2002, p. A01). An acquittal resulted and the *Handschu* litigation followed in several weeks.

Initially, the *Handschu* court considered the balance of equities; on one hand, Judge Weinfeld recognized the lawful scope of law enforcement surveillance (*Handschu*, 1972, p. 769):

> The use of secret informers or undercover agents is a legitimate and proper practice of law enforcement and justified in the public interest—indeed, without the use of such agents many crimes would go unpunished and wrongdoers escape prosecution. It is a technique that has frequently been used to prevent serious crimes of a cataclysmic nature. The use of informers and infiltrators by itself does not give rise to any claim of violation of constitutional rights.

But the Judge additionally noted (*Handschu*, 1972, p. 771) that the law also recognized that the police may not collect information by unconstitutional means for unconstitutional purposes and that if such a "pattern of unconstitutional conduct" existed, then the plaintiffs would be entitled to injunctive relief.

After fourteen years of protracted litigation and negotiation, the case ultimately settled with a stipulation and order that prohibits any investigation of "political activity" defined as "the exercise of a right of expression or association for the purpose of maintaining or changing governmental policies or social conditions" except in connection with a criminal investigation or the planning of a public event, only by the Public Security Section (PSS) of the Intelligence Division, and only under the supervision of an Authority made up of the First Deputy Commissioner, the Deputy Commissioner for Legal Matters (both police department officials), and a civilian appointed by the Mayor (*Handschu*, 1985, p. 1391). More specifically, the New York stipulation provided (*Handschu*, 1985 pp. 1421–1422):

- That the PSS may begin an investigation if it submits an investigation statement to the Authority containing "specific information ... that a person or group engaged in political activity is engaged in, about to engage in or has threatened to engage in conduct which constitutes a crime"

- That within thirty days of such initiation, the PSS must present a request for approval to the Authority, which may either terminate the investigation or permit it to continue beyond the first thirty days

- That the PSS may use undercover infiltrators in such cases only with the express approval of the Authority, but that plainclothes officers may be present at "public activities of political organizations" without separate approval if they are part of an investigation for which proper statements and applications have been filed with the Authority

- That no file may be opened solely on the basis of the political, religious, or sexual preference of an individual as well as the collection of certain types of information including the fact that a person has signed a petition, has his name on a mailing list, supports a group by contributions, or has written any political or religious work

Viewed historically, the *Handschu* litigation was important in three respects: creating limited civilian oversight, recognizing that the actions of individuals often involve mixed political and criminal aspects, and establishing the rule that police investigations should be limited to those circumstances where there exists actual or potential criminal conduct. The decisions in other jurisdictions varied in certain aspects given unique procedural issues but generally followed these parameters. For example, both the Chicago (*Alliance to End Repression v. City of Chicago*) and the Memphis litigations (*Kendrick v. Chandler*) went poorly for the police after they destroyed records. In Memphis, highly embarrassing records nevertheless surfaced in the media and drove an early settlement in 1978 that substantially influenced the Chicago settlement three years later. Specifically, the Memphis decree proscribed any actions "for the purpose of political intelligence" (*Kendrick,* 1978, p. 3) but also recognized the fact of mixed investigations and required the approval of the Director of Police for the "collection of information about the exercise of First Amendment rights" in such circumstances (*Kendrick,* 1978, p. 4). It was weakened, however, by a lack of guidelines and of civilian oversight. The Chicago decree was similar, albeit somewhat more detailed in its regulation of mixed investigations and also novel in explicitly establishing a "reasonable suspicion"—not a "probable cause"—standard for the initiation of an investigation (*Alliance,* 1982, p. 564). This is the same factual predicate as the "stop and frisk" standard established in the U.S. Supreme Court's earlier decision in *Terry v. Ohio* (1968). The public policy problem, however, with most if not all of the resolutions in these cases was three-fold—the police felt burdened by cumbersome, time-consuming, bureaucratic procedures that limited responsiveness to community threats; the courts remained involved in oversight thus increasing the burden on police; and there was a general chilling effect on police personnel in their performance of official duties.

The Congressional investigations also contributed significantly to the public understanding to the facts and legal issues surrounding domestic intelligence. For example, the 1975 Rockefeller Commission report

(Rockefeller, 1975, pp. 137–149) provided disturbing evidence that the Central Intelligence Agency (CIA) had engaged in questionable domestic intelligence and counterintelligence operations against dissidents, primarily those on the political left, during the late 1960s and early 1970s. The Special Operations Group's activities under their "Operation CHAOS" projects, although not as invasive and extensive as those of the FBI, nevertheless involved the creation of hundreds of thousands of political files on American citizens through intelligence operations on American soil that pushed the envelope of jurisdictional boundaries specified by the 1947 National Security Act. Although CHAOS operations had terminated in 1974, the revelations of the Commission were compounded by the Church and Pike Committees' investigations of the FBI. Their 1976 report shed light on many questionable FBI operations throughout the J. Edgar Hoover years, particularly COMINFIL in the 1950s and COINTELPRO during the 1960s and 1970s (Church, 1976, Books II–IV). In sum, the reports brought to light a seemingly endless litany of civil right offenses against domestic dissidents and activist groups across the ideological spectrum from far left to extreme right— including infiltration, nonconsensual seizures, intrusive surveillance, and assorted "dirty tricks" (Church, 1976, Book III, pp. 17–18).

Most damning at the time were revelations about the operations involving Martin Luther King, Jr. who most certainly was a mainstream advocate of civil rights but was targeted with highly intrusive surveillance by the government as if he presented a criminal or counterintelligence threat (Church, 1976, Book III, pp. 449–450). The details of the decades-long effort are well documented (e.g., Murphy, 2002); a number of seminal facts and two unmistakable conclusions remain relevant today. The facts include investigation predicated solely on involvement in the racial movement and later baseless suspicion of Communist infiltration as well as active efforts to discredit if not defame King through infiltration and external propaganda efforts. The conclusions include: First, as considered in more detail in the next segment, the lack of attorney general regulations setting parameters on investigations and intelligence methods was the critical factor in permitting this abuse of rights. Thus, as early as 1957, and even though then-FBI Director Hoover acknowledged that there was an "absence of any indication that the Communist Party has attempted, or is attempting, to infiltrate" King's organization, agents were ordered to "remain alert" for relevant information simply because of his involvement "in the racial field" (Church, 1976, Book III, pp. 87–88). There were simply no limitations of surveillance and investigation to a required finding of some reasonable suspicion of criminal conduct, or a criminal conspiracy, or necessity for foreign intelligence purposes. Second, and equally relevant to the future of surveillance, was the lack of internal documentation as to the decision-making process as well as the lack of extra-agency oversight. As recognized by the Church Report (1976, Book III, pp. 457–459), this failure of required regularity, documentation, and oversight renders accountability impossible.

The Early Regulation of Surveillance

The critics' arguments against unrestricted government surveillance posited a number of significant points that led to legal change. One was that these actions were inconsistent with the Fourth Amendment because they had not been predicated on any degree of reasonable and particularized suspicion of criminal conduct, but rather the exercise of First Amendment rights. Another was that these actions had occasioned identifiable harm by creating an atmosphere of fear and intimidation and by chilling the First Amendment rights of every subject. Yet another suggested that the scope of potential harm was exacerbated when the efforts moved from surveillance to more active efforts, including infiltration and active participation if not suggestion and direction.

The legal changes that resulted from the disclosures of excesses span the period from the Ford through the Reagan administrations and were intended to better regulate intelligence activities in the U.S., especially in the context of surveillance. First, Attorney General Edward Levy in 1976 (and subsequent attorneys general) adopted guidelines for the Department of Justice, including the FBI, to regulate foreign and domestic intelligence collection and to prohibit use of First Amendment protected activities as the basis for investigation. Second, the FISA was enacted in 1978 to address foreign intelligence collection in the U.S. Third, in 1981 Executive Order (EO) 12333 was issued to regulate intelligence in general both inside and outside the nation's borders. And fourth, the Electronic Communications Privacy Act (ECPA), updating domestic criminal law on electronic surveillance, passed in 1986.

Although these provisions will be considered in turn and in some detail, it is important to recognize the dichotomy between the collection of information (e.g., surveillance) for criminal law enforcement purposes and, conversely, collection for intelligence purposes. Why there should be separate law for the same type of intrusion is an intriguing Constitutional question and reflects the complicated intersection of the Fourth Amendment with the President's authority to conduct foreign relations, provide for national defense, and collect foreign intelligence. These conflicting considerations have led to judicial recognition that, even with respect to U.S. citizens, there are substantial limitations on Fourth Amendment rights in the context of foreign intelligence— whether in the U.S. or overseas—if the person is acting on behalf of foreign powers. In essence, a "foreign intelligence" exception to the Fourth Amendment warrant requirement has been held by most circuit courts but never explicitly by the U.S. Supreme Court (e.g., *United States v. Bin Laden, et al.,* 2000). As appears logical, the courts have consistently held that the Fourth Amendment does not apply to foreign nationals overseas even if the activity was conducted by U.S. government officers and the person is brought to this country for trial. This is because the Fourth Amendment protection applies at the time, location, and object of the seizure and only to "the people," meaning generally U.S. citizens (e.g., *United States v. Verdugo-Urquidez*, 1990).

Nevertheless, as the courts established from matters arising from the times and activities of the Nixon Administration, there is no domestic intelligence exception to the Fourth Amendment (*United States v. United States District Court [Keith]*, 1972). In this case, a criminal prosecution involving the bombing of a CIA office in Ann Arbor, Michigan, as an anti–Vietnam War protest, domestic warrantless electronic surveillance was utilized by the FBI, challenged at trial, and found unlawful. Although the Justice Department sought a Writ of Mandamus to compel the District Court to vacate its order, the U.S. Court of Appeals refused as did the U.S. Supreme Court, which offered this historical observation as to the conflict between individual rights and governmental power (*Keith*, 1972, p. 314):

> History abundantly documents the tendency of Government—however benevolent and benign its motives— to view with suspicion those who most fervently dispute its policies. [Constitutional] protections become the more necessary when the targets of official surveillance may be those suspected of unorthodoxy in their political beliefs. The danger to political dissent is acute where the Government attempts to act under so vague a concept as the power to protect "domestic security." Given the difficulty of defining the domestic security interest, the danger of abuse in acting to protect that interest becomes apparent.

Addressing the legal merits, the Court reached three critical findings. First, the President's Article II powers "to protect our Government against those who would subvert or overthrow it by unlawful means" provide authority for national security surveillance (*Keith*, 1972, p. 310). However, and second, the "convergence of First and Fourth Amendment values" in such surveillance cases made the Court wary of potential abuse (*Keith*, 1972, p. 313). Thus, and thirdly, "the duty of Government to protect the domestic security and the potential danger posed by unreasonable surveillance to individual privacy and free expression" must be balanced (*Keith*, 1972, pp. 314–315). As such, the Court rejected the government's argument for a domestic security exception to the general Fourth Amendment warrant requirement but noted that it was expressing no opinion on the authority for warrantless searches involving foreign governments and agents.

Included, however, in the opinion was an observation that has engendered argument until the present time (*Keith*, 1972, p. 322): "We recognize that domestic security surveillance may involve different policy and practical considerations from the surveillance of ordinary crime." This sentence, and the withholding of a judgment about surveillance of foreign agents, has led the DoJ to argue that Fourth Amendment standards may not apply fully in domestic national security cases and that the Supreme Court had "invited" the Government to set statutory standards for such surveillance. Whether there was an invitation continues to be debated. In point of fact, the Congress did enact the FISA several years later to address the issue of foreign agents and foreign powers but did not then, or thereafter, provide separate rules for domestic threats. More recently, the Bush administration has revisited this issue of an "invitation" arguing that, although the Supreme Court held that a warrant is required for domestic security surveillance, more flexible standards could apply to the issuance of such a warrant.

The Early Attorney General Guidelines

As a direct result of the noted Watergate-era abuses and the Congressional investigations and concerns, Attorney General Edward Levi, in 1976 under President Ford, issued guidelines to better define permitted federal law enforcement investigations and techniques. There were three critical factors: First, FBI inquiries and investigations required a minimal factual showing that they were predicated on suspected criminal, terrorist, or foreign intelligence activity (i.e., specific and articulable facts); second, the FBI was prohibited from engaging in disruption of protected First Amendment activity and from attempting to discredit individuals; and third, the guidelines were issued in consultation with the House and Senate Judiciary Committees, thus evidencing a broad-based balancing of citizen rights and state needs for security tools (Berman, 2002).

Seven years later, during the Reagan Administration, these guidelines were liberalized with a revision by Attorney General William French Smith (1983), subsequently reissued with only minor amendments by Attorney General Dick Thornburgh (1989), and continued by Attorney General Janet Reno (1994). The purpose of this revision—maintained under multiple Democratic and Republican administrations—was to ease certain Levi restrictions believed overly cumbersome and to clarify the circumstances under which action could be initiated: "In its efforts to anticipate or prevent crimes, the FBI must at times initiate investigations in advance of criminal conduct" (Thornburgh, Part I., para. 3, p. 3). Thus, for example, these guidelines authorized the "opening of an investigation" whenever "facts or circumstances *reasonably indicate* that two or more persons are engaged in an enterprise for the purpose of furthering political or social goals wholly or in part through activities that involve force or violence and a violation of the criminal laws of the United States" (Thornburgh, Part III.B.1.a, p. 13). Moreover, they established an even lower basis for a preliminary inquiry, which the FBI was authorized to begin based on the receipt of any information or allegation "whose responsible handling requires some further scrutiny" in light of the criminal and foreign intelligence interests of the United States (Thornburgh, Part II.B, para. 1, p. 4). Under the revised guidelines, preliminary inquiries could be conducted without any initial headquarters approval and could include overt interviews, physical surveillance, and the tasking of confidential informants, but not more intrusive activities such as wiretapping, mail opening, or mail covers (Thornburgh, Part II.B.(4)–(6), p. 5).

Examination of these revised guidelines and a comparison with the initial Levi guidelines reveals factors relevant to consideration of additional changes in current times. First, the change from requiring *specific facts* (under Levi) to *reasonable indication* (under Smith, Thornburgh, and Reno) as an investigative threshold was significant. This standard was less than probable cause of a specific crime or even reasonable suspicion; it was satisfied if there is merely an indication that some criminal activity was intended (Elliff, 1984; McConnell, 1983). Second, although the FBI was urged to respect First Amendment rights, it was clearly authorized to undertake appropriate investigatory actions when "statements advocate criminal activity or indicate an apparent intent to engage in crime, particularly crimes of violence" (Thornburgh, Part I, page 3). Third, as evidenced by multiple FBI investigations of groups ranging from white supremacists to black militants, no religious cloak has ever insulated criminal or terrorist enterprises. And fourth, there were no limitations on the source of information to initiate an inquiry; as long as a source was credible, a private citizen, a civil rights advocate, the media, or any other public source could serve as the basis.

The Foreign Intelligence Surveillance Act

In addition to the public interest litigation and Attorney General regulations, the Congress also moved to address the conflicting rights of the public and the President vis-à-vis foreign intelligence with the passage in 1978 of the FISA, 50 U.S.C. § 1801 et seq. The Act authorizes the collection of "foreign intelligence information" about foreign powers or agents of foreign powers in the United States through a scheme of Attorney General procedures and approvals (e.g., Reno, 1995a) and in most cases applications to the specially appointed U.S. Foreign Intelligence Surveillance Court (FISC). As with the legal theory of exigent circumstances in the law enforcement environment, emergency situations also permit temporary Attorney General approvals—an authority personally exercised by Attorney General Ashcroft through mid-2003 in the form of 170 emergency surveillance orders—with mandatory subsequent request for approval from the FISC within 72 hours (Goldstein, 2003).

The Act defines "foreign intelligence information" as information about 1) an actual or potential attack or other grave hostile acts of a foreign power, 2) sabotage or international terrorism by a foreign power or an agent of a foreign power, 3) clandestine intelligence activities by a foreign power or agent, or 4) concerning a foreign country that is necessary to the national defense or the security of the U.S. or the conduct of the foreign affairs of the U.S. Orders from the FISC may authorize the acquisition of electronic information, physical searches, and access to certain types of business records. The FISC is composed of eleven federal district court judges (increased by § 208 of the U.S.A. Patriot Act from seven), appointed by the Chief Justice for staggered terms and from different circuits, who review Attorney General applications for electronic surveillance, physical searches, and demands for other information such as business records. The cases are presented *ex parte* and *in camera* by attorneys from the DoJ Office of Intelligence Policy and Review and the records and files of the cases are secret and sealed; they may not be revealed even to persons whose prosecutions are based on evidence obtained under FISA warrants, except to a limited degree set by district judges' rulings on motions to suppress.

Most significant are the threshold conditions for FISA warrants. First, the sole or primary purpose of the investigation must be counterintelligence or counterterrorism; and, second, the required showing is not probable cause of criminal activity, as with traditional Fourth Amendment law, but rather probable cause that the target is a foreign power or an agent of a foreign power. Thus, these factors present significant Constitutional considerations especially because FISA orders, although seeking information about terrorism, may authorize highly intrusive electronic surveillance and physical searches, may have as a target either aliens or U.S. citizens, and may be directed to innocent third parties who are in possession of relevant information (e.g., private sector entities such as Internet service providers [ISPs] and landlords or public entities such as schools). Moreover, the transfer of FISA-acquired information to law enforcement authorities, and subsequent use in criminal prosecutions, is little restricted. For example, the retention and dissemination of information acquired in an intelligence investigation that is evidence of a "serious crime totally unrelated to intelligence matters" is fully authorized (U.S. House of Representatives, 1978, p. 62).

Accordingly, in light of these standards and provisions, and to prevent the FISA from becoming a substitute for traditional Fourth Amendment law, the FISA process has always included *minimization requirements—* intended to constrain the acquisition and retention and prohibit the dissemination of nonpublicly available information concerning U.S. persons if it does not concern "foreign intelligence." This requirement followed largely from two cases: *Keith*, where the court suppressed warrantless evidence in a strictly domestic case and *United States v. Truong Dinh Hung* (1980), involving North Vietnamese espionage, where the court also suppressed warrantless intelligence-acquired evidence after the matter became primarily a criminal investigation. Thus, when the FISA was enacted in 1978, during the pendency of *Truong*, the "primary purpose" test became the practical benchmark. After years of pragmatic applications, the test was postulated in Attorney General Reno's (1995b) formal guidelines, which included the admonition that the sharing of intelligence information with law enforcement must not result in either the fact or the appearance of law enforcement officials directing or controlling a FISA activity. Although these rules were modified slightly in later years (Holder, 2000; Thompson, 2001), effective communication between intelligence and law enforcement remained constrained.

One aspect of implementation has been the establishment of a "wall" procedure by the FISC, requiring all applications for FISA orders to include, inter alia, (a) certification that the purpose is foreign intelligence, (b) disclosure of all criminal information aspects of an intelligence case, including specifics on information sharing with law enforcement, and (c) designation of a senior official to moderate the flow of information to law enforcement, including the FISC itself, in significant overlapping criminal and intelligence cases. The purpose, in the words of the FISC, was "to preserve both the appearance and the fact that FISA surveillance and searches were not being used sub rosa for criminal investigations" and to prevent prosecutors from becoming "de facto partners in FISA searches" (*In re: All Matters Submitted to the Foreign Intelligence Surveillance Court*, 2002, p. 620).

Thus, although the law and these rules constituted an effective tool in intelligence investigations, they presented a conundrum if there were any potential for law enforcement action because the application to the FISC was required to state that the purpose was intelligence and not law enforcement. This conundrum presented serious difficulties for the government in at least 75 FISA cases where the FISC believed that the government had misrepresented matters. In such cases, the safest legal path was to proceed under traditional criminal law for the issuance of a warrant under Fourth Amendment standards (probable cause of a crime) and, in the case of an intercept warrant, under even higher standards of proof and with greater limitations. In essence, the DoJ was constantly torn between two paths and reluctant to go the intelligence route if there were any indication that a federal crime had been committed—almost always the case in terrorism. Moreover, if the Department selected the criminal route—assuming that it could meet the required showing—the law then imposed a substantial barrier to information sharing between the criminal investigation and the intelligence process, a barrier that obviously hampered the inherently time-sensitive nature of counterterrorism work. Finally, if the Department selected the intelligence route, the law still prevented surveillance of foreign nationals, even where there was a significant suspicion that terrorism might be in the works, if there were no firm evidence of links to terrorist groups or foreign states. In essence, DoJ officials faced a vexatious choice in attacking terrorist targets—a matter that would be addressed in part by the U.S.A. Patriot Act and the Attorney General interpretations; these will be considered in detail later.

The Executive Order 12333

In 1981, three years after the enactment of the FISA and to further respond to public demands for the legal control of intelligence activities, President Reagan issued Executive Order 12333 in a ground-breaking effort to provide an overall framework for the scope of allowable U.S. intelligence activities, including: electronic surveillance, unconsented physical searches, mail surveillance, physical surveillance, and the use of monitoring devices. The order, which is further implemented by classified procedures adopted by the relevant federal agencies including the FBI and CIA, remains in effect today.

First, of particular note, are the provisions of § 2.3 defining the types or subjects of information that may be collected concerning U.S. persons both in the United States and abroad; quite broadly it allows the collection of:

- Information that is publicly available or collected with consent

- Information constituting foreign intelligence or counterintelligence but not concerning the domestic activities of a U.S. person

- Information obtained in the course of a lawful foreign intelligence, counterintelligence, international narcotics, or international terrorism investigation

- Information needed to protect the safety of any persons or organizations, including those who are targets, victims, or hostages of international terrorist organizations

- Information needed to protect foreign intelligence or counterintelligence sources or methods from unauthorized disclosure

- Information concerning persons who are reasonably believed to be potential sources or contacts for the purpose of determining their suitability or credibility

- Information arising out of a lawful personnel, physical, or communications security investigation

- Information acquired by overhead reconnaissance not directed at specific U.S. persons

- Incidentally obtained information that may indicate involvement in activities that may violate federal, state, local or foreign laws

Second, the order at § 2.4 also regulates the techniques that may be used to collect information within the U.S. in general or directed against U.S. persons abroad. It does so, however, by allowing all collection techniques—including electronic surveillance, unconsented physical search, mail surveillance, physical surveillance, or monitoring devices—provided that they are utilized in accordance with Attorney General regulations. The limitations on that authority are important because they reflect consistent judicial and Congressional judgments that the President's foreign intelligence authority is to be exercised only by the President (or immediate subordinate) and properly documented. Otherwise, it is impossible to identify exactly who authorized given intelligence-related activities after the fact—was it the President or a subordinate official without the Constitutional authority? In essence, the Attorney General must approve any collection effort for which a warrant would be required if undertaken for law enforcement purposes and such approvals must be based on probable cause to believe that the technique is directed against a foreign power or an agent of a foreign power. Of course, any use of electronic surveillance is further controlled by the FISA.

This executive order is also where we find a host of interesting and very specific provisions arising from the documented excesses of executive branch agencies during the political disturbances of the 1970s, including restrictions on academic contacts, undisclosed participation, human experimentation, and assassination. First, contracts or arrangements between academic institutions and the intelligence community may be undertaken only with the consent of appropriate officials of the institution. Second, no individuals acting on behalf of agencies within the intelligence community may join or otherwise participate in any organization in the U.S. without disclosing his or her intelligence affiliation to appropriate officials of the organization, except in accordance with Attorney General procedures. Moreover, no such participation may be undertaken for the purpose of influencing the activity of the organization or its members except in the course of a lawful FBI investigation, or if the organization is composed primarily of individuals who are not U.S. persons and are reasonably believed to be acting on behalf of a foreign power. Third, human experimentation is prohibited except in accordance with Health and Human Services regulations, specifically including informed consent. And, fourth, although assassination is prohibited, it is not otherwise defined and, according to media reports, the DoJ has opined that the use of directed deadly force against a foreign official in time of war does not constitute assassination as prohibited by this order or other law.

Law Enforcement and the Electronic Communications Privacy Act (ECPA)

Following the initiatives previously discussed, it became evident to federal authorities by 1986 that the basic law applicable to surveillance of electronic communications in criminal investigations required updating largely as a consequence of technology change. That new law—the ECPA—and indeed the history of electronic surveillance, is even more complex than the law of physical searches for documentary records because of the constant evolution of technology and the limited reach of the Fourth Amendment. As the Supreme Court has recognized (*Katz v. United States*, 1967, p. 361), the Fourth Amendment protects only those individual expectations of privacy that "society is prepared to accept."

What about telephone conversations and the computer equivalents, especially because much of this information is managed and held by a third party? The U.S. Supreme Court first addressed the issue in 1928, holding that a telephone wiretap was not covered by the Fourth Amendment. However, Justice Brandeis in his dissent was eloquent as to the issue (*Olmstead v. United States,* pp. 475–476):

> The evil incident to invasion of the privacy of the telephone is far greater than that involved in tampering with the mails. Whenever a telephone line is tapped, the privacy of the persons at both ends of the line is invaded, and all conversations between them upon any subject, and although proper, confidential, and privileged, may be overheard. Moreover, the tapping of one man's telephone line involves the tapping of the telephone of every other person whom he may call, or who may call him. As a means of espionage, writs of assistance and general warrants are but puny instruments of tyranny and oppression when compared with wire tapping.

Congress first addressed the issue in the Communications Act of 1934 that, inter alia, proscribed any person not authorized by the sender to intercept any communication. But by 1968, the law had become so complex and confused that Congress enacted Title III of the Omnibus Crime Control and Safe Streets Act to address the specific conditions and procedures under which wiretaps would be authorized, and again revisited the issue in 1986 with the ECPA as electronic communications became prevalent.

The ECPA has a four-tier approach to acquiring the very broad range of electronic communications information—some falling within the scope of the Fourth Amendment (e.g., real-time intercepts of message content) and some wholly without (e.g., subscriber information held by the telephone company) and thus outside core Fourth Amendment protections. First is a *real-time intercept order* that requires a greater showing than a regular search warrant and must be authorized at the most senior levels of the DoJ. It applies to any real-time voice or data transmission; it may be used only for specific, serious crimes and only when normal investigative techniques for obtaining the information have failed, are likely to fail, or are too dangerous. The intercept order has two parts, one authorizing the law enforcement agency to conduct the intercept and the other granting a service provider necessary assistance. When a provider cannot comply, Carnivore may be deployed. Carnivore, recently renamed DCS-1000, is a special filtering tool that can be inserted into a communication network and configured to gather the information authorized by court order, and only that information. In other words, it is a Local Area Network (LAN) packet sniffer that identifies and copies packets (by IP address) so authorized. The existence and use of Carnivore does not change any Constitutional or statutory requirements for warrants and court orders, but it has been the subject of enormous civil liberties controversy because of the possibility of intentional or accidental misuse. In light of these concerns, former Attorney General Reno convened an impartial academic review; the report of the Illinois Institute of Technology verified that the system operates as stated, albeit that proper use relies on the operator's ability to configure the filter correctly and fully (Smith et al., 2000).

Second is a *search warrant* that is used for stored electronic communications (e.g., unopened e-mail at a service provider). It requires a determination by a federal judge that probable cause exists to believe that a crime has been committed and that the information sought is material to that offense. Although a regular criminal search warrant (issued under Federal Rule 41) and this ECPA search warrant are identically termed and issued under the same legal basis of probable cause, they are somewhat different in the method of execution. Ordinarily, an ECPA search warrant is served on a provider and the provider then produces the information described in the warrant.

Third, and less difficult yet is a *court order*, generally used to obtain transactional records. It includes "pen register" information which comes from devices that record the telephone number dialed by the subject as well as "trap and trace" information which comes from devices that identify an incoming telephone number. Such orders must be granted automatically if the government certifies that there are "reasonable grounds to believe data is relevant to ongoing criminal investigation" as supported by specific and articulable facts (18 USC § 2703). The court order may also be utilized to access a limited subset of electronic mail—unopened messages held by an "electronic communications service" provider for more than 180 days or opened mail held by a "remote computing service" provider (18 USC § 2703). The arcane nature of these terms, underlying the entire ECPA framework, comes from the early days of computing services when the ECPA was passed. Essentially, Congress likened an electronic communications service to the post office and required a higher standard for access, but a remote computing service provider was likened to a third party and required a lower standard (a subpoena) for access. In either case, however, notice to the subscriber is required, although "delayed notice" may be authorized by the judge (18 USC § 2705).

Least difficult of all is an *administrative or grand jury subpoena* issued by the government itself, without judicial assistance, to obtain information. Typically, in the electronic arena, such subpoenas are used to identify the subscriber. As addressed in the next section, technology advances that led to the ECPA would later lead to changes by the U.S.A. Patriot Act.

To complete the consideration of the ECPA, it should be noted that most if not all states have similar laws, some of which are even more protective of privacy. For example, Maryland's statute (Maryland Courts & Judicial Proceedings Code, Title 10, Subtitles 4 and 4A), includes a more restrictive provision on taping of telephone calls than most (§§ 10–402 and 410); it requires consent of both parties but provides a most unusual legal defense—a lack of specific knowledge of the law. By example of application, it is this statute under which Linda Tripp was prosecuted but not convicted during the Clinton administration when the state was precluded from utilizing specific testimony of Monica Lewinsky that would have established the requisite knowledge.

New Surveillance Authorities in the Age of Terrorism
The Drive for the U.S.A. Patriot Act

The events of September 11, 2001, greatly changed the operating environment, although for some time prior, the law enforcement and intelligence communities, as well as civil liberties community, had chaffed under the statutory, regulatory, and judicial decrees in place that controlled investigations in general and surveillance in particular. Among the issues: The law of electronic surveillance had lagged behind technological advances (e.g., much of the law still focused on dial telephones), the intelligence restrictions hampered terrorism investigations (e.g., they had been formulated in the days of the "evil empire" [Reagan, 1982, online] and not Al Qaeda), and there was the general perception that extant regulations and judicial decrees hamstrung preemptive and prompt law enforcement action. It thus became clear to many government officials that new authorities and the removal of old restrictions were immediately required to prevent new and imminently feared attacks.

As previously noted, the immediate result was the extraordinarily rapid passage of the U.S.A. Patriot Act in late 2001 with minimal debate after the Attorney General admonished the Senate Judiciary Committee that debate was unwarranted (CNN.com/US, 2001). This legislation began in the Senate as The U.S.A. Act and the House of Representatives as The PATRIOT Act; the Senate bill passed on October 11, 2001, but the House took until October 24 to process language substitutions, resolve differences with the Senate version, and pass a final bill. The Senate agreed to the changes the same day and the matter was sent to the President and signed on October 26, 2001. But the continuing legacy of that then-existing consensus and lack of detailed consideration has been a bitter national argument as to the proper balance between civil liberties and national security in these threatening times, including most specifically the appropriate scope of domestic intelligence surveillance. In point of fact, the Act accomplished many needed and logical changes but also other troubling changes related to individual rights.

The Act contains ten titles, extends over 131 pages, and amends numerous laws. It expanded electronic surveillance of communications in law enforcement cases by removing anomalies between different forms of communication and also authorizing the sharing of law enforcement information with intelligence. It expanded the acquisition of electronic communications as well as business records in intelligence cases. It broadened the authority of the government to combat money laundering and to address the ease with which terrorists can enter and remain in the country. It also created new crimes (e.g., terrorism against mass transit systems, harboring or providing material support to terrorists, and possession of biological weapons materials) as well as conspiracy provisions to a number of crimes including arson and murder. And, of particular interest to information professionals, it amended the Computer Fraud and Abuse Act, 18 U.S.C. § 1030, to increase penalties for hackers who damage protected computers (ten to twenty years), to clarify the *mens rea* required for such offenses (need only intend damage, not a specific type of damage), to create a new offense for damaging computers used for national security or criminal justice (without dollar threshold), to expand the coverage of the statute to include computers in foreign countries so long as there is an effect on U.S. interstate or foreign commerce, to count state convictions as "prior offenses" for sentencing enhancements, and to allow losses to several computers from a hacker's actions to be aggregated for purposes of meeting the $5,000 jurisdictional threshold.

The Changes to the ECPA

The amendments in the law enforcement arena focus to a substantial degree on and accommodate advances in technology or remove inconsistencies in the protection of information based on its particular format; it:

- Expands the authority for issuance of intercept orders to crimes related to terrorism and computer fraud

- Facilitates the current roving wiretap authority by allowing search warrants for stored messages (e-mail and voice) and court orders for transactional records (e.g., "pen register") to be valid everywhere in the nation, issued by any court with jurisdiction over the offense, and without naming specific common carriers although a carrier can request certification from the government as to specific applicability to their company

- Allows stored voice mail messages to be acquired by the slightly easier search warrant process, thus harmonizing the law for stored voice mail and stored e-mail

- Broadens the scope of orders for transactional records to include any form of electronic communication (e.g., e-mail or Web surfing), not just telephone communications

- Broadens slightly the scope of the subpoena authority for subscriber information by allowing for access to payment and type of service information rather than merely current name and address

- Makes the rules for cable company service providers the same as for all other ISPs and provides immunity for all ISPs when acting in good faith reliance on government orders of any type

- Changes the rules on dissemination of criminal investigation information by allowing the automatic sharing with intelligence; the reverse takes place now under existing law

- Authorizes, if requested, assistance by law enforcement to ISPs or businesses under computer attack, but only where person is trespasser and not in existing contractual arrangement, and limits intercepts to communications to/form the attacker

- Broadens and clarifies the rules for voluntary disclosure of information by an ISP:

  - for nonpublic providers, any data may be disclosed for any reason

  - for public providers, the content of messages *and* subscriber data 1) where immediate danger is presented or 2) if necessary to protect ISP property; this provision is significant because the previous law provided for voluntary disclosure of content (but not, anomalously, noncontent information) and only to protect property (no provision for emergency)

  - clarifies immunity because all voluntary disclosures and relevant immunity are addressed in §§ 2702 and 2707(e), respectively; good faith reliance on a court order or a statutory provision allowing disclosure is now a complete defense

These changes are in large part logical refinements of the law and, in the case of sharing between law enforcement and intelligence, required as a matter of common sense and legal standards: Why should intelligence be barred from tracking leads that exist by virtue of a lawful criminal investigation when that information has been acquired under a probable cause standard? There is, however, controversy as to several of the amendments. One is the pen/trap provision and whether it includes header information in e-mail transactions and Web surfing that implicitly reveals content; the DoJ has confirmed that this authority would not be used in this way—that neither subject lines in e-mail, nor Web surfing addresses beyond the high-level domain name corresponding to an IP address, would be acquired. Another is the obligation with respect to such information by providers (e.g., a library); there is general agreement that the U.S.A. Patriot Act neither requires providers to change their current data retention practices nor to reconfigure their systems if presented with a court order. That said, it should be noted 1) that the FBI could insist on deployment of their Carnivore system if the provider could not comply, and 2) that a provider is obligated to preserve specific information if so requested by law enforcement pending issuance of an appropriate court order or other process. Lastly, the provision for ISPs to request law enforcement assistance raises the concern that this could provide entrée by way of suggestions by law enforcement to providers; again, it is clear that this is the right of the provider and not a right of law enforcement.

The Changes to the FISA

The U.S.A. Patriot Act was specifically formulated also to facilitate foreign intelligence collection against the terrorist threat; it thus amended the FISA in several regards. Some provisions were largely administrative and included making electronic surveillance warrants "roving" (applicable to all relevant communications carriers), as was largely the case in law enforcement. Other provisions, specifically § 215, were more extensive and significantly changed the extent of access to business records. Previously, only a very limited range of records held by transportation and travel-related businesses could be acquired by a FISA court order. Now, such orders can be directed to any entity, for any "tangible thing," and with an assertion only that they are "required" for an intelligence or terrorism investigation (15 Stat. 287–288). The surveillance implications are clear because any information in the possession of third parties is now available to intelligence demands with minimal showing.

But the real issue—presented by the dichotomy in public opinion and recognized in the pending Congressional bills—is balance. For example, library records have never been inviolate to government demands and are not infrequently sought in criminal investigations. Indeed the DoJ has argued that § 215 of the U.S.A. Patriot Act is more restrictive than a federal grand jury subpoena for the same records in that a court must explicitly authorize the use, there are explicit safeguards for First Amendment activities, and for a U.S. person the government must certify that the information is sought in a terrorism investigation. The point not addressed by the Department is that the total secrecy of the FISA process, including § 215, feeds public and Congressional concerns over potential abuse.

Another change affected the little known, FISA-related provision known as the National Security Letter (NSL) authority for certain electronic communications information and financial records. The authority is found in three separate statutes: the ECPA (18 USC § 2709) for telephone toll and electronic communications transactions records; the Fair Credit Reporting Act (15 U.S.C. § 1681u) for the names and addresses of all financial institutions at which an individual maintains or has maintained an account; and the Right to Financial Privacy Act (12 U.S.C. § 3414(a)(5)(A)) for an individual's institution-specific banking and credit records. These demands, essentially the intelligence corollary to the administrative or grand jury subpoena provisions for criminal law enforcement investigations, can be issued directly by the FBI and, under the U.S.A. Patriot Act, only on an assertion that such information is required for an investigation. Moreover, as with all FISA-related provisions, there is a prohibition of any disclosure at any time of the fact that the FBI has sought such information.

In December 2003, the Intelligence Authorization Act for 2004 further broadened the NSL authority. Specifically, § 374 modified the definition of "financial institutions" in the Right to Financial Privacy Act of 1978 to be the same as in the recently expanded law addressing money laundering (31 U.S.C. § 5312): essentially any business whose cash transactions have a high degree of usefulness in criminal, tax, or regulatory matters including, for example, casinos or car dealers.

Yet another change to the FISA authority, perhaps most important from a rights perspective and certainly the most complex, involved the situation warranting the use of FISA rather than traditional criminal law enforcement authority (e.g., search warrants issued pursuant to Fourth Amendment standards of probable cause of criminal conduct). Heretofore, the focus of the FISA statutory authority was solely counterintelligence or counterterrorism and it was available only where such reasons were the *sole or primary* purpose for the order. Although the Constitutional rationale was clear, the practical issue was that threats such as terrorism most often presented dual objectives—gathering intelligence and prosecuting—yet the terms of the FISA forced a binary and often arbitrary choice (proceed as intelligence investigation with less restrictions and forego criminal prosecution—or proceed as criminal matter with full Constitutional demands), thus hampering the most effective defense to critical threats.

The U.S.A. Patriot Act addressed this issue in two ways. First, at § 218, it changed the basis for granting FISA orders to a "significant" purpose, allowing a dual criminal and intelligence investigation. Second, it clarified the rules on information sharing between intelligence and criminal law enforcement officials. The FISA (pre-U.S.A. Patriot Act) had specifically allowed intelligence officials to pass "foreign intelligence" as well as routine criminal information to law enforcement, but a perception had grown among such officials that there were significant limitations—either directly prohibiting passage or indirectly by endangering the intelligence operation if a law enforcement purpose emerged. Thus, § 504 of the U.S.A. Patriot Act expressly authorized intelligence officers who are using FISA "to consult" with federal law enforcement officers in order to "coordinate efforts to investigate or protect against" foreign threats to national security. In the opinion of the Attorney General (Ashcroft, 2002e, online) this means that "intelligence and law enforcement officers may exchange a full range of information and advice" concerning foreign intelligence investigations including "information and advice designed to preserve or enhance the possibility of a criminal prosecution." Together, the intent of these provisions was to reduce the conflict between intelligence and law enforcement personnel as to the path of an investigation and to make clear that a FISA order could be obtained even if a criminal prosecution was contemplated or indeed had been instituted during the pendency of an intelligence investigation.

The Attorney General's Interpretation of the FISA Changes

With this change of statutory language, there was little understanding among the public or Congress that there was any further grant of positive authority—beyond enhanced information sharing and dual-track investigations. But the implementing guidelines issued by the Attorney General on March 6, 2002 (Ashcroft, 2002e, online) did more than give criminal prosecutors access to "all information developed" in FISA cases and the right to consult and provide advice and recommendations to intelligence officials regarding any intelligence case including the "initiation, operation, continuation or expansion of FISA searches." The guidelines further authorized the use of FISA "primarily for a law enforcement purpose as long as a significant foreign intelligence purpose remains" (Ashcroft, 2002e, online). Clearly, the increased sharing was authorized by the terms of the U.S.A. Patriot Act. However, the direct use of the FISA by law enforcement was more problematic and was predicated solely on the "significant purpose" change—that a foreign counterterrorism intelligence purpose can be met by the fact of a criminal prosecution for terrorism and hence such a criminal prosecution may make direct use of the FISA (115 Stat. 291). The contrary argument was that the "significant purpose" change recognizes and authorizes the dual objectives of intelligence and prosecution, but that the FISA authorities are to be exercised only by intelligence officials because use by prosecutors would substantially eviscerate Fourth Amendment rights requiring warrants based on probable cause of criminal activity.

On May 17, 2002, the Foreign Intelligence Surveillance Court (FISC) found that the sharing provisions were not inconsistent with the U.S.A. Patriot Act, but that the direct use of FISA authorities by law enforcement officials represented a substantial and Constitutional issue (*In re: All Matters Submitted to the Foreign Intelligence Surveillance Court*, 2002). Specifically, the court held that the Attorney General directive would give criminal prosecutors a significant role in "directing" FISA investigations, would violate the statutory basis for FISA orders (intelligence purpose), and would effectively substitute the FISA for the Fourth Amendment. In sum, the Court recognized the appropriate nature of cooperation and sharing between intelligence and law enforcement, but rejected a process "vulnerable to abuse" (Rosenzweig, 2002, p. A16). Accordingly, the FISC revised the proposed directive in part to allow the new levels of coordination and sharing but to retain the critical aspects of the "wall" that precludes direct use or control of FISA authorities by law enforcement personnel. Hence, to be maintained pursuant to the FISC order was the bright line between criminal and intelligence investigations and the respective rights of U.S. persons under these two bodies of law.

An appeal of that decision followed on August 23, 2002, by the government to the special three-judge court that was created to review FISC decisions. This court—the U.S. Foreign Intelligence Surveillance Court of Review—which had not met previously in its two decades of existence, consists of three federal appellate court judges appointed on a rotating basis by the Chief Justice; the members at that time were Judge Ralph B. Guy of the 6th Circuit, Judge Edward Leavy of the 9th Circuit and Judge Laurence H. Silberman of the D.C. Circuit. The government in its brief (Ashcroft, 2002d, online), asserted that Senator Leahy, who it identified as the "drafter of the coordination amendment" in the U.S.A. Patriot Act, was in agreement that there is no longer a distinction between using FISA for a criminal prosecution and using it for foreign intelligence collection. However, in a statement on September 10, 2002 in the context of a Senate Judiciary Committee hearing, the Senator was clear in his renunciation of this assertion that he supports the Administration position (Leahy, 2002, online):

> That was not and is not my belief. We sought to amend FISA to make it a better foreign intelligence tool. But it was not the intent of these amendments to fundamentally change FISA from a foreign intelligence tool into a criminal law enforcement tool. We all wanted to improve coordination between the criminal prosecutors and intelligence officers, but we did not intend to obliterate the distinction between the two, and we did not do so. Indeed, to make such a sweeping change in FISA would have required changes in far more parts of the statute than were affected by the U.S.A.

> PATRIOT Act ... [and] such changes would present serious
> constitutional concerns.

The argument on appeal, *ex parte* as are all matters involving the FISA, was heard on September 9, 2002, and, consistent with a point that the author has suggested in the past (Strickland, 2003b) (i.e., the FISC and the Court of Review should be open to the receipt of public filings), the American Civil Liberties Union (ACLU) and other public interest groups filed an *amici* brief on September 20 (Beeson, Jaffer, & Shapiro, 2002). According to media reports, the Court of Review was not fully persuaded by the government submissions and ordered a supplemental brief. Contemporaneous with the hearing, and through the efforts of the Senate Judiciary Committee and particularly Senator Leahy, the Chief Judge of the Court of Review committed to making available to the public a declassified version of their decision (Lumpkin, 2002). That promise was fulfilled on November 18, 2002, when the judgment of the U.S. Foreign Intelligence Surveillance Court of Review was released (*In re: Sealed Case 02–001*). In doing so, it overturned the unanimous opinion of the FISC by rejecting the "wall" between intelligence and law enforcement but nevertheless acknowledging the paradox in the government's argument—that legally there has never been a limitation on law enforcement use of the FISA (the "wall") but that the U.S.A. Patriot Act substantially removed such limitation by changing the intelligence predicate from "primary" to "significant" purpose. Specifically, the court interpreted the FISA foreign intelligence predicate as imposing no limits on who may use the authority, provided that the information sought is broadly defined as foreign intelligence information, that the target is deemed an agent of a foreign power, and that the purpose must include something more than solely a national security-related criminal prosecution (e.g., a terrorism-type crime). What this means is that criminal prosecutors have carte blanche use of the FISA provided only that there is a scintilla of separate intelligence interest and that the investigation and prosecution do not focus solely on ordinary or street crime— although noting that such crimes (e.g., credit card fraud or bank robbery) could be inextricably intertwined with foreign intelligence and/or terrorism activity and thus qualify for FISA investigation.

In addition to construing the FISA in statutory terms, the Court of Review also considered its constitutionality—comparing the FISA with established Fourth Amendment law. In doing so, it acknowledged the differences in probable cause standard, the standard of judicial review, and the required particularity showing. In the end, however, the court avoided a definitive statement by deeming the question "close" and conceding that a FISA order may not be a warrant contemplated by the Fourth Amendment. That said, however, the court noted that this was not the end of the inquiry given that there are recognized circumstances where the government may proceed without a warrant and the question in all such circumstances is one of reasonableness. Here, the Court of

Review found that the government was presented with "special needs" and recognized, as most courts have, that the President has inherent authority to conduct warrantless searches to obtain foreign intelligence information, although the exact parameters have never been decided by the U.S. Supreme Court. As such, the Court of Review concluded that the FISA passes Constitutional muster consistent with other cases that have drawn distinctions between ordinary crime and extraordinary situations and that have led to the approval of even warrantless and suspicionless searches in schools, in apprehending intoxicated drivers, and in securing the national borders from unlawful entry.

Thereafter, the ACLU attempted to secure review of the order by the U.S. Supreme Court on behalf of unknown targets of FISA orders stating that "we are urging the Supreme Court to reject the extreme notion that Attorney General Ashcroft can suspend the ordinary requirements of the Fourth Amendment to listen in on phone calls, read e-mails, and conduct secret searches of Americans' homes and offices" (Beeson, 2003, online). That effort was denied on March 24, 2003, on the grounds of standing, thus changing, in perhaps significant degree, government surveillance authorities in criminal cases that involve foreign intelligence aspects (*American Civil Liberties Union, et al. v. United States*, 2003). The qualification is, of course, predicated on the fact that all FISA orders are secret and there are no data at present to suggest how the FISA authorities may have been utilized by criminal prosecutors.

The Attorney General's Revised Investigatory Regulations

Following the new FISA procedures, Attorney General Ashcroft also issued new "Guidelines on General Crimes, Racketeering Enterprise and Terrorism Enterprise Investigations" dated May 30, 2002 (Ashcroft, 2002b). These guidelines, accompanied by three others entitled Guidelines on Federal Bureau of Investigation Undercover Operations, Guidelines Regarding the Use of Confidential Informants and Procedures for Lawful, Warrantless Monitoring of Verbal Communications (Ashcroft, 2002a, 2002c, 2002f), replaced those of former Attorneys General Levi, et al., and have generated substantial public and media controversy largely because of their allowance of so-called public surveillance. However, this is but a small aspect of the authority granted, and the concern tends to be misdirected with a focus on privacy rather than the real issue of anonymity. Indeed, the provisions regarding public surveillance present a classic and recurring conundrum in our society. On one hand, it is only logical that the FBI should be authorized to undertake actions that ordinary citizens might do in order to ascertain the continued safety of their community (e.g., observe public events for signs of danger). But conversely, those very actions suggest the days of the FBI under J. Edgar Hoover when special agents investigated perceived threats where there were no indications of criminal malevolence.

The new guidelines address initially the purpose and scope of criminal investigations in general (Part II) as well as racketeering (Part IIIA) and terrorism-related investigations (Part IIIB)—defined as concerning enterprises that seek to further political or social goals through violence or that otherwise aim to engage in terrorism or the commission of terrorism-related crimes. They specifically provide for three levels of increasing investigative activity to include: 1) the limited checking of initial leads, 2) preliminary inquiries, and 3) full investigations. The lowest level is authorized "whenever information is received of such a nature that some follow-up as to the possibility of criminal activity is warranted" and the objective is to determine whether a higher level of activity is warranted. The next level, the preliminary inquiry, is authorized whenever "information or an allegation which indicates the possibility of criminal activity and whose responsible handling requires some further scrutiny beyond checking initial leads." Here, any lawful investigative technique is allowed—with the exception of mail opening and nonconsensual electronic surveillance—including (a) the review of FBI indices and files, of other federal, state, and local government files, of any records open to the public, or of any public sources of information; (b) interviews of the complainant, established informants, the potential subject, or any other person with potential knowledge of facts relevant to the allegations; (c) physical or photographic surveillance of any person, or (d) one-party consent electronic surveillance. In such circumstances, a preliminary inquiry may last up to 180 days, but multiple 90-day extensions are available albeit with escalating approval levels.

Thereafter, two types of full investigations may follow—a "general crimes investigation" or a "criminal intelligence investigation" that would focus on racketeering or terrorism. The difference is related to the information identified by the prior levels of inquiry and the investigative focus. A general crimes investigation is predicated on the existence of facts or circumstances "reasonably indicating that a federal crime has been, is being, or will be committed." It thus may have a focus on preventing a crime or solving a crime already committed and it should be noted that the "reasonable indication" standard is "substantially lower" than probable cause. Conversely, a "criminal intelligence investigation" is predicated on the involvement of a group rather than an individual or individuals involved in a specific act. The initial purpose of such intelligence investigations is to ascertain "information concerning the nature and structure of the enterprise—including information relating to the group's membership, finances, geographical dimensions, past and future activities, and goals—with a view toward detecting, preventing, and prosecuting the enterprise's criminal activities." As evident, intelligence investigations are long-term and have both a remedial and preventive focus. In addition, such intelligence (or enterprise) investigations are also predicated on the "reasonable indication" standard but focus on whether two or more persons are engaged in an enterprise of "furthering political or social goals wholly or in part through activities that involve force or violence and a federal crime" or engaging in terrorism or racketeering.

One immediate question is the meaning of the "reasonable indication" standard because there is not the benefit of several centuries of judicial interpretation as with probable cause. The guidelines instruct the Bureau to balance the magnitude of the threatened harm, the likelihood it will occur, the immediacy of the threat, and any danger to privacy or free expression posed by an investigation. In the opinion of the DoJ, the following would qualify: a statement that directs or advocates the use of violence or merely the demonstrated ability to carry out a violent or other criminal act as by the purchase of weapons, the stockpiling of dangerous substances, or by the demonstration of any other behavior "that appears to harm, intimidate or interfere with the rights of others." Thus, although investigations cannot be predicated on First Amendment protected activities or mere speculation, there is no shield for activities that present some indicia of violence either in terms of future action or mere advocacy and a terrorism enterprise investigation may properly ensue.

A second question is the permitted scope of a terrorism enterprise investigation. Under the Ashcroft guidelines, the Bureau may collect information including the identity of the membership (or those acting in concert) as well as the finances, geographical scope, the past and present activities, and the goals of the enterprise. In so doing, the Bureau may use any lawful investigative technique including the use of confidential informants, undercover operations, nonconsensual electronic surveillance in all of the varied forms, consensual electronic monitoring, warrantless monitoring of verbal (oral) communications and, of course, court-approved search warrants.

It is Part VI of the Guidelines that have generated the majority of public and media attention and concern because they identify a number of activities that can be carried out even in the absence of the approved investigatory efforts—checking of leads, preliminary inquiry, or full investigation. To the extent that the focus is counterterrorism, the Bureau is now authorized to operate and/or use essentially any information system (e.g., commercial database or other government agency repository) and to visit any public place or event and collect any information relevant to that event provided that the event is open to the public and that the information collected may be retained only if it relates to potential criminal or terrorist activity. And here, the definition of terrorism is broad: "for the purpose of identifying and locating terrorists, excluding or removing from the United States alien terrorists and alien supporters of terrorist activity as authorized by law, assessing and responding to terrorist risks and threats, or otherwise detecting, prosecuting, or preventing terrorist activities" (Ashcroft, 2002b, p. 21). Additionally, to the extent that the focus is counterterrorism or otherwise, the Bureau is also now authorized to carry out general topical research and to conduct online searches and access online sites and forums on the same terms and conditions as members of the general public (Ashcroft, 2002b, p. 22).

As might be expected, there is a range of logical and political opinions on this aspect of the new guidelines as reported in the media (Miller, 2002): Laurence Tribe, the noted constitutional scholar at Harvard University, has argued that there is no violation of speech or privacy rights (provided there is surreptitious infiltration), but the Cato Institute has opined that "the idea that agents cannot do what ordinary Americans can do" is indeed "odd." Conversely, Rep. Kucinich (D. Ohio) has relied on history for an assessment: "Americans have not forgotten the abuse of civil liberties which took place in the 60s and 70s under the name of law enforcement."

Perhaps the best expression of the issue with the new Ashcroft guidelines is not that they allow expansive new access to information for which there is a reasonable expectation of privacy—although this may be an area of concern with certain U.S.A. Patriot Act provisions as detailed. Rather, the issue is that the guidelines allow the collection of generally public information with minimal or even no suspicion of criminal activity. Although the guidelines prohibit the maintenance of such information unless it relates to counterterrorism, the public fear is that the standard is so imprecise and so open to interpretation, as to allow the collection of political dossiers and reopen the problems that the original Levi guidelines solved. The essential argument is well expressed by Mark Rasch (2002, online), a former chief of the Justice Department's computer crime unit:

> Privacy is the right to be left alone. Political freedom is the right to engage in vigorous discourse and exploration of ideas—even unpopular and potentially subversive ideas. We have and should expect a right to privacy even in those things that occur and are reported on in the public. The mere act of the FBI collecting newspaper clippings mentioning our name—something clearly public—has a chilling effect on our free discussion. The requirement that they do so only where there is some reason to believe that were are engaged in criminal activity is not unreasonable.

And the public concern is exacerbated by official and media releases that document the extent of such overt collection and the data mining plans. Documents released voluntarily or under the FOIA confirm that the FBI has contracted with commercial information databases (e.g., ChoicePoint or Experian) for access to information on individuals, and that local police departments have conducted intelligence collection operations against public interest groups. For example, as documented by the ACLU,[1] such surveillance, file maintenance, and dissemination of information have occurred vis-à-vis organizations that support nothing more than peace and natural resources. One case evolved in 2002 from the coordinated surveillance actions of Denver and Colorado Springs authorities against the American Friends Service Committee and the Colorado Coalition to Prevent Nuclear War, where license plates, names, and home addresses were collected and disseminated under the rubric of combating criminal extremism. On April 17, 2003, the ACLU of Colorado announced a settlement of its lawsuit against the Denver Police for these and similar actions against various advocacy groups.

Disclosures also document that data mining has become a central focus of government organizations including the Pentagon and the Transportation Security Administration—an issue that will be addressed in some depth in the next section of this chapter. This, of course, is not to suggest that data mining in general or access to publicly available information sources are per se violations of civil rights. It simply highlights the fact that relaxed guidelines and a number of critical incidents can converge to present a public relations failure that will continue to erode public support for required government response to the terrorist threat.

The Local Impact of the Changed Climate

In September 2002, following the revised Attorney General regulations allowing, inter alia, overt surveillance of public activities, the New York City Police Department moved to revise the previously considered *Handschu* regulations arguing that they unnecessarily restricted terrorism investigations, that "the safety of New Yorkers would be jeopardized if officers had to prove suspicion of criminal activity before they could investigate political groups in search of connections to terrorism," and that the proper standard was that any activity must simply serve a law enforcement purpose (Perrotta, 2002a, p. 1). In opposition, counsel argued that the changes were intended to be permanent changes in police practices that would prevent direct violations or chilling effects on the public's constitutional rights (Perrotta, 2002a, 2002b, 2002c).

On February 11, 2003, Federal Judge Charles Haight, citing "fundamental changes in the threats to public security" (*Handschu*, 2003, p. 338), modified the 1985 *Handschu* decree to conform closely with the Ashcroft guidelines but left in place the three-member oversight board with the authority to investigate specific complaints that police investigations have violated individuals' constitutional rights. In substantial part, the decision was based on affidavits filed by David Cohen, the NYPD Deputy Commissioner for Intelligence and the former Deputy Director for Operations of the CIA, wherein he detailed (Weiser, 2003, p. A17) the "changed circumstances" that legally authorized and required a change in the existing court order. Those "changed circumstances" in his informed view included the fact that numerous Islamic mosques and organizations in the United States were "radicalized" and operated in a manner "to shield the work of terrorists from law enforcement scrutiny by taking advantage of restrictions on the investigation of First

Amendment activity." But in his decision, Judge Haight (*Handschu*, 2003, pp. 348–349) was "mindful of the crucial importance of preserving both individual freedoms and public safety and balancing the legitimate demands of those two goals" and thus, although the NYPD was entitled to new authorities given "the nature of public peril," he reiterated the fact that the "Constitution's protections are unchanging" and the continuation of the oversight panel was required and would ensure that fact (CNN.com/Law Center, 2003). As considered in the conclusion of this chapter, such a mechanism of oversight is critical.

The Use of the New Powers

As briefly discussed in the introduction, a primary element of the public and political opposition to the U.S.A. Patriot Act and the Attorney General's new grants of investigative powers is the dearth of insight into their exercise. The annual report on the use of the FISA authority states merely that "all 1228 applications presented to the Foreign Intelligence Surveillance Court in 2002 were approved." (Ashcroft, 2003, online). Slightly more information was provided in two reports by the DoJ (Brown, 2003; Bryant, 2002) to the House Judiciary Committee including the following facts:

- The § 213 authority for delayed notice of the execution of a warrant has been requested 47 times and "the courts have granted every request" (Brown, 2003, online).

- The delay period is most commonly seven days but has ranged from one to 90 days and has included unspecified duration pending the unsealing of indictments.

- Extension of the delay period has been requested 248 times.

- The § 212 authority for voluntary disclosures by computer-service providers in emergency situations has been used on "many occasions" including regular criminal (not terrorism) investigations such as kidnapping, bomb threats, and death threats (Brown, 2003, online).

- No records are maintained on the use of the Attorney General authority to attend public meetings in general or conduct surveillance activities at mosques; an informal survey of FBI field offices suggests the latter number is less than ten.

- From the enactment of FISA in 1978 through September 11, 2001, the Attorneys General issued 47 emergency authorizations for electronic surveillance and/or physical searches under

FISA; in the one-year following, 113 emergency authorizations were issued.

- All other questions regarding the use of FISA-related authorities were answered with a claim of classification.

More recently as reported (Eggen, 2003, p. A02), the Attorney General augmented these disclosures in a September speech in Memphis where he characterized critics of the U.S.A. Patriot Act as "hysterics" and proceeded to assert that no § 215 FISA court orders had been issued against libraries or bookstores. This disclosure is as notable for its approach on building a national consensus as for what it did not say: It did not detail any other use of § 215 against other targets or any use of other provisions authorized by the U.S.A. Patriot Act (e.g., FISA-authorized physical or electronic surveillance, National Security Letters, or delayed notice criminal warrants) against libraries and booksellers. What exists is sparse information on which to build a public, much less a Congressional, consensus on a legal approach to the threat of terrorism including specifically the contours of domestic surveillance. It is a problem that is exacerbated when we consider the demands for new powers in the following section.

The Proposals for Additional Powers

The new legal authorities and investigatory guidelines considered previously have not presented the entire story in the contest between government needs and citizen rights. In February 2003, draft legislation prepared by the DoJ styled as the Domestic Security Enhancements Act (DSE) of 2003 (and often referred to as Patriot Act II), was leaked and received substantial media and Congressional criticism. Many of the provisions were extreme, including the authority to revoke citizenship, and others would have substantially increased the power of government surveillance. Indeed, according to media sources (Goldstein, 2003, p. A1), Rep. Sensenbrenner, Chairman of the House Judiciary Committee, and Senator Hatch, Chairman of the Senate Judiciary Committee, "deterred" the circulation and consideration of this proposed legislation and informed the Attorney General "in no uncertain terms" that such an effort "would be extremely counterproductive."

However, much like the proverbial phoenix, certain of the provisions resurfaced on the second anniversary of the September 11th attacks with the support of President Bush who called on Congress to expand the surveillance powers of the government under the 2001 U.S.A. Patriot Act and "untie the hands of our law enforcement officials" (Milbank, 2003, online). Specifically sought was a subset of the provisions in the previously proposed DSE Act, including new authority for the issuance of administrative subpoenas and broadened authority for the denial of bail and the imposition of the death penalty.

Although the latter provisions may invoke the most emotional concern, the little understood administrative subpoena authority is most significant to our discussion of government surveillance. Such subpoenas are a category of governmental authority to compel the production of documents or testimony that has been granted by Congress to various federal agencies. Unlike search warrants that are issued by federal courts and require a demonstrated showing of probable cause, administrative subpoena are issued by federal agents themselves, without any prior approval by the courts, need only relate generally to the subject matter of the investigation, and may generally be enforced through the civil and criminal contempt powers of the federal courts. Today, there are well over 300 existing administrative subpoena authorities and the government (U.S. Department of Justice, 2002, online) describes them as a "complex proliferation" of law with varying enforcement methods and varying provisions for the protection of individual rights, including privacy in each of the enabling statutes. It is thus this breadth of authority and general balance in favor of the government that is of concern with respect to the instant proposal.

Indeed, in the context of this consideration of surveillance authority, it is significant to note that the U.S. Supreme Court has consistently ruled in favor of a broad interpretation of administrative subpoena authority—using a "reasonableness" and not a "probable cause" analysis that incorporates a substantial deference to the government. The reasonableness or good faith standard requires only that 1) the investigation is conducted for a legitimate purpose, 2) the information requested is relevant to that purpose, 3) the agency does not already have the information sought, and 4) the agency followed its own administrative procedures in issuing the subpoena (*United States v. Powell*, 1964). As summarized by other courts, the federal judiciary will enforce administrative subpoena unless the information sought is "plainly incompetent or irrelevant to any lawful purpose of the [requesting official] in the discharge" of official duties (*Endicott Johnson Corp. v. Perkins*, 1943, p. 509).

But this is not to suggest that there are no limits on the authority, although the limits are generally found in the specific enabling statute, and will typically be argued in the context of a motion to quash. And, there is also the baseline question of who has knowledge of the subpoena and thus can mount an opposition—the recipient who has the information may only be a custodian for the real person in interest. In any event, one argument would be that the good faith factors in general were violated by the agency (i.e., there is no legitimate purpose to the underlying investigation). Another focus would be privacy and could be argued in several contexts. First would be the overall requirement of reasonableness where the Supreme Court on occasion has discussed the need to balance the public need and personal privacy (*Oklahoma Press Publishing Co. v. Walling*, 1946). And second would be statutorily imposed privacy rights often juxtaposed with notice rights; here, by way of example, the Right to Financial Privacy Act (RFPA), codified at 12

U.S.C. § 3401 et seq., generally prevents the release of personal banking information pursuant to an administrative subpoena (or otherwise) until required notice is provided to the customer. More specifically, 12 U.S.C. § 3405(2) requires the government notice to the customer to include 1) a copy of the administrative subpoena including a description of the records sought, 2) the purpose of the subpoena, and 3) the customer's right to file a motion to quash. There are, however, exceptions to the notice provision including delayed notice in law enforcement cases and absolute secrecy in foreign intelligence cases under existing National Security Letter authorities. But note that the delayed notice provisions, found at 12 U.S.C. § 3409, allow delayed notice if the agency petitions a federal court and establishes that providing current notice would endanger the life or safety of any individual, allow flight from prosecution, destruction of evidence, or intimidation of witnesses, or otherwise endanger the investigation; this delayed notice provision requires judicial involvement and a detailed finding—not simply the determination of the law enforcement agency.

Thus, although the exact form of the proposed anti-terrorism administrative subpoena authority is unknown at the moment, the form provided in the abortive DSE Act at § 128 allowed the production of any record or tangible item (or the testimony of witnesses), required nondisclosure (i.e., secrecy) by the recipient, provided for enforcement through the contempt powers of the federal courts, but made no provision for notice to or protection of the privacy of the subject.

The Impact of Technology on Surveillance and Citizen Rights

Technology and the Law in General

We have considered technology and its impact on surveillance in an evolutionary sense—how advances in communications technology such as e-mail had necessitated legal changes in the form of the U.S.A. Patriot Act to enable surveillance in this new environment. We also noted that the judiciary historically has been slow to address technology changes, citing the 1928 U.S. Supreme Court decision in *Olmstead v. United States* that held new surveillance technology in the form of a telephone wiretap was not covered by the Fourth Amendment—a decision not changed until 39 years later in the 1967 decision of *Katz v. United States*. That decision, as recognized today, was significant for not only extending the Fourth Amendment protections to electronic surveillance but also defining the general scope of the law as protecting one's reasonable expectation of privacy that society was prepared to accept and not turning per se on the presence or absence of a physical intrusion into any given enclosure. Thus, consistent with this theory, the courts have deemed a "plain sight" exception to the Fourth Amendment generally irrespective of the use of technology including the following actions:

observations from low-flying airplanes (*California v. Ciraolo*, 1986), recording of telephone numbers dialed (*Smith v. Maryland*, 1979), or trained dog sniffs (*United States v. Place*, 1983). The law has tended to allow biological systems as well as technology platforms for information collection through surveillance—deeming that such activities simply do not constitute a search for purposes of the Fourth Amendment and may be regulated by statute if at all.

But today, the use of hardware in lieu of biological systems is a major activity as technology products are designed and deployed to identify drugs; chemical, biological, and radiological contaminants; and many other measurable phenomena that may provide evidence of criminal activity. As might be expected, this advance of technology has presented conflict in court as cases are considered where technology has enhanced the regular senses of law enforcement officers. And rather consistently, their use has been approved provided that the sanctity of the home was not invaded. In essence, although the *Katz* court spoke of the Fourth Amendment protecting people and not places, the fact is that the courts have recognized that actions in the public sphere present a lesser expectation of privacy and thus raise lesser Fourth Amendment considerations. Thus, for example in *United States v. Knotts* (1983), the Supreme Court held that a person traveling in an automobile on public thoroughfares—with drug chemicals containing a tracking beeper—had no reasonable expectation of privacy in his movements from one place to another including the fact of whatever stops he made and the fact of his final destination when he exited from public roads onto private property. The logic: Because traditional visual surveillance from public places of such travels was not a search for purposes of the Fourth Amendment, the fact that law enforcement had relied not only on visual surveillance, but also on the use of the beeper to assist in the visual surveillance, did not alter the situation. As the court stated (*Knotts*, 1983, p. 276): "Nothing in the Fourth Amendment prohibited the police from augmenting the sensory faculties bestowed upon them at birth with such enhancement as science and technology afforded them in this case." And just a year later, the Supreme Court reiterated the rules on homes, public places, and technology-assisted surveillance. Here, in *United States v. Karo* (1984), involving a very similar factual matter to *Knotts*, the Court found that a car with a beeper installed with an informant's consent raised no Fourth Amendment considerations so long as it traveled in public spaces but did raise such issues when the beeper continued to be monitored inside a private home where the residents had a justifiable privacy interest.

But if radio frequency beepers are good, and the latest surveillance technology is better (in a functional sense), then what limits exist on the authority of the state to use new technology not only in secret but without consent and without court orders? Remember that in the cases considered previously, the beepers had been placed in the goods with the consent of the owner of the goods and the officers were augmenting their normal faculties in public spaces. But what if that were not the case—would critical civil liberties considerations be at issue? Two cases, one federal and one state, provide some guidance in our quest to define the proper balance between required surveillance and individual privacy.

The Courts Speak Today on Technology

To speak definitively on the contest between technology and privacy, the U.S. Supreme Court finally addressed the issue in 2001 in a case that concerned the use of infrared sensing devices—*United States v. Kyllo*. The issue was whether the use of this technology without a warrant constituted a search under the Fourth Amendment in that it violated a reasonable expectation of privacy. Previously, five U.S. Courts of Appeal had answered these questions in the negative and approved the use of the technology. More specifically, those courts had found that activities within a residence are not protected from outside, nonintrusive government observation, that the use of heat-sensing technology to enhance government surveillance did not turn permissible observation into impermissible search, and that the crucial inquiry was whether the technology revealed intimate details.

But in *Kyllo* (2001, p. 40), the U.S. Supreme Court, in a surprising 5 to 4 decision led by Justice Scalia, struck down the use of this technology and held that where "the Government uses a device that is not in general public use, to explore details of a private home that would previously have been unknowable without physical intrusion, the surveillance is a Fourth Amendment search and is presumptively unreasonable without a warrant." The Court continued by noting that the real question here was whether the observation was "a search" and repeated the *Katz* test—that asks in a given case whether there was a subjective expectation of privacy that society was willing to recognize as reasonable. In sum, the Court found that sense-enhancing technology, not in common use, should not be allowed to erode the traditional concepts of privacy. The dissent, however, focused on the fact that the device did not access information through the wall of the house (breaching a privacy expectation) but rather escaped emanations exposed to the public thus analogizing to a long line of plain sight cases (e.g., dog sniffs, as previously considered) as ample precedent for this use of technology.

Less than a year after *Kyllo*, the courts in Washington State in *State v. Jackson*, confronted a similar contest but involving global positioning system (GPS) locators and whether they could be utilized without a court-ordered warrant—a somewhat surprising case because the general rule of law is that the placement of a beeper requires lawful consent or a warrant. Here, in working to solve a murder of a young child, local police did secure a warrant to emplace the GPS device but subsequently, during the course of the defendant's challenges to the warrant, the trial court held that no warrant was required to install the GPS devices under the state constitution. Thereafter, the intermediate-level Court of

Appeals (*State v. Jackson*, 2002, p. 683) affirmed the decision that no warrant was required noting that the monitoring of public travels in a vehicle by use of the GPS device is "reasonably viewed as merely sense augmenting, revealing open-view information of what might easily be seen from a lawful vantage point without such aids." Derided as an "unprecedented extension of governmental power to monitor citizens" by the Seattle-King County Public Defender Association and as raising the "possibility of a police state" by the ACLU of Washington (George, 2003, online), the case was accepted for final review by the Washington Supreme Court.

On September 11, 2003, the Court strongly sided with privacy advocates holding that a warrant is required for the use of such technology but doing so under the broader state constitutional protection for privacy rather than the federal Fourth Amendment. Here, the Court (*State v. Jackson*, 2003, p. 259) held that the state constitution provides that "no person shall be disturbed in his private affairs, or his home invaded, without authority of law" and thus the fact of whether advanced technology leads to diminished subjective expectations of privacy is irrelevant to the decision. What is relevant, observed the Court (*State v. Jackson*, 2003, p. 260), is the definition of "private affairs" and what one exposes to the public is not private, but when there is a "substantial and unreasonable departure from a lawful vantage point, or a particularly intrusive method of viewing," then there is a violation. The Washington Supreme Court thus distinguished between the more typical visual surveillance by a police officer and the all-knowing, all-time GPS solution— a more encompassing definition of citizen rights than the *Kyllo* decision.

Policy or Technology?

Although these cases have considered the impact of technology on surveillance and privacy essentially in a hardware sense, there are additional aspects and policy considerations in this arena. One is that the seminal question is not merely the initial collection but equally the subsequent maintenance, accretion, analysis, utilization for other purposes, and dissemination to other entities that present civil liberties issues. This question is paramount when considering "data mining" systems that collect individual-specific information from multiple systems, although there is some controversy as to whether such systems are more properly denominated as "knowledge discovery" systems. Data mining is perhaps best defined as the automated analysis of existing data to identify patterns or relationships that can be used in a predictive or future manner. Thus, data mining is in some instances misused to describe systems that merely express existing data in different forms. Knowledge discovery is more appropriate nomenclature to represent generally the extraction of knowledge from data by various means including specific data mining tools.

It is because of this issue—the use of information without or in excess of individual consent—that the recent efforts by the Information Awareness Office (IAO) at the Department of Defense for a Total Information Awareness (TIA) program as well as the Transportation Security Administration's (TSA) Computer Assisted Passenger Prescreening System (CAPPS II) have raised substantial public opposition.

Although TIA was later renamed the Terrorist Information Awareness, thus preserving the acronym, there are two points relevant to any assessment with respect to public policy: First, both systems were based on perceived, extensive data mining of commercial and government databases collected for one purpose and then to be shared and used for another purpose (to identify national and aviation-specific threats, respectively); and second, both systems were thus sidetracked by a widely perceived threat to personal privacy. To a substantial degree this perception is correct; as recognized by many public interest organizations (e.g., Center for Democracy and Technology, 2003a, 2003b, 2003c), there are few legal restrictions on government access to personal data held by commercial entities. The U.S. has no comprehensive privacy law for commercial data and the Privacy Act of 1974 either does not apply to searches of commercial databases, or law enforcement or intelligence agencies can easily exempt themselves from the majority of provisions.

Thus, although the law provides little redress, the political process has been more responsive to public opinion. More specifically, the TIA effort was required by a unanimous Congressional action (Consolidated Appropriations Resolution, 2003) sponsored by Senator Ron Wyden (D. Ore.) to issue a comprehensive privacy analysis in May 2003 (U.S. Department of Defense, 2003) and was substantially embarrassed in July when Senator Wyden and others revealed the existence of a government-operated online betting pool on future terrorist attacks known as "FutureMAP" (Wyden, 2003). This program was terminated in September 2003 when Congress specifically eliminated funding for the Office of Information Awareness, the DARPA division that included TIA (Department of Defense Appropriations Act of 2004; Hulse, 2003). This legislation specifically allows the continuation of certain specific DARPA projects that do not concern data mining, including bio-event recognition, analytical war gaming, and foreign language processing, and also the use of any technology tools developed by TIA through the National Foreign Intelligence Program (NFIP) for foreign intelligence purposes. As of early 2004, little if any information has emerged about such continuation efforts.

The political process has also focused fierce criticism on CAPPS II for its initial privacy notice in January 2003 (Department of Transportation Privacy Act of 1974: System of Records Notice) as well as its revised privacy notice in August 2003 (Department of Transportation Interim Final Privacy Act Notice). For example, although the new notice narrowed the range of information that would be acquired and maintained, it expanded the potential use from airline security to criminal law enforcement and thus did little to quell public controversy. Congress entered the fray when it halted funding in September 2003 (Department of Homeland Security Appropriations Act of 2004) and mandated a comprehensive General Accounting Office (GAO) privacy review of the system before it could become operational. Public concerns and tensions were exacerbated the same month when Delta Air Lines ended their testing efforts (Goo, 2003) and a highly questionable data transfer between JetBlue and a Pentagon contractor came to light. (Goo, 2004). Here, Passenger Name Records (PNRs) were voluntarily provided by JetBlue, at the request of TSA, to Torch Concepts, a company doing data mining research for the Pentagon in a separate but functionally identical project. The disclosure resulted in a first-ever investigation by the Chief Privacy Officer at the Department of Homeland Security and in a report issued in February 20, 2004, found (Kelly, pp. 1, 8–9) that the transfer raised "serious concerns about the proper handling of personally identifiable information by government employees" albeit not a civil or criminal violation of the federal Privacy Act because the information in question never was in the possession of the government.

As a result, two issues, one technical and one political, will focus public attention on the CAPPS II system during 2004. First, because name-based immigration systems have already proven ineffective (with numerous false positives and false negatives), there will be an inexorable move toward biometrics, with the concomitant problem of how such data would be captured without untoward economic impact on an already troubled industry: "If you start tampering with [online booking], there's a whole section of air travel that's going to fall by the wayside" according to the Geneva-based International Airline Trade Association (Reuters, 2004, online). Second, given the level of public concern (e.g., with Delta and with JetBlue in a different but functionally similar project), it appears unlikely that any airline will now voluntarily participate, with the result that TSA rule making may well be made mandatory (Warner & Neely, 2004). As of early 2004, TSA had yet to identify which airlines would test a prototype, and the project operational deployment planned for summer 2004 was in doubt (Goo, 2003; O'Harrow, 2003).

Congress mandated an assessment of CAPPS II development, including safeguards to protect the individual privacy of the traveling public. That report (U.S. General Accounting Office, 2004) found significant project management shortcomings (e.g., incomplete requirements, schedule, and budget plan) as well as likely delays in testing and incremental deployment due in part to the inability to acquire testing data. Most significantly, however, the GAO found that the TSA had failed to meet seven of eight high priority issues identified by Congress. Other than the establishment of an internal program review board, the critical civil liberties issues of unauthorized access, data accuracy, data privacy, abuse prevention, and redress process remain unresolved—as do critical performance issues, including the adoption of operating policies, stress testing, and validation of the accuracy and effectiveness of employed search tools. Several examples of the findings are instructive. With respect to notice, although the previously cited federal Privacy Act notice proposed to exempt CAPPS II from seven requirements, no reasons (as required by law) were provided. And, with respect to redress, TSA has not determined the period of data retention, the degree of data access that will be provided to a complainant, or the process of data correction.

During the course of the review, the GAO identified three additional issues that present a major risk: international cooperation, enlargement of mission, and the implications of identity theft. Although all points are significant, the fact that CAPPS II concerns are not limited to the U.S. poses perhaps the greatest risk to effectiveness. In 1995, the Council of Ministers of the European Commission (EC) adopted a directive "on the protection of individuals with regard to the processing of personal data and on the free movement of such data" that requires member states to conform their national privacy laws (European Parliament and the Council of European Union, 1995). Known formally as the *Directive on Data Protection* (EU 95/46/EC) and with an effective date of October 25, 1998, it requires that public or private sector organizations in the European Union (EU) that process personal data must ensure that they are 1) collected for specified and legitimate purposes and not processed in a way incompatible with those purposes, 2) adequate, relevant, accurate, current, not excessive, and not retained in identifying form for any longer than necessary, 3) processed only if the data subject has consented "unambiguously" or if the processing falls within an exception (e.g., contract terms), and 4) transferred to third countries only with "adequate" privacy protection, with adequacy to be determined on a case-by-case basis in light of all the circumstances surrounding a particular data transfer. Moreover, the Directive grants EU citizens a plethora of rights including: a right of access to their personal data, a right to correct any data that are inaccurate, a right to refuse use of their data for activities such as direct marketing, and a right of recourse if unlawful processing occurs. In this latter area, it requires that EU states provide judicial remedies for any breach of the rights, adopt enforcement mechanisms and sanctions, and establish supervisory authorities to monitor the application of national law related to the Directive.

The nexus with the U.S. is twofold. First, the Directive does not apply to the processing of personal data by public agencies concerning public security, defense, state security (including the economic well-being of the State), and state criminal law enforcement. This allows, for example, the unrestricted transfer of counterterrorism information between EU states and the U.S. The second concerns the provision that prohibits the transfer of personal data to non-EU countries that do not provide an "adequate" level of privacy protection. This, for example, would restrict a private-sector company from transferring any information to the U.S. unless the EU had certified the acceptability of U.S. law. This provision led to much negotiation between the EU and the U.S. Department of Commerce—made difficult by the fact that, unlike the EU, the U.S. policy approach to privacy relies on a mix of subject-specific legislation, regulation, and self-regulation. Ultimately, in July 2000, the European Commission (the final authority in the EU) approved a "safe harbor" agreement with the U.S. that meets the "adequate privacy protection" requirement of the Data Directive. The terms of the agreement are specified in the "Safe Harbor Privacy Principles" issued by the Department of Commerce (U.S. Department of Commerce, 2000) and enforceable by the Federal Trade Commission (FTC). These principles reflect the long-established, fair information practices with specific provisions of notice, choice, onward transfer, security, data integrity, access, and enforcement. They further include a requirement to publicly declare adherence to the Principles, making a failure to comply actionable under Section 5 of the Federal Trade Commission Act that prohibits unfair and deceptive trade practices.

The onward transfer provision is critical to the implementation of CAPPS II for international air travel; foreign carriers and even U.S. carriers operating out of EU airports must comply with the Data Directive and this requires certification of U.S. data handling practices as suitably equivalent. After substantial discussions, the European Commission agreed on December 16, 2003, that passenger data could be provided to U.S. authorities (Commission of the European Communities, 2003), although there remains substantial confusion as to the exact circumstances, given conflicting statements made by EU Commissioner Bolkestein (Lettice, 2004). Based on communications from Bolkestein and statements by Nuala O'Connor Kelly, Chief Privacy Officer for the Department of Homeland Security (Statewatch.org, 2004), it appeared that an agreement was in place for a trial run of CAPPS II with sharing of Passenger Name Record data and limited retention of such data. However, in January 2004 the Article 29 Working Party, an independent European advisory body on data protection and privacy, issued an opinion (Rodota, 2004, p. 4) that there should be no agreement until there was an appropriate "adequacy decision and an international agreement." Because this body is composed of the senior data protection (privacy) officials in each of the EU member states, it appears that an agreement for EU transfers to an operational CAPPS II system will be a focus of additional negotiations in 2004—as it will for other countries such as Canada that have significant concerns about U.S. privacy policies (Strohm, 2004).

Having considered the domestic and international aspects of these surveillance systems, a concluding assessment of the civil liberties implications remains difficult. For example, TSA has stated in the *Federal Register* that only four data elements (name, address, phone number, and date of birth) would be provided to commercial data aggregators and only for passenger identity authentication, has asserted in public statements that the data aggregators would not use health or credit information, and has clarified some data maintenance policies (e.g., commercial entities could not retain any government-supplied information and the government data would be purged within an undefined number of days after travel). But conversely, the TSA also proposes to exempt CAPPS II from most of the protections provided by the federal Privacy Act of 1974, including the identity of the sources of records from which it will draw information as well as the right of an individual to have access, to contest the accuracy, or to have any judicial review of alleged violations. Moreover, there is simply no public insight into the operations of the system by which a valid civil liberties judgment can be rendered. The public does not know how the commercial data companies will compute a passenger's "authentication score" either in data source or mathematical terms. The public does not know how the TSA will conduct its risk assessment, with respect to either the government data that will be utilized or the criteria that will be applied to the data. And the public does not know the functional results of the process—what percentage of travelers will be deemed a risk, what error rate will be deemed acceptable, and what will be results for those so labeled (American Civil Liberties Union, 2003; Center for Democracy and Technology, 2003b, 2003c). In sum, the system to meet the critical need of protecting the nation's commercial aviation system is seriously troubled, from the prospective of not only program management but also public acceptance, given the lack of insight and independent oversight of operational details.

But assuming that CAPPS II eventually becomes operational, and notwithstanding the demise of TIA at least with respect to its original vision, it remains important to assess certain additional policy and operational issues presented by such systems. The first issue is to distinguish *research* from the *actual deployment* of technical tools that could be operated in a manner to surreptitiously collect and/or analyze personal information (e.g., transactional data from the daily lives of citizens). TIA was a research program, not an implementation program. But by failing to develop and publicize the necessary information policy, DARPA allowed public and media perceptions that electronic dossiers would be created based on the transactions of every American to become the generally accepted belief. In point of fact, the technologies being developed by DARPA and other federal agencies including the CIA (e.g., the data mining tool Quantum Leap [Powell, 2003]) are concerned not with information collection per se but with information use (analysis) in terms of collaboration, decision support, foreign-language translation, pattern recognition, and predictive modeling tools. And these technologies, designed to respond to the challenges presented by terrorism, have a dual focus. The first is concerned with effectively exploiting the information currently held by the government, to allow better sharing and collaboration as well as better tools for analysis. The second, a research-oriented focus, is concerned with approaches to bring the totality of the information space to bear on the terrorism problem, including any information in government, private, or public databases. The most critical problems are presented in this research environment: technical (e.g., are there terrorism signatures and can they be detected within the information space?), legal (e.g., will proposed operations violate guaranteed rights?), and policy (e.g., what constraints on access and dissemination will facilitate public support?). To date, work here has addressed only data that are lawfully in the current possession of government or simulated data (Jonietz, 2003). To suggest that data mining itself is improper fails to recognize that the critical elements include comprehensive information policy in addition to technology.

The reality, of course, is that privacy must be an intrinsic element of every law enforcement and national security activity, whether it is new legislation or new technology solutions, and whether we are concerned with information collection or analysis. In saying this, however, it should be noted that the privacy answer can derive from more than law and policy—that technology itself can assist. Consider the data mining technique of data transformation where structured or full-text data are converted to a mathematical representation that is inaccessible to humans but subject to statistical or other analysis for the identification and extraction of terrorist signature information. Or consider selective data authorization where technology, policy, and law combine to authorize the release of a subset of information to an analyst although protecting the most privacy-sensitive information unless and until evidence had been identified or deduced by the analyst that would justify (in a legal sense) additional disclosures. In essence, a privacy application (or privacy appliance, in other terminology) would be positioned between the data and the user, contain policy rules on access at both the data element and data user levels, and would ensure that only defined appropriate data were provided to specific users with the requisite authorities (e.g., given data elements would only be available to specific FBI special agents assigned to terrorism investigation with an issued FISA court order). This approach indeed reflects current Fourth Amendment practice that is iterative in nature (e.g., law enforcement receives information, corroborates through a subpoena, and thus establishes probable cause authorizing a search warrant). "In short, TIA can be safely implemented. Failing to make the effort poses grave risks and is an irresponsible abdication of responsibility" (Rosenzweig, 2003, p. A16).

Lastly, consider that technology-based collection may directly benefit privacy: Video cameras and face recognition software, biometric identification, or technical equivalents of dog sniffs all have the potential, if managed properly, to enhance our individual rights. Consider that video surveillance and detection systems are merely a slight improvement in the practices in play for more than one hundred years—where law enforcement employed printed wanted posters, hard copy distribution, viewing by police officers every morning, and then police visual scanning of crowds and individuals for matches—all accomplished through, of course, the subjective lens of the eyes of individual police officers who could bring his or her personal bias into play. Technical solutions such as face recognition systems may be a preferable approach from a perspective of both security and the protection of individual constitutional rights because they substantially eliminate bias in law enforcement. Moreover, other technical solutions (e.g., body scans at air terminals) may also serve to preserve constitutional expectations of personal privacy and dignity. In sum, technology is not the enemy of personal privacy; rather it is the lack of critical information policy regulating data from collection through conversion to information and beyond to dissemination and alternative uses.

A Proposed Scheme for the Management of Surveillance in a Representative Democracy
Balancing Needed Secrecy and Oversight

It appears clear that, to many, the primary concern with American domestic intelligence gathering (e.g., through physical and electronic surveillance) is the attendant secrecy and the lack of checks and balances—the inability to know about and challenge actions directed against individual citizens and groups. Yet, the question is not whether the law should prohibit such action because the very foundation of national security is intelligence—the ability to understand and thus proactively respond to the threats presented. And secrecy per se is not inappropriate because it is an inherent element of intelligence, as George Washington (1777/1933, pp. 478–479) recognized in his correspondence: "The necessity of procuring good intelligence is apparent and need not be further urged ... upon secrecy, success depends in Most Enterprizes ... and for want of it, they are generally defeated." Indeed, most if not all discussions of intelligence have recognized the paramount need for secrecy (e.g., *United States v. Curtiss-Wright Export Corp.*, 1936). Nor is the question whether a new agency is formed similar to the United Kingdom's MI5, or whether the FBI mission can be reformed, although many question whether another organization or simply improved information focus and sharing would be more effective (e.g., Edwards, 2002).

Rather, the critical questions are how a representative democracy concisely defines the role of domestic intelligence and brings an appropriate and effective level of checks and balances to the intelligence process. In other words, how does the U.S. avoid the political abuses of the past yet acquire the needed information to protect citizens' lives and property? And, it is a question that is not so clearly black and white as argued from the various perspectives of the political spectrum. Indeed, there are three facts that are the predicate for any discussion. First, America is at war with forces committed to the destruction of the state and the people: Osama bin Laden has warned of a new "storm of planes" (Abu-Ghayth, 2001, p. 1) and has directed the members of his terrorist network, al Qaeda, "to kill the Americans and plunder their money wherever and whenever they find it" (Bin Laden, 1998, online). Second, these forces understand the potential to exploit our system of government and personal liberties. Extremists on the right and left, internal and external, have long demonstrated an ability to use Constititutional protections to cloak criminal activity. And third, the government requires the ability—information collection and analysis tools—to meet the real challenges that threaten citizens and indeed national survival.

To examine this ultimate question further, consider three scenarios for the future where proactive, comprehensive surveillance operations would be in place and play a preventative role: first, a biological attack that avoids the current very partial deployment of sensors and does not appear via the reporting network until after there are massive incidents of infection; second, continued, discrete attacks and attempts against the multiplicity of relatively soft targets requiring substantial expenditures of federal funds for defense and resulting in substantive economic damage; and third, a return to more traditional threat patterns with relatively isolated, minor attacks. Do these scenarios suggest a requirement for different legal authorities and a concomitant change in citizen rights? Historically, as discussed previously, that has been the case but with a general return to a traditional expression of rights upon the conclusion of the threat.

But in the instant circumstances, the threat is far from traditional in terms of both structure and duration. The courts have recognized two critical legal propositions vis-à-vis national threats: First, no expression of rights in the Constitution is absolute—each, from freedom of speech to prohibitions on government intrusion, has significant limitations in order to preserve the state and its people. As the U.S. Supreme Court observed in *Haig v. Agee* (1981, pp. 307 and 309–310) and cases cited therein: "It is obvious and unarguable that no governmental interest is more compelling than the security of the Nation [and although] the Constitution protects against invasions of individual rights, it is not a suicide pact." Second, that balance of rights and needs may properly shift in appropriate circumstances. As the Supreme Court also observed, in *Schenck v. United States* (1919), the Constitution cannot be said to protect rights incompatible with the national defense in times of war. Finding that reasonable balance between government power and citizen rights in this time of asymmetric, nontraditional threat is the national task and the focus of this section.

This requirement for balance presents three questions that necessitate answers in the form of comprehensive information policy. *First, what is the threshold to focus on an individual or group and initially collect personally identifiable information?* Do we require specific external, validated information that an individual poses a specific threat? Do we expand our focus from individual threats to group threats? Do we make a political judgment that individuals or entire groups may be an appropriate target because of their views, statements, and associations? Do we agree that, although the First Amendment may preclude a criminal prosecution for even violent advocacy of anti-American views, it may still be appropriate to monitor such statements in order to prevent advocacy from developing into actual violence? Do we establish different rules for U.S. citizens and aliens? And how would such rules work effectively when we know that terrorism transcends national origin, religion, ethnicity, and even citizenship? Or do we allow law enforcement to conduct ad hoc, even suspicionless surveillance in order to most proactively identify threats?

As previously considered, Attorney General Ashcroft has moved the investigatory guidelines from a requirement of specific and articulable facts indicating potential criminal activity to the possibility of criminal activity and has generally authorized the collection of public information—the latter being the hallmark of an effective intelligence agency or, as some might argue, the beginning of a police state. The issue, it might be suggested, is less the guideline standards, because preventive policing is required in a dangerous world; rather the issue, as will be discussed in more detail, may well be the lack of oversight and redress that would erode civil liberties from within.

Second, what is the threshold to retain and disseminate information acquired through domestic surveillance? Do we severely restrict such information to data which are relevant to terrorism or a specific crime? Or do we understand that the intelligence process proceeds from the collection and subsequent evaluation of many data points where relationships and hence value may not be discerned for some time? As considered previously, the ACLU has documented what it describes as a program by the Colorado Springs Police Department to spy and disseminate reports on peaceful critics of government policy. Beyond issues of appropriateness in terms of resource utilization, this vignette focuses concern on the line between effective intelligence to protect the nation from threats and unconstitutional intrusion. Again, the issue is not so much policy as oversight. Although the U.S. Constitution prohibits the collection of information against citizens exercising their First Amendment rights (because the mere act of monitoring constitutes a chilling effect and restriction of most basic rights), it does not prevent enforcement of laws against actual or prospective criminal acts where speech is merely an incidental element.

And, third, what are the mechanisms for oversight and redress—a critical element in a representative democracy based on a system of checks and balances? There should be little debate that substantially empowered intelligence and law enforcement, necessary to protect the nation from this new threat, present an inherent danger to liberty and require oversight and balance. Indeed, this inherent conflict between security and liberty has been recognized from the founding of the Republic where James Madison (1788, online) in the *Federalist Papers* cogently argued that "the accumulation of all powers ... in the same hands ... may justly be pronounced the very definition of tyranny." For example, should the FISC continue to function in an *ex parte* environment where the Court must decide the case and thus assess the validity of the government's position without the benefit of an adversarial process? It is, of course, ludicrous to suggest that targets should be notified and represented by counsel. But it is conceivable that the FISC could and should appoint cleared counsel, perhaps as a Special Master and/or Amicus, to review factual filings and present relevant legal arguments. What is certain is that a domestic intelligence process, including a FISA court operating in substantial secrecy, may not engender the public and Congressional support necessary for intelligence activities. This observation suggests the importance of creating organizations and processes with as much openness as possible. Or should all personal information maintained or accessed by the government and, whether compiled by the government or private business, be protected by the regulatory scheme established by the federal Privacy Act of 1974 that addresses collection, maintenance, use, and dissemination of personal information? Today, specific statutory provisions exempt intelligence information from the basic provisions of the Privacy Act; although unrestricted access would obviate the value of intelligence information, models could be developed that would permit independent review of the appropriateness of the collection, maintenance, and use. In other words, why, in the discussion of the Pentagon's TIA program or TSA's CAPPS II program, have we heard so much of information collection and data mining technology and so little of information and rights protection—until very recently?

The British Model of Surveillance

Indeed, this final point of independent oversight presents the questions of whether the U.S. should consider the British model for the conduct and oversight of intelligence that includes independent review mechanisms for the legality and appropriateness of surveillance activities as well as an external tribunal empowered to hear complaints from any member of the public who is aggrieved by any conduct of the intelligence services.

To understand the British regulatory scheme, we must briefly consider the focus and operations of their domestic intelligence service. The British Security Service, more commonly referred to as MI5, is responsible for national security and assessing threats to it. MI5 operates only domestically, has no powers of arrest, but discloses information to the appropriate law enforcement body that then procures warrants through the evidentiary information developed. The statutory role of the Security Service is to (a) investigate, collect, and analyze intelligence and make assessments of threats to national security and economic well-being; (b) counter threats, meaning acting or allowing others, such as the local police force, to act in response to a specific and identified threat; (c) advise the government and fellow intelligence and law enforcement agencies of possible threats and the appropriate response in terms of action and security measures deemed necessary; and (d) assist other agencies, governmental departments, and organizations, although the scope of any assistance must fall within the standards set out in statutory law and be appropriate to the perceived threat, and be approved by the Commissioners (MI5, 2003c).

The Service, much like the FBI, operates today under the Regulation of Investigatory Powers Act (RIPA) of 2000. That legislation was introduced in the House of Commons on February 9, 2000, passed by Parliament on July 23, and received Royal Assent on July 28; it replaced an array of surveillance laws, including the Interception of Communications Act of 1985, but amended others, including the Intelligence Services Act of 1994 and the Police Act of 1997, to better regulate access to communications and property (RIPA, 2000). The Service, as authorized by the RIPA, employs four basic intelligence collection techniques in the realm of surveillance: the "interception of communications" with interception warrants; "intrusive surveillance" that involves covert collection taking place on residential property or a private vehicle that includes clandestine listening or searching and requires a property warrant but not the placement of a tracking device on a private vehicle; "directed surveillance" that includes the covert following and/or observing of targets as well as the interception of communications where there is consent of one party; and the "use of covert human intelligence sources" that generally involves "inducing, asking and assisting" a source, paid or unpaid, into the life of the target for the collection of non-public information and insulates the source from any civil liability for "incidental" conduct (MI5, 2003b, online).

The interception of communications generally requires a warrant (Regulation of Investigatory Powers Act, Part I). Exceptions, similar to U.S. law, are provided at §§ 3 and 4, if one or more of the parties have consented to the interception, if it is necessary to the rendition of a communications service (e.g., open a letter to find a return address or complete the transmission of a message), or if it is in the course of lawful business practices (e.g., at a prison or mental hospital). Such interception warrants may be issued only by the Secretary of State, only on a finding that it is necessary and proportional (§ 5(2)), and only on one of three grounds: the interests of national security, preventing or detecting serious crime, or safeguarding the national economic well-being (§ 5(3)(a)–(c)). The necessary provision is significant because it prevents the issuance of a warrant merely if the information sought would be useful (e.g., supplementing other information) or interesting—indeed the RIPA defines the word as "necessary in a democratic society" inherently balancing the interests previously considered with the requirements of Article 8 of the European Convention on Human Rights (ECHR). As is generally the practice in the U.S., all matters relating to interception warrants are secret, including the existence of a given warrant and the contents of any intercepted communication (RIPA, § 19) and there are criminal penalties for violation (§ 19(4)) but exceptions for consultation with counsel (§ 19(5)–(7)). In addition, there are emergency provisions that authorize the signing of a warrant by a designated senior official but only if expressly authorized to do so by the Secretary in the particular case—meaning that even in emergency situations, the Secretary must have given personal consideration to the application and have directed signature (§ 7).

Intrusive surveillance, conceptually equivalent to civil liberties concern as interceptions, is regulated somewhat similarly (Regulation of Investigatory Powers Act, Part II). There must be an "authorization" from the Secretary of State or from a larger group of persons known as the "senior authorizing officer" in the various security and police services (§§ 32, 41, and 42); but in such cases the authorizations by these lesser officials must be reviewed and approved by the Surveillance Commissioner—a position that will be discussed subsequently (§§ 35 and 36). The grounds or standards for intrusive surveillance are identical to the interception standards, including the necessary and proportionate provisions (§ 32).

Directed surveillance and the use of covert human sources (Regulation of Investigatory Powers Act, Part II, §§ 26–31) may be authorized for a broad array of purposes including the interests of national security, preventing or detecting crime, preventing disorder, the economic well-being of the nation, public safety, public health, the collection of any form of government revenue, or as otherwise specified by the Secretary of State (§§ 28 and 29). The authorizations include a broad array of public officials designated by the Secretary, provided it is for one or more of the stated purposes and the "activity is proportionate to what is sought to be achieved by it" (§ 30). Thus, there are no requirements for a warrant or approval otherwise at the most senior governmental levels.

The British Model of Oversight

By far the most interesting and potentially relevant provisions of the RIPA to the debate in the U.S. regarding the balance between state needs and individual rights are the mechanisms for oversight and redress: independent commissioners and a review tribunal (Regulation of Investigatory Powers Act, Part IV, §§ 57–72). Specifically, the Act (§ 57) provides for the appointment of an Interception of Communications Commissioner (IofCC) who "holds or has held a high judicial office (within the meaning of the Appellate Jurisdiction Act 1876)" and has a substantial portfolio of authorities. These include the review of the Secretary of State's role in the issuance of interception warrants and the operation of the regime for acquiring communications data (including decryption notices and the protection of encryption keys). To discharge these duties, the IofCC is granted the necessary technical facilities and staff, may require the cooperation of and access to the records of those departments utilizing the interception authorities, and reports to the Prime Minister.

The Act (§ 59) also provides for an Intelligence Services Commissioner (ISC) who also must hold or have held a high judicial office and has the responsibility to review the Secretary of State's powers under the Intelligence Services Act 1994 relating to the grants of warrants for interference with wireless telegraphy or entry and interference with property as well as the powers in general with respect to the oversight of the activities of the intelligence services.

The Act (§§ 62–64) also provides for a group of additional commissioners to provide oversight for those activities not requiring warrants—primarily surveillance through the use of agents and undercover officers—and certain aspects of decryption, including notices and protection of keys. In addition to the Chief Surveillance Commissioner, there may be Assistants (at the rank of circuit judges) and certain limited delegation of authority to members of their respective staffs.

But of import far beyond the commissioners are the provisions of the Act (§ 65) that relate to the establishment of a comprehensive independent tribunal—the Investigatory Powers Tribunal—to hear citizen complaints of intelligence activities. The tribunal came into existence on October 2, 2000, and assumed responsibility for the jurisdiction of the previous Interception of Communications Tribunal, the Security Service Tribunal, and the Intelligence Services Tribunal. The members of the Tribunal are appointed for terms of five years by letters patent issued by the Queen and are selected from the senior ranks of the legal profession, including the judiciary. The Tribunal has the jurisdiction to hear any citizen complaint brought under the European Convention on Human Rights or alleging any personal harm as a result of any use of investigatory powers under the RIPA including any entry on or interference with property, any interference with communications, or any disclosure of personal information. In its inquiry, the Tribunal may interview any member of and have access to any records of the Service; the complainants are entitled to appear directly or by counsel and make their own representations. A complaint is deemed successful if the Service has exceeded or violated its legislative mandate; in such cases the Tribunal will provide a summary of its conclusions (but must protect any information from disclosure that might be damaging or prejudicial to the state) and has the power to award compensation, quash any warrant, and require the destruction of records or information in any form. In unsuccessful cases, the complainants receive notice of that fact, but will not be informed whether they have been subject to surveillance (MI5, 2003a).

The Good and the Bad

All of this discussion does not establish, however, that the general approach of the British to domestic security is without substantial criticism, does not have a history of excessive implementation, or is a model for emulation in totality. Several seminal works (e.g., Donohue, 2001,

2003) as well as influential public policy organizations, the media, and even elements of the government (e.g., www.statewatch.org; Millar & Norton-Taylor, 2003) have detailed a history of sacrificing individual rights for state security in the United Kingdom. Today these concerns range from government demands for encryption keys for information acquired through electronic surveillance to an ever-widening circle of government officials entitled to use the extraordinary powers for reasons ever-more removed from national security. Previously, and beginning in the early 1970s with the resurgence of the "Irish troubles," these concerns mirrored many faced by the U.S. today; there, those identified as IRA terrorists or supporters faced an alleged targeted killing policy, detention for periods without trial, and the abandonment of most civil rights for those tried (e.g., suspension of right of jury, right of silence, and many rules of evidence). Even in 1922, there was enormous concern with respect to the enactment of the Civil Authorities (Special Powers) Act for Northern Ireland that enabled all emergency legislation in Northern Ireland for more than fifty years and allowed the government, delegable to any police officer, "to take all such steps and issue all such orders as may be necessary for preserving the peace and maintaining order, according to and in the execution of this Act and the regulations in it." More specifically, this Act included a 35-element "Schedule for the Regulations for Peace and Order in Northern Ireland" that, inter alia, allowed the closure of any place of business, restrictions on exercise of what would be considered First Amendment protected activities in the U.S., the right to enter or seize possession of any building without cause, and the arrest of any person who does any act to impede the preservation of peace and order under the Act (Ewing & Gearty, 2000).

From these and similar events it appears clear that the almost inexorable result of democratic governments responding with unrestricted state power to security threats is not only the loss of civil rights, but also a substantial diminution of governmental credibility with its people and of moral authority in the world (Donohue, 2003). But what is the answer? There is little question that democracy is a notably inefficient form of government but it was created intentionally with and for this flaw. For example, James Madison recognized the inefficiencies that come from the separation of powers but also the important safeguards from abuse that are a natural consequence; of course, inefficient does not equate to ineffective. Indeed, it is important to note that this concept is so central to American democracy that the Department of State's Office of International Programs[2] presents this as a central tenet of our system of government.

Thus, perhaps the most perspicuous answer to changed threat formats is the cautious introduction of state authorities with the maximum transparency. This is the lesson of reconciliation efforts in the world and, although the decisions may not meet with general approval, the open nature engenders support. It is for these reasons that the British model of an independent Tribunal represents a substantial improvement and a potential model for the management of surveillance in a democratic society. This said, it must also be noted that none of the tribunals has ever upheld a complaint. More specifically, officials of the Home Office testified in Parliament in November of 2003 that the current Investigatory Powers Tribunal had considered some 470 complaints and found none to present a contravention of the RIPA or the Human Rights Act 1998 (Wearden, 2003). Does this invalidate the conceptual model or perhaps suggest different rules of procedure (e.g., greater openness or rights of counsel)? No definitive answer can be offered because the public cannot discern whether the British pattern of decisions represents a failure in terms of undue deference to the intelligence authorities, or a success in terms of ensuring adherence to the law by the intelligence authorities. What is certain is that any mechanism of oversight, especially one that incorporates an adversarial process, only improves democratic government.

Conclusion

Regardless of the political position taken on these complex issues, two political realities are learned and relearned from history. The first is that secrecy in any organization, whether in the public, private, or religious sector, poses a real risk of abuse and must be constrained through effective checks and balances. Simply stated, information awareness can be equally critical for national security and dangerous to individual liberty. The second is that a domestic intelligence process operating in substantial secrecy will not engender the necessary public and Congressional support for the required intelligence activities to combat the very real threat of terrorism. This is not to say that secrecy in many discrete, specific national security activities is unwarranted. However, the impetus for secrecy and efficiency must be balanced against the rights of the public and their elected representatives for access to the information needed in a democratic society. If a reasonable balance between openness and secrecy is not obtained through appropriate information policy, then the political process may well act to restore the balance in ways that could harm the national security. As even Admiral Poindexter, the former head of the IAO acknowledged (O'Harrow, 2002, p. A04): "We can develop the best technology in the world and, unless there is public acceptance and understanding of the necessity, it will never be implemented." This is the lesson of the 1970s and it will prove the lesson of today as well as tomorrow.

Moreover, it is certain that information science and technology professionals are ideally situated to provide the tools and mechanisms by which the necessary domestic intelligence is collected and civil liberties are protected through established law and policy. What is different today is that the application of information analysis technology (i.e., knowledge discovery tools) must be managed. A preliminary question is whether such tools should be restricted to terrorism or used for general criminal matters because of their unusually intrusive potential and the requirements of the Fourth Amendment. Given the state of the law as considered herein, legislation is likely required but difficult to formulate with the blurring of lines between crime and terrorism as well as between domestic and external threats.

Thereafter, and to the extent the focus is on information specific to individuals, tool use must be regulated by clear policy that is consistent with the current threshold for law enforcement investigations in general (e.g., Attorney General guidelines applicable to general criminal and terrorist investigations). But when the focus is the generalized detection of patterns or relationships from the universe of information, tool use must preserve individual anonymity and permit access to and disclosures of individual identities only if the evidence required meets a discernible legal standard (e.g., reasonable suspicion). Stated differently, the examination of public or private sector databases for information suggestive of potential terrorist activity must proceed in two stages—first, the search for information must ensure anonymity; second, the acquisition of specific identity, if required, must be court authorized under appropriate standards. This point is especially critical and debatable for a number of reasons. At one level, any disclosure could be an arguable violation of the Fourth Amendment if there were no probable cause for the initial automated search, analysis, and identification. However, a more measured legal argument would be posed in terms of "special needs" or "overall reasonableness"—two rubrics that have allowed numerous exceptions to the general warrant requirement of the Fourth Amendment. Indeed, these theories arise in contexts that often present the issue of anonymity—a right that should be recognized even with respect to events that are exposed to the public (e.g., an airline flight). But the law as to when and under what circumstances the government may require (or otherwise acquire) an individual's identification is not clearly established. In one context, a stop to check identification and thereafter a prosecution for a then discovered violation of law, the U.S. Supreme Court has required at least "reasonable suspicion of an offense" (a somewhat lower standard than "probable cause") before law enforcement can require the production of identification (*United States v. Hensley*, 1985). However, the law is somewhat clouded by two previous cases on whether a person may be punished solely for refusing to produce identification. In one case (*Brown v. Texas*, 1979), the Court avoided the issue of whether an individual could be punished for refusing to identify himself in general because of the specific context of the case—a proper investigatory stop that met Fourth Amendment standards. In the other case (*Kolender v. Lawson*, 1983) the Court held that a California statute requiring loiterers to produce "credible and reliable" identification was unconstitutional on vagueness grounds but did not decide the Fourth Amendment issue.

It was for this reason that the U.S. Supreme Court announced on October 20, 2003, that it would decide the issue of the right of anonymity in the context of a Nevada law that allowed a charge of obstructing a law enforcement officer for failure to produce identification (*Hiibel v. Sixth Judicial District Court of the state of Nevada*, 2003). In this case, the Nevada court had stated that the law was critical in the war against terrorism:

> Most importantly, we are at war against enemies who operate with concealed identities and the dangers we face as a nation are unparalleled. Terrorism is 'changing the way we live and the way we act and the way we think.' During the recent past, this country suffered the tragic deaths of more than 3,000 unsuspecting men, women, and children at the hands of terrorists; seventeen innocent people in six different states were randomly gunned down by snipers; and our citizens have suffered illness and death from exposure to mail contaminated with Anthrax. We have also seen high school students transport guns to school and randomly gun down their fellow classmates and teachers. It cannot be stressed enough: 'This is a different kind of war that requires a different type of approach and a different type of mentality.' To deny officers the ability to request identification from suspicious persons creates a situation where an officer could approach a wanted terrorist or sniper but be unable to identify him or her if the person's behavior does not rise to the level of probable cause necessary for an arrest. (*Hiibel v. Sixth Judicial District Court of the state of Nevada*, 2003, p. 1206)

Quite clearly, this case may well set the legal standard by which law enforcement, assisted by information science and technology, will fight the war on terrorism.

In any event, there must be a robust system of oversight, transparency, and right of individual challenge that can address investigations alleged to be wrongful as well as errors in the surveillance system (e.g., false positives for technical reasons). Quite simply, a tribunal system as in the U.K. would be a significant step in the U.S. system of ensuring rights. But we could substantially improve the process. Although complainants or those aggrieved of determinations could not invariably be provided access to the corpus of relevant information (e.g., the government could not even confirm or deny that an individual had been the subject of terrorism-related surveillance but could, in some instances, disclose the information leading to transportation screenings), it would be possible to create a robust adversarial system through the establishment of a cleared staff of counsel serving as citizen advocates whose responsibility it would be to consider, investigate, and argue cases before the FISC or other administrative tribunals. The costs would be insignificant in terms of the public consensus that would be built and could be reduced to very minimal levels through the appointment of *pro bono* counsel (as is being contemplated with respect to the military tribunals). The essential point is that public support of government initiatives in a representative democracy must have the highest priority. This is an especially critical priority in the current environment because the failure of public support could lead the nation to clear error—sacrificing civil liberties for security or prohibiting government authorities that could prevent other catastrophic attacks. In sum, the debate is not over government authority or civil liberties. The answer is firstly balance, meaning not that we sacrifice some of both objectives, but rather that any response to the threat of terrorism is as concerned with individual rights as with providing the necessary tools. The answer secondarily is transparency, meaning that the exercise of government powers in a representative democracy must be subject to effective oversight. This, we submit, is the real challenge of the war on terrorism—engineering the collection and use of information to preserve both our safety and our rights.

Endnotes

1. www.aclu-co.org/news/pressrelease/release_spyfilescosprings.htm and www.aclu-co.org/news/pressrelease/release_spyfilessettle.htm
2. http://usinfo.state.gov/products/pubs/whatsdem/whatdm7.htm

Legal Statutory References

Civil Authorities (Special Powers) Act (Northern Ireland). (1922). Retrieved March 5, 2004, from http://cain.ulst.ac.uk/hmso/spa1922.htm

Computer Fraud and Abuse Act, 18 USC § 1030, as amended.

Consolidated Appropriations Resolution of 2003 (Subsection 111(b) of Division M), Public Law 108-7.

Department of Defense Appropriations Act of 2004 (H.R. 2658), Public Law 108-87.

Department of Homeland Security Appropriations Act of 2004 (H.R. 2555), Public Law 108-90.

Department of Transportation Privacy Act of 1974: System of Records Notice, 68 Fed. Reg. 2101, January 15, 2003.

Department of Transportation Interim Final Privacy Act Notice, Department of Transportation, 68 Fed. Reg. 45265, August 1, 2003.

Departments of Commerce, Justice and State, the Judiciary and Related Agencies Appropriation Act of 2004, H.R. 2799. Retrieved April 1, 2004, from http://frwebgate. access.gpo.gov/cgi-bin/getdoc.cgi?dbname=108_cong_bills&docid=f:h2799pcs.txt.pdf

Domestic Security Enhancement Act, proposed legislation by the Department of Justice. (2003). Retrieved June 1, 2003, from http://www.privacy.org/patriot2draft.pdf. An alternative analysis by the ACLU can be found at http://www.aclu.org/SafeandFree/Safeand Free.cfm?ID=11835&c=206

Electronic Communications Privacy Act, 18 U.S.C. §§ 2510-22 (real-time intercepts, also known as Title III of the Omnibus Crime Control and Safe Streets Act of 1968), 18 U.S.C. §§ 2701–2712 (stored communications) and 18 U.S.C. §§ 3121–27 (trap and trace); all sections were amended by the U.S.A. Patriot Act.

Executive Order (EO) 12333, 46 Fed. Reg. 59941, 1981. Retrieved April 1, 2003, from www.cia.gov/cia/information/eo12333.html

Executive Order (EO) 12958 as amended by EO 13292, 68 Fed. Reg. 15315, 2003. Retrieved June 30, 2002, from http://www.archives.gov/isoo/rules_and_regulations/executive_order_12958_amendment.html

Foreign Intelligence Surveillance Act, 50 U.S.C. §§ 1801–1862, as amended.

Freedom to Read Protection Act of 2003, H.R. 1157. Retrieved April 1, 2004, from http://frwebgate.access.gpo.gov/cgi-bin/getdoc.cgi?dbname=108_cong_bills&docid=f:h1157ih.txt.pdf

Intelligence Authorization Act for 2004 (H.R. 2417), Public Law 108-177.

Internal Security Act, 64 Stat. 987 (title II repealed by Public Law 92-128).

Library, Bookseller and Personal Records Privacy Act, S. 1507. Retrieved April 1, 2004, from http://frwebgate.access.gpo.gov/cgi-bin/getdoc.cgi?dbname=108_cong_bills&docid=f:s1507is.txt.pdf

National Security Letter authority, 18 U.S.C. § 2709 (ECPA exception for telephone toll and electronic communications transactions records), 15 U.S.C. § 1681u (Fair Credit Reporting Act exception for names and addresses of all financial institutions at which an individual maintains or has maintained an account, and 12 U.S.C. § 3414(a)(5)(A) (Right to Financial Privacy Act exception for an individual's institution-specific banking and credit records).

Protecting the Rights of Individuals Act, S. 1552. Retrieved April 1, 2004, from http://frwebgate.access.gpo.gov/cgi-bin/getdoc.cgi?dbname=108_cong_bills&docid=f:s1552is.txt.pdf

Regulation of Investigatory Powers Act (U.K.). (2000). Retrieved March 5, 2004, from http://www.legislation.hmso.gov.uk/acts/acts2000/20000023.htm and http://www.home-office.gov.uk/crimpol/crimreduc/regulation/index.html

Security and Freedom Ensured Act of 2003 (SAFE Act), H.R. 3352. Retrieved April 1, 2004, from http://frwebgate.access.gpo.gov/cgi-bin/getdoc.cgi?dbname=108_cong_bills&docid=f:h3352ih.txt.pdf

Security and Freedom Ensured Act of 2003 (SAFE Act), S. 1709. Retrieved April 1, 2004, from http://frwebgate.access.gpo.gov/cgi-bin/getdoc.cgi?dbname=108_cong_bills&docid=f:s1709is.txt.pdf

The Sedition Act of 1798, officially titled "An Act for the Punishment of Certain Crimes Against the United States." Approved 14 July 1798. 1 Stat. 596–597.

Sedition Act of 1918, amending the Espionage Act of 1917, 40 Stat. 553–554.

Smith Act, 54 Stat. 671.

U.S.A. Patriot Act, Pub. L. 107-56, 115 Stat. 272 (2001). Retrieved April 1, 2004, from http://frwebgate.access.gpo.gov/cgi-bin/getdoc.cgi?dbname=107_cong_public_laws&docid= f:publ056.107.pdf

Legal Case References

Abrams v. United States, 250 U.S. 616 (1919).

Albertson v. SACB, 382 U.S. 70 (1965).

Alliance to End Repression et al. v. City of Chicago, 561 F.Supp. 537 (N.D. Ill. 1982) (approving settlement agreements and entry of permanent injunction); 66 F.Supp.2d 899 (N.D.Ill. 1999) (denying motion of City to amend consent judgment); rev'd, 237 F.3d 799 (7th Cir. 2001) (with instructions to enter modifications to consent judgment requested by City).

American Civil Liberties Union et al. v. United States, 538 U.S. 920 (2003).

American Civil Liberties Union et al. v. U.S. Department of Justice, 265 F.Supp.2d 20 (D.D.C. 2003).

Andersen v. Sills, 256 A. 2d 298 (Ch. Div. 1969); rev'd, 265 A.2d 678 (1970); dismissed as moot, 363 A.2d 381 (1976).

Aptheker v. Secretary of State, 378 U.S. 500 (1964).

Brandenburg v. Ohio, 395 U.S. 444 (1969).

Brown v. Texas, 443 U.S. 47 (1979).

California v. Ciraolo, 476 U.S. 207 (1986).

Center for National Security Studies et al. v. U.S. Department of Justice, 215 F.Supp.2d 918 (D.D.C. 2002); rev'd., 331 F.3d 918 (D.C. Cir. 2003).

Cohen v. California, 403 U.S. 15 (1971).

Communist Party v. SACB, 367 U.S. 1 (1961).

Dennis v. United States, 341 U.S. 494 (1951).

Detroit Free Press v. Ashcroft, 303 F.3d 681 (6th Cir. 2002).

Endicott Johnson Corp. v. Perkins, 317 U.S. 501 (1943).

Gitlow v. New York, 268 U.S. 652 (1925).

Haig v. Agee, 453 U.S. 280 (1981).

Handschu v. Special Services Division (NYPD), 349 F. Supp. 766 (S.D.N.Y., 1972) (establishing initial guidelines); 605 F.Supp. 1384 (S.D.N.Y. 1985) (final decree affirming settlement agreement signed 30 December 1980); aff'd, 787 F.2d 828 (2nd Cir. 1986); 2003 U.S. Dist. LEXIS 2134 (S.D.N.Y. Feb. 11, 2003) (adopting new FBI-based guidelines); 2003 U.S. Dist. LEXIS 13811 (S.D.N.Y. Aug. 6, 2003) (ordering new guidelines previously incorporated in NYPD Patrol Guide also incorporated in Order and Judgment of Court so as to be enforceable by contempt power).

Hiibel v. Sixth Judicial District Court of the state of Nevada, 59 P.3d 1201 (Nev. 2002); cert. granted, ___ U.S. ___, 2003 U.S. LEXIS 7710 (Oct. 20, 2003).

Hoffa v. United States, 385 U.S. 293 (1966).

In re: All Matters Submitted to the Foreign Intelligence Surveillance Court, Docket Numbers Multiple (FISC, May 17, 2002). 218 F. Supp.2d 611.

In re: Sealed Case No. 02-001, 310 F.3d 717 (FIS Ct. Review, 2002).

Irvine v. California, 347 U.S. 128 (1961).

Katz v. United States, 389 U.S. 347 (1967).

Keith. See United States v. United States District Court.

Kendrick v. Chandler, Civil Action No. 76-499, unpublished Order, Judgment and Decree (W.D. Tenn., filed September 14, 1978).

Kenyatta v. Kelly, 375 F.Supp. 1175 (E.D.Pa. 1974).

Kolender v. Lawson, 461 U.S. 352 (1983).

Laird v. Tatum, 408 U.S. 1 (1972).

North Jersey Media Group v. Ashcroft, 308 F.3d 198 (3rd Cir. 2002); cert. denied, ___ U.S. ___, 2003 U.S. LEXIS 4082 (May 27, 2003).

Oklahoma Press Publishing Co. v. Walling, 327 U.S. 186 (1946).

Olmstead v. United States, 277 U.S. 438 (1928).

Philadelphia Yearly Meeting of the Religious Society of Friends v. Tate, 519 F.2d 1335 (3rd Cir. 1975).

Reporters Committee for Freedom of the Press v. AT&T, 593 F.2d 1030 (D.C. Cir. 1978); cert. denied, 440 U.S. 949 (1979).

Richmond Newspapers Inc. v. Virginia, 448 U.S. 555 (1980).

Schenck v. United States, 249 U.S. 47 (1919).

Smith v. Maryland, 442 U.S. 735, 741–46 (1979).

State v. Jackson, 111 Wn. App. 660 (2002); rev'd, 150 Wa.2d 251 (2003).

Terry v. Ohio, 392 U.S. 1 (1968).

Texas v. Johnson, 491 U.S. 397 (1989).

United States v. Bin Laden et al., 126 F.Supp.2d 264 (S.D.N.Y. 2000).

United States v. Curtiss-Wright Export Corp., 299 U.S. 304, 320 (1936).

United States v. Freitas, 800 F.2d 1451 (9th Cir. 1986).
United States v. Hensley, 469 U.S. 221 (1985).
United States v. Karo, 468 U.S. 705 (1984).
United States v. Knotts, 460 U.S. 276 (1983).
United States v. Kyllo, 190 F.3d 1041 (9th Cir. 1999) (*en banc*); rev'd, 533 U.S. 27 (2001).
United States v. Place, 462 U.S. 696 (1983).
United States v. Powell, 379 U.S. 48 (1964).
United States v. Troung Dinh Hung, 629 F.2d 908 (4th Cir. 1980).
United States v. United States District Court (Keith), 407 U.S. 297 (1972).
United States v. Verdugo-Urquidez, 494 U.S. 259 (1990).
United States v. Villegas, 899 F.2d 1324 (2nd Cir. 1990).
Yates v. United States, 354 U.S. 298 (1957).

References

ABC News. (2003a, September 10). *Rights intrusions all right*. Retrieved September 10, 2003, from http://abcnews.go.com/sections/us/World/sept11_terrorwar_poll030910.html

ABC News. (2003b, September 10). *Poll: Most Americans recognize rights losses, but say it's justified*. Retrieved September 10, 2003, from http://abcnews.go.com/sections/us/politics/Results_privacy030910.html

ABC News. (2003c, September 10). *The price of progress: Ashcroft defends Patriot Act, post 9/11 detentions*. Retrieved September 10, 2003, from http://abcnews.go.com/sections/wnt/US/PJ_Ashcroft_transcript030910.html

Abu-Ghayth, S. (2001). Statement of Al-Qa'ida spokesman Sulayman Abu-Ghayth, broadcast on Al-Jazirah satellite channel television in Arabic, 2100 hours GMT, 09 October 2001, as translated and reported by the U.S. Foreign Broadcast Information Service (FBIS RES 092211Z). Washington, DC: Foreign Broadcast Information Service, Central Intelligence Agency.

American Civil Liberties Union. (1997). *ACLU briefing paper number 10: Freedom of expression*. Retrieved on September 1, 2003, from http://archive.aclu.org/library/pbp10.html

American Civil Liberties Union. (2003). *The five problems with CAPPS II: Why the airline passenger profiling proposal should be abandoned*. Retrieved October 10, 2003, from http://www.aclu.org/SafeandFree/SafeandFree.cfm?ID=13356&c=206

Andrews, R., Biggs, M., & Seidel, M. (1996). *The Columbia world of quotations*. New York: Columbia University Press.

Andrews, W. (n.d.). A history of the NYPD: The early years: The challenge of public order. *Spring 3100 Magazine*. Retrieved July 1, 2003, from http://www.ci.nyc.ny.us/html/nypd/html/3100/retro.html

Ashcroft, J. (2002a). *Attorney General's guidelines on Federal Bureau of Investigation undercover operations*. Washington, DC: U.S. Department of Justice. Retrieved April 8, 2003, from http://www.usdoj.gov/olp/fbiundercover.pdf

Ashcroft, J. (2002b). *Attorney General's guidelines on general crimes, racketeering enterprise and terrorism enterprise investigations*. Washington, DC: U.S. Department of Justice. Retrieved April 8, 2003, from http://www.usdoj.gov/olp/generalcrimes2.pdf

Ashcroft, J. (2002c). *Attorney General's guidelines regarding the use of confidential informants*. Washington, DC: U.S. Department of Justice. Retrieved April 8, 2003, from http://www.usdoj.gov/olp/dojguidelines.pdf

Ashcroft, J. (2002d). *Brief for the United States In re [Deleted]*. Washington, DC: U.S. Department of Justice. Retrieved June 1, 2003, from http://www.fas.org/irp/agency/doj/fisa/082102appeal.html

Ashcroft, J. (2002e). *Memorandum to Director FBI, subject: Intelligence sharing procedures for foreign intelligence and foreign counterintelligence investigations conducted by the FBI*. Washington, DC: U.S. Department of Justice. Retrieved April 8, 2003, from http://www.fas.org/irp/agency/doj/fisa/ag030602.html

Ashcroft, J. (2002f). *Procedures for lawful, warrantless monitoring of verbal communications*. Washington, DC: U.S. Department of Justice. Retrieved April 8, 2003, from http://www.usdoj.gov/olp/lawful.pdf

Ashcroft, J. (2003). *Memorandum from Attorney General to Director, Administrative Office of the U.S. Courts*. Washington, DC: U.S. Department of Justice. Retrieved June 30, 2003, from http://www.usdoj.gov/04foia/readingrooms/2002annualfisareporttocongress.htm

Beeson, A. (2003, February 18). *In legal first, groups urge high court to review secret court ruling on government spying*. New York: American Civil Liberties Union. Retrieved March 1, 2003, from http://www.aclu.org/SafeandFree/SafeandFree.cfm?ID=11840&c=206

Beeson, A., Jaffer, J., & Shapiro, S. R. (2002, September 19). *Brief on behalf of amici curiae*. New York: American Civil Liberties Union. Retrieved October 10, 2003, from http://www.cdt.org/security/U.S.A.patriot/020919fiscrbrief.pdf

Berg, B. (1999). *Law enforcement: An introduction to police in society*. Boston: Allyn & Bacon.

Berman, J. (2002, June 6). *The FBI guidelines and the need for congressional oversight: Some observations and lines of inquiry*. Washington, DC: The Center for Democracy and Technology. Retrieved July 4, 2003, from http://www.cdt.org/testimony/020606berman.shtml

Berry, D. (1982, January). The First Amendment and law enforcement infiltration of political groups. *Southern California Law Review, 56*, 207.

Bin Laden, O. (1998, February 23). *Jihad against Jews and Crusaders* (Fatwa: World Islamic Front Statement) (Federation of American Scientists, trans.). Retrieved July 10, 2003, from http://www.fas.org/irp/world/para/docs/980223-fatwa.htm

Bowden, M. E., Hahn, T. B., & Willliams, R. V. (1999). *Proceedings of the 1998 Conference on the History and Heritage of Science Information Systems*. Medford, NJ: Information Today, Inc.

Broehl, W. (1964). *The Molly Maguires*. Cambridge, MA: Harvard University Press.

Brown, J. E. (2003, May 13). *Letter to Chairman James Sensenbrenner, Jr., House Judiciary Committee re U.S.A. Patriot Act administration*. Retrieved June 1, 2003, from http://www.house.gov/judiciary/patriotlet051303.pdf

Bryant, D. J. (2002, July 26). *Letter to Chairman James Sensebrenner, Jr., House Judiciary Committee re U.S.A. Patriot Act administration (with supplementary responses)*. Retrieved August 1, 2002, from http://www.house.gov/judiciary/patriotresponses101702.pdf

Center for Democracy and Technology. (2003a, September 30). *Comments of the Center for Democracy and Technology on Notice of Status of System of Records, Interim Final Notice, 68 Fed. Reg. 45265-01 (Aug. 1, 2003), Docket No. DHS/TSA-2003-01*. Retrieved on October 10, 2003, from http://www.cdt.org/security/U.S.A.patriot/030930cdt.pdf

Center for Democracy and Technology. (2003b, July 31). *Policy post: A briefing on public policy issues affecting civil liberties, vol. 9, 16*. Retrieved April 1, 2004, from http://www.cdt.org/publications/pp_9.16.shtml

Center for Democracy and Technology. (2003c, May 28). *Privacy's gap: The largely non-existent legal framework for government mining of commercial data*. Retrieved August 10, 2003, from http://www.cdt.org/security/usapatriot/030528cdt.pdf

Chevigny, P. G. (1984). Symposium: National security and civil liberties: Politics and law in the control of local surveillance. *Cornell Law Review, 69*, 735–784.

Church, F. (1976). *Final report of the Select Committee to Study Governmental Operations with Respect to Intelligence Activities, S. Rep. No. 94-755, 94th Cong., 2nd Ses.* (Books I–VI). Washington, DC: U.S. Government Printing Office.

CNN.com/Law Center. (2003, February 11). *NYPD wins motion to curb political surveillance.* Retrieved April 8, 2003, from http://www.cnn.com/2003/LAW/02/11/nypd.surveillance.limits

CNN.com/U.S. (2001, December 7). *Ashcroft: Critics of new terror measures undermine effort.* Retrieved April 8, 2003, from http://www.cnn.com/2001/US/12/06/inv.ashcroft.hearing

Cohen, C. (2002, February 20–26). Night train to Baltimore. *Baltimore City Paper Online.* Retrieved April 8, 2002, from http://www.citypaper.com/2002-02-20/charmed.html

Cole, D. (2002, Spring). The Ashcroft raids. *Amnesty Now Magazine.* Retrieved August 10, 2003, from http://www.amnestyusa.org/usacrisis/ashcroftraids.html

Commission of the European Communities. (2003, December 16). *Communication from the Commission to the Council and the Parliament: Transfer of air passenger name record (PNR) data: A global EU approach.* Retrieved December 20, 2003, from http://europa.eu.int/comm/external_relations/us/intro/apis_en.pdf

Court News. (1907, May 8). Police kill plot to build up alibi. *Brooklyn Standard Union.* Retrieved on October 1, 2003, from http://www.bklyn-genealogy-info.com/Court/1907.Court.html

Davenport, C. (n.d.). *Understanding covert repressive action: An assessment of political surveillance and anti-red squads in the U.S.* College Park, MD: Center for International Development and Conflict Management, University of Maryland. Retrieved September 1, 2003, from http://www.bsos.umd.edu/gvpt/davenport/understanding%20covert%20repressive%20action.PDF

Dives, Pomeroy, & Stewart. (1911). *History of the County of Schuylkill.* Pottsville, PA: Author.

Donner, F. (1990). *Protectors of privilege: Red squads and police repression in urban America.* Berkeley: University of California Press.

Donohue, L. K. (2001). *Counter-terrorist law and emergency powers in the United Kingdom 1922–2000.* Dublin, Ireland: Irish Academic Press.

Donohue, L. K. (2003, April 6). The British traded rights for security, too. *The Washington Post,* B1, B4.

Edwards, J. (2002, December 18). Wrong job for the FBI. *The Washington Post,* A35.

Egelko, B. (2002, September 16). FBI snooping has librarians stamping mad; local woman jailed in 70's in informant flap. *The San Francisco Chronicle.* Retrieved September 16, 2002, from http://www.sfgate.com/cgi-bin/article.cgi?file=/chronicle/archive/2002/09/16/BA229715.DTL

Eggen, D. (2003, September 19). Patriot monitoring claims dismissed: Government has not tracked bookstore or library activity, Ashcroft says. *The Washington Post,* A02.

Eisenberg, A. N. (2003, May 21). *Testimony of Arthur N. Eisenberg presented to the New York Advisory Committee to the U.S. Commission on Civil Rights.* Retrieved August 1, 2003, from http://www.nyclu.org/ny_advisory_commission_oncivilrts_052103.html

Eisenhower, D. D. (1960, May 25). *Television report after the Paris summit.* Retrieved September 1, 2003, from http://www.cia.gov/csi/monograph/firstln/eisenhower.html

Elliff, J. T. (1984). National security and civil liberties: The Attorney General's guidelines for FBI investigations. *Cornell Law Review, 69,* 785–815.

European Parliament and the Council of European Union. (1995). *Directive on data protection (95/46/EC).* Retrieved June 1, 2003, from http://europa.eu.int/smartapi/cgi/sga_doc?smartapi!celexapi!prod!CELEXnumdoc&lg=EN&numdoc=31995L0046&model=guichett

Ewing, K. D., & Gearty, C. A. (2000). *The struggle for civil liberties: Political freedom and the rule of law in Britain, 1914–1945.* New York: Oxford University Press.

Factbook on intelligence. (2003). Washington, DC: Central Intelligence Agency.

Foerstel, H. N. (1991). *Surveillance in the stacks: The FBI's Library Awareness Program.* Westport, CT: Greenwood.

George, K. (2003, May 12). Court will decide if police need warrant for GPS 'tracking'. *Seattle Post-Intelligencer Reporter.* Retrieved August 10, 2003, from http://seattlepi. nwsource.com/local/121572_gps12.html

Goldstein, A. (2003, September 8). Fierce fight over secrecy, scope of law. *The Washington Post*, A1.

Goo, S. (2003, September 9). Fliers to be rated for risk level. *The Washington Post*, A01.

Goo, S. (2004, February 21). TSA helped JetBlue share data, report says. *The Washington Post*, E01.

Heuer, R. J., Jr. (1999). *The psychology of intelligence analysis.* Washington, DC: Center for the Study of Intelligence. Retrieved March 5, 2004, from http://www.cia.gov/ csi/books/19104/index.html

Hoffman, B. (1999). Old madness, new methods: Terrorism evolves toward netwar. *Rand Review, 22*, 18–23. Retrieved June 1, 2003, from http://www.rand.org/publications/ran-dreview/issues/rrwinter98.9.pdf

Holder, E. (2000, January 18). *Memorandum re to recommend that the Attorney General authorize certain measures regarding intelligence matters in response to the interim recommendations provided by Special Litigation Counsel Randy Bellows.* Washington, DC: U.S. Department of Justice. Retrieved April 8, 2003, from http://www.fas.org/irp/ agency/doj/fisa/ag012100.html

Holland, J. J. (2003, August 16). *Washington today: Momentum growing against Patriot Act; government tries to shore support.* Washington, DC: Associated Press.

Hulse, C. (2003, September 26). Congress shuts Pentagon unit over privacy. *The New York Times*, A20.

International Association of Chiefs of Police. (2003a). *About IACP history.* Retrieved April 30, 2003, from http://www.iacp.org/about/history.htm

International Association of Chiefs of Police. (2003b). *About IACP timeline.* Retrieved April 30, 2003, from http://www.iacp.org/about/timeline.htm

Jonietz, E. (2003, July–August). Total information overload: Interview with Robert Popp, Deputy Director, U.S. Defense Advanced Research Projects Agency (DARPA), Information Awareness Office. *Technology Review, 106*(6), 68–70.

Kelly, N. O. (2004). *Report to the public on events surrounding the JetBlue data transfer.* Washington, DC: Department of Homeland Security, Privacy Office. Retrieved April 5, 2004, from http://www.dhs.gov/interweb/assetlibrary/PrivacyOffice_jetBlueFINAL.pdf

Kenny, K. (1998). *Making sense of the Molly Maguires.* New York: Oxford University Press.

Kerner, O. (1968). *Report of the National Advisory Commission on Civil Disorders.* Washington, DC: U.S. Government Printing Office.

King, M. L., Jr. (2001). *Autobiography of Martin Luther King, Jr.* (C. Clayborne, Ed.). New York: Warner Books.

Krug, J. (2004). *Alternative press needs librarians' stories of government seeking information on library users.* Retrieved April 5, 2004, from http://www.ala.org/ala/oif/ifissues/issues-relatedlinks/usapatriotactletter.htm

Leahy, P. (2002, September 10). *Statement of the Honorable Patrick Leahy before the Senate Judiciary Committee. The U.S.A. Patriot Act in practice: Shedding light on the FISA process.* Retrieved June 30, 2003, from http://judiciary.senate.gov/member_statement. cfm?id=398&wit_id=50

Lettice, J. (2004, January 15). US using EU airline data to 'test' CAPPS II snoop system. *The Register* (UK). Retrieved January 20, 2004, from http://www.theregister.co.uk/ content/6/34915.html

Lumpkin, B. (2002, September 14). *Further FISA: Halls of justice: A weekly look inside the Justice Department.* New York: ABC News.com. Retrieved September 30, 2002, from http://abcnews.go.com/sections/us/HallsOfJustice/hallsofjustice136.html

Madison, J. (1788). *The particular structure of the new government and the distribution of power among its different parts* (Federalist paper No. 47, The New York Packet). Retrieved March 5, 2004, from http://lcweb2.loc.gov/const/fed/fed_47.html

McConnell, R. A. (1983). Letter from Robert A. McConnell, Assistant Attorney General for Legislative Affairs, to Senator Walter D. Huddleston, April 7, 1983. On file at *Cornell Law Review.*

MI5 The Security Service. (2003a). *The Investigatory Powers Tribunal.* Retrieved June 1, 2003, from http://www.mi5.gov.uk/accountability_funding/accountability_6.htm

MI5 The Security Service. (2003b). *Methods of gathering intelligence.* Retrieved June 1, 2003, from http://www.mi5.gov.uk/how_we_operate/how_we_operate_2.htm

MI5 The Security Service. (2003c). *Principles of our work.* Retrieved June 1, 2003, from http://www.mi5.gov.uk/how_we_operate/how_we_operate_1.htm

Milbank, D. (2003, September 11). President asks for expanded Patriot Act: Authority sought to fight terror. *The Washington Post,* A01. Retrieved March 5, 2004, from http://www.washingtonpost.com/wp-dyn/articles/A57827-2003Sep10.html

Millar, S., & Norton-Taylor, R. (2003, March 21). Police and MI5 tapping of phones and emails doubles under Labour. *The Guardian.* Retrieved April 1, 2004, from http://www.guardian.co.uk/humanrights/story/0,7369,884275,00.html

Miller, B. (2002, May 31). Ashcroft: Old rules aided terrorists; FBI agents get freer hand; Civil liberties groups criticize new guidelines. *The Washington Post,* A13.

Milles, S. (1995, September 7). Battling the executioners and their bosses: From the Molly Maguires to Mumia. *Worker's World,* 2–3.

Moffett, C. (1894). The overthrow of the Molly Maguires: Stories from the archives of the Pinkerton Detective Agency. *McClure's Magazine, 4,* 90–100. Retrieved June 1, 2003, from http://www.history.ohio-state.edu/projects/coal/mollymaguire/mollymaguires.htm

Murphy, L. W. (2002). *The dangers of domestic spying by federal law enforcement: A case study on FBI surveillance on Dr. Martin Luther King.* Washington, DC: ACLU Washington Office. Retrieved March 5, 2004, from http://www.aclu.org/Files/Files.cfm?ID=9999&c=184

Nieves, E. (2003, April 21). Local officials rise up to defy the Patriot Act. *The Washington Post,* A01.

O'Harrow, R. (2002, November 12). U.S. hopes to check computers globally: System would be used to hunt terrorists. *The Washington Post,* A04.

O'Harrow, R. (2003, July 31). Surveillance proposal expanded: CAPSS II would look at more air passengers. *The Washington Post,* E01.

Palmer, A. M. (1920). The case against the "Reds." *Forum, 63,* 173–185.

Perrotta, T. (2002a, November 27). NYPD Defends Bid to Relax Spying Rules. *The New York Law Journal, 228,* 1. Retrieved Apri 1, 2004, from http://www.law.com/jsp/article.jsp?id=1036630501833

Perrotta, T. (2002b, November 6). Police ask court to view secret papers. *The New York Law Journal, 228,* 1. Retrieved April 1, 2004, from http://www6.law.com/lawcom/displayid.cfm?statename=NY&docnum=162944&table=news&flag=full

Perrotta, T. (2002c, September 26). Police request broader surveillance rights. *The New York Law Journal, 228,* 1. Retrieved April 1, 2004, from http://www6.law.com/lawcom/displayid.cfm?statename=NY&docnum=154916&table=news&flag=full

Pinkerton, A. (1973). *The Molly Maguires and the detectives.* New York: Dover. (Original work published in 1877)

Powell, B. (2003, October 13). How George Tenet brought the CIA back from the dead. *Fortune Magazine,* 129–135.

Powell, M. (2002, November 29). Domestic spying pressed. *The Washington Post*, A01.

Powers, R. G. (1987). *Secrecy and power: The life of J. Edgar Hoover*. New York: The Free Press.

Rasch, M. D. (2002, June 24). The domestic spying renaissance. Securityfocus.com. Retrieved June 1, 2003, from http://www.securityfocus.com/columnists/90

Reagan, R. (1982). Evil empire speech, June 8, 1982, President Reagan: Speech to the House of Commons. *Modern history sourcebook*. Retrieved April 1, 2004, from http://www.fordham.edu/halsall/mod/1982reagan1.html

Reno, J. (1994). *Attorney General guidelines on general crimes, racketeering enterprise and domestic security/terrorism investigations*. Washington, DC: U.S. Department of Justice. Retrieved April 8, 1994, from http://www.fas.org/irp/agency/doj/fbi/generalcrime.htm

Reno, J. (1995a). *Attorney General guidelines for FBI foreign intelligence collection and foreign counter intelligence investigations (with government-authorized redactions)*. Washington, DC: U.S. Department of Justice. Retrieved April 8, 2003, from http://www.usdoj.gov/ag/readingroom/terrorismintel2.pdf

Reno, J. (1995b). *Memorandum re: Procedures for contacts between the FBI and the Criminal Division concerning foreign intelligence and foreign counterintelligence investigations*. Washington, DC: U.S. Department of Justice. Retrieved April 8, 2003, from http://www.fas.org/irp/agency/doj/fisa/1995procs.html

Reuters. (2004, January 12). Net flight bookings may be next terror casualty. Retrieved January 20, 2004, from http://news.com.com/2100-1038_3-5139631.html

Rockefeller, N. (1975, June). *Report to the President by the Commission on CIA Activities within the United States*. Washington, DC: U.S. Government Printing Office.

Rodota, S. (2004). Opinion 2/2004 on the adequate protection of personal data contained in the PNR of air passengers to be transferred to the United States' Bureau of Customs and Border Protection (US CBP). Brussels, Belgium: Article 29 Data Protection Working Party. Retrieved March 1, 2004, from http://europa.eu.int/comm/internal_market/privacy/docs/wpdocs/2004/wp87_en.pdf

Rosenbaum, D. (1976, January 30). House prevents releasing report on intelligence. *New York Times*, A1.

Rosenzweig, P. (2002, October 14). Prosecutions or "I Spy"? *National Law Journal*, A16.

Rosenzweig, P. (2003, August 7). Proposals for implementing the Terrorism Information Awareness system. Washington, DC: The Heritage Foundation. Retrieved August 15, 2003, from http://www.heritage.org/Research/HomelandDefense/lm8.cfm

Schmidt, S. (2003, October 22). Patriot Act misunderstood, senators say: Complaints about civil liberties go beyond legislation's reach, some insist. *The Washington Post*, A4.

Simic, A. (1995, November). *The workers' movement in Serbia and ex-Yugoslavia. Conference: Crisis, war and the world economy—the prospects of the organized working class in the countries of former Yugoslavia*. Berlin, Germany. Retrieved August 1, 2002, from http://www.alter.most.org.pl/iam/ex-yugo.htm

Smith, S. P., Crider, J. A., Perritt, H. H., Jr., Shyong, M., Krent, H., Reynolds, L. L., et al. (2000, December 8). *Independent review of the Carnivore System: Final report*. Lanham, MD: IIT Research Institute. Retrieved June 1, 2003, from http://www.usdoj.gov/jmd/publications/carniv_final.pdf

Smith, W. F. (1983, March 7). Attorney General's guidelines on general crimes, racketeering enterprise and domestic security/terrorism investigations. *Criminal Law Reporter, 32*, 3087.

Statewatch.org. (2004, January). *USA to use EU PNR data for CAPPS II testing despite assurances the agreement would not cover it*. Retrieved February 10, 2004, from http://www.statewatch.org/news/2004/jan/04eu-us-pnr-cappsII.htm

Stern, S. (2002, August 16). Pressure up to balance rights and security in terror war. *The Christian Science Monitor*. Retrieved April 1, 2004, from http://www.csmonitor.com/2002/0816/p02s01-usju.html

Strickland, L. S. (2002a). Information and the war against terrorism (Part I). *Bulletin of the American Society for Information Science and Technology, 28*(2), 12–17.

Strickland, L. S. (2002b). Information and the war against terrorism (Part II): Were American intelligence and law enforcement effectively positioned to protect the public? *Bulletin of the American Society for Information Science and Technology, 28*(3), 18–22.

Strickland, L. S. (2002c). Information and the war against terrorism (Part III): New information-related laws and the impact on civil liberties. *Bulletin of the American Society for Information Science and Technology, 29*(3), 23–27.

Strickland, L. S. (2002d). Information and the war against terrorism (Part IV): Civil liberties vs. security in the age of terrorism. *Bulletin of the American Society for Information Science and Technology, 28*(4), 9–13.

Strickland, L. S. (2002e). Information and the war against terrorism (Part V): The business implications. *Bulletin of the American Society for Information Science and Technology, 28*(6), 18–21.

Strickland, L. S. (2002f). Updates in the war against terrorism (Part I): Civil liberties developments. *Bulletin of the American Society for Information Science and Technology, 28*(5), 20–21.

Strickland, L. S. (2002g). Updates in the war against terrorism (Part II): Terrorist plotting or privileged information? The secret conspiracy of the blind sheik and the radical defense attorney. *Bulletin of the American Society for Information Science and Technology, 28*(5), 21–22.

Strickland, L. S. (2003a). Breaking developments in domestic intelligence. *Bulletin of the American Society for Information Science and Technology, 29*(3), 20–22.

Strickland, L.S. (2003b). Civil liberties vs. intelligence collection: The secret FISA Court speaks in public. *Government Information Quarterly, 20*, 1–12.

Strickland, L. S. (2003c). Spying and secret courts in America: New rules and insights. *Bulletin of the American Society for Information Science and Technology, 29*(2), 8–10.

Strohm, C. (2004, January 30). *U.S., Canada launch talks on sharing citizen data*. GovExec.com. Retrieved February 1, 2004, from http://www.govexec.com/dailyfed/0104/013004c1.htm

Thompson, L. D. (2001, August 6). *Memorandum re intelligence sharing*. Washington, DC: U.S. Department of Justice. Retrieved April 8, 2001, from http://www.fas.org/irp/agency/doj/fisa/dag080601.html

Thornburgh, D. (1989, March 21). *The Attorney General's guidelines on general crimes, racketeering enterprise and domestic security/terrorism investigations*. Washington, DC: U.S. Department of Justice. Retrieved April 8, 2003, from http://www.usdoj.gov/ag/readingroom/generalcrimea.htm

Treverton, G. F. (2001). *Reshaping national intelligence for an age of information*. New York: Cambridge University Press.

U.S. Department of Commerce. (2000, July 21). *Safe harbor privacy principles*. Retrieved August 1, 2003, from http://www.export.gov/safeharbor/SHPRINCIPLESFINAL.htm

U.S. Department of Defense. (2003, May 20). *Report to Congress regarding the Terrorism Information Awareness Program*. Retrieved June 1, 2003, from http://www.darpa.mil/body/tia/tia_report_page.htm

U.S. Department of Justice. (1997). *Report on the availability of bombmaking information, the extent to which its dissemination is controlled by federal law, and the extent to which such dissemination may be subject to regulation consistent with the First Amendment to the United States Constitution*. Washington, DC: Office of Legal Counsel. Retrieved April 1, 2004, from http://www.cybercrime.gov/bombmakinginfo.html

U. S. Department of Justice. (2002). *Report to Congress on the use of administrative subpoena authorities by executive branch agencies and entities.* Washington DC: Office of Legal Counsel. Retrieved March 5, 2004, from http://www.usdoj.gov:80/olp/intro.pdf

U.S. General Accounting Office. (2004, February). *Aviation security: Computer-assisted passenger prescreening system faces significant implementation challenges* (GAO-04-385). Retrieved February 13, 2004, from http://www.gao.gov/new.items/d04385.pdf

U.S. House of Representatives. (1978). *House Report on the Foreign Intelligence Surveillance Act (FISA), H.R. Rep. No. 95-1283, Part I, 95th Cong. 2d Sess.*

U.S. Joint Chiefs of Staff. (2001). *Department of Defense dictionary of military and associated terms (as amended through 17 December 2003)* (Joint Publication 1-02). Washington, DC: Department of Defense. Retrieved April 1, 2004, from http://www.dtic.mil/doctrine/jel/new_pubs/jp1_02.pdf

U. S. Senate. Committee on Education and Labor. (1940). *A resolution to investigate violations of the right of free speech and assembly and interference with the right of labor to organize and bargain collectively: Part 64, documents relating to Intelligence Bureau or Red Squad of Los Angeles Police Department.* Hearings on S. Res. 266, 74th Cong., 2d Sess. and 76th Cong., 3d Sess. 23507.

U.S. Senate. Committee on the Judiciary. (2003, October 21). *Hearing on protecting our national security from terrorist attacks: A review of criminal terrorism investigations and prosecutions.* Washington, DC: United States Senate. Retrieved October 22, 2003, from http://judiciary.senate.gov/hearing.cfm?id=965

Vila, B., & Morris, C. (1999). *The role of police in American society: A documentary history.* Westport, CT: Greenwood.

Warner, B., & Neely, J. (2004, January 12). *Net flight bookings may be next terror casualty.* Reuters. Retrieved January 13, 2004, from http://www.reuters.co.uk/newsArticle.jhtml?type=internetNews&storyID=4111803&section=news

Washington, G. (1933). Letter to Col. Elias Dayton, July 26, 1777. In J. C. Fitzpatrick (Ed.), *The writings of George Washington from the original manuscript sources, 1745–1799* (pp. 478–479). Washington, DC: U.S. Government Printing Office.

Watson, B. (2002, February). Crackdown! When bombs terrorized America, the Attorney General launched the "Palmer Raids." *Smithsonian Magazine,* 51–53.

Wearden, G. (2003, November 5). *Data surveillance complaints have zero success rate.* ZDNet UK.com. Retrieved December 1, 2003, from http://news.zdnet.co.uk/0,39020330,39117640,00.htm

Weiser, B. (2003, February 12). Rules eased for surveillance of New York groups. *The New York Times,* A17.

Williams, R. V. (n.d.). *The Chronology of Information Science and Technology.* University of South Carolina, College of Library and Information Science. Retrieved on August 1, 2003, from http://www.libsci.sc.edu/bob/istchron/ISCNET/ISCHRON.HTM

Wyden, R. (2003, July 28). *Wyden, Dorgan call for immediate halt to tax-funded "terror market" scheme* [Internet home page of Senator Ron Wyden]. Retrieved August 10, 2003, from http://wyden.senate.gov/media/2003/07282003_terrormarket.html

Theory

CHAPTER 12

Managing Social Capital

Elisabeth Davenport
Napier University

Herbert W. Snyder
North Dakota State University

Introduction

Social capital has been defined as "the wealth (or benefit) that exists because of an individual's social relationships" (Lesser, 2000, p. 4). It includes the relationships that form a network and the resources that are made available through that network (Nahapiet & Ghoshal, 2000). In an attempt to integrate these relationships and resources, Adler and Kwon (2000, p. 93) offer the following definition: "Social capital is a resource for individual and collective actors created by the configuration and content of the network of their more or less durable social relations." Social capital has been well explored in sociology (Portes, 2000) where analysts distinguish two perspectives: the egocentric, focusing on contacts among individual actors, and the sociocentric, focusing on relative position within a given network. These distinct forms have been invoked to explain the power of networks or relationships in improving personal position and building social cohesion at many different levels of social aggregation, from studies of international and regional development (Woolcock & Narayan, 2000) to dog fanciers (Syrjänen & Kuutti, 2002). It is an important concept in discussions of community, most recently in attempts to define a "sociability index" for online communities (Preece, 2000).

In this review our focus is narrow: We explore ways in which social capital may be managed with information and communication technologies (ICTs). We do this because claims have been made by knowledge management analysts (such as Huysmans 2002a, 2002b; Huysmans & Wulf, in press) that social capital analysis skills should be part of a standard resource kit for knowledge sharing in organizations. To examine this claim, we look at instances of formalisms of different types to see how they are created and exploited in management frameworks for social capital. We consider, for example, ways in which technology makes social capital more visible via tools that track and visualize interaction at a variety of levels and contexts, thus allowing relational assets to be modeled and managed in ways that were not possible before. Such implementations raise issues about technological choice and political interests; we discuss these at the end of the chapter. The ground we cover is not virgin territory. In 1994, a special issue of the *Communications of the ACM* on social computing (Schuler, 1994) presented many of the issues discussed here: Grudin (1994), for example, has raised a number of caveats (or issues) about an overly technological emphasis on groupware development.

In keeping with the management focus of this chapter, we have drawn on sources in the following domains: knowledge management, information management, computer-supported cooperative work, social shaping of technology, and management science. We have based the chapter on a commonly used management framework that models social capital as a tripartite phenomenon, all of whose elements are needed, though the relative position of each may vary across contexts. Nahapiet and Ghoshal (2000) articulate this view of social capital in terms of a structural dimension (network analysis), a cognitive dimension (shared codes), and a relational dimension (norms, trust). (Adler & Kwon [2002] proffer the three areas of networks, shared norms, and shared beliefs.)

The sections that follow consider each of the elements of social capital in sequence, discussing models and descriptions that are offered by different analysts as supports for managerial decision making about resources, and examining a number of relevant case studies. There are two main themes. We provide examples of the types of "objects" and "rules" (broadly defined) that allow elements of social capital to be configured into management support systems (including financial management systems). But we also address the contexts in which such objects are embedded, and the "collisions and convergences" (O'Day, Adler, Kuchinsky, & Bouch, 2001, p. 399) that may be observed. This leads us to consider social capital as a sociotechnical phenomenon (Resnick, 2002), and raises a number of implications for managers. The issue is complicated by two aspects of social capital that we have mentioned: the difference between personal and group social capital, and the range of levels of social aggregation at which the concept is invoked.

We suggest that the topic is timely for a number of reasons. The theme contributes to an ongoing exploration of the social dimension of information science, a subject that has attracted a number of recent *ARIST* authors (see, for example, Davenport & Hall, 2002; Jacob & Shaw, 1999; Pettigrew, Fidel, & Bruce, 2001). It also contributes to the understanding of human–technology interaction in ways that go beyond simple process models, by drawing attention to the management of metalevel or "second order" organizational phenomena—structures, routines, and norms and forms—by means of systematic analysis and representation. It thus provides a stronger empirical basis for those who seek to account for intangible assets in organizations.

The Structural Dimension: Models and Mechanisms

The links between social structure and organizational performance have been explored comprehensively in a corpus of work by Burt (2000a, 2000b), whose thinking is invoked in a number of studies of social capital. In his exposition of "the networked structure of competition," Burt (1992, p. 57) demonstrates that social networks vary in their shape and thus their effects, and that simple assumptions are misplaced in analysis of the relationship between network density, efficiency, and effectiveness. Many networks show what he labels "structural holes" (Burt, 2000b, p. 353) or dense nuclei characterized by tightly packed bonds (a source of bonding capital) that are separated by sparse, weak links (a source of innovation and bridging capital). The context of much of Burt's discussion is interallied network commerce or the world of networked organizations; in a number of studies (e.g., Burt, 1992) he discusses the relative efficiency of different types of clusters. A classic and much cited case is presented by Putnam (1993), contrasting patterns of small firm interaction in northern and southern Italy, where differences in network structure (sociocentric versus egocentric) correlate strongly with regional differences in economic performance.

Burt draws on earlier work on social networks, specifically Granovetter's (1973) seminal paper, "The Strength of Weak Ties," which suggested that significant exchanges happen on the intersecting edges of groups and that analysts should not neglect such (weak tie) spokes in favor of (strong tie) hubs. The insights derived by Burt and others from regional economic analysis are replicated at the microlevel of the organization, where a robust corpus of work on "knowledge networking" (reviewed in Teigland, 2003) provides insight into the working of strong and weak ties. Brown and Duguid (1994), for example, in a discussion of the importance of margins, observe that strong bonds that define groups are needed for continuity and cohesion, while weak bonds across groups are a source of innovation and adaptation. The periphery is thus as deserving of attention as the core. A similar point is made by Wenger (1998, 2000) in his analyses of communities of practice.

The early work of Burt and Granovetter analyzes the relationship between networks and social capital in environments that are not mediated by ICTs. Do their findings hold in the world of networked commerce? How do the social capital effects of strong and weak ties intersect with technology configuration? We consider these questions from two perspectives: microlevel work on organizational patterns of association, and more recent high-level work on patterns of relationship on the World Wide Web. Teigland, a microanalyst, addresses these issues in a series of systematic and intensive studies of multinational, networked organizations. With her co-author she suggests (Wasko & Teigland, 2003, pp. 354–355) that a variant of the community of practice, the electronic network of practice or ENOP, draws much of its strength from weak ties. In an exemplary application of network analysis techniques co-authored with colleagues (Schenkel, Teigland, & Borgatti, 2003), she identifies five measures that managers may use to assess centrality and reciprocity and to observe patterns of internal and external trading in different groups.[1] Teigland (2003) links creative performance to loose structure, and efficient performance to tight structure.

The instrument for analysis used in these studies is Krackplot, one of the earliest publicly available software applications designed by Krackhardt (Krackhardt, Blythe, & McGrath, 1994) for human resource management. (A full review of this and other methods is provided in the practical manual by Scott [2000].) It has been the basis of a number of studies of organizational communication by library and information science (LIS) analysts (Grosser, 1991; Haythornthwaite, 1996). Krackhardt's analysis uses standard matrix techniques to derive node and arc maps that chart reciprocal interactions of different types. By qualifying these interactions in terms of values such as gift-giving, consultancy, and "looking out for," Krackhardt blends formal modeling and qualitative analysis to support those making decisions about human resources. In a convincing case study (Krackhardt & Hanson, 1997), he demonstrates that social network analysis can reveal latent problems in interpersonal communication. Krackplot is one of a number of instruments reviewed by Garton, Haythornthwaite, and Wellman (1999), in an extensive body of work that ranges across different social domains. Decisions about media to support social networking are an important theme: To what extent may weak ties be fostered for their recognized benefits (Teigland & Wasko, 2003), and in what ways may they be converted to strong ties? Haythornthwaite (2001) suggests that media richness theory (Trevino, Daft, & Lengel, 1990) should be linked to social tie theory. This would assist those who make decisions about infrastructures that involve multiple media by helping them identify which configurations are likely to foster strong ties and which may convert the weak to the strong.

High-level structural analyses of the World Wide Web suggest that a dominant pattern is visible in network analysis across domains—that of the power curve, which indicates that networks tend toward a stable configuration with many links focused on a few powerful hubs. This pattern is clearly visible in simulations of interactions on the Web undertaken in recent years by Huberman (2001) and by Barabási and his colleagues whose work is summarized in his recent popular science volume *Linked* (Barabási, 2002). Each of these analysts relies on computational and graphical representation of relationships (analogous to those that underpin the citation maps of science that were developed by Small and his colleagues [reported retrospectively by Small in 1999]). A readable review of the techniques involved is given by Buchanan (2003, pp. 27–28) who discusses the related phenomenon of "Erdös numbers," a measure of graph-theoretic distance between a core and peripheral entities. Barabási offers examples of how his techniques may be exploited *in vivo*, drawing on, among other sources, the cases described by Gladwell (2000) who illustrates, in layman's terms, the managerial implications of intersecting structures, and the power of the hubs and connectors who participate in multiple networks. This layman's version of topological analysis has been invoked by consultants such as Prusak (2001) as a guide to social choices about roles and associates. Gladwell's (2000) popular monograph, *The Tipping Point*, instantiates the power of hubs in a number of narratives or parables. Hubs, says Gladwell, are strong attractors in terms of social capital as their membership of multiple networks makes them major resources in terms of both strong and weak ties. Gladwell (2000, p. 30) distinguishes "hubs" from two other types of agents of social capital formation, "mavens" and "salesmen." In a high-level account of the Internet and its shaping of social structures, Castells (2000) suggests that the repertoires of ties shaping personal and corporate social interactions are complex and that the Web allows this complexity to be more clearly understood. This has a number of implications: It allows individuals to manage their participation in the multiple networks that shape their lives, and, as a corollary, allows others to observe and capitalize on these networks. Media richness is only one of many contingencies shaping different types and mixes of social capital. (The mobility, fluidity, accessibility, and availability of e-mail make it an important channel in the formation of strong ties, for example.) In some contexts an actor's visible membership of the right group or list will provide a profile that is sufficient social capital in itself to lead others to engage with him or her (see, for example, Kollock, 1999). This is, of course, the principle that underlies computerized referral systems of different types, or collaborative recommender systems (Davenport & Cronin, 2001; Furner 2002; Munro, Hook, & Benyon, 1999; Resnick & Varian, 1997). (If the pattern is indeed a general phenomenon, then it raises questions about the need for the cognitive modeling that characterizes much information-seeking and information retrieval research, as the individual utility maximization involved in such cases relates to status rather than anomalous knowledge states.) Resnick (2002) discusses the ambivalent role of recommender systems in social capital formation at some length; in some cases, this mechanism supports and sustains cliques, rather than stimulating an expansion of the relationships in a network.

Several mapping techniques have been developed to visualize and interpret the microlevel interactions that are indicative of the structural dimension of social capital. Examples may be found in studies of turn-taking and discourse dynamics, where visualization of conversational moves reveals microlevel interaction dynamics (Donath, 2002; Erickson, Halverson, Kellogg, Laff, & Wolf, 2002; Nardi, Whittaker, Isaacs, Creech, Johnson, & Hainsworth, 2002). These may be stored as objects in configurable "knowledge bases." The emerging generation of micro-sensors radio frequency identification (RFID) tags and other miniature tracking devices (such as "smart dust" and "mica motes") will greatly extend our ability to map patterns of interaction and exploit such representations in an array of pervasive systems, such as customer relationship management applications, road traffic control systems, and "advanced" recommenders (Ananthaswamy, 2003, p. 26). It may be noted that Grudin (2002) has questioned the desirability of such trends in a discussion of the implications of ubiquitous computing.

Analysts of learning communities have shown considerable interest in ways in which structural capital may be managed to encourage participation (e.g., Riel, 1995). Haythornthwaite (1996) has for a number of years advocated social network analysis (SNA) as a useful general method for information science, and has subsequently undertaken a number of studies of distance and flexible academic learning environments (e.g., Haythornthwaite, 2000), where the visualization of differential bonding patterns can fine tune an instructor's grasp of classroom dynamics. Such a focus is timely, given the continuing shift in education away from teaching to learning support by means of managed learning objects. This concept reflects a reconfiguration of classroom dynamics to an environment where a structural analysis revealing the power bases (and, thereby providing insight into likely levels of social capital) among students may be a useful managerial tool (Koku, Nazer, & Wellman, 2001; Koku & Wellman, 2004). Such understanding has always been part of the tacit knowledge portfolio of experienced teachers, who have understood the importance of classroom dynamics and have managed classes as micro-organizations where a group (or groups) work together to achieve outcomes that are assessed according to mutually agreed performance criteria.

Relationships and relative roles in this environment are not yet clearly understood—is the teacher a "master" of apprentices, a "manager," or simply a "facilitator"? This lack of clarity may partly explain the failure of many implementations of e-learning platforms that have not met the expectations of those who designed and invested in them. Critical analyses of online and electronic learning environments are offered by Kling and his colleagues (e.g., Hara & Kling, 2001; Kling & Courtright, 2003). Visions of the class of learners as a computer-supported community may be further compromised by inappropriate training of teacher/managers, notably in cases where this predates the adoption of technology-mediated learning (e.g., Renninger & Shumar, 2002), and by the expectations of learners that they should receive instruction. Various learning platforms, or mechanisms, are available that support structural analysis, monitoring, and tracking to enhance both instructor and learner experience (Hardless & Nulden, 1999).

These and cognate issues are discussed in a recent issue of *The Information Society* (Barab, 2003) and in a monograph co-edited by contributors to the special issue (Barab, Kling, & Gray, 2004). In addition, a useful review of social networking and social capital in a number of learning communities is provided in Wellman and Haythornthwaite's (2002) edited monograph. Wellman's oeuvre provides the most substantial account of the area and has developed from earlier studies of social networks (Wellman, 1982) to more recent studies that are explicitly focused on the structural dimensions of social capital (Wellman & Frank, 2001; Wellman, Haase, Witte, & Hampton, 2001).

Networks of Science: The Archetype of Structural Capital

As we have noted, the social networks that define scientific communication have been a focus of interest in information science since its inception. The pioneers of this analytic approach worked as a small and highly efficient cadre (readers are referred to the review by Borgman & Furner, 2002; and see also Small, 1999). Recently, visualization studies that provide insight into social dynamics have emerged in the domain of computer science (e.g., Börner and Chen, 2003; Börner, Chen, & Boyack, 2003; Kirschner, Shum, & Carr, 2003). The work of the pioneers of citation mapping was closely coupled with work on the sociology of science, where the pattern of the power law was identified and elegantly described in terms of the "Matthew effect" (Merton, 1968, p. 55). Historically, the raw material of traditional (or bibliometric) citation analysis was limited in range (journal article references), exactly specified (according to the presentation norms of different domains), and inherently sociometric. The resulting maps were intended as a guide to the dynamics of subject domains and the distribution, rather than the evaluation, of expertise (the latter was managed in the covert network of the "invisible college" [Crane, 1969, p. 335]). The use of citation analysis to rank individuals, groups, and institutions (with serious consequences in terms of resource allocation) is a strong example of a computational technique that has powerful implications for the management of structural capital (Sosteric, 1999).

The relationship between the published structures of science and their underpinnings, varied networks of interactions among scientists, have been scrutinized since the early days of bibliometrics. In the past decade, bibliometricians have considered a wider set of textual artifacts, moving beyond the original citation to consider phenomena such as acknowledgments (Cronin, 1995; Cronin, Shaw, & La Barre, 2004) that are not totally commensurate with citation patterns. And several studies (Almind & Ingwersen, 1997; Björneborn & Ingwersen, 2001; Cronin, Snyder, Rosenbaum, Martinson, & Callahan, 1998; Thelwell, Vaughan, & Björneborn, in the present *ARIST* volume) have explored linking on the Web. Recognizing that the citation, and its Web equivalent the outward link, is only one of a number of social practices that may contribute to the formation of social capital in scholarly regimes, analysts have studied an extended set of communication channels and genres (Web pages, teaching materials, contributions to lists, and so on) to acquire a rich picture of what Kling and his colleagues describe as "socio-technical interaction networks" (Kling, McKim, & King, 2001, p. 1; see also Kling, Spector, & McKim, 2002; Lamb & Kling, 2002). The visibility of contributions to, and participation in, a wider range of publicly accessible channels has changed the nature of communication in many disciplines, notably those driven by rapid dissemination of empirical results. These issues were discussed by Kling and Callahan (2003) in their *ARIST* review.

As we have noted, traditional citation analysis (see Cronin, 1984, 1995) has focused largely on sociocentric network analysis (concerned with the relative position of individual texts, authors, and institutions within a relevant disciplinary network). White suggests that egocentric maps are of interest; he recently, with Wellman and Nazer (White, Wellman, & Nazer, 2002), addressed the relationship between structural analysis of networks and patterns of social behavior. This line of exploration was also followed by Cronin and Shaw (2001, p. 127) in a discussion of "identity" citation (based on whom I cite) and "image" citation (based on who cites me). Cronin has further developed the theme of ecology of participation to suggest that authorship should be perceived as a form of distributed cognition (Cronin, 2004). Such analyses, added to the Web-based maps of the multiple networks that constitute participation in academic life, may transform ways in which academics manage their career trajectories. In many institutional contexts, academic status is determined by an individual's position in the multiple networks that are typical of a domain: Bourdieu (1988, 1990), for example, provides detailed and comprehensive understanding of this phenomenon in analyses of different types of capital; of particular pertinence here is his account of the French higher education system. The management potential of a visualization of concurrent participation on multiple network applications (e-mail, lists, blogs, and so on) is as yet unclear. Haythornthwaite (2002) touches on this issue in a discussion of multiple channels in the online classroom. Kling's account of academic apprenticeship provides a similar analysis; his "guild model" candidates with high potential are entrained into an expanding series of networks (Kling et al., 2002, p. 1).

The Cognitive Dimension: Models and Mechanisms

In the previous sections we explored the structural dimension of social capital and some of its intricacies. ICTs contribute to the management of structural capital in different ways: They are the means by which relationships are articulated (in the case of networked organizations), and they also embody formalisms and rules that allow structures to be identified, visualized, and exploited—both overtly and covertly—in the form of manageable objects. Although network structures are revealed in the form of patterns (such as power curves) that are manifest across domains, locally they depend on mutually recognized artifacts of varying types such as codes, procedures, instruments, and maps (like the network topologies themselves). These constitute what Nahapiet and Ghoshal (2000) call the cognitive dimension; they may be formalized in different ways. A simple view considers such artifacts to be the common ground of collaboration—but this is contested in a number of empirical studies that considers objectification to be misleading or even oppressive, as shared artifacts are indeterminate and undergo continuous negotiation (Koschman & LeBaron, 2003). This divergence of perspective between formalists and social analysts is long standing (Suchman, 1993). In 1989, Star and Greisemer proposed that deep investigation was needed of what they described as boundary objects, or negotiable entities whose fluidity is acknowledged, that are accessible by participants in shared work who come from different backgrounds.[2] Boundary objects have proved to be an important focal point in discussions of online communities of practice (Davenport & Hall, 2002) and in discussions of collaborative work across domains. The concept has been refined by some analysts of situations where the boundary objects are not yet stable: O'Day et al. (2001, pp. 404–406) talk of "boundary-objects-in-the-making" in their account of micro-array work in bioinformatics.

Codes, Contents, and Contests

The cognitive dimension is the low-hanging fruit of social capital. Shared codes are objects managers believe they understand, and can do something about, by standardizing existing, scattered codified material, for example, or by "codifying" previously "uncoded" knowledge, a mantra in much of the knowledge management literature. Failure to develop social capital is often attributed to imperfect content management or lack of "shared codes" (often in the shape of shared procedures), something that features in many accounts of interventions to overcome underdevelopment and social exclusion. In this section of the chapter we review a number of cognitive elements that are supported by ICTs—classification schemes, ontologies, expertise models, and intranets. As we have indicated, the management implications are more complex than they first appear.

Classification and ontology schemes feature in many accounts of collaborative work; Bowker and Star (2000) provide a critical review of some of these in their socially oriented series of studies of classification structures, ontologism, and digital infrastructure. A thoughtful exposition of the issues is offered by Simone and Sarini (2001) in a discussion of the adaptability of classification schemes in cooperative work. They evaluate a number of systems specifically designed as boundary objects and conclude that flexible formalisms (a desideratum in the work of Bowker and Star) can be derived, but that these will operate at a high level of abstraction, thus offering partial solutions to the problems of shared systems. "Partial" is an advantage here, as these formalisms can evolve to meet the contingencies of local situations. This account is compatible with O'Day et al.'s (2001) notion of "boundary-objects-in-the-making."

We suggest that work on documentary (and electronic) genres can provide further examples of this phenomenon of partial, high level formalisms. A body of work on documentary genres offers an alternative approach to formalizing the cognitive dimension by codifying at a higher or more social level. There is evidence from work on business genres that inferences can be made about compatibility and confidence on the basis of shared articulation of practice and expression. Davenport and Rosenbaum (2000) have suggested that genre profiles (a surrogate for the formation of social capital) might serve as the basis of matching individuals and groups for collaborative work. Spinuzzi (2003), in a recent monograph, provides further examples of the workings of genres and the negotiations and conflicts that lead to their evolution.

Various studies have described the negotiations that convert shared data into a workable resource. Harper (1997, p. 363), in an extensive ethnographic study of the International Monetary Fund, describes the process of converting numbers into information as one of giving them "a voice." O'Day et al. (2001, p, 406) describe this as "making numbers count." O'Day and colleagues frame the process of negotiation in terms of stories that must be aligned. In their account of a group of bioinformatics specialists (computer scientists and geneticists) working with micro-arrays (a "disruptive" technology [O'Day et al., 2001, p. 423] that disturbs the "rhythm" of work [O'Day et al., 2001, p. 414]), they dissect the interplay of "computational stories" (O'Day et al., 2001, p. 408) and "biological stories" (O'Day et al., 2001, p. 407) and observe that "the complexity of trust" (O'Day et al., 2001, p. 406) is a central theme of the interdisciplinary collaboration between ecologists and computational experts. "We do not mean to imply that people are suspicious of each other, but rather that both biologists and computational experts are still trying to bring their problems and methods into alignment so they can both feel confident of their results" (O'Day et al., 2001, p. 406). Similar observations are made by Sonnenwald, Maglaughlin, and Whitton (in press) in a recent longitudinal study of ecologists. A further example of the difficulties of adequately managing shared meanings is provided in an account of water planners by Schiff, Van House, and Butler (1997). In this interdisciplinary, planning project, sharing of documents in an online repository was compromised by some scientists' concerns that documents prepared by experts within their own disciplinary tradition (and thus embedding much implicit understanding of the entities and procedures presented in the documentation) would be misappropriated by readers whose expertise lay in a different domain. Expertise is characterized by implicit (or embedded or tacit) understanding; the assumption that shared codes or other common objects can work across groups of experts in different domains may be invalid.

Within some domains, similar processes of collision and convergence may be observed—see the detailed accounts of the high-energy physics world by Knorr Cetina (1999) and Kling and his colleagues (Kling et al., 2002; Meyer & Kling, 2002). In others, where the environment is less volatile or the focus of cognitive effort less complex, the contribution of shared codes to cognitive capital appears less problematic. In the high-technology sector, some analysts suggest that cognitive prowess is the primary driver of association in a domain where success depends on expertise or knowing one's job (Cohen & Fields, 2000; Collins, Miller, Spielman, & Wherry, 1994; De Carolis, 2002). This is corroborated in a number of recent studies exploring the Linux community (Berquist & Ljungberg, 2001; Moody, 2002; Ye & Kishida, 2003). In this group, knowledge of the code is to be taken literally as an explanation of the working of social capital. This knowledge defines status in the group, and as members participate to learn, those with greater knowledge are strong attractors. Entry barriers are high and participants jointly accomplish programming outcomes that cannot be achieved individually. The process of legitimation in this context does not depend on a traditional master/apprentice relationship (one way of ensuring continuity in a traditional craft), but on establishing one's reputation rapidly. In other communities (professions such as law or clinical medicine), the acquisition of appropriate shared codes and meanings (the apprenticeship or familiarization process) may be an important and protracted element of practice (Lave & Wenger, 1990).

Such environments are examples of tight structures, and there are specific issues to be addressed in the management of cognitive capital in such circumstances. These are illustrated in a case study by Baxter (2000) of a problematic software engineering project. In this case, managers were not masters of the code (a specific software program) that was the working medium of those whom they managed. Developers were able to sustain their own employment on a deliberately and continuously delayed project by hiding behind their expertise (or knowledge of the code). In this case, lack of shared codes did not impede collaboration, but impeded control.

There is considerable debate about the legitimacy of formalisms in the context of the modeling of expertise. Various frameworks are available to managers (e.g., Zachmann Institute for Framework Advancement, n.d.). Ackerman and Malone's Answer Garden (Ackerman, 1994) is a computer-based expertise locator that works on a referral basis and is designed to direct questions to those with appropriate levels of expertise. Groth and Bowers (2001, p. 298), making a plea for "lightweight approaches which add value to existing systems," suggest that the heuristic that underlies Answer Garden fails to take account of the local embeddedness of work "in the face of contingency" by "artful" workers who try to meet the "demands of getting work collectively done." Pipek and Wulf (2003, p. 19), in a study of maintenance engineers, contest this critique because it treats Answer Garden as a "replacement architecture" or substitute for existing practice, which it is not intended to be. As a "supplement architecture" or extension of existing practice, it can be a useful managerial tool. McDonald and Ackerman's (2000) extension of Answer Garden, the Expertise Recommender, is based on the latent semantic analysis techniques (Dumais, 2004) that underlie other recommender systems (Resnick & Varian, 1997). This and other systems are reviewed by Becks, Reichling, and Wulf (2002) and their application to social capital formation is discussed. These techniques raise serious issues of control and ownership.

Cognitive Capital in the Political Process

The European Commission has recently funded attempts to embed knowledge management architectures into the political process in order to increase citizen participation. Recent initiatives to promote inclusion across Europe (Gronlund, 2002; Malkia, Anttiroiko, & Savolainen, 2002) have been premised on the provision of shared protocols for debate and reporting, based on standardized architectures and ontologies. A parallel set of developments has attempted to establish standards for government transaction services (Adams, Fraser, Macintosh, & McKay-Hubbard, 2002), with the assumption that shared codes will contribute to transparency and trust and thereby persuade citizens of the benefits of participating in the policy-making and resource-allocating networks that constitute government (Whyte & Macintosh, 2001).[3] Initial validations of the systems that have been developed under such initiatives have raised as many problems as they have solved. Code sharing is difficult to achieve above a certain level of aggregation and where there is an asymmetry on the process of code-making (governments, or other "outsiders" set the codes; citizens are "trained") social capital formation may be compromised rather than facilitated. Briggs (1998), for example, describes an intervention in an area of urban deprivation where the failure to select the right "connector" led to a mismatch of understandings and the failure of the project. A further reported inhibitor is the time lag between citizen input and outcomes that leads citizens to perceive codes as ineffective. Lack of accountability is a problem in current implementations of e-democracy and e-government platforms (Whyte & Macintosh, 2001): It is not clear why processes and decisions work as they do, and citizens cannot fathom the consequences of their input (Nissenbaum, 1994). Technical fixes for such problems have been proposed, including visualization, again, in the form of argumentation tools that can make the relationship between input and output into policy making visible (Kirschner et al., 2003), or e-consultation tools.

Intranets and Cognitive Capital

Intranets have proved to be a most seductive technology for managers who wish to promote social capital by means of knowledge sharing, and an extensive literature of case studies is available (Davenport & Hall, 2002; Newell, Scarbrough, Swan, & Hislop, 2000). Intranets are considered both as repositories for shared protocols, procedures, and best practices, and as providers of access to collaborative work spaces, although these, as Detlor (2004) observes, are often the least exploited components of such platforms. Studies have demonstrated that the effects of local implementations are contingent and emergent and that developers with naïve and over-ambitious expectations are often insensitive to contextual issues (see, for example, Adler & Kwon, 2000; Arnold & Kay, 2000; Portes & Landolt, 1996). Groth and Bowers (2001) comment on the use of an intranet in their study of mechatronics engineers: The intranet was not useless, but it was not the fundamental bedrock for collaboration envisaged by those who commissioned it—it was used as a handy source of factual information such as telephone numbers and train times. An exemplary study (because it is an instance of a common pattern of failed expectations) of an e-bank intranet is provided by Edelman, Bresnen, Newell, Scarbrough, and Swan (2002). They describe an outsourced project that was institutionally unable to appropriate/exploit the social capital on which the subsequent use of the system would depend. An intranet implementation in the consultancy sector described by Hall (2002) involved a top-down initiative which also proved problematic because the decision makers involved failed to take account of the social networks that shaped exchange patterns in different groups and focused on providing repositories of shared and formatted information. A parallel economy developed, with much of the effective exchange undertaken outside the official forum.

Managing the Relational Dimension: Models and Mechanisms

In the previous sections we reviewed a number of cases where ICTs support the management of structural and cognitive capital. They do this in diverse ways, both as technical infrastructures and as superstructures, providing the means for description, manipulation, and analysis. We suggest that both of these levels intersect and involve complex issues of legitimacy, ownership, and choice. Nahapiet and Ghoshal (2000) suggest that the relational dimension of social capital addresses areas such as norms, cultural institutions, social bonds, and trust. It is thus overtly concerned with ethical and political issues and deeply intertwined with the other dimensions of social capital (Dibben, 2000, states that it emerges from the other two). In this section we focus on ICTs and the management of trust, following a number of analysts such as Preece (2002, p. 36), who equates trust with social capital, on the grounds that it is "the glue that holds organizations together" (a view that is shared by the anonymous author of an article in *The Economist* (Economics focus: A question of trust, 2003). Like structural and cognitive social capital, the elements of the relational dimension (trust and related areas) may be formalized to a greater or lesser extent. Much of our recent understanding of how this may be done has emerged in two areas: computer development work on intelligent agents and behavioral studies of groups (supported to a greater or lesser extent by computers) at different levels of organization. For a comprehensive review of these issues, readers may consult the *ARIST* chapter by Marsh and Dibben (2003), and the seminal text by Kramer and Tyler (1996), an edited monograph that presents a number of models of trust.

Modeling Trust

Trust (or "learned trust") has been the focus of much design work on intelligent agents, as it may be deconstructed and analyzed in terms of game theory. This reduces the phenomenon to a set of rules that may be programmed. Marsh (1994; Marsh & Dibben, 2003), building on earlier work in the digital library environment (where "socially intelligent" agents make decisions on cooperative retrieval moves on the basis of game theory), has designed "socially adaptive agents," which can acquire understanding of what moves are appropriate in typical contexts. Their cumulating knowledge is stored in a library of routines that is activated as appropriate to recognized sets of circumstances; we suggest that is an example of formalized social capital. This theme is developed in an edited volume on trust in virtual organizations (Castelfranchi & Tan, 2001). Falcone and Castelfranchi (2001), for example, explore social trust in terms of a continuum; Rea (2001) considers how trust may be engendered in electronic environments; and Weigand and van den Heuvel (2001) discuss trust in terms of speech acts and workflow modeling (an approach that depends on formal analysis and representation). An additional pertinent series of studies at the Electrical Engineering Department at Imperial College of Science, Medicine and Technology, London, can be accessed at http://alfebiite.ee.ic.ac.uk/Templates/papers.htm.

Dibben (2000), inspired by Marsh's early work, has deconstructed interpersonal trust in the material or physical world of business transactions. He provides a template to capture the stages and events that characterize interactions between investment angels and entrepreneurs and demonstrates that trust has recognizable and observable focal points, such as past performance, current competence, and mutually understood practice. Trust is, in effect, a form of tacit knowledge that can be apprehended by observing instances of reciprocal behavior. The focal points of trust overlap to a great extent with Nahapiet and Ghoshal's (2000) structural dimension (past performance and reputation dictate one's position in a network), and cognitive dimension (mutually understood forms and practices). This implies that the relational element of the tripartite model emerges from the first two, although not always from both.

A parallel approach to representing the relational dimension can be extracted from the literature on emotional intelligence in organizations (Bartel & Saavedra, 2000). Druskat and Wolff (2001) distinguish different facets: Emotion is a brief reaction to particular persons or events and implies the existence of a specific goal; mood is distinguished from emotion by longer duration and lack of a specific focus—it involves beliefs about future affective states and may be a byproduct of emotion; affect is a broad and inclusive label that refers to both mood and emotion and reflects "value" changes. Emotional intelligence implies awareness of all of these, and ability to regulate them, and thus to monitor one's own and others' emotions, to discriminate among them, and to use the information to guide one's thinking and action (Cherniss & Goleman, n.d.).

Instances of behavior can be classified and the classification can be the basis of an observational instrument, such as Larsen and Diener's (1992, p. 25) "circumplex model." Behavior in this context is to some extent self-organizing, as moods are generated by emotional comparison and emotional contagion; the "thermostat" is group norms, for perspective taking, for handling confrontation, and for caring behavior. Norms, say Druskat and Wolff (2001), build "emotional capacity."

Emotional capacity, defined in this way, is similar to the "collective mind" in organizations described by Weick and Roberts (1993, p. 367) as "patterns of heedful interrelating of actions in a social system" that contributes to "heedful performance," (Weick & Roberts, 1993, p. 362) or the "outcome of training and experience that weave together thinking, feeling and willing" (Weick & Roberts, 1993, p. 362). This is a significant factor in group cohesion, as, in the right circumstances, "newcomers will learn this style of responding ... whether mind gets renewed may be determined by the candor and narrative of insiders and the attentiveness of newcomers" and the collective mind may be cultivated where it is "visible, rewarded, modeled, discussed and preserved in vivid stories" (Weick & Roberts, 1993, p. 367). Insiders who narrate richly remind themselves of forgotten details as they reconstruct previous events. The downside of failing to be expansive is "newcomers acting without heed, as they have only banal conversations to internalize" (Weick & Roberts, 1993, p. 368). The issue of how to present and represent such narratives and their relation to social capital is discussed in depth by Snowden (2000).

Managing Relational Capital in Online Teams

Recent work on the formation and maintenance of online teams raises a number of issues about collective mind and the time in which it may be developed. The concept of "swift trust," developed by Meyerson, Weick, and Kramer (1996, p. 167) to account for the emergence of trust relations in situations where the individuals have a limited history of working together, is relevant here. According to Jarvenpaa and Leidner (1998, p. 6), Meyerson et al. coined the term for temporary teams "whose existence ... is formed around a common task, with a finite life span." For many, the term social capital implies cumulation over time (capital is cumulative), and history is a condition of the development of structural capital. Social capital in swift trust environments appears to depend heavily on cognitive capital with additional elements of heedfulness, or awareness that emerge from the current situation; "swift trust" neatly encapsulates the world of high-technology heroes described in the previous section. Jarvenpaa and Leidner (1998, p. 6), for example, list the requisites of successful virtual team performance: a clear definition of roles and responsibilities, clarity in order to avoid confusion and disincentive, effective handling of conflict, and "thoughtful exchange of messages at the beginning of the team's existence." Such factors can serve as indicators to distinguish virtual teams with high trust from those with low trust. These insights are corroborated by recent accounts of collaboration in bioinformatics teams by Sonnenwald, Whitton, and Maglaughlin (2003). Awareness is not a distinctive property of online work: Groth and Bowers (2001, p. 288) playfully suggest that an "availability concierge" would be as useful as an "expertise concierge" (advocated by the designers of Answer Garden) in ensuring that work is completed in the material environment of his study.

Iacono and Weisband (1997, p. 1) describe a project with distributed electronic teams, who must "quickly develop and maintain trust relationships with people that they hardly know, and may never meet again, with the goal of producing interdependent work." In this situation, say the authors, trust is less about relating than doing, as swift trust is "less an interpersonal form than a cognitive and action form" (Iacono & Weisband, 1997, p. 1). Temporary systems require quick mutual adjustments so that people can innovate as required; in online work, technology must support these adjustments. Good communication habits are key; these include the ability to handle multiple tasks and remote requests while attending to local demands.

Active participation may be seen, say the authors, as a system of initiations and responses. Initiations involve trust, because they "make one's preferences public" (which may incur risk) (Iacono & Weisband, 1997, p. 2); each initiation strengthens participants' perceptions that trust is reasonable and incurs more initiations. The making of responses "signals and inspires trust" in the group (Iacono & Weisband, 1997, p. 2), and action moves forward in an initiation–response cycle. Drew (1995) suggests that the anticipatory interactive planning (AIP) supported in initiation–response sequences is a defining characteristic of social intelligence. Iacono and Weisband (1997) observed initiations, which they categorized as getting together, work-process, work-content, work-technical, needing-contact, and fun-talk. Work-process and work-content initiations correlated with high performance, as did number of total initiations, and the pattern of timing.

Weisband (2002) draws on extensive work by Steinfield, Jang, and Pfass (1999) on the design of a collaborative platform for teamwork. This suggests that transparency, presence, and awareness are critical components in successful online work.[4] Weisband (n.d.) has summarized a subsequent study (15 teams in two universities): Low performing teams rely on their perceptions of others as a predictor of good performance; high performing teams rely on what people do and say as a predictor of good performance; teams who may not engage in the hard work of doing distant collaboration may feel good about the process and each other, but such perceptions do not lead to successful outcomes. It appears that microlevel "shared situational awareness," (borrowing a term from macrolevel studies of teambuilding in the U.S. defense forces [Loughran, 2000]) based on Weisband's work, contributes more to high performance than warm feelings about others. This is corroborated in a study by Jarvenpaa, Knoll, and Leidner (1998), who suggest that perceptions of others' ability and integrity are important to initial trust and that perceptions of benevolence are least important. It may be noted, however, that a subsequent empirical study (Spring & Vathanophas, 2003) suggests that adding an awareness function to computer platforms for collaboration does not, in itself, add to the development of social capital.

Awareness is a key feature of situated trust in the work of the bioinformatics specialists described by Sonnenwald et al. (in press), who derived their own local template for addressing the situational dimensions of collaboration. Their approach resonates with material in Cool's (2001) *ARIST* chapter that reviews the concept of "situation" in library and information science.

Davenport, Graham, Kennedy, and Taylor (2003) have attempted to build a probe to support empirical investigation of the interplay of different dimensions of social capital; used in the longer term, the probe may allow analysts and managers to assess the effect of intervening at the formation stage by balancing the elements of a team to meet the requirements of different types of projects. They have drawn on prior studies of social capital and organizational trust and based their work on the studies described previously and on "use cases" developed with a number of commercial partners. The focus is social capital and trust in specific organizational situations. They have exploited Nahapiet and Ghoshal's (2000) social capital model to construct a framework for assessing the social capital potential of prospective partners. To accommodate business partners, the dimensions are described in terms that managers can use. "Structural dimension" is represented in a "competence" layer that captures track record or "hindsight" and by a social browser, or spreading graph, that incorporates dynamic judgments about the cognitive dimension and the relational dimension.

This system accommodates both egocentric and sociocentric profiles; it is intended to support judgments about social capital in cases where there is little time for it to develop: Judgments must rely on surrogates and cues. In this case, managers are invited to score the attributes of potential partners in what is in effect a social capital index. Such an approach allows the elements of social capital (in the form of profiles of different types) to be modeled as objects in different configurations. The data for the probe are stored in an Extensible Markup Language (XML) server and can be manipulated with a visualization tool that provides flexible support for those who seek to establish viable teams and partnerships.

Measuring Social Capital: Financial Management Systems

That the human relations which comprise social capital have a value is generally agreed upon by most accountants and economists, even those who find fault with accounting models of intellectual capital.

Several studies have demonstrated that social relations both inside and outside the firm have value (e.g., Dzinkowski, 2000; Fukuyama, 1995; Hazelton & Kennan, 2000; Huang, 1998; Jacques, 2000; Onyx & Bullen, 2000), and theories have been proposed for understanding the value of social capital in both economic and noneconomic terms. (Bontis, 2000; Day, 2002; Klamer, 2002; Teece 2000).

If we accept the idea of social capital as a resource for making organizations operate more effectively (whether making a profit or not), then there needs to be some method of managing it. Successful organizations and effective managers have, of course, always recognized the value of social connections and made attempts of more or less formal natures to use them (viz., Mintzberg, 1979). However, as organizations have come to rely less on physical assets and more on knowledge assets, a more rigorous means of identifying and valuing these assets becomes necessary (Bassi & VanBuren, 2000).

Many social capital theorists have noted the difficulties both in valuing social capital and in establishing quantitative connections between social capital expenditures and organizational outcomes. In some cases, the outcomes are expressed in terms of what does not occur rather than in gains. Fukuyama (1995), for example, identifies increased transaction costs (e.g., decline in trust, increased legal costs, and discrimination suits) that occur in the absence of social capital. Other researchers such as Hazelton and Kennan (2000) and Bourdieu (1985) point to the difficulty in identifying the value of social capital, the outcomes from it, and the connection to the outcomes.

That such social capital can, or should, be valued in economic terms is by no means universally accepted. Day (2002, p. 1079) takes a strongly Marxist critique of economic models of social capital and suggests that the notion of capital is inherently inadequate to capture value because "the social doesn't occur as capital" and exists in a larger context than capital can measure. Similarly, Klamer (2002) notes that economic capital alone fails to account for the value that social and cultural capital bring to the life of humans.

However, whether economic measures such as those used in traditional accounting are adopted or not, the notion of measurement is inherent in the idea of control. Most organizations have only a vague understanding of the nature and value (in any terms) of their social capital. Regardless of whether a monetary value can be applied, researchers have identified the measurement of social capital is an integral step in realizing the value of these assets (Bontis, 2000; Bassi & VanBuren, 2000). The next section examines a variety of classification and measurement schemes for intangible assets and their effectiveness in capturing social capital; it concludes with an examination of information technology (IT) systems that attempt to capture not only financial transactions but also the internal and external actor-networks that facilitate their occurrence.

Traditional Accounting Treatment of Intangible Assets

Within the context of professional standards in accounting (known as generally accepted accounting procedures or GAAP), social capital falls under the rubric of intangible assets. That is, social capital is a resource with no physical embodiment. Although GAAP admits of the existence of intangible assets, the recording (or recognition) of social capital within traditional financial statements (i.e., in monetary values) is problematic. Under GAAP in most countries, assets must be owned by a company and have a clearly recognizable historical cost in order for them to be recognized. Because social capital resides in relationships among employees and those outside the firm, it clearly fails to meet the first criterion. Similarly, because the relationships are internally developed and have no obvious means of being sold to test their market value (a test of "separability"), they also clearly fail to meet the second criterion (Bontis, 1999; Snyder & Pierce, 2002; Sveiby, 2000).

This is not to say that the value of social relationships has gone completely unvalued in traditional accounting systems. The concept of a "going concern" whose market value is significantly greater than the sum of its tangible assets is well recognized and is easily discernible in the high book-to-market values for companies such as Microsoft or Coca-Cola. In a market-based economy, differences between selling price and tangible assets are generally subsumed under the umbrella of goodwill, a term that acts as a catchall for differences between sales price and valued assets. Although goodwill has the advantage of being simple to calculate, it has been criticized for coming into existence only after a transaction, for significantly undervaluing many businesses, and for lumping together a wide range of assets that may require different management strategies.

In response to the shortcomings of goodwill, a number of measures such as Tobin's Q, market-to-book, and calculated intangible value have been devised that purport to measure the cumulative effect on intangible assets, such as social capital, in a single metric. The measures have the advantage of being standardized and applicable across many businesses and industries, and they do not necessarily rely on market transaction to supply a value. However, they also share the criticism with goodwill for lumping diverse assets together (Snyder & Pierce, 2002).

Intellectual Capital and Social Capital

Both the failure to recognize intangible assets and the use of single, undifferentiated accounts as a catchall for intangible assets became problematic as industries increasingly relied on intellectual assets in the creation of wealth (Brooking, 1996; Osborne, 1998; Teece, 2000). In response to the shortcomings, the concept of intellectual capital (IC) together with the means for classifying and measuring it, gained interest during the 1990s and continues to the present.

IC admits of a variety of definitions; among the more common are "intellectual material-knowledge, intellectual property, experience—that can be put to use to create wealth" (Stewart, 1997, p. xx), knowledge transformed into something of value for the company (Lynn, 2000), and the entire stock of knowledge-based equity that a firm possesses (Dzinkowski, 2000). What all the definitions imply is that IC can be both the end result of a knowledge transfer process and the knowledge itself. It is within the former part of the definition that social capital is more easily accommodated because it is more properly a means to obtain resources as a result of membership in social structures rather than the resources themselves (Hazelton & Kennan, 2000).

The advantage of placing social capital within the construct of IC is that it inherits both a recognized structure for identifying intangible resources and also methods for attaching value to these resources. IC systems do not necessarily include social capital as an explicit type of IC, but, to a large extent, they do accommodate the concept of value in social relationships and access to resources.

Classifying Social Capital Within IC Models

Several IC models have been proposed over the last decade. However, most conform to the schemes put forth by Stewart (1997), Sveiby (1997), Edvinnson and Malone (1997), Edvinsson and Sullivan (1996), Saint-Onge (1996), and Bontis (1999). In an effort to tie IC to actual strategic objectives, the models move beyond simple definitions to create an IC management framework (Larsen, Bukh, & Mouritsen, 1999; Snyder & Pierce, 2002). Allowing for differences in wording, the schemes divide IC into three inter-related categories of capital: human, customer (or external relations), and organizational (including internal relations).

Human capital is the accumulated value of competence training and knowledge that resides within an organization's members. Although social capital has a strong social component (Adler & Kwon's [2000] sociocentric element), a case can be made for including at least some part of human capital because it is impossible to conceive of a network or organization without people (Sveiby, 1997).

Customer capital is the value derived from connections outside the firm. This includes some of the structural dimension proposed by Nahapiet and Ghoshal (2000) and Hazelton and Kennan (2000) and particularly the relational dimension as it relates to trust and perceptions of a firm's reputation. Much has been written concerning the role of the external environment on organizational success (e.g., Dzinkowski, 2000; Huang, 1998), particularly reputation and perceptions of trust, and various tools such as market-conceived quality profiles have been developed to understand customers and their perceptions.

Organizational capital subsumes all the remaining measures of IC and includes internal networks and the knowledge embedded in the routines of a company.

Part of the difficulty in applying such existing schemes for valuing IC is inherent in the nature of social capital, which, as noted, has both an individual (egocentric) and a social (sociocentric) component (Adler & Kwon, 2000). In the former case, the capital resides within individuals (e.g., Sveiby's professional competence and Edvinssons's human capital) and leaves the organization if the individual leaves. In contrast, sociocentric capital resides within the organization (both internal and external structure in Sveiby's system and organizational capital in Edvinsson's system). As Sveiby (1997) has noted, however, this is not a problem unique to social capital, but rather a larger difficulty that results when systems attempt to account for the value of relationships among individuals.

Similarly, conventional IC schemes require internal and external relationships to be accounted for separately. Although this results in more information with concomitantly higher costs, it is probably inevitable given the differences between the two assets. Internal relations such as those used by a firm's managers to better accomplish their agendas (cf. Mintzberg, 1979) are very different from those with suppliers and customers and thus need to be identified and managed separately.

Measurement and Reporting Using IC

Practices for measuring and reporting IC range widely and to date the accounting profession has not produced standards that can be reliably applied to compare one organization with another. However, several measurement schemes have been developed that rely on the three-part IC model described earlier. The measurement schemes are characterized by the underlying theory that money is only one of many possible proxies for measuring human action.

Such schemes as Skandia Navigator (based on Edvinsson's work) and Intangible Assets Monitor (based on Sveiby's work) combine standard financial metrics such as profit with IC metrics (customer satisfaction, customers per employees). In addition, they also include nonmetric or nonfinancial reporting such as narratives concerning customer value, the nature and value of organizational culture, and relationship capitals between employees and customers (Larsen et al., 1999).

These nonmetric reports have come under considerable criticism for their "fuzzy" nature (e.g., Rutledge, 1997), and even among proponents there is agreement that difficulties exist in applying a standard set of reporting measures across different companies (Edvinsson & Malone, 1997; Sveiby, 1997). However, other researchers have observed that only through nonfinancial measures can the total value of the firm's assets be conveyed (Bontis, 2000; Sveiby, 2000). Moreover, the use of narrative is consistent with not only the production of value through collaboration, but also a more accurate representation of the process (Guthrie, Petty, & Johanson, 2001; Mouritsen, Larsen, & Bukh, 2001). As Roberts (1999, p. 14) observes,

"The combination of narration, collaboration and social construction also works together to produce accountability."

Capturing Relationship Assets in IT-Based Accounting Systems

In addition to the formal accounting practice of classifying and measuring the value of social relationships, various IT-based accounting systems have attempted to capture these relationships, albeit without necessarily placing a monetary value on them. Guthrie et al. (2001) observe that assets come in bundles or networks and cannot be separated without loss of value. At the same time, the proliferation of information technologies has allowed both the implementation of conceptually designed business models (Walden & Nerson, 1995) and the closer coupling of enterprise systems. (Bradley, 1997; Geertz & McCarthy, 2002; Hoque, 2000). Enterprise Resource Planning (ERP), for example, standardizes the actual interactions while also making the nature and extent of the relationships transparent (Roberts, 1999, p. 16).

McCarthy (1982) proposed the Resource-Event-Agent (REA) model, an accounting system expanded to collect both financial information and the social network that creates value; it was later expanded by Geertz and McCarthy (1999). The REA model decomposes the business processes of an enterprise value chain (or more accurately a value network [Stabell & Fjelstad, 1998]) into primitives called resources, events, and agents, all of which are subsequently linked through their relationship in an economic event. Unlike standard accounting systems, which capture information only after a transaction and only information of a monetary nature, the REA model incorporates all of the information necessary for an event to occur, including the individuals who create the event.

Conclusion

Our review is exploratory and the examples indicative rather than exhaustive. We have identified a number of models, ranging from high-level frameworks from Nahapiet and Ghoshal's (2000) tripartite model to more grounded models of structure and performance such as those provided by Teigland and her colleagues (Schenkel, Teigland, & Borgatti, 2003). We have also provided examples of different mechanisms for managing social capital with ICTs; these may be broadly categorized as infrastructure (the network and database applications that support collaborative work in organizations and institutions), and superstructure (the representations of such work in the form of descriptions, and maps, largely visual, that are manipulated by managers). We have also discussed some ways of measuring the elements of social capital, in the form of indexes (and indicators), grids, typologies, and ranking mechanisms that may support those who make decisions, including financial decisions. But we have not synthesized all of these into a grand narrative or model for a number of reasons. First, the corpus we have used to make our points is an emerging one and there are as yet few robust, longitudinal studies of technology and social capital. Wellman and Haythornthwaite (2002, p. 24), reflecting on the question "Does the Internet increase social capital?" observe that there "is no clear position." Quan-Haase, Wellman, Witte, and Hampton (2002, p. 320) in the same volume can identify "no single effect" that accounts for the "changing composition" of social capital. Resnick (2002, p. 10) also observes that "no comprehensive theory of how social capital grows or erodes has come to my attention and I have not yet been able to develop an adequate one;" he redresses this by scoping out a research agenda. We suggest, however, that a grand, general narrative may not emerge for the reasons that follow.

The empirical studies we have reviewed indicate that social capital formation is a local process, and a complex one, contingent on evolving social practices (Suchman, Blomberg, Orr, & Trigg, 1999). Under such circumstances, a deterministic view of what technology can achieve may disappoint. In addition, the technology associated with the formation of social capital is itself evolving, a set of components which designers continuously reconfigure to meet the demands of practice in the workplace (MacKenzie & Wajcman, 1999) where those affected strive to establish a working social order that accommodates the intervention (Haythornthwaite, 2002, provides an example of this in the online classroom). The result is a process that may be described more accurately as "managing ICTs with social capital" than the reverse. Drawing on empirical work described in this review, we have exposed a number of "myths" about management in the discourse of social capital in organizations. These, like other management narratives articulate an agenda for control; they are countered by the stories of resistance that populate the case studies we have discussed.

Management Myths About Social Capital and ICTs

The first may be called the "field of dreams" myth, often expressed as "if we build it they will come." The use of "we" is important. The case studies of intranet development we have discussed are stories of unintended consequences, some more deleterious than others. Many are premised on design principles that take little account of diversity of work practice and the multiple channels that characterize "getting work done." Social capital effects are more evident in "working around" the implementation, than working with it.

The second is the myth that common ground can be achieved easily with technical fixes such as ontologies, process models, shared document repositories, and so on. This is often articulated in terms of technology to support the provision of "boundary objects." These are described in a simplistic way that belies the original authors' description of entities that are a complex, and negotiated, element of infrastructure. To re-emphasize the contingent and open-ended nature of these entities, O'Day et al. (2001, p. 405) offer the term "objects-in-the-making." In several of the case studies that we reference, "boundary objects" may have unforeseen consequences—having a role more of "hot potato" than a shared dish at the feast. Like the previous example, boundary objects do not support social capital, but serve as focal points for the social capital that is needed to make them effective.

A third myth is that a convergence of interests will lead to rapid adoption of technologies conceived by managers; the inverse of this myth is articulated in many of the studies that we cover: the dark side of social capital, or the downside of knowledge management. In many accounts, cliques are seen simply as dissent, rather than as a natural part of the structural landscape that provides insight into the local workings of social capital. A range of motives leads individuals to participate in networks—we have discussed cases where participation in multiple networks (by means of weak ties) is a source of innovation and creativity. We also considered examples of industry sectors and company cases where visibility rather than mutuality is a driving force and individual association with high achievers enhances reputation. Although ample research evidence suggests that a rich ecology with diverse tie strengths is desirable, the management of such diversity is difficult; but a wise manager will work with such phenomena, not against them.

Endnotes

1. The five measures are: connectedness, graph-theoretic distance (the number of points that separate one entity from another—a measure that may be compared with those that characterize small-world theory [Buchanan, 2003; Milgram, 1967; Watts, 2003]), density, core-periphery structure, and coreness.
2. It may be noted that the term "boundary" (like its correlate "community") has been used so extensively that the complex entity described by the authors of the construct has been lost and the term has been subsumed into a catchall for any externalization of concepts or processes that are referenced by parties in an interaction, social or otherwise.
3. It should be noted that to homogenize at this level of aggregation may be counterproductive. If, as we have suggested, innovation and adaptation happen at the edges of groups, diversity is a prerequisite of adaptive viability in volatile environments. (See, for example, Lievrouw, 2003; Lievrouw & Farb, 2003.)
4. Weisband (n.d.) lists four types of awareness: 1) activity awareness, or knowing what actions are ongoing at any given moment; 2) availability awareness or knowing whether others can meet or take part in an activity; 3) process awareness that allows people to see where they fit at any give time and how the project is moving along; 4) perspective awareness that gives information (about beliefs and values, for example) that is helpful for making sense of actions.

References

Ackerman, M. (1994). Augmenting the organizational memory: A field study of Answer Garden. *Proceedings of the 1994 ACM Conference on Computer Supported Cooperative Work, CSCW 94*, 243–252.

Adams, N., Fraser, J., Macintosh, A., & McKay-Hubbard, A. (2002). Towards an ontology for electronic transaction services. *International Journal of Intelligent Systems in Accounting, Finance and Management, 11*(3), 173–181.

Adler, P., & Kwon, S.-W. (2000). Social capital: The good, the bad and the ugly. In E. Lesser (Ed.), *Knowledge and social capital* (pp. 89–115). Oxford, UK: Butterworth-Heinemann.

Adler, P. S., & Kwon, S.-W. (2002). Social capital: Prospects for a new concept. *Academy of Management Review, 27*(1), 17–40.

Almind, T., & Ingwersen, P. (1997). Informetric analysis on the World Wide Web: A methodological approach to "webometrics." *Journal of Documentation, 53*(4), 404–426.

Ananthaswamy, A. (2003, August 23). March of the motes. *New Scientist, 179*(2409), 26–31.

Arnold, B., & Kay, F. (2000). Social capital, violations of trust and the vulnerability of isolates: The social organization of law practice and professional self-regulation. In E. Lesser (Ed.), *Knowledge and social capital* (pp. 201–222). Oxford, UK: Butterworth-Heinemann.

Barab, S. (2003). Introduction to the special issue. *The Information Society, 19*(3), 197–201.

Barab, S., Kling, R., & Gray, J. (Eds). (2004). *Designing for virtual communities in the service of learning.* Cambridge, UK: Cambridge University Press.

Barabási, A.-L. (2002). *Linked: How everything is connected to everything else and what it means for business, science, and everyday life.* Cambridge MA: Perseus Publishing.

Bartel, C., & Saavedra, R. (2000). The collective construction of work group moods. *Administrative Science Quarterly, 45*(2), 197–231.

Bassi, L., & VanBuren, M. (2000). New measures for a new era. In D. Morey, M. Maybury, & B. Thuraisingham (Eds.), *Knowledge management* (pp. 355–374). Cambridge, MA: MIT Press.

Baxter, L. (2000). Bugged. In C. Prichard, R. Hill, M. Chumer, & H. Wilmott (Eds.), *Managing knowledge: Critical investigations of work and learning* (pp. 37–48). Houndmills, UK: MacMillan Press.

Becks, A., Reichling, T., & Wulf, V. (2002, May). *Expertise finding: Approaches to foster social capital.* Paper presented at the Workshop on Social Capital and IT, Amsterdam.

Bergquist, M., & Ljungberg, J. (2001). The power of gifts: Organizing social relationships in open source communities. *Information Systems Journal, 11*(4), 305–320.

Björneborn, L., & Ingwersen, P. (2001). Perspectives of webometrics. *Scientometrics, 50*(1), 65–82. Retrieved January 30, 2004, from http://www.kluweronline.com/oasis.htm/337310

Bontis, N. (1999). Intellectual capital: An exploratory study that develops measures and models. *Management Decision, 36*(2), 63–76.

Bontis, N. (2000). Managing organizational knowledge by diagnosing intellectual capital. In D. Morey, M. Maybury, & B. Thuraisingham (Eds.), *Knowledge management* (pp. 375–402). Cambridge, MA: MIT Press.

Borgman, C., & Furner, J. (2002). Scholarly communication and bibliometrics. *Annual Review of Information Science and Technology, 36*, 3–72.

Börner, K., & Chen, C. (Eds.). (2003). *Visual interfaces to digital libraries.* New York: Springer.

Börner, K., Chen, C., & Boyack, K. (2003). Visualizing knowledge domains. *Annual Review of Information Science and Technology, 37*, 179–255.

Bourdieu, P. (1985). The forms of capital. In J. Richardson (Ed.), *Handbook of theory and research for the sociology of education* (pp. 241–258). New York: Greenwood.

Bourdieu, P. (1988). *Homo academicus.* Cambridge, UK: Polity Press.

Bourdieu, P. (1990). *The logic of practice.* Stanford, CA: Stanford University Press.

Bowker, G. C., & Star, S. L. (2000). *Sorting things out: Classification and its consequences.* Cambridge MA: MIT Press.

Bradley, K. (1997). Intellectual capital and the new wealth of nations. *Business Strategy Review, 8*(1), 53–62.

Briggs, X. (1998). *Doing democracy up-close*: Culture, power and communication in community building. *Journal of Planning Education and Research, 18*, 1–13.

Brooking, A. (1996). *Intellectual capital: Core assets for the third millennium enterprise.* London: Thomson Business Press.

Brown, J. S., & Duguid, P. (1994). Borderline issues: Social and managerial aspects of design. *Human-Computer Interaction, 9*(1), 3–36.

Buchanan, M. (2003). *Small world.* London: Orion Books.

Burt, R. (1992). The social structure of competition. In N. Nohria & R. Eccles (Eds.), *Networks and organizations: Structure, form and action* (pp. 57–91). Cambridge, MA: Harvard Business School Press.

Burt, R. (2000a). The contingent value of social capital. In E. Lesser (Ed.), *Knowledge and social capital* (pp. 255–287). Oxford, UK: Butterworth-Heinemann.

Burt, R. (2000b). The network structure of social capital. In R. I. Sutton & B. M. Straw (Eds.), *Research in organizational behavior,* (Vol. 22, pp. 345–423). Greenwich, CT: JAI Press.

Castelfranchi, C., & Tan, Y.-H. (2001). *Trust and deception in virtual societies.* Dordrecht, The Netherlands: Kluwer Academic Publishers.

Castells, M. (2000). *The Internet galaxy.* Oxford, UK: Oxford University Press.

Cherniss, C., & Goleman, D. (n.d.). *Bringing emotional intelligence to the workplace.* Retrieved January 30, 2004, from http://www.eiconsortium.org

Cohen, S., & Fields, G. (2000). Social capital and capital gains in Silicon Valley. In E. Lesser (Ed.), *Knowledge and social capital* (pp. 179–200). Oxford, UK: Butterworth-Heinemann.

Collins, W. R., Miller, K. W., Spielman, B. J., & Wherry, P. (1994). How good is good enough?: An ethical analysis of software construction and use. *Communications of the ACM, 37*(1), 81–91.

Cool, C. (2001). The concept of situation in information science. *Annual Review of Information Science and Technology, 35*, 5–42.

Crane, D. 1969. Social structure in a group of scientists: A test of the "invisible college" hypothesis. *American Sociological Review, 34*, 335–352.

Cronin, B. (1984). *The citation process: The role and significance of citations in scientific communication.* London: Taylor Graham.

Cronin, B. (1995). *The scholar's courtesy: The role of acknowledgement in the primary communication process.* London: Taylor Graham.

Cronin, B. (2004). Bowling alone together: Academic writing as distributed cognition. *Journal of the American Society for Information Science and Technology, 55*(6), 557–560.

Cronin, B., & Shaw. D. (2001). Identity-creators and image-makers: Using citation analysis and thick description to put authors in their place. *Proceedings of the 8th International Conference on Scientometrics and Informetrics* Vol.1, 127–138.

Cronin, B., Shaw, D., & La Barre, K. (2004). Visible, less visible, and invisible work: Patterns of collaboration in twentieth century chemistry. *Journal of the American Society for Information Science & Technology, 55*(2), 160–168.

Cronin, B., Snyder, H., Rosenbaum, H., Martinson, A., & Callahan, E. (1998). Invoked on the Web. *Journal of the American Society for Information Science, 49*, 1319–1328.

Davenport, E., & Cronin, B. (2001). The citation network as a prototype for representing trust in virtual environments. In B. Cronin & H. B. Atkins (Eds.), *The web of knowledge: A Festschrift in honor of Eugene Garfield* (pp. 517–534). Medford, NJ: Information Today, Inc.

Davenport, E., Graham, M., Kennedy, J., & Taylor, K. (2003). Managing social capital as knowledge management: Some specification and representation issues. *Proceedings of* the 66th Annual Meeting of the American Society for Information Science and Technology, 101–108.

Davenport, E., & Hall, H. (2002). Communities of practice and organizational learning. *Annual Review of Information Science and Technology, 36*, 171–228.

Davenport, E., & Rosenbaum, H. (2000). A system for organizing situational knowledge in the workplace that is based on the shape of documents. *Proceedings of the Sixth International Society for Knowledge Organization Meeting*, 352–358.

Day, R. (2002). Social capital, value and measure: Antonio Negri's challenge to capitalism. *Journal of the American Society for Information Science and Technology, 53*(12), 1074–1082.

De Carolis, D. M. (2002). The role of social capital and organizational knowledge in enhancing entrepreneurial opportunities in high-technology environments. In C. W. Choo & N. Bontis (Eds.), *The strategic management of intellectual and organizational knowledge* (pp. 699–709). Oxford, UK: Oxford University Press.

Detlor, B. (2004). *Toward knowledge portals: From human issues to intelligent agents*. Dordrecht, The Netherlands: Kluwer Academic Press.

Dibben, M. R. (2000). *Exploring interpersonal trust in the entrepreneurial venture*. London: MacMillan Press.

Donath, J. (2002). A semantic approach to visualizing online conversations. *Communications of the ACM, 45*(4), 45–50.

Drew, P. (1995). Interaction sequences and anticipatory interactive planning. In E. Goody (Ed.), *Social intelligence and interaction* (pp. 111–138). Cambridge, UK: Cambridge University Press.

Druskat, V. U., & Wolff, S. B. (2001). Building the emotional intelligence of groups. *Harvard Business Review, 79*(3), 81–90.

Dzinkowski, R. (2000). Managing the brain trust. *CMA Management, 73*(8), 14–18.

Dumais, S. T. (2004). Latent semantic analysis. *Annual Review of Information Science and Technology, 38*, 189–230.

Economics focus: A question of trust. (2003, February 22). *The Economist*, 92.

Edelman, L., Bresnen, M., Newell, S., Scarbrough, H., & Swan, J. (2002, April). *The darker side of social capital*. Paper presented at the 3rd European Conference on Organizational Knowledge, Learning, and Capabilities, Athens laboratory of Business Administration, Athens, Greece. Retrieved January 30, 2004, from http://www.alba.edu.gr/OKLC2002/Proceedings/pdf_files/ ID469.pdf

Edvinsson, L., & Malone, M. (1997). *Intellectual capital: Realizing your company's true value by finding its hidden brain power*. New York: Harper Collins.

Edvinsson, L., & Sullivan, P. (1996). Developing a model for managing intellectual capital. *European Management Journal, 14*, 356–364.

Erickson, T., Halverson, C., Kellogg, W., Laff, M., & Wolf, T. (2002). Social translucence: Designing social infrastructures that make collective activity visible. *Communications of the ACM, 45*(4), 40–44.

Falcone, R., & Castelfranchi, C. (2001). Social trust: A cognitive approach. In C. Castelfranchi & Y.-H. Tan (Eds.), *Trust and deception in virtual societies* (pp. 55–90). Dordrecht, The Netherlands: Kluwer Academic Publishers.

Fukuyama, F. (1995*). Trust: The social virtues and the creation of prosperity*. New York: Free Press.

Furner, J. (2002). On recommending. *Journal of the American Society for Information Science, 53*(9), 747–763.

Garton, L., Haythornthwaite, C., & Wellman, B. (1999). Studying on-line social networks. In S. Jones (Ed.), *Doing Internet research: Critical issues and methods for examining the Net* (pp. 75–105). Thousand Oaks, CA: Sage.

Geertz, G., & McCarthy, W. (1999, July/August). An accounting object infrastructure for knowledge-based enterprise models: Expert opinion. *IEEE Intelligent Systems*, 1–6.

Geertz, G., & McCarthy, W. (2002). *Using object templates from the REA accounting model to engineer business processes and tasks.* Retrieved October 26, 2003, from http://www.msu.edu/~mccarth4

Gladwell, M. (2000). *The tipping point: How little things can make a big difference.* Boston: Little, Brown.

Granovetter, M. (1973). The strength of weak ties. *American Journal of Sociology, 78*, 1360–1380.

Gronlund, A. (Ed.). (2002). *Electronic government: Design, applications, and management.* London: Idea Group Publishing.

Grosser, K. (1991). Human networks in organizational information processing. *Annual Review of Information Science and Technology, 30*, 77–109.

Groth, K., & Bowers, J. (2001). On finding things out: Situating organizational knowledge in CSCW. *Proceedings of the Seventh European Conference on Computer Supported Cooperative Work, ECSCW 2001*, 279–298.

Grudin, J. (1994). Groupware and social dynamics: Eight challenges for developers. *Communications of the ACM, 37*(1), 92–105.

Grudin, J. (2002). Group dynamics and ubiquitous computing. *Communications of the ACM, 45* (12), 74–78.

Guthrie, J., Petty, R., & Johanson, U. (2001). Sunrise in the knowledge economy. *Accounting, Auditing & Accountability Journal, 14*(4), 365–382.

Hall, H. (2002, April). Sharing capability: The development of a framework to investigate knowledge sharing in distributed organizations. Paper presented at the 3rd European Conference on Organizational Knowledge, Learning, and Capabilities, Athens, Greece. Retrieved January 30, 2004, from http://www.alba.edu.gr/OKLC2002/Proceedings/pdf_files

Hara, N., & Kling. R. (2001). *Students' distress with Web-based distance education course* (SI Working Paper No. 00-01). Retrieved January 30, 2004, from http://www.slis.edu/csi/papers.html#00-01

Hardless, C., & Nulden, U. (1999). Visualizing learning activities to support tutors. *In Proceedings of CHI 99, Extended Abstracts, Student Posters.* New York: ACM, 312–313.

Harper, R. (1997). Gatherers of information: The mission process at the International Monetary Fund. *Proceedings of the Fifth European Conference on Computer Supported Cooperative Work*, 361–376.

Haythornthwaite, C. (1996). Social network analysis: An approach and technique for the study of information exchange. *Library & Information Science Research, 18*, 323–342.

Haythornthwaite, C. (2000). Online personal networks: Size, composition and media use among distance learners. *New Media and Society, 2*(2), 195–226.

Haythornthwaite, C. (2001). Tie strength and the impact of new media. *Proceedings of the 34th Hawaii International Conference on System Sciences.* Retrieved January 30, 2004, from http://alexia.lis.uiuc.edu/~haythorn/HICSS01_tiestrength.html

Haythornthwaite, C. (2002). Strong, weak and latent ties and the impact of new media. *The Information Society, 18*(5), 385–401.

Hazelton, V., & Kennan, W. (2000). Social capital: Reconceptualizing the bottom line. *Corporate Communications, 5*(2), 81–86.

Hoque, R. (2000). *XML for real Programmers.* San Francisco: Morgan Kaufmann

Huang, K.-T. (1998). Capitalizing on intellectual assets. *IBM Systems Journal, 37*, 578–583.

Huberman, B. (2001). *The laws of the Web: Patterns in the ecology of information.* Cambridge, MA: MIT Press.

Huysman, M. (2002a, May). *Design requirements for knowledge sharing tools: A need for social capital analysis.* Paper presented at the Workshop on Social Capital and IT, Amsterdam.

Huysman, M. (2002b). *Knowledge sharing in practice.* Dordrecht, The Netherlands: Kluwer Academic Press.

Huysman, M., & Wulf, V. (in press). *Social capital.* Dordrecht, The Netherlands: Kluwer Academic Press.

Iacono, C. S., & Weisband, S. (1997). Developing trust in virtual teams. *Proceedings of the 30th Hawaii International Conference on System Sciences, Virtual Communities Minitrack,* (CD-ROM).

Jacob, E., & Shaw, D. (1999). Sociocognitive perspectives on representation. *Annual Review of Information Science and Technology, 33,* 131–186.

Jacques, R. (2000). Theorising knowledge as work: The need for a "knowledge theory of value." In C. Prichard, R. Hill, M. Chumer, & H. Wilmott (Eds.), *Managing knowledge: Critical investigations of work and learning* (pp. 199–215). Houndmills, UK: MacMillan Press.

Jarvenpaa, S. L., Knoll, K., & Leidner, D. (1998). Is anybody out there? Antecedents of trust in global virtual teams. *Journal of Management Information Systems, 14*(4), 29–64.

Jarvenpaa, S. L., & Leidner, D. E. (1998). Communication and trust in global virtual teams. *Journal of Computer-Mediated Communication, 3*(4). Retrieved January 30, 2004, from http://www.ascusc.org/jcmc/vol3/issue4/jarvenpaa.html

Kirschner, P., Shum, S. J. B., & Carr, C. S. (Eds.). (2003). *Visualizing argumentation: Software tools for collaborative and educational sense-making.* London: Springer-Verlag.

Klamer, A. (2002). Accounting for social capital and cultural values. *De Economist, 150*(4), 453–470.

Kling, R., & Callahan, E. (2003). Electronic journals, the Internet, and scholarly communication. *Annual Review of Information Science and Technology, 37,* 127–177.

Kling, R., & Courtright, C. (2003). Group behavior and learning in electronic forums: A sociotechnical approach. *The Information Society, 19*(3), 221–235.

Kling, R., McKim, G., & King, A. (2001). *A bit more to IT: Scholarly communication forums as socio-technical interaction networks.* Retrieved February 24, 2004, from http://www.slis.indiana.edu/CSI/wp01-02.html

Kling, R., Spector, L., & McKim, G. (2002). Locally controlled scholarly publishing via the Internet: The guild model. Retrieved February 24, 2004, from http://www.slis.indiana.edu/CSI/WP/WP02-01B.html

Knorr Cetina, K. (1999). *Epistemic cultures: How the sciences make knowledge.* Cambridge, MA: Harvard University Press.

Koku, E., Nazer, N., & Wellman, B. (2001). Netting scholars: Online and offline. *American Behavioral Scientist, 44*(10), 1750–1772.

Koku, E., & Wellman, B. (2004). Scholarly networks in learning communities: The case of TechNet. In S. Barab, R. Kling, & J. Gray (Eds.), *Designing for online communities in the service of learning* (pp. 299–337). Cambridge, UK: Cambridge University Press.

Kollock, P. (1999). The economies of online cooperation: Gifts and public goods in cyberspace. In M. Smith and P. Kollock (Eds.), *Communities in cyberspace* (pp. 220–239). London: Routledge.

Koschman, T., & LeBaron, C. (2003). Reconsidering common ground: Examining Clark's contribution theory in the OR. *Proceedings of the Eighth European Conference on Computer Supported Cooperative Work, ECSCW 2003,* 81–98.

Krackhardt, D., Blythe, J., & McGrath, C. (1994). KrackPlot 3.0: An improved network drawing program. *Connections, 17*(2), 53–55.

Krackhardt, D., & Hanson, J. (1997). Informal networks: The company. In L. Prusak (Ed.), *Knowledge in oganizations* (pp. 37–49). Oxford, UK: Butterworth-Heinemann.

Kramer, R., & Tyler, T. (1996). *Trust in organizations: Frontiers of theory and research.* Thousand Oaks, CA: Sage.

Lamb, L., & Kling, R. (2002). From users to social actors: Reconceptualizing socially rich interaction through information and communication technology. Retrieved February 24, 2004, from http://www.slis.indiana.edu/CSI/WP/WP02-11B.html

Larsen, H., Bukh, N., & Mouritsen, J. (1999). Intellectual capital statements and knowledge management: "Measuring," "reporting," "acting." *The Australian Accounting Review, 9*(3), 15–26.

Larsen, R. J., & Diener, E. E. (1992). Promises and problems with the circumplex model of emotion. In M. S. Clark (Ed.), *Review of personality and social psychology: Emotion and social behavior* (Vol. 114, pp. 25–59). Newbury Park, CA: Sage.

Lave, J., & Wenger, E. (1990). *Situated learning: Legitimate peripheral participation.* New York: Cambridge University Press.

Lesser, E. (Ed.). (2000). *Knowledge and social capital.* Oxford, UK: Butterworth-Heinemann.

Lewicki, R. J., & Bunker, B. B. (1996). Developing and maintaining trust in work relationships. In R. Kramer & T. Tyler (Eds.), *Trust in organizations: Frontiers of theory and research* (pp. 114–139). Thousand Oaks, CA: Sage.

Lievrouw, L. (2003, September). *When users push back: Oppositional new media and community.* Paper presented at the Communities and Technology Conference, Amsterdam.

Lievrouw, L., & Farb, S. (2003). Information and equity. *Annual Review of Information Science and Technology, 37,* 499–540.

Loughran, J. (2000). *Working together virtually: The care and feeding of global virtual teams.* Retrieved January 30, 2004, from http://www.dodccrp.org/2000ICCRTS/cd/papers/Track4/009.pdf

Lynn, B. (2000, January/February). Intellectual capital: Unearthing hidden value by managing intellectual assets. *Ivey Business Journal, 79,* 45–52.

MacKenzie, D., & Wajcman, J. (Eds.). (1999). *The social shaping of technology* (2nd ed.). Philadelphia: Open University Press.

Malkia, M., Anttiroiko, A., & Savolainen, R. (2002). *eTransformations in governance: New directions in government and politics.* London: Idea Group Publishing.

Marsh, S. (1994). Trust in distributed artificial intelligence. *Artificial Social Systems: Proceedings 4th European Workshop on Modelling Autonomous Agents in a Multi-Agent World,* 95–112.

Marsh, S., & Dibben, M. (2003).The role of trust in information science and technology. *Annual Review of Information Science and Technology, 37,* 465–498.

McCarthy, W. E. (1982). The REA accounting model: A generalized framework for accounting systems in a shared data environment. *The Accounting Review, 57,* 554–578.

McDonald, D., & Ackerman, M. (2000). Expertise recommender: A flexible recommendation system and architecture. *Proceedings of the 2000 ACM Conference on Computer Supported Cooperative Work, CSCW 2000,* 231–240.

Merton, R. K. (1968). The Matthew effect in science. *Science, 159*(3810), 56–63.

Meyer, E., & Kling, R. (2002). Leveling the playing field or expanding the bleachers? Sociotechnical interaction networks at arXiv.org. Retrieved January 30, 2004, from http://www.slis.indiana.edu/CSI/WP/WP02-10B.html

Meyerson, D., Weick, K. E., & Kramer, R. M. (1996). Swift trust and temporary groups. In R. M. Kramer & T. R. Tyler (Eds.), *Trust in organizations: Frontiers of theory and research* (pp. 166–195). Thousand Oaks, CA: Sage.

Milgram, S. (1967). The small-world problem. *Psychology Today, 1,* 62–67.

Mintzberg, H. (1979). *The nature of managerial work.* New York: Prentice Hall.

Moody, G. (2002). *Rebel code: Linux and the open source revolution.* London: Penguin.

Mouritsen, J., Larsen, H., & Bukh, P. (2001). Valuing intellectual capital and the future: Intellectual capital supplements at Skandia. *Accounting, Auditing & Accountability Journal, 14*(4), 327–336.

Munro, A., Hook, C., & Benyon, D. (1999). *Designing information spaces: The social navigation approach.* London: Springer.

Nahapiet, J., & Ghoshal, S. (2000). Social capital, intellectual capital and the organizational advantage. In E. Lesser (Ed.), *Knowledge and social capital* (pp. 119–155). Oxford, UK: Butterworth-Heinemann.

Nardi, B., Whittaker, S., Isaacs, I., Creech, M., Johnson, J., & Hainsworth, J. (2002). Integrating communication and information through ContactMap. *Communications of the ACM, 45*(4), 89–95.

Newell, S., Scarbrough, H., Swan, J., & Hislop, D. (2000). Intranets and knowledge management: De-centered technologies and the limits of technological discourse. In C. Prichard, R. Hull, M. Chumer, & H. Willmott (Eds.), *Managing knowledge: Critical investigations of work and learning* (pp. 88–106). Basingstoke, UK: MacMillan Press.

Nissenbaum, H. (1994, January). Computing and accountability. *Communications of the ACM, 37*(1), 72–80.

O'Day, V., Adler, A., Kuchinsky, A., & Bouch, A. (2001). When worlds collide: Molecular biology as interdisciplinary collaboration. *Proceedings of the Seventh European Conference on Computer Supported Cooperative Work, ECSCW 2001,* 399–418.

Onyx, J., & Bullen, P. (2000). Measuring social capital in five communities. *The Journal of Applied Behavioral Science, 36*(1), 23–42.

Osborne, A. (1998). Measuring intellectual capital: The real wealth of companies. *Ohio CPA Journal, 57*(4), 37–38.

Pettigrew, K., Fidel, R., & Bruce, H. (2001). Conceptual frameworks in information behavior. *Annual Review of Information Science and Technology, 35,* 43–78.

Pipek, V., & Wulf, V. (2003). Pruning the Answer Garden: Knowledge sharing in maintenance engineering. *Proceedings of the Eighth European Conference on Computer Supported Cooperative Work, ECSCW 2003,* 231–240.

Portes, A. (2000). Social capital: Its origins and applications in modern sociology. In E. Lesser (Ed.), *Knowledge and social capital* (pp. 17–41). Oxford, UK: Butterworth-Heinemann.

Portes, A., & Landolt, P. (1996). Unsolved mysteries: The Tocqueville Files II: The downside of social capital. *The American Prospect, 7*(26). Retrieved January 30, 2004, from http://www.prospect.org/print/V7/26/26-cnt2.html

Preece, J. (2000). *Online communities: Designing usability, supporting sociability.* New York: Wiley.

Preece, J. (2002). Supporting community and building social capital. *Communications of the ACM, 45*(4), 36–39.

Prusak, L. (2001, September). Keynote address presented at the Seventh European Conference on Computer Supported Cooperative Work.

Putnam, R. (1993). *Making democracy work: Civic traditions in modern Italy.* Princeton, NJ: Princeton University Press.

Quan-Haase, A., Wellman, B., Witte, J. C., & Hampton, K. (2002). Capitalizing on the Net: Social contact, civic engagement and sense of community. In B. Wellman & C. Haythornthwaite (Eds.), *The Internet in everyday life* (pp. 291–324). Oxford, UK: Blackwell.

Rea, T. (2001). Engendering trust in electronic environments. In C. Castelfranchi & Y.-H. Tan (Eds.), *Trust and deception in virtual societies* (pp. 221–236). Dordrecht, The Netherlands: Kluwer Academic.

Renninger, K. A., & Shumar, W. (Eds.). (2002). *Building virtual communities: Learning and change in cyberspace.* Cambridge, UK: Cambridge University Press.

Resnick, P. (2002). Beyond bowling together: Sociotechnical capital. In J. Carroll (Ed.), *HCI in the new millennium* (pp. 247–272). New York: Addison-Wesley.

Resnick, P., & Varian, H. (1997). Recommender systems [Special issue]. *Communications of the ACM, 40*(3).

Riel, M. (1995). Cross-classroom collaboration in global learning circles. In S. L. Star (Ed.), *The cultures of computing* (pp. 219–242). Oxford, UK: Blackwell.

Roberts, H. (1999, May). *The control of intangibles in the knowledge-intensive firm.* Paper presented at the 22nd Annual Congress of the European Accounting Association. Retrieved February 20, 2004, from http://www.fek.su.se/home/bic/meritum/download/EAA99_Roberts.pdf

Rutledge, J. (1997, April 7). You're a fool if you buy into this. *Forbes ASAP Supplement, 159*(7), 42–46.

Saint-Onge, H. (1996). Tacit knowledge: The key to strategic alignment of intellectual capital. *Strategy & Leadership, 24*(2), 10–14.

Schenkel, A., Teigland, R., & Borgatti, S. (2003). Theorizing structural properties of communities of practice: A social network approach. In R. Teigland (Ed.), *Knowledge networking: Structure and performance in networks of practice* (pp. 280–303). Stockholm, Sweden: Institute of International Business.

Schiff, L., Van House, N., & Butler, M. (1997). Understanding complex information environments: A social analysis of watershed planning. *Proceedings of the ACM Digital Libraries Conference*, 161–186.

Schuler, D. (Ed.). (1994). Social computing [Special issue]. *Communications of the ACM, 37*(1).

Scott, J. (2000). *Social network analysis.* Thousand Oaks, CA: Sage.

Simone, C., & Sarini, M. (2001). Adaptability of classification schemes in cooperation. What does it mean? *Proceedings of the Seventh European Conference on Computer Supported Cooperative Work, ECSCW 2001*, 19–38.

Small, H. (1999). Visualizing science by citation mapping. *Journal of the American Society for Information Science, 50*, 799–813.

Snowden, D. (2000). The social ecology of knowledge management. In C. Despres & D. Chauvel (Eds.), *Knowledge horizons: The present and the promise of knowledge management* (pp. 237–265). Boston: Butterworth-Heinemann.

Snyder, H., & Pierce, J. (2002). Intellectual capital. *Annual Review of Information Science and Technology, 36*, 467–502.

Sonnenwald, D. H., Maglaughlin, K. L., & Whitton, M. C. (in press). Designing to support situational awareness across distances: An example from a scientific collaboratory. *Information Processing & Management.*

Sonnenwald, D. H., Whitton, M. C., & Maglaughlin, K. L. (2003). Evaluating a scientific collaboratory: Results of a controlled experiment. *ACM Transactions on Computer Human Interaction, 10*(2), 150–176.

Sosteric, M. (1999). Endowing mediocrity: Neoliberalism, information technology, and the decline of radical pedagogy. *Radical Pedagogy, 1*(1). Retrieved February 25, 2004, from http://radicalpedagogy.icaap.org/content/issue1_1/sosteric.html

Spinuzzi, C. (2003). *Tracing genres through organizations: A sociocultural approach to information design.* Cambridge, MA: MIT Press.

Spring, M., & Vathanophas, V. (2003). Peripheral social awareness information in collaborative work. *Journal of the American Society for Information Science and Technology, 54*(11), 1006–1014.

Stabell, C., & Fjelstad, O. (1998). Configuring value for competitive advantage: On chains, shops and networks. *Strategic Management Journal, 19*, 413–437.

Star, S. L., & Griesemer, J. R. (1989). Institutional ecology, "translations" and boundary objects: Amateurs and professionals in Berkeley's Museum of Vertebrate Zoology. *Studies of Social Science, 19*, 387–420.

Steinfield, C., Jang, C.-Y., & Pfass, B. (1999.). *Supporting virtual team collaboration: The TeamSCOPE system.* Retrieved October 22, 2002, from http://cscw.msu.edu/reports/scope.htm

Stewart, T. (1997). *Intellectual capital.* New York: Bantam Doubleday.

Suchman, L. (1993). Do categories have politics? The language/action perspective reconsidered. *Proceedings of the Third European Conference on Computer Supported Cooperative Work, ECSCW '93*, 1–14.

Suchman, L., Blomberg, J., Orr, J., & Trigg, R. (1999). Reconstructing technologies as social practice. *American Behavioral Scientist, 43*(3), 392–408.

Sveiby, K.-E. (1997). *The new organizational wealth: Managing and measuring knowledge-based assets.* San Francisco: Barrett-Kohler.

Sveiby, K.-E. (2000). *Sveiby knowledge management.* Retrieved October 10, 2003, from http://www.sveiby.com.au

Syrjänen, A., & Kuutti, K. (2002, May). *Trust, acceptance and alignment: The role of IT in redirecting a community.* Paper presented at the Workshop on "Social Capital and IT" Amsterdam.

Teece, D. (2000). *Managing intellectual capital: Organizational, strategic, and policy dimensions.* Oxford, UK: Oxford University Press.

Teigland, R. (2003). *Knowledge networking: Structure and performance in networks of practice.* Stockholm: Institute of International Business.

Teigland, R., & Wasko, M. (2003). Extending richness with reach: Participation and knowledge exchange in electronic networks of practice. In R. Teigland (Ed.), *Knowledge networking: Structure and performance in networks of practice* (pp. 350–362). Stockholm: Institute of International Business.

Trevino, L., Daft, R., & Lengel, R. (1990). Understanding managers' media choices: A symbolic interactionist perspective. In J. Fulk & C. W. Steinfield (Eds.), *Organizations and communication technology* (pp. 71–94). Newbury Park, CA: Sage.

Walden, K. & Nerson, J.-M. (1995). *Seamless object-oriented software architecture.* New York: Prentice Hall.

Wasko, M., & Teigland, R. (2003). The provision of online public goods: Examining social structures in a network of practice. In R. Teigland (Ed.), *Knowledge networking: Structure and performance in networks of practice* (pp. 306–318). Stockholm: Institute of International Business.

Watts, D. I. (2003). *Six degrees: The science of a connected age.* New York: Norton.

Weick, K., & Roberts, K. (1993). Collective mind in organizations: Heedful interrelating on flight decks. *Administrative Science Quarterly, 38*, 357–381.

Weigand, H., & van den Heuvel, W.-J. (2001). Trust in electronic commerce. In C. Castelfranchi & Y.-H. Tan (Eds.), *Trust and deception in virtual societies* (pp. 237–257). Dordrecht, The Netherlands: Kluwer Academic.

Weisband, S. (2002). Maintaining awareness in distributed team collaboration: Implications for leadership and performance. In P. Hinds & S. Kiesler (Eds.), *Distributed work* (pp. 311–333). Cambridge, MA: MIT Press.

Weisband, S. (n.d.). *Maintaining awareness in distant team collaboration.* Retrieved May 4, 2002, from http://misdb.bpa.arizona.edu/~lzhao/brownbag/suzie-abstract.html

Wellman, B. (1982). Studying personal communities. In P. Marsden & N. Lin (Eds.), *Social structure and network analysis* (pp. 61–80). Beverly Hills, CA: Sage.

Wellman, B., & Frank, K. (2001). Network capital in a multi-level world: Getting support in personal communities. In N. Lin, K. Cook, & R. Burt (Eds.), *Social capital: Theory and research* (pp. 233–273). Chicago: Aldine de Gruyter.

Wellman, B., Haase, A. Q., Witte, J., & Hampton, K. (2001). Does the Internet increase, decrease or supplement social capital? Social networks, participation and community commitment. *American Behavioral Scientist, 45*(3), 437–456.

Wellman, B., & Haythornthwaite, C. (2002). *The Internet in everyday life*. Oxford, UK: Blackwell.

Wenger, E. (1998). *Communities of practice: Learning, meaning, and identity*. New York: Cambridge University Press.

Wenger, E. (2000). Communities of practice. In C. Despres & D. Chauvel (Eds.), *Knowledge horizons: The present and the promise of knowledge management* (pp. 205–224). Boston: Butterworth-Heinemann.

White, H., Wellman, B., & Nazer, N. (2002). *Does citation reflect social structure? Longitudinal evidence from the "Globenet" interdisciplinary research group*. Retrieved January 30, 2004, from http://www.chass.utoronto.ca/~w/publications/index.html

Whyte A., & Macintosh A. (2001). Transparency and teledemocracy: Issues from an "e-consultation." *Journal of Information Science, 27*(4), 187–198.

Woolcock, M., & Narayan, D. (2000). Social capital: Implications for development theory, research and policy. *World Bank Research Observer, 15*(2), 225–249.

Ye, Y., & Kishida, K. (2003). Toward an understanding of the motivation of open source software developers. *Proceedings of 2003 International Conference on Software Engineering, CSE2003*, 419–429.

Zachmann Institute for Framework Advancement. (n.d.). ZIFA [home page]. Retrieved February 13, 2004, from http://www.zifa.com

CHAPTER 13

Labor in Information Systems

Julian Warner
The Queen's University of Belfast, UK

Introduction

Labor is a condition of human existence in the Judeo-Christian tradition. Once out of Eden, we are condemned to work:

> cursed *is* the ground for thy sake; in sorrow shalt thou eat *of* it all the days of thy life; Thorns also and thistles shall it bring forth to thee; and thou shalt eat the herb of the field; In the sweat of thy face shalt thou eat bread, till thou return unto the ground; for out of it was thou taken: for dust thou *art* and unto dust shalt thou return. (Genesis 3:17–19)

Labor is the punishment for false choice, and, having eaten of the tree of knowledge, we are compelled to choose further:

> The World was all before them, where to choose
> Their place of rest, and Providence their guide:
> They hand in hand, with wandering steps and slow,
> Through Eden took their solitary way. (Milton, 1674/2003, Paradise Lost, Book 12, lines 645–648)

Following Genesis, labor has often been conceived as physical rather than mental labor and regarded as imposed, but has less often been connected with choice.

Marx can be located in the Judeo-Christian tradition that regards labor as inescapable:

> The labour process ... is purposeful activity aimed at the production of use-values. It is an appropriation of what exists in nature for the requirements of man. It is the universal condition for the metabolic interaction (*Stoffwechsel*) between man and nature, the everlasting nature imposed condition of human existence, and it is therefore independent of every

form of that existence, or rather it is common to all forms of society in which human beings live. (Marx, 1976, p. 290)

Labor for Marx is, then, the activity by which men make their own history by modifying the natural environment and their social, cultural, and environmental inheritance. Physical control over the environment, rather than intellectual control over data or development of means of communication, is emphasized by Marx, in accord with the mid- to late-nineteenth century context. The concepts of individual and communal *mental* labor are seen as significant to obtaining *physical* control over the environment, particularly in the treatment of science and technology (Marx, 1973, 1976; Warner, 2002, 2004). Science and technology can reduce direct human labor in the transformation of the environment into useful goods and offer progressive liberation from toil.

Economics, partly in its early development as a branch of political economy, developed the labor theory of value in the late eighteenth century and has progressively departed from it since the 1870s. The set of *activities* understood as labor was similar to Marx's understanding, focusing on human physical work in the transformation of natural resources into manufactured products. The *role* given to labor was less extensive and not connected with humans making their own history. The labor theory of *value* was understood as the costs of the human labor involved in the manufacture of a product determining, or at least strongly influencing, the exchange value of that product. From the 1870s onward the labor theory of value was increasingly displaced by market considerations. From an encompassing historical perspective, the labor theory of value can be seen to develop concurrently with manufacture and industrialization in Britain and then to be eroded as the overall rate of production was enabling increased leisure, the formation of a world market, and the diffusion of message transmission technologies (Warner, 2004). Labor and market determinants and theories of value might then be less antithetical than modern economics has tended to assume.

Economics has given limited attention to labor for informational purposes. Information processes have been recognized as increasingly economically significant (Lamberton, 1984; Stiglitz, 2000). Attempts have also been made to assimilate information goods to established economic models (Kahin & Varian, 2000). One crucial difficulty lies in the altered relation between selling and the exchange- and use-value of a product. Classically, in selling a product, including labor, the use-value of that good is alienated and its exchange value obtained. In selling a copy of an information product, the use-value of the product (apart from the aspect which is associated with exclusive ownership) is retained while its exchange value is still realized. The concept of information has been highly significant to the study of markets, with models incorporating concepts reminiscent of classical information theory (Shannon, 1993, pp. 5–83).

Information science has given only limited consideration to concepts of labor and more strongly in its classic antecedents than its current practice. Charles Babbage, connected to modern information science through the gestalt of the computer (V. Rosenberg, 1974) and also obtaining a fugitive existence in economics (Schumpeter, 1961, p. 541), discussed both mental labor and copying technologies (Babbage, 1963, 1989). Zipf founded his study of the dynamics of language on the principle of least effort. Zipf's (1936) law is concerned with the influence of this principle on the statistical distribution of word forms in spoken and written language. Modern studies have addressed concepts of labor in the restricted sense of workforce requirements and also begun to give some attention to the human labor involved in the making of records for catalogs and databases (Hayes, 2000). The related, although seldom fully intersecting, field of communication studies has noted the absence of explicit considerations of labor combined with their simultaneous, pervasive, implicit presence and influence and has made some incomplete steps toward formalizing relevant concepts (Schiller, 1996).

Discussions of the information society (a significant but not necessarily fully mutually communicating context for information science [Brown, 1987; Mattelart, 1996; Webster, 2002]) have often counterposed capitalist and informational modes of development (Webster, 2002). Human labor and technology as embodied human labor have accordingly often been excluded from consideration. More recent discussions have developed the concept of informational labor (Dyer-Witheford, 1999), but without the further distinctions to be articulated here.

A coherent tradition of attention to labor in information systems in potentially relevant and contributing disciplines does not, then, exist to be reviewed. Nor is even a scattered set of considerations available to assemble. Rather, synthesis from implied concepts revealed in other considerations and patterns of activity must be attempted. The assumption articulated by Zipf (1936) of resistance to labor can be simultaneously carried forward but changed. Rather than being received as a universal aspect of human behavior, it is transformed into an empirically supported observation of a widespread preference for economy of labor, theoretically connectable with the high costs of direct human labor.

The combination of a relative absence of explicit consideration with an implicit and pervasive presence suggests the possibility of constructing a powerful analytical framework, by transforming the implicit into the explicit. Labor and the costs of labor, particularly the high costs of direct human labor, have greatly influenced patterns of activity central to information science, for instance, in the depth of humanly assigned description given to documents and records for databases and catalogs. A promise of robustness for the analysis can also be derived from the existence of analogous concerns with labor in ordinary discourse and in economics. The analysis should be relevant to fields other than information science, most obviously the cognate subject but separately developed discipline of information systems (Ellis, Allen, & Wilson, 1999).

The analysis must be approached progressively, with cumulative development of concepts. First, the idea of technology as a human construction will be introduced through studies of productive technology and extended to information technology, with information differentiated from productive technology. Then the different forms of labor embodied in information technologies and in systems enabled by information technology will be considered. Information technologies can be regarded as the embodied product of communal labor, the cooperation of humans working together, building on universal labor or the general intellect, historically accumulated human knowledge. A distinction between syntactic and semantic levels and processes held in ordinary and some scholarly discourses can be extended to syntactic and semantic labor, with information technologies regarded as capable of syntactic labor. The distinctions of universal from communal labor and of syntactic from semantic labor can then be made to yield insights into domains crucial to information science—into the information theory formalized by Shannon in 1948 (Shannon, 1993, pp. 5–83)—and information retrieval, with systems dealing with written language differentiated from those concerned with oral speech and images. Finally, the productivity of the concepts introduced will be reviewed.

Technology as a Human Construction

A view of technology as a radical human construction will be taken as the basis for subsequent discussion. Classically, this view was developed by Marx (N. Rosenberg, 1976, 1982, 1994) primarily, although not exclusively, with regard to industrial rather than information technologies:

> Nature builds no machines, no locomotives, railways, electric telegraphs, self-acting mules etc. These are products of human industry; natural material transformed into organs of the human will over nature, or of human participation in nature. They are *organs of the human brain, created by the human hand*; the power of knowledge, objectified. The development of fixed capital indicates to what degree general social knowledge has become a *direct force of production*, and to what degree, hence, the conditions of the process of social life itself have come under control of the general intellect and been transformed in accordance with it. (Marx, 1973, p. 706)

Control mechanisms ("self-acting mules") and message transmission technologies ("electric telegraphs") are mentioned in this passage, but they are not its primary focus. The idea of technology capable of performing autonomous labor as exclusively industrial technology would have been broadly true of Marx's historical period:

Only in large-scale industry has man succeeded in making the product of his past labour, labour which has already been objectified, perform gratuitous service on a large scale, like a force of nature. (Marx, 1976, p. 510)

Information technologies for message transmission were increasingly diffused from the mid-1860s, and these are acknowledged by Marx (Haye, 1980; Warner, 1999b) in a later passage that takes an inclusive view of communication:

the last fifty years have brought a revolution that is comparable only with the industrial revolution of the second half of the last century. On land the Macadamized road has been replaced by the railway, while at sea the slow and irregular sailing ship has been driven into the background by the rapid and regular steamer line; the whole earth has been girded by telegraph cables. (Marx, 1981, p. 164)

The industrial technologies of the nineteenth century, such as the steam-hammer "that can crush a man or pat an egg-shell" (Dickens, 1946, p. 150), would have contained control mechanisms for variation in force (in one instantiation of the steam-hammer, a hand-controlled steam valve, enabled the "workman ... [to] *think in blows*" [Nasmyth, 1885, p. 263]). Such mechanisms are not fully acknowledged in the classic concept of the simple machine (Minsky, 1967, p. 7). Primitive logic machines, such as Jevons' logic piano, were also developed in the late nineteenth century (Gardner, 1958).

More recently, the Marxian conception of technology as a radical human construction has been extended to information technologies, understood currently rather schematically as a form of knowledge concerned with the transformation of signals from one form or medium into another (Warner, 2004). From this perspective, the language—including the written language—used by Marx can be seen as a cumulative creation of the "general intellect." Congruently with the growth of message transmission technologies, the late nineteenth century also witnessed the diffusion of nonverbal and abbreviated forms of writing, in logical notations, telegraphic codes, and shorthand.

The extension of a concept describing industrial technologies to include information technologies implies a continuity from industrial to information societies, with both potentially subsumed under capitalism. Familiarly, within discussions of the information society, continuities are counterposed to disjunctions with industrial and capitalist eras (Webster, 2002). A perspective derived from Marx can again be both novel and informative in this context:

It is not what is made but how, and by what instruments of labor, that distinguishes different economic epochs. ... The writers of history have so far paid very little attention to the development of material production, which is the basis of all social life, and therefore of all real history. But prehistoric times at any rate have been classified on the basis of the investigations of natural science, rather than so-called historical research. Prehistory has been divided, according to the materials used to make tools and weapons, into the Stone Age, the Bronze Age and the Iron Age. (Marx, 1976, p. 286)

Developments in the instruments of informational labor must be acknowledged, with the computer as a universal information machine displacing calculation and, increasingly, writing by hand, as well as special purpose information machines. Yet an underlying and underpinning continuity also exists, strikingly revealed in the theoretical development of the computer from an account of mathematical operations as the writing, erasure, and substitution of symbols (Warner, 1994). It is questionable whether modern information technologies constitute a transformation in material production rather than a significant addition (Warner, 1999a). An understanding of information as a perspective rather than as a disjunction from pre-existing forms of social organization is, then, preferred here (Warner, 1999b).

Awakening of Dead Labor

Classically, living labor is required to reawaken the dead labor embodied in machinery and thereby to confer use- and exchange-value on inert stuff (Marx, 1976, p. 527; Warner, 2004). The fictional or mythic analogue to this process is supplied by Frankenstein giving life to his creation:

> With an anxiety that almost amounted to agony, I collected the instruments of life around me that I might infuse a spark of being into the lifeless thing that lay at my feet. It was already one in the morning; the rain pattered dismally against the panes, and my candle was nearly burnt out, when, by the glimmer of the half-extinguished light, I saw the dull yellow eye of the creature open; it breathed hard, and a convulsive motion agitated its limbs. (Shelley, 1998, pp. 38–39)

The awakening of dead physical or industrial labor by human action has analogies in the use of information technologies, specifically, to one interpretation of nondeterminism in automata theory, where human intervention moves the machine on from a halted state, corresponding to the awakening of dead physical or industrial labor.

Universal and Communal Labor

Marx separates universal from communal labor:

> We must distinguish here, incidentally, between universal labour and communal labour. They both play their part in the production process, and merge into one another, but they are each different as well. Universal labour is all scientific work, all discovery and invention. It is brought about partly by the cooperation of men now living, but partly also by building on earlier work. Communal labour, however, simply involves the direct cooperation of individuals. (Marx, 1981, p. 199)

Universal labor, understood as science, discovery, and invention, could be regarded as an aspect of the general intellect that transforms the process of social life. Communal labor is crucial to the awakening and use of universal labor, as embodied in both technologies and written texts. In the narrative of Frankenstein, universal labor would be represented by the learning used by Frankenstein and by the instruments of life, and communal labor, here mediated through a single individual, in the application of that learning and those instruments.

With regard to "building on earlier work," disciplines are understood to differ in the extent to which they are cumulative. Disciplines marked by the extensive use of syntactic operations, most obviously mathematics, are regarded as more strictly cumulative than the human sciences, and, even more the texts and artifacts studied in the human sciences (consider the reduction of Shannon's [1938] seminal work on analogies between Boolean logic and switching circuits to material for secondary education, over the subsequent fifty years).

Semantic and Syntactic Labor

A distinction between semantics and syntax is made in ordinary discourse in literate Western societies. Semantics would be concerned with the issues of meaning. Syntax, by contrast, would be concerned with the form of messages and usually understood to include the grammar of spoken and, particularly, written language. The resonance of Searle's (1980) critique of claims for the intelligence of computers, that syntax is not semantics, may derive from its ordinary discourse roots and also points to more formalized distinctions.

In semiotics, in particular, a four-level distinction has been constructed: from pragmatics or the intentions of the senders of messages and their effects on recipients, through semantics or issues of meaning, to syntactics or the form of statements (including formal logic), and to empirics or message transmission (including information theory in the Shannon sense). The distinctions have been brought to the study of information systems (Liebenau & Backhouse, 1990). In that context, the distinction of considerations of intention and meaning (pragmatics and semantics) from those of form and message transmission (syntactics and empirics) has proved sharper than the distinction between intention and meaning or between form and message transmission.

Issues connected with labor at each level of analysis have not received much attention but are still pervasive. They can be recovered and explicitly reconstructed from ordinary and from scholarly discourse, particularly from logic and discussions of mathematics.

The origins of the distinction between syntax and semantics in ordinary discourse can be traced to the transition from oral to oral-and-written verbal communication. In primarily oral communications, we receive not self-identical signs but variable and mutable signs (Vološinov, 1986). Primarily oral societies tend not to have concepts of grammar (and, self-evidently, not of orthographic correctness) and may differentiate good from bad speech by the effects it has on communal welfare. With the introduction of written language, a distinction between voice and speech develops (Aristotle, 1981, pp. 59–61). Speech, embodied in written language, can be detached from its producers and made an object for grammatical and logical study (Harris, 1989). Transformations can then be carried out on statements; and it is these transformations that we understand as demanding labor. That labor is both physical and material, involving direct human physical effort and the material technologies for writing; the labor can be syntactic, concerned only with the form of statements, or semantic, engaging with considerations of meaning, in character.

An idea of syntactic labor is embodied in ordinary discourse and experience, although it is not necessarily made fully explicit. For instance, in nineteenth-century legal practice in Britain and the U.S., scriveners would mechanically copy documents and compare (technically, collate) them for accuracy by listening to an oral reading of the original or primary source. The lawyer in the practice would be responsible for the semantic labor or understanding and interpretation of documents (Melville, 1997). Babbage, in the mid-19th century, gives a detailed analysis of devices for copying (Babbage, 1963, pp. 69–113; 1989; Hyman, 1982), although the devices, such as the stencil duplicator, which would displace hand copying, were not brought into wide use until the late nineteenth century (Day, 1996, p. 683; Ohlman, 1996). For 19th century scriveners, syntactic labor is intimately bound up with physical labor (consider the effort of copying documents by hand) and is performed directly by humans, assisted by the established technologies of writing. Direct human labor has high costs, even under 19th century capitalism, where wages might be limited to the reproduction cost of that labor (Marx, 1976). In the diffusion of copying devices in the late 19th century, we can see the beginnings of a dynamic where machine labor, which has lower direct costs, is substituted for direct human syntactic labor.

Semantic labor, then, is concerned with transformations motivated by the context, meaning, or, in semiotic terms, the signified of the message. Syntactic labor, by contrast, is concerned with transformations determined by the form, expression, or signifier of the message. The aim of both forms of labor may be the production of further messages, for instance, a description of the original message or a dialogic response. Syntactic labor is better understood than semantic labor and the primitive operations possible (the writing, erasure, and substitution of symbols) can be recovered from accounts in formal logic, mathematics, and automata theory. These discourses, particularly automata theory and its concept of nondeterminism, also contain the basis for a distinction of semantic from syntactic transformations.

The mid- to late 19th century was a crucial period for the emerging formalization of syntactic labor in formal logic. Boole (1854) demonstrated that the predominantly verbal form of the Aristotelian syllogism could be replaced by notational forms, with rule-governed transformations possible between expressions. Boole conceived his project as being concerned with the laws of thought and would not have explicitly differentiated syntactic from semantic transformations. A crucial restriction was that a symbol must retain an unaltered meaning (signifier inextricably linked to a single signified) during the course of a single argument. Semiotics itself, particularly in its North American manifestation in Charles Peirce (1991), developed in its modern form in the late 19th century. Mechanical devices for carrying out syntactic transformations, from Jevons' logic piano to the more widely diffused Hollerith tabulating machine, also proliferate in this period. Modern message transmission technologies, such as the telephone and telegraph, develop from the early to the mid-19th century and particularly intensively in the 1870s (Warner, 2004). Message transmission technologies precede their theoretical modeling in Shannon's 1948 information theory. Practical technologies and understandings often developed in advance of scientific theory, before the development of scientific research as a corporate enterprise in the late 19th century and the complexity of modern technology. Analogously, Boole's logic and Peirce's semiotics had a delayed impact on relevant academic discourses (Collins, 1998).

Following the convergence of mathematics and logic in the late 19th and early 20th centuries (Whitehead & Russell, 1962), discussions of mathematical logic clearly isolated the primitive operations of mathematics. These were the writing, erasure, and substitution of symbols (Ramsey, 1990) (there may be an implicit limitation to a discrete alphabet of symbols, analogous to the restriction of information theory, in its primary form, to discrete information sources [Shannon, 1993, pp. 5–83]). From the perspective here, the writing, erasure, and substitution of symbols would be regarded as the primitive operations possible on discrete messages and labor as the work expended in these operations. Syntactic labor occurs when transformations are determined by the form or signifier and not directly motivated by the meaning or signified.

The models of the computational process developed in the 1930s by Turing, for example, may well have been influenced by the preceding isolation of the primitive operations of mathematics (see Davis, 1965). Turing (1937, p. 231), whose model subsequently became dominant, began by comparing "a man in the process of computing a real number to a machine which is only capable of a finite number of conditions." The primitive operations of such a machine (the Turing machine) are the writing, erasure, and substitution of symbols, identical with those previously isolated in discussions of mathematics. The Turing machine is regarded as capable of imitating the operations of any information machine (with, again, a possible implicit limitation to discrete sources). The universal Turing machine, subsequently embodied in the computer, can imitate the actions of any given Turing machine. Recent evidence has revealed a greater continuity between Turing's conceptualization of the computer in 1936 and its subsequent invention, or demonstration of technical feasibility, in the early 1940s, particularly in dialogue between Turing and von Neumann (Davis, 2000, p. 192). In the derivation of the conceptualization of the computer from an account of writing or graphic inscription, there is a curious, although suggestive, analogy with Marx's (1976, p. 500n) remark, that, "it is not labour, but the instrument of labour, that serves as the starting point of the machine."

A distinction between syntactic and semantic transformations and the labor involved in those transformations can be opened up from within mathematical logic and, particularly, from automata theory. A critique of the reduction of mathematics to the writing, erasure, and substitution of symbols acknowledges that these are the primitive operations of mathematics but questions whether that is all a mathematician does; this turns attention to the meaning of symbols and their connection with analogous terms and concepts (for instance, of number) in ordinary discourse (Ramsey, 1978). The distinction between syntax and semantics would also be analogous to the classic mathematical distinction between form and interpretation.

A crucial concept in automata theory is that of nondeterminism. Nondeterminism is understood in a number of senses, but relevant here is the classic sense of a Turing machine or algorithm that reaches a configuration at which it halts and can only be moved on by choice from a human operator. Transformations in the deterministic periods between human intervention are conducted syntactically, with substitutions made on the basis of form of symbols (Turing, 1937, pp. 131–132). Human intervention may then be made on the basis of the meaning of symbols or semantically, including selecting between different, but legitimate, possibilities for writing, erasure, or substitution permitted by the syntax of the expression. Between the deterministic process and the nondeterministic intervention, a distinction between syntactic and semantic processes, and of labor, can be opened up.

Syntactic and semantic transformations are, then, formally similar, with a common set of primitive operations. For syntactic transformations, writing, erasure, and substitution of symbols are determined by the form of symbols alone (consider the current state and symbol scanned of a Turing machine determining the symbol written and next state). For semantic transformations, choice is motivated by the meaning of symbols. In some instances, semantic transformation may involve selection from syntactically legitimate possibilities. For instance, in the syntagmatic sequence, "the method by which mathematics arrives at its equations is the method of substitution" (Wittgenstein, 1981, § 6.24), other selections from the paradigm could legitimately replace *substitution* but might be less semantically informative or even semantically dissonant (everyday practice with spell checkers would confirm this). In historical practice, both syntactic and semantic transformations involved direct human labor, although labor was of different types and costs (clerical as contrasted with intellectual work). Because of the intensive development of modern information technologies from the late 19th century, syntactic transformations can be automatically executed, whereas semantic transformations continue to require direct human intervention. Syntactic transformations have lost of some of their material character and dispense with physical effort (contrast copying an electronic file with hand copying of a manuscript), and semantic labor has emerged more clearly as a separable category. Classically, but not necessarily successfully, computer science has been concerned with modeling semantic as syntactic processes, with far less explicit attention to the labor involved.

The automation of syntactic labor has substantial analogies with the replacement of physical labor by machine labor. For Marx (1976, p. 389), "with the help of machinery, human labour performs actions and creates things which without it would be absolutely impossible of accomplishment." Analogously, graphic inscription and not just oral utterance was crucial to the development of mathematics, and, for Russell, enabled the construction and consideration of regions of thought that would have otherwise been impossible to contemplate (Whitehead & Russell, 1962, p. 2). Modern information technologies—particularly, although not exclusively, the computer—have enhanced the possibilities of exactness classically associated with writing (Warner, 2001, pp. 33–46). The industrial machine was "a mechanism that, after being set in motion, performs with its tools the same operations as the worker formerly did with similar tools" (Marx, 1976, p. 495). With information technologies, there has been a greater change in the tools of labor, reflecting and embodying the partial dematerialization of syntactic processes, but a comparable continuity in the primitive operations remains possible. With mechanization of physical labor, the "number of tools that a machine can bring into play simultaneously is from the outset independent of the organic limitations that confine the tools of the handicraftsman" (Marx, 1976, p. 495). Similarly, a modern database or catalog enables a control of complexity over a larger amount of data than would be remotely possible for a single human mind, not enabled by modern technologies. Large-scale industry had to produce machines by means of machines (Marx, 1976, p. 506) and the construction of modern information technologies, for instance the design of logic gates, is similarly dependent on existing information technologies. The threat to direct human syntactic labor offered by modern information technologies has not necessarily been fully recognized within relevant disciplines, such as mathematics.

Semantic and syntactic processes are formally similar, both involving the writing, erasure, and substitution of symbols; it is these similarities that have enabled the modeling of semantic processes as syntactic transformations. Semantic and syntactic processes differ, not in their form, but in their motivation, with syntactic transformations determined by the form of symbols and semantic processes motivated by their meaning or signified. Accordingly, they also differ in the labor involved. An analogy is discernible with the critique of the view of mathematics as consisting solely of the erasure, writing, and substitution of symbols and the questioning of whether that is all a mathematician does (Ramsey, 1990, pp. 164–224): A mathematician's thought, for instance, selecting between synchronically legitimate alternatives, would correspond to semantic labor. To pursue the analogy with Marx's comment on the instrument of labor as the starting point for the machine, regarding syntactic and semantic transformations as identical would be to confuse the instrument of labor and its autonomous operations when embodied in a machine with the whole labor process, which includes elements of direct human intervention. The modeling of semantic processes as syntactic transformations rests on the formal similarities between the processes but has also exposed the differences in motivation and the difficulty of modeling semantic as syntactic processes (Searle, 1980; Warner, 2002).

Both syntactic and semantic labor are costly when performed directly by humans. The costs of that labor can be related to its production costs in education into literacy and in the acquisition of knowledge for particular semantic domains, as well as to the markets for educated human labor. In relation to the specific concerns of information science, the direct human labor required to describe documents for catalogs is known to be costly (to describe a document to the standard required for OCLC's World Cat is estimated to cost in the region of US$40); Hayes (2000, 2001) discusses the costs to cataloging while noting the lack of attention to these costs and to the monetary value of the resources created in both the literature of information science and for accounting purposes. Labor delegated to information technologies, by contrast, is relatively and increasingly less costly than direct human labor. For instance, the costs of automatically creating an index to a record would be minimal, once the information technologies for this (in both their hardware and software aspects) are formalized and robust (these technologies can be regarded as the products of communal labor working on accumulated universal labor or the general intellect).

The contrasting costs of direct human and delegated syntactic labor create the possibility of a dynamic similar to that between physical or material and industrial labor. Dead labor embodied in machinery is substituted for direct human labor, first industrial for physical labor and then, in modern practice, intellectual for human intellectual labor, both reducing the direct costs of the processes and enabling processing on a scale previously impossible. Syntactic processes—such as copying, creation of indexes where the meta-language of description is directly derived from the object language described, and message transmission—which in historical practice were directly performed by humans, can be increasingly delegated to information technologies. One approach, emerging in practice before being formalized in theory, in a number of areas, such as information retrieval and citation analysis for research assessment (Warner, 2000; 2003b), is to combine syntactic and semantic transformations, using information technologies to manipulate data but reserving human judgment for the interpretation of results.

Summary

Distinctions, then, have been created between syntactic and semantic labor, formalizing the distinctions from their ordinary discourse analogues and adding the idea of labor to that of transformations and levels of analysis. A powerful dynamic, continuous with the dynamic of the substitution of dead for living labor under capitalism, has been detected, but in relation to intellectual and not physical labor. The dynamic may have predictive value as well as current and retrospective application, anticipating patterns in future developments.

Information Theory

Information theory was influential in the early development of information science (Bar-Hillel, 1964; Brown, 1987; Roberts, 1976, pp. 1–27; V. Rosenberg, 1974; Shannon, 1993, pp. 5–83; Weaver, 1949; Wiener, 1954), providing models for communication that were also adapted to the understanding of information retrieval. Its early promise as a metaphysic for the field of information science was not fulfilled, but there are current indications of a more informed revival and a subtler recognition of its continuing relevance to communication (Cornelius, 2002; Warner, 2003a).

Concepts of labor are both implicit and, to some extent, explicit in information theory. Historically accumulated intellectual labor is embodied in the coding systems (telegraph codes, systems of shorthand, Morse code, and alphabetic written language itself) that preceded Shannon's formalization of information theory. These coding systems can be described in terms of information theory and, both historically and biographically (Horgan, 1990; Warner, 2003a), may have impelled its formalization. Information theory is primarily adapted to discrete rather than continuous information sources, for instance, to written language rather than oral speech. In this sense, it deals with the congealed products of communication rather than with forms of communication where process and product are inseparable. There is a continuity with Zipf (1936) in the statistical perspective on communication and, specifically, in the understanding of a word as "a cohesive group of letters with strong internal statistical influences" (Shannon, 1993, pp. 197–198). Ideas of labor emerge with regard to selection of messages from the information source, in work done in encoding and decoding, and in the search for economy in the use of the transmission channel. Dialectical relations between these specific forms of labor are also discernible.

Labor in selection (*selection labor*) is implied in the choice of messages from the source to accord with combinative constraints of the message sequences. Selection labor can include both physical and intellectual components. For instance, the messages for selection could be the individual letters of the Roman alphabet and the combinative constraints those of the English language lexicon. If the number and variety of the messages for selection are increased to accord with the anticipated combinative constraints (for instance, with printers' ligatures), recurrent labor in selection is reduced, but additional intellectual labor is expended in learning how to choose between the increased set of messages (the additional labor expended in learning could be regarded as nonrecurrent capital cost [Warner, 2003a]). Coding practices existing before the formalization of information theory can, then, embody both descriptive understandings of the principles of information theory and a preference for economy in the total labor to be expended (Cherry, 1978).

Encoding labor by the transmitter is more explicitly acknowledged rather than merely implied within information theory, although as delays or time consumed rather than directly as labor expended (Shannon, 1993, pp. 5–83; Verdú & McLaughlin, 2000). Labor in encoding is understood as the work done on the message to produce the signal and would often be performed with the aim of reducing demands on channel capacity (corresponding to a preference for economy in the use of the channel as a product of labor). A common strategy formalized within information theory would be to reduce the redundancy of the message when transforming the message into the signal. The reduced redundancy in the signal can render it more vulnerable to corruption by noise in a manner that complicates reconstruction of the message from the signal by the receiver. This strategy would be exemplified by systems of shorthand and by Morse code: Systems of shorthand enable operations on messages, for instance, sequences from the English lexicon, to transform them into reduced signal sequences by replacing redundant characters by a single symbol; Morse code uses short signal sequences for frequently occurring characters in the message. Other post-Shannon coding systems, such as those for the compression of text files, use more deliberately theoretically informed techniques to reduce redundancy in the signal (Verdú & McLaughlin, 2000). The receiver transforms the received signal into the message and the amount of labor expended in this operation tends to be directly rather than inversely correlated with encoding labor (the production of the message sequences produced by decoding operations may require further physical or material labor). The receiver then passes the message to the destination and any distortions produced by uneliminated noise can complicate interpretation by the destination. Historically, for instance, in mid- and late 19th century practice, encoding and decoding were conducted by direct human labor (for instance, by a human telegrapher), but, in modern practice, are likely to be delegated to information technologies.

Redundancy has various effects on the amounts of labor required at different points in the process of communication modeled in information theory. Redundancy in message sequences increases physical and material labor in selection and composition but may reduce intellectual labor in selection (contrast the physical or material with the intellectual labor required to complete the message sequence, *afterw*, in the English language lexicon, once combinative constraints have been mastered). Redundancy of the signal transmitted counteracts noise in the channel but uses channel capacity (channel capacity is either the product of labor or, in historical instances, involves direct human labor). Redundancy in the signal received may reduce labor in decoding at the receiver and in interpretation by the destination.

Concepts of labor, understood primarily as direct human labor, are more explicit and developed in the related field of cryptography than information theory, possibly due to the immediate experience of constructing and deciphering systems. For a cryptanalyst, a secrecy system is strongly analogous to a noisy communication system and the cryptogram to the distorted signal (Shannon, 1993, p.113). Redundancy of the signal could assist the transformation of the signal into the message by the receiver, and, similarly, redundancy in the original messages enciphered makes a solution possible (Shannon, 1993, p. 117). Labor expended in encoding is broadly correlated with labor expended in decoding by intended receivers and in deciphering by interceptors. The balance between labor in encoding and decoding would also be implicitly understood in ordinary discourse references to the complexity of coding systems. Specifically, within cryptography, a unicity point is distinguished, after which there will usually be a unique solution. Data beyond the unicity point can reduce labor in deciphering but additional data may not reduce labor further. At a trans- rather than individual system level, a dialectic over time can be detected between the introduction of new systems resistant to known methods of solution and the development of cryptanalytic techniques for deciphering such systems (Shannon, 1993, p. 132). From the distinctions established here, this process can be seen as the transformation of communal into universal labor. For individual systems, it is recognized that perfect secrecy is possible, for instance, where the number of possible messages is small, but that the key must then be equivalent in amount (and, by implication, in labor expended on agreeing and transmitting the key) to the messages for selection (Shannon, 1993, p. 111); Warner (2003a) gives an historical example of equivalence between labor expended in the key and the message, with only two messages for selection.

Complex but still comprehensible patterns for the distribution of labor can, then, be discovered in information theory. A principal aim of information theory was to enable economy in the use of a channel for transmitting signals, corresponding to a preference for economy in use of the products of labor. Achieving economy in transmission tends to involve delays or labor expended in encoding by the transmitter and decoding by the receiver. If redundancy in the signal is greatly reduced with the aim of economy in transmission, reconstruction of the message by the receiver and interpretation by the destination may be complicated by the effects of noise. The distribution of labor between the components in the communication process, between selection, encoding, transmission, decoding, and interpretation, can be expected to reflect the costs of direct human labor and of the technologies in which accumulated labor is embodied. Information theory itself was developed by individuals working in communal contexts, building on the theoretical (preceding analogues to information theory [Cherry, 1978]) and practical (working coding systems) products of historically accumulated labor. With publication and the transition from the relatively closed context of wartime cryptography, information theory itself becomes part of universal labor, although with delayed diffusion to public consciousness and the design of coding systems.

Information Retrieval Systems

I wish, in this context, to confine attention to systems predominantly concerned with written language. Oral and nonverbal forms of graphic communication, which have undergone less clearly marked historically accumulated forms of coding, present different issues for retrieval system design. Most obviously, they do not necessarily offer readily distinguishable syntactic units with potential semantic significance.

Two antithetical, if not always clearly distinguished, traditions can be detected in information retrieval system design and evaluation. The idea of query transformation, understood as the automatic transformation of a query into a set of relevant records, has been dominant in information retrieval theory (the literature of classic information retrieval research is extensive, and reference to reviews of the literature is made in the interest of economy [Blair, 2003; Ellis 1996; Harter & Hert, 1997]). A contrasting principle of selection power has been valued in ordinary discourse, librarianship, and, to some extent, in practical system design and use (Ellis, 1984; Wilson, 1996). Philosophical antecedents to the idea of selection power can also be found (Warner, 2000) (the derivation of the term, *intelligence*, from *inter-legere*, or to choose between, would be relevant). The debate between query transformation and selection power may not be resolvable within either paradigm, but in this context, I wish to treat selection power as the founding principle for system design, evaluation, and use.

Selection power may be the design principle, but *selection labor* could be regarded as the primary concept from which selection power is derived. Let us assume that a certain quantity of selection labor associated with the number and variety of objects for selection is distributed between system producer and searcher, with the possibility for variation of the distribution between the producer and searcher.

Selection power is valued by searchers as it reduces their selection labor (and an exhaustive serial search may not be a practical possibility). *Description labor* by the system producer tends to aim to increase the selection power of the searchers and reduce their selection labor (description labor is understood to include cataloging, or document description, and classification, or subject categorization, incidentally revealing the congruence between their aims). The semantic and syntactic intellectual labor embodied in objects or documents described is here treated as a given. The description labor of the system producer can contain elements of syntactic labor, for instance, transcription or algorithmic transformation of the object-language of documents described into the meta-language of index representations, and of semantic labor, for instance, the application of thesaural terms derived from a controlled vocabulary or of cataloging codes to the description of documents. In the 19th century, both syntactic and semantic labor might have involved continuous human intervention (consider the creation of *Palmer's Index to The Times* and the primarily syntactic labor of transcribing newspaper headlines as index entries); in modern Western practice, syntactic labor is delegated to humanly constructed technologies, and, accordingly, human intellectual labor becomes almost exclusively semantic.

Universal labor is understood as information technologies in both their hardware and software aspects, and communal labor as the awakening or use of those technologies, including semantic record description.

A diagram may clarify the application of the distinctions between semantic and syntactic and communal and universal labor to information retrieval systems (see Figure 13.1). The classification of systems from highly to loosely structured is tautological in that it is derived from the objects described and the framework of description, but may still be informative.

The *Financial Times,* in its various searchable manifestations, provides a peculiarly pure example of the distinction between syntactic and semantic labor. It is available as a Web resource without payment at the point of use, with largely syntactically generated search facilities that operate on identifiable units of the source. It is also available with additional description, generated from human semantic labor (which could be syntactically assisted), from a number of vendors. For instance, the file available on Dialog labels articles by geopolitical region and product/industry names, including NAICS (North American Industry Classification System) code. Direct payment at point of use is made for

Degree of structure	Example systems	Syntactic labor	Semantic labor	Universal intellectual labor	Communal intellectual labor for system construction and maintenance	Selection work of searcher
High	OCLC World Cat; *Financial Times* with additional indexing	Delegated to technology in all cases	Human semantic labor (different degrees of intensity)	Information technologies	Record creation (standards for cataloging and subject description)	Reduced
Medium	Open Directory Project; Yahoo! Web site directory			Algorithms for automatic (syntactic) processing of written language	Description of resources; creation and maintenance of directory structure	Intermediate
Loose	Google; Newspaper Web site; Palmer's Index to the Times		Human semantic labor not applied		Automatically generated descriptions	Intensive

Figure 13.1 **Forms of labor in information system construction and searching**

the resources that embody additional semantic labor. The continuity of such sources is market testimony to a readiness to pay for additional selection power (and further evidence for the congruence of the concept of selection power with ordinary discourse understandings and everyday practice). Provision of both types of resource involves similar access to the universal labor embodied in information technologies and comparable communal labor to reinvigorate those technologies.

The costs of human labor in description can be more specifically considered. For instance, Duval, Hodgins, Sutton, and Weibel (2002) briefly discuss the costs of human creation of metadata for Internet resources. The labor in description may contain syntactic elements, for instance, in transcription, but will be predominantly semantic. Costs of syntactic labor, by contrast, in storage, manipulation, and transmission of records have diminished historically and continue to diminish as communal human labor is transformed into universal labor. Labor invested in record description increases the selection power and reduces the selection labor of the searcher.

Returning to the overall schema embodied in the diagram, we can see that producers of information systems, from highly to loosely structured, have comparable access to universal intellectual labor and to its products, embodied in the language they use and, specifically, in the information technologies available. Comparable, although contrasting, levels of communal labor would be required for system design and maintenance. Strikingly different levels of direct human labor are given to document description: For records in library and union catalogs, intense semantic labor is required (the intensity of which could be related to the exactness required); for Internet directories, selection and description of resources, although to less exacting standards; for Internet search engines, very little, if any, additional semantic labor. The communal labor invested in the description of resources reduces the selection labor of the searcher (with both forms of labor reflecting the high costs of direct human employment).

The model can be validated, from macro- to microlevels. At a macrolevel, syntactically based systems proliferate (consider the variety of Internet search engines), whereas semantically enriched systems, such as World Cat, may occupy unique market positions. Simultaneously, the search facilities of syntactically based and semantically enriched systems, products of universal labor, are converging in appearance and power. At an intermediate level, the function of library cooperatives has changed over time, moving along the horizontal axis of the diagram, from adapting universal labor to a concern with sharing the descriptive labor of cataloging (from awakening Frankenstein's monster to distributing its limbs). At a more microlevel, the relative costs of communal and universal labor, considered in relation to market demand, form the decision framework for the conversion of historical resources from paper to electronic form (including *Palmer's Index to The Times* [Chadwyck-Healey, 1998]). For information retrieval systems, the communal labor invested in description at production reduces the labor required at use; proposals for coding in the semantic Web could be understood as part of this dialectic (Berners-Lee, Hendler, & Lassila, 2001). The distribution of direct human labor between producer and searcher may depend on the nature of the market for the product.

Information retrieval systems, then, can be seen to exhibit the fundamental dynamic of capitalism, the substitution of dead for living labor, although semiotic rather than physical labor. The specific and already known dynamic of bibliography between order and chaos is accentuated (Grogan, 1987; Roberts, 1997). Chaos is further enabled by the reduced costs of making information public. Possibilities for order are enhanced by the availability of delegated syntactic labor (although the limitations of such labor are becoming painfully known [Blair, 2003; Warner, 1992, p. 231]). The resources giving control themselves contribute to overall disorder (consider Search Engine Watch, at http://www.searchengine watch.com in relation to Theodore Besterman's [1940] *A World Bibliography of Bibliographies* and classic concerns with bibliographic proliferation).

Conclusion

Attention to labor in information systems has, then, made some powerful forces and implicit concepts more explicit. An established distinction between universal and communal labor has been adapted to information technology and systems. The concept of labor has been added to more familiar distinctions between syntactic and semantic levels of analysis and processes. These distinctions have been used to clarify issues and patterns of activity in information theory and information retrieval. Application to other domains within information science would be possible. A dynamic involving the substitution of dead for living human labor, continuous with the dynamic of capitalism, has been detected. The dynamic may have predictive value as well as explanatory power and could be used to inform information policy decisions. The social and technological aspects of information science, often divided from each other, have been brought together and the divide between the social and the technical partly dissolved. Technology has been humanized, explicitly recognized as a human construction, and the human user of technology also humanized, with full recognition given to human judgment and choice.

References

Aristotle. (1981). *The politics* (T. A. Sinclair, Trans.; T. J. Saunders, Ed.). Harmondsworth, UK: Penguin Books.

Babbage, C. (1963). *On the economy of machinery and manufactures* (4th ed.). New York: A. M. Kelley. (Original work published in 1835)

Babbage, C. (1989). *Science and reform: Selected works of Charles Babbage* (A. Hyman, Ed.). Cambridge, UK: Cambridge University Press.

Bar-Hillel, Y. (1964). *Language and information: Selected essays on their theory and application*. Reading, MA: Addison-Wesley.

Berners-Lee, T., Hendler, J., & Lassila, O. (2001). The semantic Web. *Scientific American*. 279, 5. Retrieved January 21, 2004, from http://www.sciam.com/article.cfm?articleID= 00048144-10D2-1C70-84A9809EC588EF21&catID=2

Besterman, T. (1940). *A world bibliography of bibliographies*. Printed for the author at Oxford, UK: Oxford University Press.

Blair, D. C. (2003). Information retrieval and the philosophy of language. *Annual Review of Information Science and Technology*, 37, 3–50.

Boole, G. (1854). *An investigation of the laws of thought, on which are founded the mathematical theories of logic and probabilities*. London: Walton and Maberly.

Brown, A. D. (1987). *Towards a theoretical information science: Information science and the concept of a paradigm*. Sheffield, UK: Department of Information Studies, University of Sheffield.

Chadwyck-Healey. (1998). *Palmer's full text online 1785–1870*. Cambridge, UK: Chadwyck-Healey.

Cherry, C. (1978). *On human communication: A review, a survey, and a criticism* (3rd ed.). Cambridge, MA: MIT Press.

Collins, R. (1998). *The sociology of philosophies: A global theory of intellectual change*. Cambridge, MA: Harvard University Press.

Cornelius, I. (2002). Theorizing information for information science. *Annual Review of Information Science and Technology, 36,* 393–425.

Davis, M. (1965). *The undecidable: Basic papers on undecidable propositions, unsolvable problems and computable functions.* Hewlett, NY: Raven Press.

Davis, M. (1988). Mathematical logic and the origin of modern computing. In R. Herken (Ed.), *The universal Turing machine: A half-century survey* (pp. 149–174). Oxford, UK: Oxford University Press.

Davis, M. (2000). *The universal computer: The road from Leibniz to Turing.* New York: W. W. Norton.

Day, L. (1996). Language, writing, and graphic arts. In I. McNeil (Ed.), *An encyclopaedia of the history of technology* (pp. 665–685). London: Routledge.

Dickens, C. (1946). *Great expectations.* London: Oxford University Press.

Duval, E., Hodgins, W., Sutton, S., & Weibel, S. L. (2002). Metadata principles and practicalities. *D-Lib Magazine, 8,* 4. Retrieved January 21, 2004, from http://www.dlib.org/dlib/april02/weibel/04weibel.html

Dyer-Witheford, N. (1999). *Cyber-Marx: Cycles and circuits of struggle in high-technology capitalism.* Urbana, IL: University of Illinois Press.

Ellis, D. (1984). Theory and explanation in information retrieval research. *Journal of Information Science, 8,* 25–38.

Ellis, D. (1996). *Progress and problems in information retrieval.* London: Library Association Publishing.

Ellis, D., Allen, D., & Wilson, T. (1999). Information science and information systems: Conjunct subjects disjunct disciplines. *Journal of the American Society for Information Science, 50,* 1095–1107.

Gardner, M. (1958). *Logic machines and diagrams.* Brighton, UK: Harvester.

Grogan, D. (1987). *Grogan's case studies in reference work: 3: Bibliographies of books.* London: Clive Bingley.

Harris, R. (1989). How does writing restructure thought? *Language and Communication, 9,* 99–106.

Harter, S. P., & Hert, C. A. (1997). Evaluation of information retrieval systems: Approaches, issues, and methods. *Annual Review of Information Science and Technology, 32,* 3–94.

Haye, Y. de la (1980). *Marx & Engels on the means of communication (the movement of commodities, people, information & capital): A selection of texts.* New York: International General.

Hayes, R. M. (2000). Assessing the value of a database company. In B. Cronin & H. B. Atkins (Eds.), *The web of knowledge: A Festschrift in honor of Eugene Garfield* (pp. 73–84). Medford, NJ: Information Today, Inc.

Hayes, R. M. (2001). *Models for library management, decision-making, and planning.* San Diego, CA: Academic Press.

Horgan, J. (1990). Profile: Claude E. Shannon: Unicyclist, juggler and father of information theory. *Scientific American, 262*(1), 16–17.

Hyman, A. (1982). *Pioneer of the computer.* Oxford, UK: Oxford University Press.

Kahin, B., & Varian, H. R. (Eds.). (2000). *Internet publishing and beyond: The economics of digital information and intellectual property.* Cambridge, MA: MIT Press.

Lamberton, D. M. (1984). The economics of information and organization. *Annual Review of Information Science and Technology, 19,* 3–30.

Liebenau, J., & Backhouse, J. (1990). *Understanding information: An introduction.* Houndsmill, UK: Macmillan Education.

Marx, K. (1973). *Grundrisse: Foundations of the critique of political economy (rough draft).* (M. Nicolaus, Trans.). London: Penguin Books.

Marx, K. (1976). *Capital: A critique of political economy* (Vol. 1) (B. Fowkes, Trans.). Harmondsworth, UK: Penguin Books.

Marx, K. (1981). *Capital: A critique of political economy* (Vol. 3) (D. Fernbach, Trans.). Harmondsworth, UK: Penguin Books.

Mattelart, A. (1996). *The invention of communication* (S. Emanuel, Trans.). Minneapolis: University of Minnesota Press.

Melville, H. (1997). Bartleby: A tale of Wall Street. In H. Melville, *The complete shorter fiction* (pp.18–51). New York: Alfred A. Knopf.

Milton, J. (2003). *The major works* (S. Orgel & J. Goldberg, Eds.). Oxford, UK: Oxford University Press. (Original work published 1674)

Minsky, M. L. (1967). *Computation: Finite and infinite machines*. Englewood Cliffs, NJ: Prentice-Hall.

Nasmyth, J. (1885). *James Nasmyth engineer: An autobiography* (S. Smiles, Ed.). London: John Murray.

Ohlman, H. (1996). Information: Timekeeping, computing, telecommunications and audiovisual technologies. In I. McNeil (Ed.), *An encyclopaedia of the history of technology* (pp. 686–758). London: Routledge.

Peirce, C. S. (1991). *Peirce on signs: Writings on semiotic by Charles Sanders Peirce*. (J. Hoopes, Ed.). Chapel Hill: University of North Carolina Press.

Ramsey, F. P. (1990). *Philosophical papers* (D. H. Mellor, Ed.). Cambridge, UK: Cambridge University Press.

Roberts, N. (1976). Social considerations towards a definition of information science. *Journal of Documentation, 32*(4), 249–257.

Roberts, N. (1977). *Use of social sciences literature*. London: Butterworths.

Rosenberg, N. (1976). *Perspectives on technology*. Cambridge, UK: Cambridge University Press.

Rosenberg, N. (1982). *Inside the black box: Technology and economics*. Cambridge, UK: Cambridge University Press.

Rosenberg, N. (1994). *Exploring the black box: Technology, economics and history*. Cambridge, UK: Cambridge University Press.

Rosenberg, V. (1974). The scientific premises of information science. *Journal of the American Society for Information Science, 25*, 263–269.

Schiller, D. (1996). *Theorizing communication: A history*. New York: Oxford University Press.

Schumpeter, J. (1961). *History of economic analysis*. London: George, Allen & Unwin.

Searle, J. R. (1980). Minds, brains and programs. *Behavioral and Brain Sciences, 3*, 417–457.

Shannon, C. E. (1938). A symbolic analysis of relay and switching circuits. *Transactions of the American Institute of Electrical Engineers, 57*, 713–723.

Shannon, C. E. (1993). *Collected papers* (N. J. A. Sloane & A. D. Wyner, Eds.). Piscataway, NJ: IEEE Press.

Shelley, M. (1998). *Frankenstein or the modern Prometheus*. Oxford, UK: Oxford University Press.

Stiglitz, J. E. (2000). The contributions of the economics of information to twentieth century economics. *Quarterly Journal of Economics, 115*(4), 1441–1478.

Turing, A. M. (1937). On computable numbers, with an application to the Entscheidungs problem. *Proceedings of the London Mathematical Society, 42*, 230–265.

Verdú, S., & McLaughlin, S. W. (2000). *Information theory: 50 years of discovery*. New York: IEEE Press.

Vološinov, V. N. (1986). *Marxism and the philosophy of language*. New York: Seminar Press.

Warner, J. (1992). Retrieval performance tests in relation to online bibliographic searching. *Proceedings of the 55th ASIS Annual Meeting*, 231–241.

Warner, J. (1994). *From writing to computers*. London: Routledge.

Warner, J. (1999a). Information society or cash nexus? A study of the United States as a copyright haven. *Journal of the American Society for Information Science, 50*, 461–470.

Warner, J. (1999b) An information view of history. *Journal of the American Society for Information Science, 50*, 1125–1126.

Warner, J. (2000). In the catalogue ye go for men: Evaluation criteria for information retrieval systems. *Aslib Proceedings, 52*, 76–82.

Warner, J. (2001). *Information, knowledge, text.* Metuchen, NJ: Scarecrow.

Warner, J. (2002). Forms of labour in information systems. *Information Research, 7*(4). Retrieved January 12, 2004, from http://informationr.net/ir/7-4/paper135.html

Warner, J. (2003a). Information and redundancy in the legend of Theseus. *Journal of Documentation, 59*(5), 540–557.

Warner, J. (2003b). Citation analysis and research assessment in the United Kingdom. *Bulletin of the American Society for Information Science and Technology, 30*(2), 26–27.

Warner, J. (2004). *Humanizing information technology.* Lanham, MD: Scarecrow Press.

Weaver, W. (1949). Recent contributions to the mathematical theory of communication. In C. E. Shannon & W. Weaver, *The mathematical theory of communication* (pp. 1–28). Urbana, IL: University of Illinois Press.

Webster, F. (2002). *Theories of the information society* (2nd ed.). London: Routledge.

Whitehead, A. N., & Russell, B. (1962). *Principia mathematica to *56* (2nd ed.). Cambridge, UK: Cambridge University Press.

Wiener, N. (1954). *The human use of human beings: Cybernetics and society* (Rev. ed.). New York: De Capo Press.

Wilson, P. (1996). Interdisciplinary research and information overload. *Library Trends, 45*, 192–203.

Wittgenstein, L. (1981). *Tractatus logico-philosophicus.* London: Routledge and Kegan Paul.

Zipf, G. K. (1936). *The psycho-biology of language: An introduction to dynamic philology.* London: Routledge.

CHAPTER 14

Poststructuralism and Information Studies

Ronald E. Day
Wayne State University

Poststructuralism

The meaning of the term "poststructuralism" is not without controversy. Culler (1982) has pointed to some of these problems, notably how different theorists have attempted to distinguish "structuralism" from "poststructuralism" using different criteria. Chalmers (1999, p. 1111) states that "structuralism focused on a notion of a shared, commonly understood language that individuals would use as a basis for each utterance or action. This aspect of the theory was later criticized as unrealistic ... To take account of such issues, there was a move to 'post-structuralist' theories of knowledge and interpretation." Framing the discussion in terms of literary analysis, Culler (1982, p. 22) writes, "In simplest terms, structuralists take linguistics as a model and attempt to develop 'grammars'—systematic inventories of elements and their possibilities of combinations—that would account for the form and meaning of literary works; post-structuralists investigate the way in which this project is subverted by the workings of the texts themselves." Another problem in making such a division between structuralism and post-structuralism is that late structuralists such as Roland Barthes (for example in *S/Z* [Barthes, 1974]) adopted poststructuralist modes of analysis that subverted stronger structuralist claims.

In information studies, the term "poststructuralism" is commonly associated with Michel Foucault's work under the term, "discourse analysis." In critical legal studies, poststructuralism was discussed in terms of Derridean deconstruction (see, for example, *Deconstruction and the Possibility of Justice*, 1990; and *On the Necessity of Violence for Any Possibility of Justice*, 1991). Farmer (1993), following critical legal studies, has broadly addressed poststructuralism in the legal research process and in information studies. Farmer (1993, p. 392) states that poststructuralism may be characterized by its rejection of "master narratives" and "foundational claims that purport to be based on science, objectivity, neutrality, and scholarly disinterestedness." It is also characterized by a deep interest in the notion of "text," as both a literal and an analogical form. Yet as Farmer (1993, p. 392) states, "in spite of the theoretical storm sweeping through universities and law schools, the field of library and information science seems to have remained largely apart from the maelstrom." Farmer is not incorrect in pointing to the poststructuralist critique of "master narratives" (though this phrase comes from Jean-François Lyotard [1984], who is not always positioned within poststructuralism). Farmer's definition is a rather over-encompassing framework, however, since such a definition can also apply to many other types of social critique in addition to poststructuralist (see, for example, Benoît's [2002] discussion of Habermas's theory of communicative action).

One way of forming and analyzing a category or genre of poststructuralist writings, however, is to look at theorists whose work is sometimes associated with the term, especially French theorists whose works had a huge impact on the humanities and the social sciences in Anglo-American universities and culture beginning in the 1970s, an impact that continues today. The names of theorists such as Michel Foucault, Jacques Derrida, and Gilles Deleuze are certainly prominent. Behind these thinkers, however, lies a variety of theoretical strands and writers. For Derrida, the most important must be the writer Maurice Blanchot and the philosopher Martin Heidegger (Rapaport, 1989). For Deleuze, the philosophical influences range from Spinoza to stoicism to English and Scottish empiricism; next to these philosophical roots stands the influence of baroque architecture, American literature, and his collaboration with the radical psychoanalyst Félix Guattari (Deleuze & Guattari, 1983, 1987, 1994). Foucault's writings, in some ways, parallel and anticipate both Derrida's and Deleuze's work.

Common to these figures and their influences, however, is an insistence upon the empirical. Such a statement might strike those involved in traditional research in information studies as odd, since the work of these theorists is discursive in nature and lacks mathematical tools of analysis. But it is this insistence that forms the break with structuralism, understood as transcendental contexts for evaluation and analysis. These writers insist that the site-specific and time-valued *repetition* of what can be generally called "signs" gives the possibility of structure, or what is commonly referred to as "context." It is this insistence upon language and repetition, as well, which links poststructuralism to kindred approaches in philosophy, ranging from Heideggerian phenomenology's emphasis on language as the dwelling site of being to Wittgenstein's (1958) interest in statements and what he termed, "language games." Repetition stands, too, at the heart of what is often termed in poststructuralism "events"—moments of complex repetition that construct possibilities of historical futures and of structures and identities. Such a conception of linguistic repetition in Deleuze's work, as Delanda (2002) has shown, also related to understandings of material forms of repetition and creation—involving genetic development, for example. (In

Deleuze's work, the concepts of repetition and event form the link between the "signifying semiotics" of linguistics and the "a-signifying semiotics" [Guattari, 1972] of biological information systems, and consequently, make the genetic notion of "information" not simply a metaphor of linguistics events.)

The "empirical" for these writers carries an emphasis upon temporality, historicity (i.e., historical agency), and a differential complexity that is intrinsic to any identity. Such an emphasis upon differential complexity and the potential powers of emergence within any identity, that is, upon the general economy that defines any restrictive economy or self-same identity, leads not only to a contrast between structuralist and poststructuralist writings, but also (contra Chalmers, 1999) classical hermeneutics. This is not to deny structure, but it is to insist that structure both is created and has the potential of being transformed by each material repetition. Each phenomenon can bring about the modification of "structure," whether that word is read in terms of language, social habits, culture, or aesthetic form. We must acknowledge that actual performances both (re)create structures and also may modify them. Structures are not simply historically contingent, but also historically produced. Poststructural analyses are deeply materialist—and in that sense, they are "empirical" analyses—in contrast to reified notions of "information," "language," "users," and cognitive models that oftentimes circulate in information studies theory. With poststructural analyses, one begins with the material fact of the sign or being as an event and then descriptively expands from that point onward.

Meaning and the Correspondence Theory of Truth

The relation of poststructuralism to information studies, and particularly to information science, thus appears both problematic and intriguing. It is problematic because the epistemological or ontological form of information science's notion of information is often that of self-identity (auto-affective "facts," "information," or "data") or that of being an effect of sociocultural "contexts" (a viewpoint that is often, somewhat too generally, referred to as "constructivist" arguments). Previous *ARIST* chapters by Jacob and Shaw (1998), Cornelius (2002), and Capurro and Hjørland (2003) have all pointed to two dominant strains in information science since the end of the Second World War. These two strains are, first, the information retrieval tradition where "information" (whether understood as concepts or documents, cf. Capurro and Hjørland, 2003) is treated as data. And, second, the user-studies tradition where "the user"—or subject—is either theorized as mental content to be brought into correspondence with a target "text" (itself sometimes taken as a surrogate for an author) or where the user is seen as an agency situated, in various ways, within sociocognitive contexts (Belkin, 1977; Pettigrew, Fidel, & Bruce, 2001; Shera, 1961, 1970).

As Frohmann suggests (1992, p. 369), despite the seeming break, especially, of the information user tradition from Warren Weaver's (1949) social reading of Shannon's engineering model of information theory according to a correspondence model of truth or a conduit model for communication (see Day, 2000, 2001), the attempt in information studies to theorize the subject in representational terms, along with the belief that the singular purpose of the field is to create "effective retrieval" (Jacob & Shaw, 1998), returns us to the truth claims and the communicational model with which Weaver begins. Even though the Weaver/Shannon model for *communication* premised the correspondence of content between the subject and another subject, and the cognitive model for *information* premises the correspondence of content between the subject and the object vis-à-vis the fulfillment of the subject's lack by informational content, the same formal correspondence structure is present in both the communicational and the informational models.

Once again, recent user studies have attempted to address this problem, at least on the side of the "seeking" subject, by appealing to sociocultural context; but there has been an ethnological or anthropological tendency in information studies to then view the contextualized subject as itself a "content" for correspondence. This does nothing to address the representational epistemology, the correspondence theory of truth, and the conduit metaphor, which are different aspects of the same formal model that need to be addressed. Further, many articles now refer to a sociocultural "context" without any attempt to specify the material horizons for production that constitute such contexts, thus leaving the notion of "social" or "cultural" horizons rather vague, or simply substituting an ethnological narrative for a much more needed theoretical account. (A stellar counter-example to these tendencies may be seen, however, in Belton [2003], where theoretical discourse acts to illuminate the conditions of production for the objects under discussion.) The central problems here are three. First, the attempt to use "context" as a causal explanation for agency; second, the temptation to see issues of *embeddedness* as issues of objects in structural contexts; and third, a tendency either to reify or to over-generalize subjects and objects or to trivialize them in the name of a narrative structure common to anthropological or ethnographic realism. Together, these add to the overall impression that, traditionally, information studies research has not accounted very well for the *construction* and *production* of not only the concepts of "users" and "information" (and for the various "scientific" methods that accompany these concepts), but it has not effectively addressed the historical, political, and philosophical construction and production of subjects and epistemic objects in general. Because it is unlikely that such a "critical" address can occur within the tradition of quantitative methods of "science" in information studies (where subjects and objects are already addressed as given presences for method), the term "science" sometimes sociologically acts to buffer the field from foundational critiques within it. Not only is a better, more theoretically informed practice needed, but an understanding of the actual history and practice of the different sciences outside of library and information science is desperately needed because the term "science" still seems to be used with wanton generality even among its foremost researchers (for counter-examples to this tendency, however, see Bowker and Star [1999], Lievrouw [2003], and Van House [2004]).

The correspondence model of truth within communication theory, called by Reddy (1993) the "conduit metaphor," has long been the dominant cultural trope for describing the correspondence between a speaker's or "transmitter's" meaning and the hearer's or "receiver's" reception of that meaning. The connection between information studies and communication as fields is based on this model. Some traditional, common understandings of language are confirmed by the father of structural linguistics, Ferdinand de Saussure's reference to the "speaking circuit" involved in speech acts (*parole*) in his foundational lectures (later published as his *Course in General Linguistics* [Saussure, 1916/1966]). Already in his lectures, however, Saussure had begun breaking away from the folk tradition of the conduit metaphor, arguing that linguistic value and identity were socially constructed by speech acts, and that such acts were constituted not by linguistic identity and value resting in the sign, in the mental psychology of the speaker, or even in its referents, but rather that identity and value were products of the differential play of signs in language (Saussure, 1966). It seems remarkable that such a fundamental insight—that identity is constituted by plays of difference—and Saussure's work itself are relatively unfamiliar to information studies researchers, especially in Anglo-American universities.[1]

Information studies theory has remained a positivist enterprise, squarely within the metaphysical tradition of Western Philosophy, in so far as it reifies meaning and understanding in language acts, replacing variable pragmatics with idealistic models. The philosophical dream of correspondence has a strong hold over information studies and its practice-based students, teachers, and even researchers. Terms such as "structuralism," "structural linguistics," and not the least "poststructuralism" are, remarkably, relatively unknown in much of the practicing and teaching of information studies. Lacking such terms and analysis, the conduit metaphor and organic or "auto-affective" notions of identity and value—in regard to language and in other domains—remain the guiding light. This tends to force all other types of research to return to these central concepts and models (or really, metaphors) in the name of a "science" whose ontological and epistemological claims have, arguably, been long since abandoned or demystified throughout the physical sciences and the social sciences.

The irony of library and information science's (LIS) historical tendency toward scientism is that the conduit metaphor, as a model of representation, is the model of philosophical and rhetorical representation that held such power over the humanities until the advent of "theory" in the American universities of the 1970s. Paling (2002) suggests that we must justify the intersection of LIS and rhetoric studies. Historically, however, the two have clearly—though often ironically—overlapped through both the governing epistemological model and disciplinary purpose of LIS and the central role of vocabulary in LIS research and practice. Ironically, as a metaphor, the conduit model has always allowed LIS to be an object of rhetorical analysis (as Frohmann's work has shown). Further, rhetoric has, a priori, infused LIS as a discipline: in terms of both its formal construction and guiding model and its concentration on problems of linguistic elements and units (vocabularies, discourses, and statements). Information science's concentration on a vague, metaphysical object termed "information" (Capurro & Hjørland, 2003; Schrader, 1984) along with "information retrieval" (largely, *document retrieval* [Capurro & Hjørland, 2003]), may be seen as an attempted break from a focus on language and its social functions and affects, as was the case, for example, in European documentation.

Correspondence, Structure, Repetition

Library science has historically addressed the problem of creating stable contexts for the correspondence of meaning through controlled vocabularies. The successful imposition of transcendental vocabularies as mediating devices for the stabilization of meaning, however, has always been difficult to achieve, because users tend to deploy their own terms for relating signifiers and signifieds (words and concepts). More recently, with digital technology, post-coordinate vocabularies have become more common, although problems of overload and low precision persist.

The central issue, from a poststructuralist viewpoint, however, is that there is no way to *assure* the correspondence of meaning between a speaker/text and listener/reader for two reasons: language and identity. From the aspect of repetition, vocabulary and phrases may be grafted onto new discursive regimes that are distributed in time and space. For Derrida (1976), this issue of grafting was intrinsic to Saussure's claims that identity and value in language were not established by a relation of correspondence or representation between signifier and signified, but rather by a system of differences. The *spacing* of signification, that is, the understanding that a stable meaning for signs is always *differed* by the constant play of *difference* between signs (what Derrida terms *différance*)—that the relation between signifier and signified is not only arbitrary, but also a result of the play of forces both within and beyond discourse—constitutes one of Saussure's deepest insights.

In terms of critiquing the conduit metaphor, there is no reason to see a speaker or text as the primary origin of a semantic "transmission" and a hearer or reader as the "receiver" of this transmitted meaning; and there is no reason to equate successful communication with transmission. For, first of all, meaning frequently occurs in the very "failure" (or in reality, nonexistence) of this correspondence; this is easily demonstrated by observing that dialog forms our fundamental approach to communication and even to information retrieval (e.g., through repeated retrieval attempts). We engage in dialog in order to approach a pragmatic degree of clarity on what we think the other means when he or she speaks or writes. Second, we are faced with difficulties in proposing mental intentions as the cause of received understanding, or even worse, of performed actions by the "receiver." (There is the impossibility of objectively identifying what the speaker or writer *intends* to transmit—introspective statements of intent are notoriously problematic, and using these as the basis of effects sets up a very problematic causal model.) Third, if we begin with the Saussurian arbitrariness of the sign and its constitution in meaning by plays of differences and powers, the identity of the classical subject in terms of being an auto-affective self is itself displaced by its being a linguistic element within a play of differences. In sum, as Reddy (1993) points out, the conduit metaphor is a poor model for language, much less for communication. In this light, poststructuralism, "critical theory," the avant-garde arts, and other, "critical" activities during the twentieth century have formed alternative horizons, outside of representation, for viewing language, knowledge, and communication.

Poststructuralism is intriguing in relation to information studies because it aims at destroying the metaphysical assumptions of positivism and bringing into question the stronger claims of structuralism; and it does this in regard to contemporary notions of knowledge, information, and popular notions of language. It thus opens up other, older understandings of knowledge, information, and language, such as those based on affect and events (Day, 2001; Peters, 1988). This reopening of language to something other than auto-affective meaning or data, this reopening of knowledge to something other than certain mental content, and this reopening of information to something other than representation, fact, or "true belief" constitute a challenge to the metaphysical and epistemological assumptions that have, for so long, dominated not only information studies research and even practice, but also popular conceptions of the materials it studies.

"Information," Historicity, and the Fragment

Peters (1988) and Day (2001) see historical and political implications in the forgetting of this previous conception of information as affect[2] and event. At worst, there is the appropriation or collapse of all areas of knowledge and affect into an ideology of information as being self-evident knowledge and commodity. Arguably, this transformation has led to the hyperinflation of the value of what is now considered information throughout areas of culture and society that were previously unaffected by such a definition of their intellectual materials and "assets" (e.g., economics, education, and the arts); another consequence is the erasure or reduction of other values for knowledge and affect (Day, 2001; Marazzi, 1997). The recent dominance of an "information" or "symbolic" economy (that is, an economy of mixed semiotic and financial capital—what Franco Berardi [2001] has termed, "semiocapital") is evidence of the cultural power of reducing knowledge and affects to information.

The "information age" and post-Fordist reduction of knowledge and affect to information essentially transform the totality of the life-world to a form of production easily manipulated by financial capital, what Marx termed, real subsumption (Day, 2002; Dyer-Witheford, 1999, Marx, 1992; Negri, 1989, 1996, 1999). This was, as Belton (2003) has pointed out, prefigured by sixteenth century and later European colonial appropriations of newly "discovered" lands and the codification of such discoveries and explorations in informational terms that were readily understood in terms of European capital appropriations and calculations, communicational channels, and, in general, wealth. One might argue, however, that the appearance of knowledge explicitly understood not only in terms of production in general, but also in terms of capital in particular, is a twentieth-century phenomenon, concurrently increasing with the "informalization" of historical thought in public space, where the past is erased as a problematic of interpretation, rhetoric, and power, leaving the present and future to be seen as givens *within* history.

For the philosopher Martin Heidegger, writing in 1966 (Heidegger, 1977, p. 370-392), the latest encroachment of the informational model (as the embodiment of metaphysics in the form of cybernetics) threatened the radical critique of modern culture and metaphysics promised in the interpretative and affective activities of art. Art, understood according to cultural heritage and through what can be termed "the museum affect," for example, reduces art's critique of modernity's teleological mode of production to that same production, symbolized in the form of a timeline of "art history" and the art historical categories of author, period, and genre. Through such frames of cultural value and historical transmission, the *work* of art, even when it is constructed against such values, becomes easily transmitted and speaks communicationally between temporal periods and geographical and social sites. The discourse and practice of "cultural heritage" and the "museum affect," thus, are also preliminary activities to the reproduction of art as a content of historical tradition and a commodity of cultural value transmitted through digital reproductions of the artwork. For Heidegger, at the historical moment when the "work" of art, qua *work*, becomes understood as the content of historical tradition and the commodity or object of cultural value, the modern ideology of information can be seen as having appropriated its most radical opponent.

Capurro (2000, p. 80) in a less pessimistic manner, however, sees information as "the shape of knowledge at the end of modernity," reading contemporary understandings of the concept of "information" in a somewhat "postmodern" manner. According to Capurro, information in late modernity comes in the form of: 1) the abandonment of universality in favor of fragmentariness (including the privileging of science as the privileged form of discourse); 2) the abandonment of subjectivity and objectivity as privileged sites for analysis and their replacement by an initial ontological state of commonality; and 3) knowledge as mediated rather than transcendental *episteme*. Capurro (2000), through a reading of Heidegger, warns of the same historical and political dangers as Day and Peters, but he also argues that the necessity for information to be interpreted can act as a brake on modernity's technological and teleological interpretation of "information" and, in fact, can lead to a form of thought marked by greater creativity.

Capurro's reading of knowledge in terms of the informational fragment suggests, in some ways, Lyotard's (1984) view that information is a form of fragmented knowledge, although Lyotard unambiguously ties that knowledge to science, namely, the pragmatics of scientific method. Lyotard reads the postmodern valorization of scientific knowledge in terms of method as a major change over the "grand narratives" offered by nineteenth-century and early twentieth-century research in universities and even in industry and the state. The foundation of scientific method for Lyotard lies in the role of negative proof for knowledge. Science is based on a "language game" of challenging claims and disproving findings ("agnostics"). This challenge is built into whatever positive claim may be made in the discourse of science. Knowledge in science, according to Lyotard, is thus built up of these fragments of knowledge whose reliability is due not to their being historically passed on as accepted truths but rather to their inability to be disproved. He adds, however, that the method of science has now been co-opted by criteria of efficiency and production—instrumentalized—under the direction of capitalism and the state; it has become the guiding method to unfold knowledge in instrumentalized fashions through big science and the funding of educational institutions. Narrative has thus returned to science, but now under the guise of efficiency and capitalist production rather than internal conditions for truth.

In her keynote address to the Document Academy's first meeting, held at the University of California at Berkeley's School of Information Management and Systems, Johanna Drucker (2003), Director of Media Studies at the University of Virginia, discussed the concept of the document in terms of the historical conception of the fragment, citing the discussion of the fragment in nineteenth-century Romantic movements as key moments for the emergence of such a discourse. According to Drucker, the fragment in Romantic poetry is posed in contrast to totalizing or unified knowledge, although the fragment also demonstrates a level of entanglement involving a whole (which itself is not, however, reducible to a concept of totality). Such a position is aligned with Derrida's interests in paratextual elements, Paling's (2002) interest in paratextuality and classification, and with Cronin's (1995) work on the social significance of traditionally "marginal" or "paratextual" textual elements such as acknowledgments.

The fragment in early documentation represented both a material specificity and a cultural completeness for documentation as a vanguard cultural activity. The notion of the fragment was treated as a metonymic element in culture, for example, by the father of European documentation, Paul Otlet, in his *Traité de Documentation* (Otlet, 1934), where the term "the Book" both specifies exemplary documentary materials and a total theory for knowledge and historical progress (akin to the Christian *logos*). The theorization of documents, marginalia, or even "information" as fragments is a contentious issue, but it points to the importance of that which is outside the body of the text proper in the construction of meaning in the text.

Cronin (2004), for example, points out that published articles have an almost endless stream of what generally can be referred to as coauthors. Some of these crowd up contemporary authorship headings in articles, many others are partially or altogether absent from acknowledgments. Practically speaking, such a view challenges, or at least mucks up, a whole variety of reward mechanisms for authorship in academe and other areas of textual production based on the quantitative appearance of one's name as an "author," whether one has done any actual writing on the article or not.

Historically, Drucker (2003) has suggested, authors such as the German Romantic Friedrich Schlegel in his *Philosophical Fragments* (Schlegel, 1991) pointed to the importance of the fragment in reflecting the whole without claiming or even attempting to represent a totality. One example of the fragment operating in this way today might be the document in open hypertext systems where citation or "linkage" can exceed what can be embodied in traditional citational and bibliographical forms (Day, 2003). As we have pointed out, the understanding that each sign has meaning only through chains of signification was fundamental to Saussure's structuralism; this was brought to bear in Roland Barthe's famous article "Death of the Author" (Barthes, 1977, p. 142–148).

In each of these cases, what is important about the fragment—"document," "information," "sign" (whatever may be taken as the signifying unit)—is the networked relations that are embodied in the very existence of the unit and that give both meaning and value to that unit. This view seems to stand in direct opposition to views of information as an entity unto itself, its value and meaning auto-affectively produced from, and as, its "self."

Historicity in Information Studies

The problem of historicity—historical specificity and agency—has a direct impact on information studies in terms of the field's historiography. As Buckland and Hahn (1998, p. 1) have suggested, library and information science, particularly information science, has historically been a field that has suffered "widespread amnesia" of its own history.

Today, through the efforts of Buckland, R. W. Williams, and others, the field's history is being narrated, although historical concerns remain at the margins compared to its more "practical" matters. As we have mentioned, this "forgetting" may be part of the epistemology of the modern conception of information and not a problem only for LIS. The problem of the historicity of information involves the active or passive construction of history by agents that see or do not see themselves as historical. The primary concern is the presence or absence of an historical horizon through which persons *can* see themselves as historical agents capable of shifting history, leading to questions as to the form of historical horizons as well. If the historical horizon is transcendental, as is the case with historical narratives arising from nineteenth-century historical approaches, persons tend to see themselves as passive elements *within* history. If that horizon is seen as constructed, persons may see themselves as active agents in the construction of "history"—both in the sense of shaping the possibilities of future events and in terms of actively offering, explaining, and interpreting past events that differ from the ideological norm. Thus, the problem of historiography is extremely important in any account of historicity—including in regard to "information ages" and "information societies"—for it shapes the form and the substance of historical agency.

W. Boyd Rayward (1998) touched upon some of these issues in his article, "The History and Historiography of Information Science: Some Reflections." Here Rayward points to the difficulty of defining "information science" both as a conceptual entity and as a social field. Referencing the work of Ferdinand Braudel, Rayward has pointed to the usefulness of applying categories of periodicity in order to help the historian "tell the story" (Rayward, 1998, p. 17) of information and information science. This historical task of "telling the story" was further commented on by Buckland, when he led a workshop on the history of information science at the fourth Conceptions of Library and Information Science conference at the University of Washington, Seattle (Buckland, 2002). Buckland reminded the audience that "history" is constructed narrative, so one must pay attention to the rhetorical elements of this "telling." History is not a story, even if a given historiography assumes that narrative form.

The problem of historiography has been broached extensively outside of information science. In terms of "poststructuralist"-influenced readings of history that emphasize the constitutive role of rhetoric in the construction of historical narrative, Hayden White's work (see White, 1987) is perhaps the best known example of the poststructuralist linguistic turn in historical studies. An earlier example that is often elicited in poststructuralist readings, however, is the work of Walter Benjamin (1968, 1978), where history is discussed both in terms of messianic time and of material residues to capitalist production (the decaying Paris arcades, for example). Another, more literary example, is the work of Paul de Man (1986, p. 54–72) where problems of historiography and history are deeply embedded in literary and rhetorical analyses. Academic historical studies and other humanities fields, as well as the various areas of science and technology studies, have seen huge shifts from classical historiography to the linguistic turn, to the reassertion of narrative in various, changed forms. Information studies historiography has, however, remained rather classical throughout the turbulent decades of the last third of the twentieth century, not only largely unaffected by these shifts, but seemingly, knowing little of their existence. Information studies has centered on unambiguously narrating little known but great men (and sometimes great women), great inventions, great funding projects, and establishing these narratives as "unifying perspectives" (Rayward, 1998, p. 15).

As early as 1969 in the introduction to *The Archaeology of Knowledge* Michel Foucault (1972) had discussed a "new history." According to Foucault, with this "new history," inclusive of historiography, "attention has been turned ... away from vast unities like 'periods' or 'centuries' to the phenomena of rupture, of discontinuity" (Foucault, 1972, p. 4). According to Foucault, historical investigations had shifted away from totalities, away from creating narratives of actions around particular, traditional, unifying identities (such as biographical figures; particular assumed disciplines; particular, privileged technological devices; collective mentalities), and away from using historical writing to establish "lasting foundations" (Foucault, 1972, p. 5). Instead, historical investigation had shifted toward identifying series of acts and their thresholds, the transversal movements of series (further explored by Deleuze & Guattari, 1987), and displacements and transformations of concepts, all of which lead to momentary stabilities or "identities" rather than those identities being treated as historiographical origins. Foucault's notion of "archaeological," rather than strictly "historical" investigation, of the "archive" as a set of founding assumptions and statements that lead history to unfold in certain directions rather than others and for the real to appear in such and such a form—both in the present and in the future— are preeminent in this "new history." (Interestingly, from the viewpoint of this "new history," the totality of LIS's traditional historiographical forms in their sociological and historical contexts [i.e., as foundations] precisely constitutes an archive, at least as much as, *if not more so than*, it constitutes the history of information and information technologies.) As Foucault (1972, p. 5) wrote of the "new history," "what one is seeing, then, is the emergence of a whole field of questions, some of which are already familiar, by which this new form of history is trying to develop its own theory: How is one to specify the different concepts that enable us to conceive of discontinuity (threshold, rupture, break, mutation, transformation)? By what criteria is one to isolate the unities with which one is dealing; what is *a* science? What is *an oeuvre*? What is *a* theory? What is *a* concept? What is *a* text?"

In this same book, Foucault centers his discussion on the problem of *documents*. Traditionally, as Buckland has remarked (1991a, 1991b), documents are understood in terms of being "evidence." The concept of "evidence," here, resembles the vulgar conception of "information" as it is commonly understood in such terms as "information-seeking behavior" in information studies, that is, as unitary elements that help fulfill a lack or "need" in an information structure (sometimes synonymous with the terms "mental model" or "image" [Belkin, 1977]). This structure may be metaphorically understood as literally a structure that needs filling or, more prescriptively, in a more Wittgensteinian manner as an element that allows us to proceed in a prescriptive act (like a note that helps us sing a song). But the term "evidence" always begs the question of evidence of *what*? Traditionally, within historical studies, documents are understood to be "lost," but now found, elements of an original structure, or they are thought to be elements that are lacking and are now fulfilling for a narrative of history that they help reconstitute. Historical research treated documents as "pointed to one critical aim: the reconstitution on the basis of what the documents say, and sometimes merely hint at, of the past from which they emanate and which has now disappeared far behind them; the document was always treated as the language of a voice since reduced to silence, its fragile, but possibly decipherable trace" (Foucault, 1972, p. 6). Documents were, in Derridean terms, treated as *graphic* traces of a more original historical identity, the *voice* or *"presence"* of history "behind" the traces.

Foucault (1972, p. 6–7) asserts, however, a different understanding of documents in the "new history": Documents are treated more as texts, that is, as produced elements that have roles and powers to shape present events and future historical thought:

> history now organizes the document, divides it up, distributes it, orders it, arranges it in levels, establishes series, distinguishes between what is relevant and what is not, discovers elements, defines unities, describes relations. The document, then, is no longer for history an inert material through which it tries to reconstitute what men have done or said, the events of which only the trace remains; history is now trying to define within documentary material itself unities, totalities, series, relations. Foucault (1972, p. 7)

compares the status of documents to archaeological artifacts: less as speaking traces of the voice of a lived past and more as "inert traces" from which pasts and futures are created. For Foucault (1972, p. 10), we have shifted from a "total history" to a "general history" (possibly here he is echoing the notion of *écriture générale*, that is, a general economy of signification): "A total description draws all phenomena around a single centre—a principle, a meaning, a spirit, a world-view, an overall shape; a general history, on the contrary, would deploy the space of dispersion." Here, total history is the legacy of the metaphysical subject (Foucault, 1972, p. 14), the defining of history and historiography by "anthropological constraints" (Foucault, 1972, p. 15). A general history, however, recognizes "centers" as products and productions of power.

Historical centers, as Latour (1996) has argued, are not only evidence of the real, but more importantly, they are constructors of the real. Libraries are important centers for accumulating "evidence"; far from being neutral, historically, they have acted as centers for the deployment of power by constructing subjects and objects of power (Latour, 1996). (Modern public libraries have never denied this. In fact, they herald this civic duty that they perform.) Day (2000, 2001) has analyzed the textual production and deployment of history, particularly that of an informational history, through professional and other authoritative texts about information. Belton (2003), in the first chapter of his book detailing the historical European construction and exploitation of the Orinoco region in Venezuela, provides an outstanding example of a contemporary historiographical and ethnological approach that recognizes the power of documents not just to *be* "information" about the past, but rather more, to be social and historical forces that construct imaginations of real and historical units (i.e., to be archives *for* certain histories). Belton examines the Orinoco region of Venezuela as an "object" for social, economic, political, and historical production and analysis, appropriated and produced within European discourse formations, knowledge disseminations, and spatial production for exploitation and wealth since the fifteenth century. The region, as a form of "biopower" (Foucault, 1990; Hardt & Negri, 2000), was opened as a *Grundrisse*, formed and produced as an information resource in a chain of appropriation/redesign, management, exploitation, and wealth production much as biological organisms are today designed at the molecular level in the biotech industry for further movements down the chain of production, consumption, reproduction (both of products and consumers), and wealth (Lievrouw, 2003). The natural and cultural "history" of the region was retroactively inscribed to follow this archival form, reforming its "natural" qualities so as to tighten relations within and between actors and stages in the economic chain.

Foucault's and Belton's (2003) approach does not simply take historical units as pregiven facts or as conveniences of historical narratology, rather it accounts for them as symbolically encoded spatial and temporal sites or *topoi*, constructions of "conscious" or "unconscious" political, economic, and social acts and desires. Foundational historical acts, such as political imperialism, historicist gestures, or the social implantation of a common noun (e.g., "information" [Day, 2001]) are attempts to form a political unconscious whose foremost concern and product will be its own formal repetition. It must be added, however, that such acts are not just predatory, but also must make use of existing fertile conditions (as Suzanne Briet hoped to see after the Second World War with the spread of the profession of documentation to address the medical needs of the third world [Day, 2000, 2001]). Nor are historical acts strictly bound up with imperialist desires; concepts may become reappropriated and redeployed—as Foucault emphasized throughout his works, much depends on extra-discursive forces for maintaining archival order (e.g., police and military powers, social organizations [schools, libraries, media], social pressures, and moral orders).

The units of analysis in the archaeological uncovering of the discursive archives of history are not simply "facts" of history, but rather products of social and historical forces and further producers of forces. The notion of "archive" is that of a conceptual whole with breaks, bifurcations, and other particular, temporal characteristics, not just a set of documents that is chosen for its "intrinsic" historical value: "The idea of the archive that I develop here is that of a chronologically layered set of notions about the world, each succeeding layer becoming dominant for a time, not superseding those beneath it but rather adding another layer of world interpretation to the archive as a whole" (Belton, 2003, p. 7).

Foucault's method, and that of later writers such as Belton (2003), displaces historical narrative as the empirical and formal origin for historical studies and as the foundation for other humanities studies. It understands the historical object to be a set of complex potencies, constituted by conceptual and empirical powers as certain identities, values, and meanings in accordance with the goals of production and reproduction, and it understands that produced object itself as an event that produces further historical events and historiographical series. By displacing historical narrative, and thus, historicism, as both the origin ("the story") and the method ("telling the story") of historical studies and the humanities, Foucault's "new history"—i.e., his archaeology—helped give rise to a type of historical investigation that was more critically reflective on the epistemological categories, the rhetorical techniques, and the political and social intentions of traditional nineteenth- and twentieth-century historiography. Given the absence of work such as Belton's (2003), and given the unreflectively classical form of history in the field, it would not be unfair to write that thirty years after it first appeared in English, the appearance of or implications of *The Archaeology of Knowledge* are little known by the vast majority of library historians and information historians.

Discourse Analysis

Foucault's work, of course, is best known in information studies as an origin for "discourse analysis." Information studies and other fields also use other sources for discourse analysis where it differs from the intent of Foucault's work (see Budd & Raber, 1996). Indeed, discourse analysis as a method in information studies sometimes appears as the type of structural or scientific "method" that Foucault was disavowing in *The Archaeology of Knowledge*: an inventory of communicative statements taken as evidence for information or information behaviors. Foucault's demand that we look at evidence less as evidence *of* and more as evidence *toward*, as well as a host of other factors, such as his continuous critiques of scientism and structuralism, do not warrant interpretation of discourse analysis as a scientific method.

Discourse analysis in Foucault's work is less a method and more a series of engagements with historical events and documents that identify techniques and structures through which power is captured, intensified, and directed toward certain productive and profitable ends. It is not a neutral method at all. Rather, it is "critical" in the sense that it attempts to identify discursive and organizational assemblages that create value, hierarchies, historical and social bodies, and habits of behavior. It is concerned with the transformation of potentialities or potencies into possibilities, possibilities for structure, wealth, and the reproduction of dominant powers and forms. These may occur through the social and conceptual formation of the sexualized body, the "disciplining" of learning and knowledge, or the discursive construction of "man" as a central ideal for modern values. It is concerned with the disciplining emergence. The criticality of Foucault's work relies not simply on an opposition to repressive forces, but rather goes beyond the Marxist dialectics of its day to discuss "power" not primarily in terms of *pouvoir*, but in terms of *puissance*. *Pouvoir* (a structural or repressive force) in Foucault's work is the shaping of *puissance* (a direction or sense of force—what we have here termed potentialities or potencies—*potentia* [Negri, 1999]). All foundational activities eventually wish that potencies construct their own masochistic structures for self-discipline, incorporating *pouvoir* into the very possibility of *puissance* or *potentia* as the only framework for expression (e.g., Freud's introjection of the superego [*Über-Ich*] into the ego or "I" [*Ich*], the law's desire for its own reification in custom and habit, or workers' self-discipline and the conditions of subsumption [Day, 2002] in post-Fordism, Knowledge Management).

Frohmann, whose work is often cited in articles that take discourse analysis in various directions, has used the term in a more Foucaultian sense than many others in information studies. What aligns Frohmann's writings with Foucault's is not that discourse is the object of inquiry, but rather that discourse is seen as a material basis for the construction of social and epistemic power and structures. Frohmann's work is "critical" in that its accounts of language acts are of language as locations of and for power, not simply of the empirical occurrences of given phrases or vocabulary. Consequently, Frohmann's discourse analysis cannot be divorced from his institutional critique. In Foucault's discourse analysis, language and power are not separable.

Frohmann's (1994, p. 127) "Discourse Analysis as a Research Method in Library and Information Science" correctly states, "to study *discourse in action* shows the authorized range of possibilities for quite specific exercises of power over information, its users, and its uses" (Frohmann's italics). We might add that to show "discourse in action" means to engage not simply in conversation analysis, but also in textual analysis. Discourse in action means the construction of subjectivity and institutions by both speech and writing. "Action" can occur in any place and time. As Deleuze (1988) points out in relation to Foucault, action means the production of affects. Second, we may add that power in information studies theory, as with other institutional discourses, is not so much exercised *over*, but *through* the construction of such concepts as "information," "users," and the notion of information "use." These terms are products of the discursive and institutional machinery of not only information studies but also the "complex institutional environments" (Frohmann, 1994) in which the field is situated and situates itself.

It is this element of constructing identity based on the consistent reproduction of words, technical procedures, and institutional forms ("discursive assemblages") that marks classical disciplines and their products in modernity, particularly in professional fields. Frohmann (1994) justly points to the constant historical anxiety in information studies over its own disciplinary status, manifested in an obsession over creating a uniform sense of theory. Such a struggle for modernist foundations (which we have previously noted) seems deeply ironic given the interdisciplinary nature of the physical sciences today, and the mutually supportive and dialogic role that interdisciplinary discourses (such as Foucault's) have played in the humanities and the social sciences for the past thirty years. Theory is not unimportant in supporting university disciplines, but today theory is expressed in the sciences in terms of discourses and their technical expressions and in the humanities and much of the social sciences in terms of discourses and their analytic expressions. These discourses are not foundational for any one discipline; their importance emerges in their ability to cross and join disciplines—their interdisciplinarity creates new extensions and even new forms of power (as both disciplinary and research forces). The notion of science that information studies assumes in its understanding of defined disciplinary foundations and its theory/practice representational framework is, fundamentally, a nineteenth- and early twentieth-century concern.

Frohmann's work attempts to account for information studies within larger discursive assemblages than the field's traditional methods and scope usually allow. Foucaultian discourse analysis, as an attempt to account for bodies and powers within the forms of general history, is, by necessity, interdisciplinary, so it is doubtful if such critical analyses of modernist bodies can then act as unequivocal positive foundations or methods in the modernist way that has been so traditionally desired in information studies. As Wittgenstein argued at the end of his *Philosophical Investigations*, foundational work upon a field must involve analyses that are not fully proper to that field. Fields that define themselves according to what they claim to be "the scientific method" cannot be analyzed by those methods; foundational work on mathematics does not contain calculations (Wittgenstein, 1958, section II, xiv). More likely, one aspect of Frohmann's analysis is that it opens information studies to a form of critical and cultural analysis that has spanned the humanities and qualitative social sciences for the past thirty years but has largely been ignored by information studies.

The humanities and the qualitative social sciences constitute the educational background of many library and information management practitioners, but these areas have not been well applied by library science in its ideal of science since Dewey and the founding of the first library school (the School of Library *Economy*—i.e., of library efficiency) or by information science since the Second World War. One might suggest that the neglect of these areas has led to an overemphasis on productivity in librarianship (e.g., information *services*) and an overemphasis upon quantitative analyses in both library and information studies. On the other hand, both library science and information science have been so heavily quantitative since their beginnings that one could argue they do not engage work from the humanities and the qualitative social sciences. Although both views have their support, Frohmann's work suggests something that goes beyond either argument: namely, that information studies must be studied beyond its own restrictive economy both for fully accounting for that economy's production and in order to redirect it within real, not imagined, social and historical contingencies. Accounting for the field's social and historical emergence, its internal contradictions, restrictions, and potentials, cannot be done with terms of the discourse alone, but only in relation to "outsides" within which it both situates itself and produces.

A discourse analysis of information studies as a field and as specific topics must, by necessity, engage "general history" and general social and cultural production (though the analyses may not be general at all). Further, the topic of Foucault's analysis is not just discourse as a topic of analysis, but is discourse as a *product and a force of production—as restrictive economies generated out of more general economies*, without losing focus on the material and historic specificity of this restriction and its drive toward specific productions and reproductions. Discourse analysis is not a method or a theory in the sense of being a pre-established framework applied to discourse in order to find its structural or empirical foundations. Rather, discourse analysis is an examination of discourse as the product and the possibilities of differential relations of two types of materials (language and things) in order to find out how discursive assemblages extend themselves in space and time; how they create meaning by folding into one another and back upon themselves; how they reproduce themselves in terms of structures; and how discourses break apart, bifurcate, become transformed, and even disappear in events.

Discursive formations and their surfaces of production may be found in texts, institutional formations, social practices, and physical formations (e.g., classroom structures, or architectural structures). They may compose sensual beings and the nature and character of sexual acts (Foucault, 1990); and they may be discerned in particular speech acts that lead to particular and recognizable affects and effects. Discursive surfaces have direction (or what Frege termed "sense"), they become folded in certain ways so as to form meaning and structures for reproduction (Deleuze,

1990b, 1993); they construct objects and even subjects by maintaining regulated differences and by forming norms, orders, and hierarchies. Their principle of self-same reproduction is maintained by habit; educational institutions and techniques (particularly "training"); other discursive surfaces; "communication" and "information" ideologies, technologies, agencies, and agents; and if necessary by the State's monopoly upon violence (Benjamin, 1978, p. 277–300) and its ability to produce virtual or actual states of emergency and subsequent claims of historical exception (Benjamin 1968, p. 253–264). Discursive surfaces are material and "real" in so far as they pose resistance (Bowker & Star, 1999: "texture") to redirecting their sense, meaning, and possibility, and in so far as they allow certain beings and events to regularly emerge or to be expressed, while not allowing others to do so or while allowing others to emerge as perspectives, approaches, or minority or marginal "points of view" (see Frohmann, 2001). Materialist analyses begin with an account of beings in terms of their reproduction: that is, in terms of the forces and materials that allow production to take certain turns rather than others; that allow certain consistencies, bifurcations, and breaks to occur; and that allow certain events to happen in formally predictable manners. In other words, materialist analyses begin with analyzing how the restrictive economy of predictable possibility *emerges* from the general play of forces that make up difference and potentialities. Discursive surfaces are the physical and semiotic assemblages that allow language games—qua linguistic activity—to proceed; the material and semiotic "chessboard" upon which linguistic moves can be made. Such surfaces can be smooth or rough, illusional or self-admitting of their materiality; they can break into planes or continue throughout as smooth spaces of ideology, custom, or habit.

Hermeneutics

One aspect of discourse theory of interest in accounting for information and the various institutional structures that claim this term in modernity and today is the commonality of language and how that commonality imposes structure and shapes action. "Hermeneutics," as it has been understood in the nineteenth and twentieth centuries, is the study of the grounds for understanding through issues of interpretation. It has seen some use in information studies, often as an approach or method, competing with the dominant technical information retrieval or cognitive paradigms in the field. But, once again, as with analyses of discursive regimes and formations, what we have here is not another claim for foundations, but rather a group of writings from philosophy, textual analysis, and social analysis that goes outside and beyond information studies and helps situate the field and its internally defined voices and language in relation to larger discursive and institutional movements and powers.

The influence of hermeneutics may be located in two strands of information studies theory. The first is domain analysis, due largely to Birger Hjørland's (1997, 2000) work. Hermeneutics influences domain analysis because the meaning of language is constructed by how language is used—and for hermeneutics, "understood"—by groups of people. Domains are defined by the uniform construction and deployment of concepts through vocabulary and statements. Domain structures then may be premised, bottom-down, based on the analysis of language within already presumed categories (such as academic or research fields) or bottom-up, by the statistical measurement of language (often, in information studies, at the level of vocabulary) forming academic or research fields (or other socio-epistemological categories). Birger Hjørland's oeuvre provides a deep and broad analysis of this area of study that we will not attempt to duplicate or summarize here.

The second influence of hermeneutics is related to the notion of domain, but it approaches the problem from the perspective of "the understanding"—as the fusion of interpretative horizons—instead of from the perspective of vocabulary. Daniel Benediktsson's (1989) article attempted to introduce the term *hermeneutics* to information studies through a relatively complex review of certain modern strains in the hermeneutic tradition. Benediktsson's article remained committed to presenting hermeneutics in terms of being a possible alternative research program to more positivist and quantifiable traditions of meaning and understanding in information studies.

Capurro (2000) developed the problem of interpretative frameworks in terms of various concepts in Martin Heidegger's (1996) *Being and Time* that deal with the forestructuring of knowledge in terms of language ("preunderstanding") and in Hans-Georg Gadamer's (1985) *Truth and Method*, where understanding is discussed in terms of the fusion of horizons of understanding through linguistic negotiations. Where, in Benediktsson's article, interpretation is approached in terms of intersubjective transmission and correspondence (thus conforming to the dominance of the conduit metaphor in information studies), Capurro's work (2000) stresses commonality in language (for example, vocabulary within indexes), through which subjects may then arise.

Gary Burnett (2002) has developed from an examination of Paul Ricoeur's (e.g., 1991) work a very important argument that integrates, but also suggests, something more than a simple hermeneutic perspective for information studies. Burnett argues that virtual online communities make visible both the mutually dependent manner of writing and reading in textual production and how dissonance within texts leads to a generative power in the production of further textual materials.

Burnett's first argument shows that the context that constitutes community in textual production begins with the material elements of the text itself and the various interpretations brought to bear upon it by various readers. This does not suggest that there were no contextual preunderstandings at work in the original formation of an online community (G. Burnett [2002, pp. 165–166] outlines various types), but rather, as Burnett argues, that the analysis of a textually mediated community cannot be grounded in metanarrative conventions or in obscure references to context. The continual process of reading and writing in online communities shows, first of all, that such communities are constituted by *performative* acts generated out of material texts, and second, that reading, as a process of interpretation, is actualized in the process of writing in response to texts, (and, conversely, that writing is an act of reading other "texts"). This reversal of the structuring connotations of the term *context* by showing textual performance as the vehicle for "con-text" (here, e.g., community) is very important, *because it demonstrates that context is not an enabling condition for text, but instead, it is texts that give rise to "contexts."* We would be better off to think of texts as always already embedded in other texts and their temporally emerging from one another according to allowance conditions, which then may produce such events as selection, conjunction, and bifurcation. The emergence of texts as context or as text depends upon the existential and rhetorical horizons for emergence and their manifestation of what is written or said (this understanding of "text" follows the poststructural notion of writing as *écriture générale*). Such horizons themselves have different types of material manifestations so that texts may appear as unsaid or unwritten social or personal assumptions (what is often termed "context" in social science literature) or as more "visible" statements.

Although Burnett develops such a theme through an analysis of Ricoeur's (1991) texts, the unending hermeneutic circle of reading/writing forms an important thread in literary analyses following the writing of Maurice Blanchot (for example, Blanchot, 1982), a major influence on the French theorist Jacques Derrida. For Derrida, as for Blanchot, the threads of textuality extend through various discursive, material, and institutional structures forming knots of identity (beings, books, understandings, concepts) and loose ends of potentiality (texts, events, dialogue, friendship). Personal and social identity, "understanding," and the text or information object itself gain their value and meaning only within and through temporal, relational processes.

Burnett's second argument regarding dissonance in texts is important because it suggests that texts, and in fact linguistic units as a whole, have a degree of semantic surplus or excess intrinsic to them; restrictions of this excess give rise to more *restrictive* economies of meaning production and thus to closer possibilities of showing or demonstrating understanding as we often conceive this term (i.e., as intersubjective correspondence). General economies precede restrictive economies, as the latter are constructed restriction and regulation of the former, where certain elements and relations in the world are allowed to be expressed and others not. The types of restrictions by which general economies of signs and social relations turn into restrictive economies constitute not only the linguistic realm, but also the social and cultural realm, and make up what Harré (1984) has termed, "moral orders." The notion of "general," here, does not signify a closed system; the system of relations in general systems is governed by mobile powers and pragmatic, rather than strictly calculable, relations. It is governed by principles of difference and emergent actualization, rather than by identity and calculable combination.

To be fair, G. Burnett (2002) does not fully develop Ricoeur's (1991) argument in this direction, writing within the more traditional constraints of notions of authorial intention, public texts, and readerly reintegration, seeing textual production as mimicking physically present interlocutions. The problem remains, however, as Saussure (1966), Heidegger (for example 1971, p. 111–136), and Benjamin (1978, p. 314–332) argued, that language exists before the speaker (or writer) uses it, and that, as Derrida (1976, 1982, p. 307–330) has argued, physical presence cannot act as a guarantor, or even as a primary model, for the notion of intention or even of meaning in writing or in language as a whole. The entire notion of preunderstanding is, in fact, based on this priority of language. This does not mean that "intention" is a vapid concept, but rather that writing shows that the excess of meaning embodied in language precedes any intention and that understanding is a process of negotiating this excess within pragmatic and never exact ends. (Domain analysis proceeds according to this same understanding.) No concept of "mind" or of authorial intention or readerly mental processes is necessary to account for the pragmatic production of texts from other texts or signs from other signs, and no correspondence model of truth is necessary to account for understanding as pragmatically conceived.

Further, as Derrida (for example 1982, p. 307–330) has argued, written language makes manifest an important characteristic of language: that the same words and phrases can be "grafted" onto other words and phrases and that generation in language proceeds in this manner. (For this reason, Foucaultian statements constitute only one form of language, namely, those constructed toward "disciplinary" production—in both the negative and positive moral senses of "discipline.") Grafting is a by-product of the essential excess of language, which is inherent in any linguistic utterance. For this reason, Derrida argues, all but the most controlled of utterances are potentially available to processes of negotiation or, in such a manner, interpretation (art works and poetry may show us this). Interpretation thus *follows* the repetition, generation, and production of language, although this repetition may or may not be in productivist manners. Thinking about the notion of excess in language leads beyond thinking language, thought, experience, and the real, according to the traditional conceptual horizons of hermeneutics—which is a more or less mentalistic and traditionally subject-inscribed conception of understanding and interpretation. Instead, it leads to seeing hermeneutic concepts such as understanding and interpretation as products of repetition.

Events

The process of reading/writing in virtual communities makes explicit the emergence of meaning from conceptual complexities; this emergence marks communities of practice as performative communities of interpretation. Such communities are formed through interpretative activity around objects that inform both the nature of community and the meaning of the objects themselves. As doubly signifying, we will call these objects "informational objects." The informational object, here, may be seen as posing a question around which the community forms itself, not only as an answer to that question, but more as a response—responses that give birth to further conceptual problematics, which are treated as questions. The discursive series—or literally, rhetorical lines—that develop out of these trains of questions and responses between concepts and persons have different "half-lives"—some "live" for the duration of a discussion group's existence, others die relatively early, many others reemerge with different groups of participants.

This process of the emergence of meaning may be understood according to "moments" and "movement" in events: phases in which meaning arises out of complexities, then begins lines of development, and then when certain lines of meaning seemingly close or die. Like musical phrases, they have rhythms, tonal ranges, and various degrees of cacophony. Gilles Deleuze's work has addressed this phenomenon in terms of the concept of "event." As has been suggested, one can see this phenomenon on online discussion lists.

Deleuze's work has not been heavily used or addressed in information studies. Some exceptions are Kathleen Burnett's (K. Burnett, 1993; Burnett & Dresang, 1999) use of Deleuze and Guattari's (1987) notion of "rhizome" from *A Thousand Plateaus* and David Snowden's (2002) plenary address to the 2002 annual meeting of the American Society for Information Science and Technology, where Deleuze's name was mentioned several times and one could identify a modified reading of Deleuze in Snowden's discussion of organizational complexity. Yet, several concepts in Deleuze's work may be useful to information studies. Because this has not been well developed in information studies, we will limit ourselves to sketching these concepts and their applicability.

The understanding of information in terms of the concept of "event" has been approached by, among others, Michael Buckland (1991a, 1991b). In his article "Information as Thing," for example, Buckland distinguishes between intangible and tangible information objects, arguing that examples of the first may be seen in terms of both knowledge and events. Tangible information, on the other hand is the representation of knowledge or events. As Buckland writes, "knowledge, however, can be represented, just as an event can be filmed. However, the representation is no more knowledge than the film is the event" (Buckland, 1991b, p. 352). We may, after Buckland, perhaps wish to speak of knowledge and events in general as "information-as-event" because they constitute a sense of information understood as general potentialities not yet representationally fixed. This more phenomeno-logical approach to information has been generally avoided in informa-tion studies because the field has traditionally seen itself as more narrowly concerned, that is, solely with the manipulation and retrieval of representations of events.

Several problems emerge if we narrow our vision of information and information studies to document and data retrieval. If we take the study of information seriously as our scope, we must be concerned with many, less formalized, modes of information such as occur in oral speech and everyday interaction. Also, information studies has not been very good at accounting for value in documents and data. It has assumed value where it may be constituted by the more general production and shap-ing of informational events rather than being internal to those events. Films select particular elements and views of events; abstracts select areas of documents to represent and then rephrase those selections; doc-uments, themselves, are selections of larger discourses. Each of these forms, too, may constitute events. The argument that information stud-ies is not concerned with the value of documents and data, but rather solely with their manipulation, flies in the face of the purposive and exis-tential environments of professional information practice where work is humanly oriented, so one must not only demand semantic value in deliv-erables but also account for the wider contexts of that value in order to determine reliability, or in the case of archival work, make claims for the authenticity of materials. Last, categories such as "the work" in art chal-lenge the very notion of value in document and data retrieval, which has been, traditionally, rooted in representation. Many art "works" are site-specific and time-valued. Avant-garde art works during the twentieth century, for example, *work* specifically against their replication over time and space and *work against* museums' desire to preserve such works as "timeless" cultural objects whose value can be spatially and temporally transferred, according to concepts of tradition or "cultural heritage." The "work" of art, here, is equivalent to the category of "event," and its most absolute example was in the "happenings" of 1960s performance art.

In each of these cases, a more restrictive economy of values attributed to information is formed out of general, phenomenological events that have informational value intrinsic to their structure. The concept of "event" is interesting in information studies because it forms both a limit and an originary, but unacknowledged and unexplored, horizon for the field's tradition, a tradition which has confined itself within restrictive readings of the concept of information according to metaphysics and epistemologies of representation.

Deleuze's discussion of events is conceptually difficult, as it is rooted in both a critical rereading of canonical texts of the Western philosophi-cal tradition and a reading of complexity and chaos theory originating from the mathematical, physical, and biological sciences. Whereas Derrida's work begins and ends with the concept of the play of forces according to *différance*, Deleuze examines difference within notions of being qua complexity. (Delanda [2002] gives an excellent account of Deleuze's scientific sources.) This aspect of Deleuze's work runs parallel to certain discourses in second order cybernetics (for example Luhmann, 1995). It is beyond the scope of this chapter to discuss the relation between these "French" and "German" readings of emergence, other than stating that the former tend to view structure and time itself as a product of affective relations rather than seeing structure "in" time and affects as relations within and between systems. It would not be unfair to state that the centrality of the term *system* is much weaker in the French tradition than the German, and that this difference could, perhaps, be viewed in terms of the influence of Nietzsche's writings upon French poststructuralism. There is also an English language philosophical literature on emergence, starting from the work of the eighteenth-century Scottish philosopher, Thomas Reid, such as the work of Rom Harré (see, for example, Harré, 2001; Harré & Madden, 1975) that runs parallel in some ways to Deleuze's concerns, including a common critique of the equation of subjectivity with traditional, metaphysically laden, understandings of "self."

Generally, "events" in Deleuze's works are phase spaces of repetition within the unfolding and refolding of complex, differential relations according to conditions of allowance. Repetition is a fact of all organisms, but repetition occurs in new space-time environments with each turn, so repetition may introduce new complexities and new forms of being or subjectivity into the organism and its emergence. Furthermore, conditions of allowance for expression may or may not vary over time (physical entities have fairly fixed conditions for allowance in the known physical universe, whereas social entities have much broader allowances, but of greatly varying types and degrees).

Being is viewed in Deleuze's work not in terms of enlightenment concepts of the self—that is, in terms of the classical subject of will and representation—but rather as a multiplicity of experiences and modes of being. These multiplicities may reach points of stability called, "singularities." Singularities are composed of multiplicities and multiplicities are made up of various singularities in processes of becoming.

Deleuze discusses the multiple series of past and future being that inhabit organs, organisms, and organizations in terms of complexities and potential powers or potencies that are actualized in affective relationships with others. Furthermore, "strange attractors" may "attract" and draw out potentialities in certain manners and ways or may lead to the bifurcation or splitting of certain lines of development (they may also let certain potentials "die"). Series can also be characterized as "strings," and strings can become knots such as single and group identities, vocabulary structures, and organized social relationships. The environment, as well as processes within the organism, allows certain powers to develop, others to be split off, and others not to develop.

The implication for such a view of individuals—in terms of potencies—has implications for organizational theory and "knowledge management" in so far as it replaces theories of person, mind, and language grounded in metaphysical dichotomies and hierarchies with theories of person, mind, and language grounded in emergence and expression. Such a view points to the difficulty in drawing strict distinctions between self and other, particularly in regard to semiotic modes of expression. It also problematizes more traditional understandings of self and mental models that are evident in information studies' cognitive approach and user studies approaches that define the user as an autonomous individual (Touminen, 1997). Last, we can see in Cronin (1995) and Paling (2002), as well as Gary Burnett's (2002) work, discussed earlier, that complex entities such as "texts" and even "authors" are composed of semantic excesses, synchronic differences, and historical multiplicities. Poststructural and emergent models for various types of notions of identity (for example, users, vocabulary structures, documents) are useful to information studies in that they show persons, information, and knowledge as not simply relational and temporal, but dynamically so. Current notions of information, users, and even knowledge are based on rather static pictures of the interaction of these concepts conceived as actually existent entities (particular information and users). Closer examination reveals that users are anything but atomic and simple individuals and that information, when spoken of as an entity, is simply based upon reifying tendencies due to misleading grammatical forms (e.g., "I have information" as akin to "I have a car"). When their correspondence is seen as proof of events of information, we have only compounded conceptual errors, and many library and information science studies, modeled on similar conceptual errors in the social sciences, simply define and demonstrate phenomena in terms of those errors (see Davenport, Day, Lievrouw, & Rosenbaum, 2003).

Certain problematics in Deleuze's work, such as not only a materialist account of subjects, objects, and events, but the material nature of meaning, and, the importance of expression and emergence in meaning and value, are useful for further investigation in information studies. Deleuze's work (particularly 1990b and 1993), develops a theory of meaning that begins from conceptions of material surfaces and production, challenging the Aristotelian tradition of categories and form/matter analysis that has a determinate role throughout library and information science, particularly in the area of classification (Bowker & Star, 1999), but also in cataloging, information architecture, the general topic of taxonomy, and, arguably, even in the field's imagination of its own disciplinary nature, pedagogy, and sociological inscription. Although Bowker and Star (1999) have ably discussed the social and historical specificity of epistemic infrastructure, more work needs to be done on the temporal nature of infrastructure, particularly as we have suggested in online environments, and the relation of this temporal nature to social and historical emergence.

Last, Deleuze's works, along with Deleuze and Guattari's (1983, 1987) joint works, have been a major influence on recent Italian political and economic theories of production. Most notably, their reading of psychological and other social bodies as "molar captures" of molecular flows and life powers, and their reading of this mechanism of capture in terms of capitalism and representation (along with Deleuze's [1990a] parallel reading of Spinoza) have greatly influenced the Italian political theorist Antonio Negri. Negri's solo works and his work with Michael Hardt (Hardt & Negri, 2000), together with his and their works' influence on new social movements, have been discussed in information studies by Dyer-Witheford (1999), Day (2002), and Wright (2002) in the context of so-called "social capital" and in the context of new social movements' use of digital information and communication technologies.

Conclusion

In this chapter we have reviewed some issues in poststructuralism and how these intersect with, and have been treated by, information studies. Let us summarize the impact that poststructural problems may have upon the field in theory and in practice.

First, the poststructural critique of identity puts into question the classical subject of Western metaphysics, which appears, for example, in the guise of the "user" in information studies and the knowledge worker in knowledge management. Agency is repositioned in poststructuralism from being autonomous and auto-affective to being socially and historically constituted and constituting. Social and historical constitution here, however, is not a grounding context for agency. Rather, social and historical series are multiple and their actualization is determined by existential agency. The "individual" is both multiple and singular: its singularity is that of multiple identities and series expressed in unique events. Such subjectivity and its expressions are complex, not a priori given or fixed, particularly in the case of social, rather than purely physical, entities.

Second, poststructuralism stresses the centrality of language in information studies theory and professional practice. The centrality of language for the field can be seen in terms of the importance of actual entities, such as vocabulary (Buckland, 1999) and discursive statements (Frohmann, 2001), but also in terms of larger notions of language, such as nonlinguistic semiotic signs and affects. Further, structural linguistics' model of identity and value in terms of difference (Saussure, 1966) is extended in poststructuralism (particularly in Derrida's writing) to problems of identity, knowledge, and value. In Deleuze's work the notion of difference more fully encompasses complexity and emergence theory than linguistics per se, and particularly more than structural linguistics—that is, according to what Guattari (1972) termed, "asignifying semiotics" rather than only "signifying semiotics." Thus, through poststructuralism's emphasis on language and semiotic signs in general, information studies' ontological, epistemological, ethical, and political concerns are deepened.

Third, the poststructural displacement of the subject outside of the framework of representation applies to objects in information studies research as well. Poststructural approaches to the object stress its inscription within discursive and affective series and the relation of these discursive and affective series to the production and reproduction of powers. Such approaches do not dissolve the status of objects to being semiotic inscriptions only (e.g., "texts"), but rather, they point to the necessity of speaking of objects in terms of both their semiotic production and their resistance to, and displacement of, certain types of semiotic and physical gestures.

Fourth, critically oriented "discourse analysis" in information studies has very successfully addressed the problematic nature of "information" as an object for "scientific" inquiry in the field. Within a theoretical discourse, at least, library and information science, and particularly information science, has sometimes since the Second World War characterized itself: 1) by a vague and almost mystical definition or conception of "information" (Capurro & Hjørland, 2003; Schrader, 1984); 2) by a misnomer and mystification of the focus of library, and particularly information, science ("information retrieval" when what was largely meant was document or document representation retrieval); 3) by a folkloric trope that historically, and to some extent still today, governs the overall theory and practice of LIS by acting as a conceptual model (the conduit metaphor); and, 4) by turning away from more material approaches in library studies and documentation claims of being *in all its aspects*, a "science."

Despite such theoretical weaknesses, and—one cannot say otherwise—sheer theoretical disasters, the practice of information science has produced successes. Theoretical discourse and practical activity are not, and may not be in any given practice, causally linked; as with medieval medicine, mystical theoretical foundations and practical successes may coexist within what is claimed to be a field of inquiry. Further, despite attempts to unify the field, it is today rather more diverse than either critics or defenders may be willing to concede; claims of "science" for such a diverse field really must be accounted for in equally diverse manners (this, however, remains to be studied). Also, with this diversity, the criteria and evidence for what constitutes success vary greatly. Last, as Frohmann (1992, 1994, 2001) has suggested, the difference between theoretical disasters and some practical successes points to the importance of viewing theoretical discourses in information studies as having been, at least in part (and in some cases, a large part), produced for both internal and external institutional purposes rather than toward either the advancement of theoretical discourses—or even of professional practice (a point emphasized by Powell, Baker, & Mika's [2002] claim that among library and information practitioners, only half those surveyed admit applying published research results to professional practice). In brief, not only the theoretical foundations, but even the meaning and relation of theory to practice in the field need to be rethought.

Fifth, poststructuralism stresses the historicity of the subject, which means that the subject is simultaneously located between necessity and freedom; "necessity" in terms of the material possibilities for its being and "freedom" in terms of its potentiality toward being (Day, 2001, Chapter 5). This historicity remains detached from anthropological narratives of context and ontic accounts of being. The power of the historical agent remains a potentiality that exceeds narrative measure and representational categories and tropes. Human agency is intrinsic to the very possibility of structure and context. The concept of "context" in information studies needs to be clarified in relation to notions of structure, and "context" and concepts of personal and social agency in the field need to be freed from anthropological and ethnological narrative conventions. Further, information studies needs to become better attuned to nonhistoricist understandings of historical agency and historical specificity (historicity) than has been the case so far, even if this jeopardizes the scientific claims of the field as a whole or relegates such claims to only part of the field.

Last, poststructuralism challenges certain assumed central topics in information studies that are founded on models of representation. For example, it challenges (as did the twentieth-century avant-garde) the digital reproduction and distal transmission of the meaning and value of art with the concept of the *work* of art. It challenges common understandings of reading as simple mental representation or reproduction ("literacy" in terms of correctness and correspondence) with a notion of reading that is interlinked with writing and is grounded in notions of semantic excess and bibliographical overabundance. Through its critique of representation, poststructuralism challenges the concept of social capital in capitalist-based discourses with notions of social and personal being that exceed quantitative measures of value, just as it challenges the temporal and ontological privilege of simple, restricted economies over abundant general economies in the production of meaning, value, power, and wealth (Negri, 1999; see also Day, 2002; Social Capital Workshop, 2002).

In sum, poststructuralism warrants further attention by information studies researchers, not only because its discourses and their progeny surround the field throughout the university structure, and not only because it points to serious areas of confusion and inattention in information studies theory, research methods, and professional practice, but also because it points to the limits and "beyond" of traditional information studies theory and thus suggests not only what we have been doing, but where we must still need to explore. Jesse Shera (1970, p. 86) suggested that LIS should not simply be a field of restricted techniques, but rather one that studies the general problems of knowledge and information. Buckland (1996) has argued for a "liberal arts" of LIS. Poststructuralism entered the intellectual stage with a concept of *écriture générale*—writing in general—that came to signify a critique of linguistic, and therefore social, production in general. *Écriture générale* was what Derrida termed a "quasi-concept": a concept of writing or inscription (*écriture*) that was the infrastructure of all the sciences and fields of scholarship and thus escaped itself becoming both an object of, and a field of study for, any particular science (including, grammatology [Derrida, 1976]). "Information," or perhaps an extended notion of "document," may be seen to play a similar role as *écriture générale*, or perhaps these terms operate as forms of *écriture générale*, whose material, social, and technological forms still must be historically and socially accounted for in terms related to, but after, what was specific to poststructuralism in the twentieth century. Poststructuralist concepts and critiques may now be engaged with considerable specificity in relation to new information and communication technologies and new forms of knowledge, identity, and sociality; they may be expanded further by new theoretical, social, and technical practices today and new forms of discourses and subjectivity. It is hard to imagine these critiques not playing a new role in both theory and practice. Certainly, such an event is taking place in media studies, communication studies, and cultural studies, as well as in areas of sociology, philosophy, and literature that study technology and culture. And, it is beginning to appear in information studies as well.

Acknowledgment

The author would like to thank Elisabeth Davenport, Ian Cornelius, Michael Buckland, Blaise Cronin, and Debora Shaw for their suggestions and work on this chapter.

Endnotes

1. Several decades later, Ludwig Wittgenstein picked up many of the same terms of Saussure's functionalist analysis, including the important metaphor of chess (compare Saussure [1966] to Wittgenstein [1958]) without, however, recuperating such games within a totalizing notion of a linguistic "science" (which Derrida [1976] critiques Saussure for doing). It may be noted, since ordinary language readings of Wittgenstein are not unknown in the information studies literature (for example, Blair, 2003), that for the same reason the Wittgensteinian notion of "language games" cannot be used to support suppositions of "ordinary language" or "ordinary language philosophy"—one can always imagine different language games and one can always position language *between* language games and use the imagination (Wittgenstein, 1958) or art technique (as in the modern avant-garde) to create other or new meanings. The metaphor of "game" in both Saussure and Wittgenstein is meant to suggest a *pragmatically* functionalist (rather than "organic") basis for value and identity in language. It is not being used to suggest that language acts have fixed—or even normative—boundaries, as a chessboard does. In other words, to use Derrida's terms (after Bataille, 1988), language must be understood economically—according to "open" or "general" economies of plays of differences—not according to closed or "restrictive" economies. Ordinary language accounts (which lean heavily toward functionalist readings) of Wittgenstein's later work (admirably done, as in Blair, 2003) work well enough for arguing some pragmatic issues of information

retrieval, but they fail to address issues of marginality and rupture as discussed in Wittgenstein's later work, particularly issues of the imagination, the ability of common linguistic usage to fool us, and the role of doubt. Such a functionalist orientation follows the positivist orientation of information science in marginalizing issues of art, the category of the art "work," and other engagements with the creation of new meaning, because the interests of information science have been with information retrieval—issues of reproduction, transmission, "aboutness," and representation in general.

2. "Affect" here should be read not simply in terms of a reductivist psychology of "feelings," but, more generally, as one body's influence upon another, although not necessarily within determined or known causal relationships. The term, thus, is conceptually related to the ancient Greek *aisthitikos* (signifying feeling) and is the root for "aesthetics," which should be understood here in a pre-Kantian manner of generalized "feelings" or "feeling effects." Viewing information in this manner, generally as affect, and exemplified in one realm of cultural relations by certain forms of artistic production and meaning, challenges contemporary understandings of information in terms of signifiers with self-evident meaning and clear, causal, and systemic relationships between such signifiers. This challenge to positivism can be seen in the twentieth century through the practices and theory of the avant-garde where the notion of "art" is freed from the Kantian confines of subjective contemplation and the notion of "information" is politically reinvigorated.

References

Barthes, R. (1974). *S/Z: An essay*. New York: Hill and Wang.

Barthes, R. (1977). *Image music text*. New York: Noonday Press.

Bataille, G. (1988). *The accursed share: An essay on general economy*. New York: Zone Books.

Belkin, N. (1977). Internal knowledge and external information. In M. deMay, R. Pinxten, M. Poriau, & F. Vandamme (Eds.), *International Workshop on the Cognitive Viewpoint* (pp. 187–194). Ghent, Belgium: Communication & Cognition.

Belton, B. K. (2003). *Orinoco flow: Culture, narrative, and the political economy of information*. Lanham, MD: Scarecrow Press.

Benediktsson, D. (1989). Hermeneutics: Dimensions toward LIS thinking. *Library & Information Science Research, 11*, 201–234.

Benjamin, W. (1968). *Illuminations*. New York: Harcourt Brace Jovanovich.

Benjamin, W. (1978). *Reflections: Essays, aphorisms, autobiographical writings*. New York: Harcourt Brace Jovanovich.

Benoît, G. (2002). Toward a critical theoretic perspective in information systems. *Library Quarterly 72*(4), 441–471.

Berardi, F. 2001. *La fabbrica dell'infelicità: New economy e movemento del cognitariato*. Rome: DeriveApprodi.

Blanchot, M. (1982). *The space of literature*. Lincoln: University of Nebraska Press.

Blair, D. C. (2003). Information retrieval and the philosophy of language. *Annual Review of Information Science and Technology, 37*, 3–50.

Bowker, G., & Star, S. L., (1999). *Sorting things out: Classification and its consequences*. Cambridge, MA: MIT Press.

Buckland, M. K. (1991a). *Information and information systems*. New York: Praeger.

Buckland, M. K. (1991b). Information as thing. *Journal of the American Society of Information Science, 42*(5), 351–360.

Buckland, M. K. (1996). The "liberal arts" of library and information science and the research university environment. *Proceedings of the Second International Conference on*

Conceptions of Library and Information Science, 75–84. Retrieved December 31, 2003, from http://www.sims.berkeley.edu/~buckland/libarts.html

Buckland, M. K. (1999). Vocabulary as a central concept in library and information science. *Proceedings of the Third International Conference on Conceptions of Library and Information Science*, 3–12.

Buckland, M. K. (2002). Workshop on the history of information science presented at Concepts of Library and Information Science 4 (COLIS4) conference. University of Washington, July 21, 2002.

Buckland, M. K., & Hahn, T. B. (Eds.). (1998). *Historical studies in information science*. Medford, NJ: Information Today, Inc.

Budd, J. M., & Raber, D. (1996). Discourse analysis: Method and application in the study of information. *Information Processing & Management 32*(2), 217–226.

Burnett, G. (2002). The scattered members of an invisible republic: Virtual communities and Paul Ricoeur's hermeneutics. *Library Quarterly, 72*(2), 155–178.

Burnett, K. (1993, January). Towards a theory of hypertextual design. *Post Modern Culture 3*(2). Retreived December 31, 2003, from http://muse.jhu.edu/journals/postmodern_culture/toc/pmcv003.html#v003.2

Burnett, K., & Dresang, E. T., (1999). Rhizomorphic reading: The emergence of a new aesthetic in literature for youth. *Library Quarterly, 69*(4), 421–446.

Capurro, R. (2000). Hermeneutics and the phenomenon of information. *Research in Philosophy and Technology, 19*, 79–85. Retrieved December 31, 2003, from http://www.capurro.de/ny86.htm

Capurro, R., & Hjørland, B. (2003). The concept of information. *Annual Review of Information Science and Technology, 37*, 343–409.

Chalmers, M. (1999). Comparing information access approaches. *Journal of the American Society for Information Science, 50*(12), 1108–1118.

Cornelius, I. (2002). Theorizing information for information science. *Annual Review of Information Science and Technology, 36*, 393–425.

Cronin, B. (1995). *The scholar's courtesy: The role of acknowledgement in the primary communication process*. London: Taylor Graham.

Cronin, B. (2004). Bowling alone together: Academic writing as distributed cognition. *Journal of the American Society for Information Science and Technology, 55*(6), 557–560.

Culler, J. (1982). *On deconstruction: Theory and criticism after structuralism*. Ithaca, NY: Cornell University Press.

Davenport, E., Day, R. E., Lievrouw, L., & Rosenbaum, H. (2003). Death of the user. *Proceedings of the 66th Annual Meeting of the American Society for Information Science and Technology*, 429–430.

Day, R. (2000). The "conduit metaphor" and the nature and politics of information studies. *Journal of the American Society for Information Science, 51*(9), 805–811.

Day, R. (2001). *The modern invention of information: Discourse, history, and power*. Carbondale, IL: Southern Illinois University Press.

Day, R. (2002). Social capital, value, and measure: Antonio Negri's challenge to capitalism. *Journal of the American Society for Information Science and Technology, 53*(12), 1074–1082.

Day, R. (2003). Critical theory and bibliography in cross-disciplinary environments. In D. W. Foster (Ed.), *International bibliography* (pp. 89–104). Jefferson, NC: McFarland.

Deconstruction and the possibility of justice. (1990, July/August). *Cardozo Law Review, 11*(5–6), 919–1733.

Delanda, M. (2002). *Intensive science & virtual philosophy*. London: Continuum.

Deleuze, G. (1988). *Foucault*. Minneapolis: University of Minnesota Press.

Deleuze, G. (1990a). *Expressionism in philosophy: Spinoza*. New York: Zone Books.

Deleuze, G. (1990b). *The logic of sense*. New York: Columbia University Press.

Deleuze, G. (1993). *The fold: Leibniz and the baroque*. Minneapolis: University of Minnesota Press.

Deleuze, G., & Guattari, F. (1983). *Anti-Oedipus: Capitalism and schizophrenia* (R. Hurley, M. Seem, & H. R. Lane, Trans.). Minneapolis: University of Minnesota Press.

Deleuze, G., & Guattari, F. (1987). *A thousand plateaus: Capitalism and schizophrenia* (B. Massumi, Trans.). Minneapolis: University of Minnesota Press.

Deleuze, G. & Guattari, F. (1994). *What is philosophy?* New York: Columbia University Press.

de Man, P. (1986). *The resistance to theory*. Minneapolis: University of Minnesota Press.

Derrida, J. (1976). *Of grammatology*. Baltimore: Johns Hopkins University Press.

Derrida, J. (1982). *Margins of philosophy*. Chicago: University of Chicago Press.

Drucker, J. (2003, August 13). Keynote address to the meeting of the Document Academy (DOCAM '03), University of California at Berkeley.

Dyer-Witheford, N. (1999). *Cyber-Marx: Cycles and circuits of struggle in high-technology capitalism*. Urbana: University of Illinois Press.

Farmer. J. A. (1993). A poststructuralist analysis of the legal research process. *Law Library Journal, 85*, 391–404.

Foucault, M. (1972). *The archaeology of knowledge*. New York: Pantheon Books.

Foucault, M. (1990). *The history of sexuality*. New York: Vintage Books.

Frohmann, B. (1992). The power of images: A discourse analysis of the cognitive viewpoint. *Journal of Documentation, 48*(4), 365–386.

Frohmann, B. (1994). Discourse analysis as a research method in library and information science. *Library & Information Science Research, 16*, 119–138.

Frohmann, B. (2001). Discourse and documentation: Some implications for pedagogy and research. *Journal of Education for Library and Information Science, 42*(1), 12–26.

Gadamer, H.-G., (1985). *Truth and method*. New York: Crossroad Publishing.

Guattari, F. (1972). *Molecular revolution: Psychiatry and politics*. London: Penguin Books.

Hardt, M., & Negri, A. (2000). *Empire*. Cambridge, MA: Harvard University Press.

Harré, R. (1984). *Personal being*. Cambridge, MA: Harvard University Press.

Harré, R. (2001). Active powers and powerful actors. *Philosophy, 48*(supp.), 91–109.

Harré, R. & Madden, E. H. (1975). Causal powers: A theory of natural necessity. Oxford, UK: Basil Blackwell.

Heidegger, M. (1971). *On the way to language*. New York: Harper & Row.

Heidegger, M. (1977). *Basic writings, from being and time (1927) to the task of thinking (1964)*. New York: Harper & Row.

Heidegger, M. (1996). *Being and time*. Albany: State University of New York Press.

Hjørland, B. (1997). *Information seeking and subject representation: An activity-theoretical approach to information science*. Westport, CT: Greenwood Press.

Hjørland, B. (2000). Library and information science: practice, theory and philosophical basis. *Information Processing & Management, 36*, 501–531.

Jacob, E. K., & Shaw, D. (1998). Sociocognitive perspectives on representation. *Annual Review of Information Science and Technology, 33*, 131–185.

Latour, B. (1996). Ces réseaux que la raison ignore: laboratoires, bibliothèques, collections. In M. Baratin & C. Jacob (Eds.), *Le pouvoir des bibliothques: La mémoire des libres en Occident* (pp. 23–46). Paris: Albin Michel.

Lievrouw, L. A. (2003). Biotechnology, intellectual property, and the prospects for scientific communication. In S. Braman (Ed.), *Biotechnology and communication: The meta-technologies of information* (pp. 145–172). Mahwah, NJ: Lawrence Erlbaum.

Luhmann, N. (1995). *Social systems*. Stanford, CA: Stanford University Press.

Lyotard, J.-F. (1984). *The postmodern condition: A report on knowledge*. Minneapolis: University of Minnesota Press.

Marazzi, C. (1997). *La place des chaussettes: Le tournant linguistique de l'économie et ses conséquences politiques*. (Trans. of Il Posto dei calzini: La svolta linguistica dell'economia e i suoi effetti nella politica.) Paris: Éditions de l'Éclat.

Marx, K. (1992). *Karl Marx: Early writings*. New York: Penguin Books.

Negri, A. (1989). *The politics of subversion: A manifesto for the twenty-first century*. Cambridge, UK: Polity Press.

Negri, A. (1996). Twenty theses on Marx: Interpretations of the class situation today. In S. Makdisi, C. Casarino, & R. Karl (Eds.), *Marxism beyond Marxism* (pp. 149–180). New York: Routledge.

Negri, A. (1999). Value and affect. *Boundary 2, 26*(2), 77–88.

On the necessity of violence for any possibility of justice. (1991, December). *Cardozo Law Review, 13*(4), 1081–1417.

Otlet, P. (1934). *Traité de documentation: Le livre sur le livre: Théorie et pratique*. Brussels, Belgium: Editiones Mundaneum, Palais Mondial.

Paling, S. (2002). Thresholds of access: Paratextuality and classification. *Journal of Education for Library and Information Science, 43*(2), 134–143.

Pettigrew, K. E., Fidel, R., & Bruce, H. (2001). Conceptual frameworks in information behavior. *Annual Review of Information Science and Technology, 35*, 43–70.

Peters, J. D. (1988). Information: Notes toward a critical history. *Journal of Communication Inquiry, 12*, 10–24.

Powell, R. R., Baker, L. M., & Mika, J. J. (2002). Library and information science practitioners and research. *Library & Information Science Research 24*(1), 49–72.

Rapaport, H. (1989). *Heidegger & Derrida: Reflections on time and language*. Lincoln: University of Nebraska Press.

Rayward, W. B. (1998). The history and historiography of information science: Some reflections. In T. B. Hahn & M. Buckland (Eds.), *Historical studies in information science* (pp. 7–21). Medford, N.J.: Information Today, Inc.

Reddy, M. J. (1993). The conduit metaphor: A case of frame conflict in our language about language. *Metaphor and thought* (2nd ed.) (pp. 164–201). Cambridge, UK: Cambridge University Press.

Ricoeur, P. (1991). *A Ricoeur reader: Reflection and imagination* (M. J. Valdes, Ed.). Toronto: University of Toronto Press.

Saussure, F. de. (1966). *Course in general linguistics* (C. Bally, A. Sechehaye, & A. Riedlinger, Eds.; W. Baskin, Trans.) New York: McGraw-Hill. (Original work published 1916)

Schlegel, F. (1991). *Philosophical fragments*. Minneapolis: University of Minnesota Press.

Schrader, A. M. (1984). In search of a name: Information science and its conceptual antecedents. *Library & Information Science Research, 6*, 227–271.

Shera, J. (1961). Social epistemology, general semantics, and librarianship. *Library Bulletin*, 767–770.

Shera, J. (1970). *Sociological foundations of librarianship*. New York: Asia Publishing House.

Snowden, D. (2002, November 20). Plenary speech presented at the Annual Meeting of the American Society for Information Science and Technology, Philadelphia, PA.

Social Capital Workshop (2002). Cornell University, September 12–14, 2002. Retrieved December 31, 2003, from http://www.einaudi.cornell.edu/europe/social_capital

Touminen, K. (1997). User-centered discourse: An analysis of the subject positions of the user and the librarian. *Library Quarterly, 67*(4), 350–371.

Turing, A. (1937). On computable numbers, with an application to the entscheidungs problem. *Proceedings of the London Mathematical Society, ser. 2, 42*, 230–265.

Van House, N. (2004). Science and technology studies and information studies. *Annual Review of Information Science and Technology, 38*, 3–86.

Weaver, W. (1949). *The mathematical theory of communication*. Urbana: University of Illinois Press.

White, H. (1987). *The content of the form: Narrative discourse and historical representation*. Baltimore: Johns Hopkins University Press.

Wittgenstein, L. (1958). *Philosophical investigations*. London: Basil Blackwell.

Wright, S. (2002). Pondering information and communication in contemporary anti-capitalist movements. *The Commoner*. Retrieved December 31, 2003, from http://www.commoner.org.uk/01-7groundzero.htm

Index

4GW (Fourth Generation Warfare), 422–423

A

Abbadi, A. E., 67
Abbas, 347–348, 355
ABC News, 440
Abelson, R., 269
Abraham, R. H., 82
Abrams v. United States, 446
Abstract Syntax and Notation One (ASN.1), 196
Abu-Ghayth, S., 492
academic status, structural capital and, 523–524
access
    to computers by children and teenagers, 352–354
    metadata for, in electronic records management, 244
    to Web, social aspects of, 318–322

accounting treatment of social capital, 535, 538
Ackerman, M., 527
Acland, G., 220, 236, 239, 243
ActiveWorlds (online community), 322
activism, electronic, 324–326
Adamic, L., 115, 120
Adamic, L. A., 117
Adams, J., 422
Adams, N., 528
Adar, E, 115
Adler, A., 518
Adler, D. A., 185
Adler, P., 528, 536–537
Adler, P. S., 518
administrative subpoena authority, 480–482
ADMs (Alternate Document Models), 91–92, 103–105
ADP (automated data processing), 231
aesthetic issues, in intercultural interface design, 281–282, 286

611

O

Further Reading in Information Science & Technology

Proceedings of the 67th Annual Meeting of the American Society of Information Science & Technology (ASIS&T)

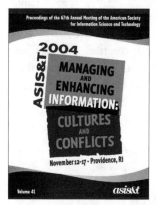

ASIS&T 2004–Managing and Enhancing Information: Cultures and Conflicts addresses the increasing tension between forces that encourage and discourage integration and cooperation in the global information society. A major focus of the conference is on conflicts involving diverse information cultures, with an emphasis on social, professional, educational, and technological issues. Particularly relevant to recent events are presentations on national security, privacy, scientific collaboration, and developing countries. Experts from a range of professional and academic settings around the world share their ideas and successes in nearly 200 papers, panels, and posters.

2004/646 pp/softbound/ISBN 1-57387-222-9
ASIST Members $55.90 • Nonmembers $69.90

Information Representation and Retrieval in the Digital Age

By Heting Chu

This is the first book to offer a clear, comprehensive view of Information Representation and Retrieval (IRR). With an emphasis on principles and fundamentals, author Heting Chu first reviews key concepts and major developmental stages of the field, then systematically examines information representation methods, IRR languages, retrieval techniques and models, and Internet retrieval systems. Chu discusses the retrieval of multilingual, multimedia, and hyperstructured information; explores the user dimension and evaluation issues; and analyzes the role and potential of artificial intelligence (AI) in IRR.

2003/250 pp/hardbound/ISBN 1-57387-172-9
ASIST Members $35.60 • Nonmembers $44.50

Knowledge Management for the Information Professional

Edited by T. Kanti Srikantaiah and Michael E. D. Koenig

Written from the perspective of the information community, this book examines the business community's recent enthusiasm for Knowledge Management (KM). With contributions from 26 leading KM practitioners, academicians, and information professionals, editors Srikantaiah and Koenig bridge the gap between two distinct perspectives, equipping information professionals with the tools to make a broader and more effective contribution in developing KM systems and creating a Knowledge Management culture within their organizations.

2000/608 pp/hardbound/ISBN 1-57387-079-X
ASIST Members $35.60 • Nonmembers $44.50

Knowledge Management Lessons Learned

Edited by Michael E. D. Koenig and T. Kanti Srikantaiah

The editorial team of Koenig and Srikantaiah have followed up their groundbreaking *Knowledge Management for the Information Professional* with this important book. While the earlier work offered an introduction to KM, the new book surveys recent applications and innovations. Through the experiences and analyses of more than 30 experts, the book demonstrates KM in practice, revealing what has been learned, what works, and what doesn't. Practitioners describe projects undertaken by organizations at the forefront of KM, and top researchers and analysts discuss KM strategy and implementation, cost analysis, education and training, content management, communities of practice, competitive intelligence, and more.

2003/550 pp/hardbound/ISBN 1-57387-181-8
ASIS Members $35.60 • Nonmembers $44.50

For a complete catalog, contact:
Information Today, Inc.
143 Old Marlton Pike, Medford, NJ 08055 • 609/654-6266
email: custserv@infotoday.com • Web site: www.infotoday.com